B1 Ref

X

DICTIONARY

Geography

METROPOLITAN BOROUGH OF WIRRAL

THIS BOOK IS FOR REFERENCE
USE ONLY, AND MUST NOT BE
TAKEN FROM THE BUILDING.

EDUCATION & CULTURAL
SERVICES DEPARTMENT

WITHDRAWN
FROM
STOCK

HarperCollins*Publishers*

HarperCollins*Publishers*
Westerhill Road, Glasgow G64 2QT

www.collins.co.uk

First published 2004

© 2004 Research Machines Plc
Helicon Publishing is a division of Research Machines
Additional text © HarperCollins*Publishers* 2004

Printing 10 9 8 7 6 5 4 3 2 1

ISBN 0-00-716595-1

A catalogue record for this book is available from the British Library.

Typeset by Davidson Pre-Press Graphics Ltd, Glasgow

Printed and bound in Great Britain by Clays Ltd, St Ives plc.

INTRODUCTION

The Collins Dictionary of Geography has been created to provide a guide to the vocabulary of modern geography, and, by giving clear explanations of terms and concepts, it aims to enhance the reader's understanding of the subject.

A feature of this dictionary is the inclusion of long explanatory entries on a range of key topics – such as **atmosphere**, **climate** or **earthquake**. These review entries provide the reader with fuller understanding of the significance of these important subjects.

The book is organized alphabetically, with each headword appearing in bold. The order is decided as if there are no spaces between words, for example:

> **green consumerism**
> **greenfield site**
> **greenhouse effect**
> **greenmail**
> **green movement**
> **Greenpeace**
> **green pound**

Cross references are indicated by an asterisk * in front of the entry being cross-referenced. Cross referencing is selective; a cross reference is shown when another entry contains material directly relevant to the subject matter of an entry, and to where the reader may not otherwise think of looking.

At the back of the book is a selection of useful geographical websites. Accessing these sites will help you further your understanding of a topic or subject area.

The text and illustrations for this book have been prepared by Helicon Publishing, a division of Research Machines Plc. The entries have been taken from their major educational encyclopedic database. These entries have been boosted by the addition of entries from the *Collins Dictionary of Environmental Science* and from *Collins Gem Geography*.

A

aa   in earth science, type of lava that has a rough surface, which may be jagged or spiny.

abiotic factor   non-living variable within the *ecosystem, affecting the life of organisms. Examples include temperature, light, and water. Abiotic factors can be harmful to the environment, as when sulphur dioxide emissions from power stations produce acid rain.

ablation   the loss of snow and ice from a *glacier by melting and evaporation. It is the opposite of *accumulation.

Ablation is most significant near the snout (foot or end) of a glacier, since temperatures tend to be higher at lower altitudes. The rate of ablation also varies according to the time of year, being greatest during the summer. If total ablation exceeds total accumulation for a particular glacier, then the glacier will retreat.

aborigine   (Latin *ab origine* 'from the beginning') any indigenous inhabitant of a region or country. The word often refers to the original peoples of areas colonized by Europeans, and especially to Australian Aborigines.

abrasion   in earth science, type of erosion in which rock fragments scrape and grind away a surface. It is also known as *corrasion. The rock fragments may be carried by rivers, wind, ice, or the sea. *Striations, or grooves, on rock surfaces are common abrasions, caused by the scratching of rock by debris carried in glacier ice.

abrasion platform   see *wave-cut platform.

abrasive   (Latin *abradere* 'to scratch away') substance used for cutting and polishing or for removing small amounts of the surface of hard materials. There are two types: natural and artificial abrasives, and their hardness is measured using the *Mohs scale. Natural abrasives include quartz, sandstone, pumice, diamond, emery, and corundum; artificial abrasives include rouge, whiting, and carborundum.

absolute age   in earth science, actual age of a fossil, rock, or stratum, normally expressed in years. It is usually measured by radiometric dating or dendrochronology (study of fossilized tree rings).

absolute humidity   in physics, amount of water vapour in air expressed as the mass of vapour (in kilograms) per cubic metre.

absolute temperature   temperature given by an absolute scale which is independent of the properties of thermometric substances. Absolute thermodynamic temperature, proposed by William Thomson, Lord Kelvin (1824–1907), is defined according to the principles of thermodynamics alone, but for experimental work is closely approximated by the International Temperature Scale of 1990 (ITS-90). It is usually expressed in units of degrees Celsius (or centigrade) from absolute zero, and freezing point is then 273.15 K.

abyssal plain   broad, relatively flat expanse of sea floor lying 3–6 km below sea level. Abyssal plains are found in all the major oceans, and they extend from bordering continental rises to mid-oceanic ridges. Abyssal plains are covered in a thick layer of sediment, and their flatness is punctuated by rugged low abyssal hills and high sea mounts.

Underlain by outward-spreading, new oceanic crust extruded from ridges, abyssal plains are covered in deep-sea sediments derived from continental slopes and floating microscopic marine organisms. The plains are interrupted by chains of volcanic islands and seamounts, where plates have ridden over hot spots in the mantle, and by additional seamounts, which were originally formed in oceanic ridge areas and transferred to the deep as the oceanic crust subsided. Otherwise, the abyssal plains are very flat, with a slope of less than 1:1000.

abyssal zone dark ocean region 2000–6000 m deep; temperature 4°C. Three-quarters of the area of the deep-ocean floor lies in the abyssal zone, which is too far from the surface for photosynthesis to take place. Some fish and crustaceans living there are blind or have their own light sources. The region above is the bathyal zone; the region below, the hadal zone.

accessibility the ease with which a place may be reached. An area with high accessibility will generally have a well-developed transport network and be centrally located or at least at a route centre. Many economic activities, such as retailing, commerce, and industry, require high accessibility for their customers and raw materials.

acculturation extensive culture change due to contact between societies. The term is most often used to refer to adaptation of subordinate or tribal cultures to Western culture.

accumulation in earth science, the addition of snow and ice to a *glacier. It is the opposite of *ablation. Snow is added through snowfall and avalanches, and is gradually compressed to form ice. Although accumulation occurs at all parts of a glacier, it is most significant at higher altitudes near the glacier's start where temperatures are lower.

acid mine drainage the seepage of sulphuric acid solutions (ph 2.0–4.5) from mines and their removed wastes dumped at the surface. These solutions result from the interaction of groundwater and percolating precipitation with sulphide minerals exposed by mining.

acid rain acidic precipitation thought to be caused mainly by the release into the atmosphere of sulphur dioxide (SO_2) and oxides of nitrogen (NO_x), which dissolve in pure rainwater making it acidic. Sulphur dioxide is formed by the burning of fossil fuels, such as coal, that contain high quantities of sulphur; nitrogen oxides are produced by various industrial activities and are present in car exhaust fumes.

Acidity is measured on the pH scale, where the value of 0 represents liquids and solids that are completely acidic and 14 represents those that are highly alkaline. Distilled water is neutral and has a pH of 7. Normal rain has a value of 5.6. It is slightly acidic due to the presence of carbonic acid formed by the mixture of CO_2 and rainwater. Acid rain has values of 5.6 or less on the pH scale.

Acid deposition occurs not only as **wet precipitation** (mist, snow, or rain), but also comes out of the atmosphere as dry particles (**dry deposition**) or is absorbed directly by lakes, plants, and masonry as gases. Acidic gases can travel over 500 km a day, so acid rain can be considered an example of trans-boundary (international) pollution.

Acid rain is linked with damage to and the death of forests and lake organisms in Scandinavia, Europe, and eastern North America. It is increasingly common in countries such as China and India that are industrializing rapidly. It also results in damage to buildings and statues. According to the UK Department of the Environment figures, emissions of sulphur dioxide from power stations would have to be decreased by 81% in order to stop the damage.

Reductions of UK emissions are being sought by using *flue-gas desulphurization plants in power stations and by fitting more efficient burners; by using gas instead of coal as a power station fuel; and, with road transport rapidly becoming recognized as the single most important source of air pollution, the compulsory fitting of *catalytic converters to all new vehicles.

A 1995 study found Manchester to be the European city worst affected by acid rain. Its rainfall was the most acidic, causing building stone to be destroyed faster there than anywhere else in Europe (other cities affected included Athens, Copenhagen, and Amsterdam). The other British test site, at Liphook,

Hampshire, also fared badly. Emissions in the UK of the two main causes of acid rain, SO_2 and nitrogen oxides, were projected to fall 60% and 30%, respectively, by the year 2003.

effects of acid rain The main effect of acid rain is to damage the chemical balance of soil. It leaches out important minerals including magnesium and aluminium. Plants living in such soils, particularly conifers, suffer from mineral loss and become more prone to infection. The minerals from the soil pass into lakes and rivers, disturbing aquatic life, for example by damaging the gills of young fish and killing plant life. Lakes affected by acid rain are virtually clear due to the absence of green plankton. Lakes and rivers also suffer more direct damage because they become acidified by rainfall draining directly from their catchment.

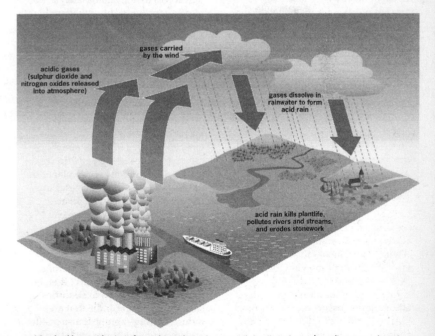

acid rain How acid rain is formed in industrial areas and distributed over long distances, where it can kill trees and damage buildings and statues.

acid rock   igneous rock that contains more than 60% by weight silicon dioxide, SiO_2, such as a granite or rhyolite. Along with the terms **basic rock** and **ultrabasic rock** it is part of an outdated classification system based on the erroneous belief that silicon in rocks is in the form of silicic acid. Geologists today are more likely to use the descriptive term *felsic rock or report the amount of SiO_2 in percentage by weight.

aclinic line   magnetic equator, an imaginary line near the Equator, where a compass needle balances horizontally, the attraction of the north and south magnetic poles being equal.

acre   traditional English land measure equal to 4840 square yards (4047 sq m / 0.405 ha). Originally meaning a field, it was the area that a yoke of oxen could plough in a day.

Acritarch   division of organic microfossils that has remained in existence from at least the Precambrian period up until the present time.

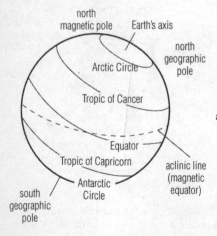

aclinic line The magnetic equator, or the line at which the attraction of both magnetic poles is equal. Along the aclinic line, a compass needle swinging vertically will settle in a horizontal position.

action area   any area within a town or city that is selected for priority treatment within a larger plan which has been prepared for the town or city. The area may require total renewal or rehabilitation by reason of its physical condition, or may have been awarded special priority because of the concentration of an environmentally related social problem, such as crowding.

activated sludge process   a biological treatment of sewage waters in which microorganisms are encouraged to grow under favourable conditions of oxygenation and nutrients, and in so doing, cleanse the waters. Dissolved or suspended organic matter in the polluted water acts as a nutrient base for the microorganisms (chiefly bacteria) which feed on the organic pollutants and secrete enzymes to digest and oxidize the absorbed material and so purify the wastes.

The actual removal of the pollutants from the liquid and the absorption by the activated sludge occurs within a few minutes of contact between sewage and the activated sludge while the oxidation of the absorbed material takes much longer. Normally the process of absorbtion and oxidation take place within the same sewage tank, although numerous variations can be incorporated into the process.

active layer   the upper layer of soil in *permafrost regions which thaws during the summer. The depth of the active layer is dependant on such features as the duration of the summer, the air temperature, the nature and extent of surface vegetation, the soil moisture content, and the mineralogical and organic composition of the soil itself. Active layers rarely exceed depths of 5 m and are often poorly drained due to the presence of permanently frozen underlying ground.

Human activities in permafrost regions, especially during the summer, can damage the delicate ecological balance of the active layer, turning it into a quagmire and causing the movement of the permafrost table through thermal erosion.

activity node   any place at which people congregate to pursue a particular activity, such as sport or shopping.

adaptive radiation   in evolution, the formation of several species, with adaptations to different ways of life, from a single ancestral type. Adaptive radiation is likely to occur whenever members of a species migrate to a new habitat with unoccupied ecological niches. It is thought that the lack of competition in such niches allows sections of the migrant population to develop new adaptations, and eventually to become new species.

The colonization of newly formed volcanic islands has led to the development of many unique species. The 13 species of Darwin's finch (named after English naturalist Charles Darwin 1809–1882) on the Galapagos Islands, for example, are probably descended from a single species from the South American mainland. The parent stock evolved into different species that now occupy a range of diverse niches.

added value   or **value added**, the sales revenue from selling a firm's products less the cost of the materials or purchases used in those products. An increasingly used indicator of relative efficiency within and between firms, although in the latter case open to distortion where mark-up varies between standard and premium-priced segments of a market.

adding value   increasing the perceived value of a product or service. A company may increase the value of a product by improving the presentation and packaging, or by providing an after sales service. Added value may cost little to the company providing the product or service but can be enough to confer a competitive advantage in a crowded market. Adding value is an essential element of the branding process.

adiabatic process   a thermodynamic process in which a change of temperature occurs in a mass of gas without a transfer of heat between the gas and its surrounding environment. The rate at which this change of temperature takes place is termed the lapse rate. The adiabatic expansion and compression of air is responsible for the cooling of rising air and the warming of descending air respectively. See *adiabatic wind, *dry adiabatic lapse rate, *saturated adiabatic lapse rate.

adiabatic wind   the movement of air which results either from the expansion of rising and cooling air (a process called *adiabatic cooling*) or from the compression of descending and warming air (a process called *adiabatic heating*). The characteristics of some local winds found in mountainous terrain such as the Föhn, the Chinook and the Southerner are associated with this phenomenon.

administrative region   a defined area in which governmental, commercial and other organizations carry out administrative functions. Examples include the regions of local health authorities and water companies, and commercial sales regions.

adobe   a mixture of clay and silt, very similar to loess, found in the southwestern United States and Mexico. It dries to a hard, uniform mass, and its use for brickmaking dates back thousands of years.

adult literacy rate   a percentage measure showing the proportion of an adult population that can read.

It is one of the measures used to assess the level of development of a country.

advanced gas-cooled reactor (AGR) a type of *nuclear reactor in which the fuel source comprises enriched uranium dioxide and the coolant comprises gaseous carbon dioxide. Graphite control rods are used to limit the rate of the nuclear reaction. Operating conditions in AGRs differ from the earlier magnox reactor design in that reactor temperatures are substantially higher at 675°C. Unlike most other designs of nuclear reactors, fuel can be loaded while the reactors continue to provide power for electricity generation. Each reactor in British AGRs contains 332 vertical fuel channels and on average five channels will be replaced each month.

advance factory any factory built to a standard design for rental to an incoming business user as part of the larger redevelopment of an area. Advance factories are not built with any particular industrial process in mind but are designed to accommodate a wide range of both type and size of enterprise. The existence of advance factories, ready for immediate occupation, has been a valuable asset in the efforts to regenerate the local economy of older industrial areas. Those in the new towns especially have attracted a wide range of entrepreneurs, often from overseas.

advection fog *fog formed by warm moist air meeting a cold ocean current or flowing over a cold land surface. It is common in coastal areas where warm air from the sea meets the colder land surface, particularly at night or during the winter.

aeolian deposits in geology, sediments carried, formed, eroded, or deposited by the wind. Such sediments include desert sands and dunes as well as deposits of windblown silt, called loess, carried long distances from deserts and from stream sediments derived from the melting of glaciers.

aerated water water that has had air (oxygen) blown through it. Such water supports aquatic life and prevents the growth of putrefying bacteria. Polluted waterways may be restored by artificial aeration.

aerial photograph a photograph taken from above the ground. There are two types of aerial photograph – a vertical photograph (or 'bird's eye view') and an oblique photograph where the camera is held at an angle. Aerial photographs are often taken from aircraft and provide useful information for map making and surveys. Compare *satellite image.

aerobic in biology, describing those organisms that require oxygen in order to survive. Aerobic organisms include all plants and animals and many micro-organisms. They use oxygen (usually dissolved in water) to release the energy contained in food molecules such as glucose in a process called aerobic respiration. Oxygen is used to break down carbohydrates into carbon dioxide and water, releasing energy, which is used to drive many processes within the cells.

aerogenerator wind-powered electricity generator. These range from large models used in arrays on wind farms to battery chargers used on yachts.

aerosol particles of liquid or solid suspended in a gas. Fog is a common natural example. Aerosol cans contain material packed under pressure with a device for releasing it as a fine spray. Most aerosols used *chlorofluorocarbons (CFCs) as propellants until these were found to cause destruction of the *ozone layer in the stratosphere.

affluent society society in which most people have money left over after satisfying their basic needs such as food and shelter. They are then able to decide how to spend their excess ('disposable') income, and become 'consumers'. The term was popularized by US economist John Kenneth Galbraith.

afforestation   planting of trees in areas that have not previously held forests. (**Reafforestation** is the planting of trees in deforested areas.) Trees may be planted:
(1) to provide timber and wood pulp;
(2) to provide firewood in countries where this is an energy source;
(3) to bind soil together and prevent soil erosion; and
(4) to act as windbreaks.

Afforestation is a controversial issue because while many ancient woodlands of mixed trees are being lost, the new plantations consist almost exclusively of conifers. It is claimed that such plantations acidify the soil and conflict with the interests of *biodiversity (they replace more ancient and ecologically valuable species and do not sustain wildlife).

Between 1945 and 1980 the Forestry Commission doubled the forest area in the UK, planting, for example, Kielder Forest, Cumbria. In 1991, woodland cover was estimated at 2.3 million hectares, around 10% of total land area. This huge increase has been achieved mainly by planting conifers for commercial use on uplands or on land difficult to farm. In spite of such massive afforestation, around 90% of the UK's timber needs are imported at an annual cost of some £4 billion.

African Development Bank organization founded 1963 to promote and finance economic development across the African continent. Its members include 51 African and 25 non-African countries. Its headquarters are in Abidja'n, Côte d'Ivoire.

African geography   Africa is connected with Asia by the isthmus of Suez, and separated from Europe by the Mediterranean Sea. The continent is an enormous plateau with terraced tablelands rising one above the other, terminating in the rugged mountains of the east, where the Nile and the Congo rivers rise. The interior plateau is bordered by mountain ranges which run parallel with the coast and descend in terraces to it. The Sahara is shut in between the Atlas Mountains on the north and the southern plateau, and the Congo basin occupies the western part of the peninsula. The southern plateau is much higher than the northern, having average elevations of nearly 1220 m.

relief Africa is bounded north by the Mediterranean Sea, west by the Atlantic Ocean, and east by the Indian Ocean and the Red Sea. Its greatest length from north to south is 8000 km and its breadth from Cape Verde to Ras Hafun is 7440 km. Its area, including Madagascar and the other adjacent islands, is 30,097,000 sq km, or three times that of Europe. The coastline is regular, with no deep seas, bays, or river estuaries of any size to afford climatic or commercial advantages; so that in proportion to its size Africa has less coastline than any other continent, its total length being about 25,600 km.

mountains The mountains of Africa may be divided into three distinct systems:
(1) the Atlas;
(2) the west coast; and
(3) the east coast.
The **Atlas Mountains** occupy the northern portion between the sea and the Sahara, from Wadi Daa to Cape Bon. The eastern Atlas, 1830–2440 m high, consist of two parallel ranges enclosing a plateau where salt lakes called chotts are found. The western Atlas, known as the Great Atlas of

Morocco, have an average elevation of 3050 m, and the highest peaks are over 4275 m. The **west coast system** consists of the Cameroon Highlands, 4000–4300 m high, and the highlands of Lower Guinea, known as the Kong Mountains. The **east coast system**, which is the most important, consists of:

(a) the southern section, containing the Drakensberg with Thabane Ntlenyana, 3482 m; and the Sneeuberge with Kompasberg, 2504 m; in Eastern Cape Province is the enormous plateau called the Great Karoo, a region of arid plains, the total area of which is 260,000 sq km.

(b) the section between the Zambezi and Ethiopia, containing the highest peaks in Africa and the Great Lakes; Kilimanjaro, 5895 m, and Mount Kenya, 5199 m, are extinct volcanic peaks. The Livingstone Range, near Lake Nyasa, is 3355 m high; the Ruwenzori Range ('Mountains of the Moon'), between Lake Albert and Lake Edward, is 4880–6100 m high; and Mount Mfumbiro, between Lake Albert and Lake Victoria, is 4507 m high.

(c) the Ethiopian system rises abruptly from the coast and gradually descends towards the Nile, and contains Ras Dashan and Abba Yared, both 4620 m.

deserts There are three great deserts: the Sahara, the largest in the world, in the north, and the Kalahari and Namib, both sandy regions with low rainfall, in the south. The Libyan and Nubian deserts are continuations of the Sahara. As well as extensive areas of sand, the Sahara, Kalahari, and Namib deserts have thousands of square kilometres of rocky expanses with scrubland that can support small antelopes, ostriches, and camels.

rivers Considering its size, and compared with other continents, Africa has relatively few rivers, and its commercial prosperity has been greatly retarded by the want of navigable rivers with good harbours. Most rivers are impeded by cataracts. The most important are the Nile, Congo, Zambezi, and Niger.

The **Nile** is of political, historical, and commercial importance, and its overflow is of vital importance to Egypt. This flow is now controlled by the Aswan High Dam. The great lakes connected with the Nile are Victoria, Albert, and Edward.

The **Congo**, which drains an equatorial rains region, has a constant water supply, and between Boyoma Falls and Pool Malebo the river is navigable for 1600 km. Lake Tanganyika supplies the Congo River with a considerable volume of water.

The **Zambezi** is the chief river in the east, and though navigable in some parts, its course is impeded by cataracts and rapids. The Victoria Falls, the largest in the world, are situated on this river.

The **Niger** is of great commercial importance, being navigable almost entirely from its source to its mouth in the Gulf of Guinea.

The **Orange** River is not useful for navigation, though imaginative schemes have been embarked upon to use the water for irrigation and industrial purposes; but the **Limpopo**, with its mouth just north of Delagoa Bay, is navigable for about 100 km.

The **Senegal**, **Gambia**, and **Ogooué** flow into the Atlantic on the west, and provide navigable waterways for some distance from their mouths.

islands Africa has very few islands, and they are all small with the exception of Madagascar (590,500 sq km) which is one of the largest in the world. In the North Atlantic are the Madeira Islands, the Canary Islands, and the Cape Verde Islands. Bioko and Annobón are two volcanic islands in the Gulf of Guinea; other nearby islands are São Tomé and Príncipe. St Helena and Ascension are solitary rocks in the Atlantic. To the east in the Indian Ocean are Madagascar, Mauritius, Réunion, the Seychelles, British Indian Ocean Territory, Comoros, and Zanzibar. Further north are Socotra and Kamaran, and in the Red Sea are Perim and Dahlak.

climate Nearly three-quarters of the total area of Africa lies within the tropics, so that there is almost perpetual summer with definite seasons of rain and drought. The variations in climate are caused by prevailing winds and height. Mounts Ruwenzori and Kenya, almost on the Equator, are covered with perpetual snow for 600–900 m downwards from their summits, and there is also perpetual snow on many peaks in Ethiopia. Many mountain slopes are very fertile, yielding different vegetation according to altitude. The prevailing winds are from the northeast and the southeast. The northeasterly winds, having come across Asia, bring no rain to North Africa. The southeasterly winds bring moisture to coastal districts, but it falls as rain on the mountains and does not reach the interior, hence the Kalahari Desert. The region of the tropical rains extends from 18° N–20° S.

African geology the continent of Africa is made up of a vast stable crystalline basement of very old rocks, mainly of *Precambrian age. Superimposed on this basement are later, largely flat-lying cover successions; along the east, north, and west coasts there are sediments of *Mesozoic and *Tertiary age, deposited in marginal marine basins.

Other major features are the Atlas Mountains in the extreme northwest, a part of the Alpine fold belt of Europe, and the *Palaeozoic Cape fold belt in the extreme south. In the east, the continent is rifted north–south along the lines of the East African Rift System, and there are huge outpourings of volcanic rocks associated with the Great Rift Valley.

The **Precambrian basement** can be divided into three large masses or *cratons; these are the Kalahari, Congo, and West African cratons. They are separated from each other by a number of mobile belts active in late Precambrian and early Palaeozoic times.

The **Kalahari craton** occupies much of southern Africa and contains some of the oldest known rocks and microfossils in the world. The oldest rocks occur in the **Transvaal province**; in this area the basement consists of *granites, *gneisses, and migmatites around 3400 million years old. Within this granitic basement are a number of greenstone belts where the rocks have been less highly metamorphosed and contain many primary features. These rocks comprise the Swaziland system. They form a thick pile of basic volcanics and cherts, which pass up into turbidites and sandstones. Gold ores (the first to be worked in South Africa) occur in fracture systems in the greenstone belts. The whole of this ancient basement became stabilised around 3000 million years ago, and was then covered by a thick sequence of shallow-water sediments, lavas, and igneous intrusions. Although of great age (deposited between 2800 and 2000 million years ago), the rocks are remarkably well preserved, and are of great economic importance. The **Witwatersrand system** at the base is the world's most productive source of gold; the younger **Transvaal system** contains important iron ores, and the bushveld igneous complex contains platinum, chromium, titanium-iron, and tin ores.

To the north of the Transvaal province, and separated from it by the younger Limpopo belt, is the **Rhodesian province**. This is composed of a similar granitic basement with narrow greenstone belts; within the greenstone belts, the sequence consists of basic volcanics, covered by greywackes, *shales, and *conglomerates. The volcanic rocks are the main gold-bearing rocks of Zimbabwe.

The **Limpopo belt** runs east–northeast separating the Rhodesian and Transvaal massifs, and is a belt of high-grade metamorphic rocks involved in a long cycle of metamorphism and deformation that ended 2000 million years ago, after the stabilisation of the adjacent massifs.

The final component of the Kalahari craton is the much younger **Orange River belt**, which loops around the southern margin of the Transvaal massif. This belt of high-grade metamorphic rocks became stabilised around 1000 million years ago.

The **Congo craton** occupies a large part of central southern Africa; its oldest rocks occur in the Tanzania province, an area of granitic basement and greenstone belts similar in structure to the Rhodesian province. Two younger belts of metamorphic rocks, the **Ubendian and Toro belts**, represent mobile belts active about 1800 million years ago. Another distinctive but younger mobile belt is the **Karagwe-Ankole belt**, which runs northeast–southwest for at least 1500 km from Uganda to Zambia. It was active 1400–1000 million years ago and suffered several periods of deformation.

The **West African craton** comprises practically the whole of western Africa from the Gulf of Guinea to the Anti-Atlas Mountains of Morocco. It is bounded east, west, and north by much younger mobile belts, while in the Sahara area it is often covered by later Phanerozoic sediments. The oldest rocks occur as scattered masses of highly metamorphosed rocks, metamorphosed during the period 2900–2500 million years ago. Rocks of the younger **Eburian province** outcrop over a wide area, occurring in Ghana, Côte d'Ivoire, and Sierra Leone. This province consists of a group of sediments and volcanics, the Birrimian group, which has been folded, metamorphosed, and intruded by a large suite of granites, the Eburian granites (1900–2200 million years old). Some of these granites in Ghana are associated with gold-quartz reefs. The final stages of activity in the craton were marked by the deposition of late Precambrian sediments, laid down unconformably on the stabilised craton.

Late Precambrian and early Palaeozoic mobile belts A number of well-defined mobile belts became established in the continent in late Precambrian times; these were the Mozambique belt, which stretches north–south along the east side of the continent from Egypt to Mozambique; the Katanga belt, which runs east–west between the Congo and Kalahari cratons; and two other less well-defined belts running through the Cape province and along the west coast of the Democratic Republic of Congo (formerly Zaire). At this time Africa was joined to the other southern continents as part of the super-continent of *Gondwanaland, and these late Precambrian belts can be traced across into the other continents.

The **Mozambique belt** is a 4000-km long belt of high-grade metamorphic rocks, representing a mobile belt which began forming in Precambrian times and underwent several phases of folding and metamorphism before its final phase of regional metamorphism some 600 million years ago at the end of the Precambrian.

The **Katanga belt** is a broad curved zone of late Precambrian orogenic activity (see *orogeny), which passes through the copper belt of Zambia and the mining district of Katanga (Shaba). The rocks forming the Katanga system are a thick succession of shallow-water sediments, up to 10 km thick. The copper deposits are sulphide ores lying mainly within shale horizons. The copper-bearing rocks have only suffered low-grade metamorphism, but towards the centre of the mobile belt metamorphism is of a higher grade. Activity in this belt ceased around 570 million years ago, with the intrusion of mantled gneiss domes, reactivated in the final stage of the orogenic cycle.

The **West Congo belt** is a sedimentary succession, similar to the Katanga system, which runs north–south for 1000 km along the west side of the Congo craton. The orogenic activity in all these belts had ceased by about 500 million years ago, and by this time the whole continent of Gondwanaland, with Africa at its core, had become a single stabilised craton encircled by zones of continuing mobile belt activity.

The **Mauritanide belt** in northwestern Africa is a mobile belt active in Palaeozoic times and having its final phase of folding in Hercynian times. Much of it is covered by younger sediments. The **Cape fold belt** cuts across the southern tip of Africa, and is another belt of Palaeozoic activity, stabilised in Upper Palaeozoic times.

glaciation Towards the end of the Upper Palaeozoic new basins of deposition were initiated within the continent and on its margins. This phase began with a widespread glaciation in Permo-Carboniferous times. Huge thicknesses of glacial deposits – tillites, *varved clays, and sandstones – were laid down; the Dykwa glacial sequence of the Karoo basin is 800–900 m thick, suggesting a lengthy glaciation. Sedimentation in the continental Karoo basin continued without interruption through Permian, Triassic, and Jurassic times. The Permian glacial deposits were closely followed by the deposition of coal measures (of economic importance in the Wankie basin of Zimbabwe), followed in turn by shales, sandstones, and finally the Karoo basalts. Large volumes of plateau basalts, up to 1 km thick, poured out in late Triassic and Jurassic times, and marking the beginning of the splitting up of the continent of Gondwanaland.

During the initial phases of **continental disruption** in late Jurassic and early Cretaceous times, narrow marine troughs developed at the sites of the future continental margins, and faulting and flexing in the basement brought about the formation of the East African Rift System. At the same time, basic and alkaline igneous activity was widespread.

In **Cretaceous** times the continents began to move apart, and since that time Africa has remained a major land mass, the only orogenic activity being that on the extreme northwestern corner of the continent, in the Atlas belt, which is of Alpine age. Fluviatile and lacustrine sediments accumulated in a number of interior basins, while thin cover sequences were laid down on the continental margins in some regions.

The **East African Rift System**, which is part of a 5000-km fracture zone extending from the Limpopo valley in the south to the Jordan valley in the north, came into existence in its southern part in the late Mesozoic, and was associated with voluminous igneous activity. Further extensive igneous activity is seen in the Tertiary to recent volcanics of the northern part of the rift, which originated in mid-Tertiary times. Fault movement forming the rift valley took place mainly in Miocene and Pleistocene times, and the area is still marked by seismic activity, volcanism, and high heat flow through the crust.

the Red Sea During Oligocene and Miocene times a new marine trough was developed on the site of the present Red Sea; *clays, *marls, *evaporites, and *limestones accumulated in Miocene and Pliocene times, and then the two sides began to split apart as new sea floor was generated at the floor of the trough. The Red Sea therefore seems to be a new ocean basin in the making.

African Union AU; formerly **Organization of African Unity** (1963–2001), association established in 1963 as the Organization of African Unity (OAU) to eradicate colonialism and improve economic, cultural, and political cooperation in Africa. Its headquarters are in Addis Ababa, Ethiopia. There are 53 members representing virtually the whole of central, southern, and northern Africa.

agate cryptocrystalline (with crystals too small to be seen with an optical microscope) silica, SiO_2, composed of cloudy and banded *chalcedony, sometimes mixed with *opal, that forms in rock cavities.

Agenda 21 non-binding treaty, signed by representatives of 178 countries in 1992, which sets out a framework of recommendations designed to protect the environment and achieve sustainable development. The treaty highlights the importance of international cooperation, but also discusses the role of individuals, communities, and local authorities in achieving the goals it sets out.

age–sex graph graph of the population of an area (see *population pyramid) showing age and sex distribution.

agglomerate a mass of coarse rock fragments or blocks of lava produced during a volcanic eruption.

agglomeration the clustering of activities or people at specific points or areas; for example, at a route centre. Firms or individuals that cluster together can often share facilities and services, resulting in lower costs (**agglomeration economies**, including *economy of scale).

agribusiness modern intensive farming which uses machinery and artificial fertilizers to increase yield and output. Thus agriculture resembles an industrial process in which the general running and managing of the farm could parallel that of large-scale industry. In the UK, many farms in East Anglia could be called agribusinesses. There small family farms have been bought up by large food-processing companies located near the farms, where the farm produce is canned, processed, frozen, etc.

A 'branch' of agribusiness is factory farming, i.e. the intensive rearing of animals such as pigs, cattle and poultry. Many of these animals spend their lives penned in special rearing units where food, water, light and temperature are carefully controlled. Thus factory farming can be seen as a production line, the sole aim being to increase output and profit. Many people disagree with this form of farming, but it is argued that less intensive methods mean higher prices for the consumer.

agrochemical artificially produced chemical used in modern, intensive agricultural systems. Agrochemicals include nitrate and phosphate fertilizers, pesticides, some animal-feed additives, and pharmaceuticals. Many are responsible for pollution and almost all are avoided by organic farmers.

agriculture human management of the environment to produce food. The numerous forms of agriculture fall into three groups: commercial agriculture, subsistence agriculture and peasant agriculture. See *agribusiness.

agroforestry agricultural system which involves planting and maintaining trees on land where crops and animals are also raised. The trees may be used for timber, fuel, food, or animal fodder. Trees may be interplanted with other crops in order to provide shade, reduce soil erosion, and increase moisture retention in the soil. Leaves dropped by the trees also improve the fertility of the soil.

A horizon see *soil.

aid financial or other assistance given or lent, on favourable terms, by richer, usually industrialized, countries to war-damaged or developing states. It may be given for political, commercial, or humanitarian reasons, or a combination of all three. A distinction may be made between **short-term aid** (usually food and medicine), which is given to relieve conditions in emergencies such as famine, and **long-term aid**, or **development aid**, which is intended to promote economic activity and improve the quality of life – for example, by funding irrigation, education, and communications programmes.

In 1970, all industrialized United Nations (UN) member countries committed to giving at least 0.7% of their gross national product (GNP). However, by 2000 only five had reached this target: Denmark, the Netherlands, Norway, Sweden, and Luxembourg; the actual average among the industrial countries in the same year was around 0.32%. The four largest donors to poor countries in

2000 were Japan, which spent $13 billion on official development assistance, the USA ($9.6 billion), Germany ($5 billion), and the UK ($4.5 billion/ £2.94 billion). Each country spends more than half its contribution on direct bilateral (by agreement with another country) assistance to countries with which they have historical or military links, hope to encourage trade, or regard as strategically important – Russia or Indonesia, for example. The rest goes to international organizations such as UN and *World Bank agencies, which distribute aid multilaterally. The World Bank is the largest dispenser of aid.

air the mixture of gases making up the Earth's *atmosphere.

airglow faint and variable light in the Earth's atmosphere produced by chemical reactions (the recombination of ionized particles) in the ionosphere.

air mass large body of air with particular characteristics of temperature and humidity. An air mass forms when air rests over an area long enough to pick up the conditions of that area. When an air mass moves to another area it affects the *weather of that area, but its own characteristics become modified in the process. For example, an air mass formed over the Sahara will be hot and dry, becoming cooler as it moves northwards. Air masses that meet form **fronts**.

There are four types of air mass. **Tropical continental** (Tc) air masses form over warm land and **Tropical maritime** (Tm) masses form over warm seas. Air masses that form over cold land are called **Polar continental** (Pc) and those forming over cold seas are called **Polar** or **Arctic maritime** (Pm or Am).

The weather of the UK is affected by a number of air masses which, having different characteristics, bring different weather conditions. For example, an

Arctic air mass brings cold conditions whereas a Tropical air mass brings hot conditions.

air pollution contamination of the atmosphere caused by the discharge, accidental or deliberate, of a wide range of toxic airborne substances. Often the amount of the released substance is relatively high in a certain locality, so the harmful effects become more noticeable. The cost of preventing any discharge of pollutants into the air is prohibitive, so attempts are more usually made to reduce the amount of discharge gradually and to disperse it as quickly as possible by using a very tall chimney, or by intermittent release.

One major air pollutant is sulphur dioxide (SO$_2$), produced from the burning of *fossil fuels. It dissolves in atmospheric moisture to produce sulphurous acid (*acid rain):

$$SO_2 + H_2O = H_2SO_3$$

The greatest single cause of air pollution in the UK is the car, which is responsible for 85% of the carbon monoxide and 45% of the oxides of nitrogen present in the atmosphere. According to a UK government report 1998, air pollution causes up to 24,000 deaths in Britain per year.

other major pollutants Oxides of nitrogen (NO$_x$), produced by car exhausts, also contribute to acid rain. The poisonous gas carbon monoxide (CO) results from the partial combustion of fuels and is particularly prevalent where there is heavy traffic, contributing to **photochemical smog**. The air pollution caused by lead compounds from leaded petrol, particularly tetraethyl lead (Pb(C$_2$H$_5$)$_4$), can cause learning impairment in children living near big roads. The air pollution caused by small particles (*PM10s), which results mainly from vehicle emissions, is also a major health risk.

recent events Possibly the world's worst-ever air pollution disaster caused by people occurred in Indonesia in September 1997. It was caused by forest clearance fires. Smoke pollution in the city of Palangkaraya reached 7.5 mg per cu m (nearly 3 mg more than in the London smog of 1952 in which 4000 people died). The pollutants spread to Malaysia and other countries of the region.

The 1997 *Kyoto Protocol committed the industrialized nations of the world to cutting their levels of harmful gas emissions.

A 1997 survey of contamination in mosses showed that Slovakia and southern Poland are the most polluted areas in central Europe, with high levels of heavy-metal pollutants including cadmium, lead, copper, and zinc caused by chemical and metal smelting industries in Slovakia, the northeastern Czech Republic, and Poland.

air pressure in earth science, pressure of the atmosphere at any place on Earth; see *atmospheric pressure.

alabaster naturally occurring fine-grained white or light-coloured translucent form of *gypsum, often streaked or mottled. A soft material, it is easily carved, but seldom used for outdoor sculpture.

albedo the proportion of solar radiation which reaches the Earth's surface and is immediately reflected back into the atmosphere. The average albedo for the Earth's surface is 40% but this figure varies greatly according to the nature of the surface. For example, the albedo for new snow is around 80% whilst the figure for forests and wet soil ranges between 5% and 10%.

alexandrite rare gemstone variety of the mineral chrysoberyl (beryllium aluminium oxide, BeAl$_2$O$_4$), which is green in daylight but appears red in artificial light.

algal bloom   the temporary, rapid growth of algae in fresh water. During the early part of the growing season an abundance of nutrients in the surface waters combined with an increase in water temperature allows algae (and other aquatic plants) to rapidly multiply until one nutrient becomes scarce or limiting, usually nitrogen or phosphorus. The phase of rapid growth is replaced by a massive and sudden mortality of algae which rise to the surface of the lake forming a green scum. The decomposing algae consume large quantities of oxygen dissolved in the water causing the lake waters to become anaerobic, killing most other aquatic life (primarily fish). The waters gradually undergo a self-cleansing process as oxygen levels are re-established. It is possible for a second bloom to occur under favourable autumn conditions.

The occurrence of algal blooms has increased in recent years due to enhanced levels of nitrogen and phosphatic materials leaching from agricultural fields into water courses. Reservoirs and shallow lakes located in intensive agricultural areas are susceptible to algal blooms. Reservoirs must sometimes be drained and pipework and filters carefully cleaned to remove the green slime produced by the algal bloom.

alienation   sense of isolation, powerlessness, and therefore frustration; a feeling of loss of control over one's life; a sense of estrangement from society or even from oneself. As a concept it was developed by German philosophers G W F Hegel and Karl Marx; the latter used it as a description and criticism of the condition that developed among workers in capitalist society.

alkylphenolethoxylate   APEO, chemical used mainly in detergents, but also in herbicides, cleaners,

packaging, and paints. Nonylphenol, a breakdown product of APEOs is a significant river pollutant; 60% of APEOs end up in the water. Nonylphenol, and other APEO breakdown products, have a feminizing effect on fish: male fish start to produce yolk protein and the growth of testes is slowed.

allotment   a small portion of ground rented annually by an individual for the purpose of growing flowers or vegetables for personal use or recreation. Allotments are most commonly found in the UK and are worked by amateur gardeners who do not have land attached to their own house.

alluvial deposit   layer of broken rocky matter, or sediment, formed from material that has been carried in suspension by a river or stream and dropped as the velocity of the current decreases. River plains and deltas are made entirely of alluvial deposits, but smaller pockets can be found in the beds of upland torrents.

Alluvial deposits can consist of a whole range of particle sizes, from boulders down through cobbles, pebbles, gravel, sand, silt, and clay. The raw materials are the rocks and soils of upland areas that are loosened by erosion and washed away by mountain streams. Much of the world's richest farmland lies on alluvial deposits. These deposits can also provide an economic source of minerals. River currents produce a sorting action, with particles of heavy material deposited first while lighter materials are washed downstream.

Hence heavy minerals such as gold and tin, present in the original rocks in small amounts, can be concentrated and deposited on stream beds in commercial quantities. Such deposits are called 'placer ores'.

alluvial fan roughly triangular sedimentary formation found at the base of slopes. An alluvial fan results when a sediment-laden stream or river rapidly deposits its load of gravel and silt as its speed is reduced on entering a plain.

The surface of such a fan slopes outward in a wide arc from an apex at the mouth of the steep valley. A small stream carrying a load of coarse particles builds a shorter, steeper fan than a large stream carrying a load of fine particles. Over time, the fan tends to become destroyed piecemeal by the continuing headward and downward erosion levelling the slope.

alluvial soil **fluvent** or **fluvisol**, a soil type of highly variable nature, found in river channels or floodplains on recently deposited *alluvium. Many alluvial soils are highly productive, with a nutrient content maintained at very high levels due to the regular flushing in of nutrients by flood waters. The lower courses of large rivers frequently contain alluvial soils.

alluvium sediments deposited by a river in its middle and lower course where the water's velocity is too low to transport the river's load – for example, on the inside bend of a *meander. Alluvium comprises silt, sand and coarser debris eroded from the river's upper course and transported downstream. Alluvium is deposited in a graded sequence: coarsest first (heaviest) and finest last (lightest). Regular floods in the lower course create extensive layers of alluvium which build up to considerable depth on the *flood plain. Some of the world's most fertile lands are found on alluvial flood plains.

The loose, unconsolidated material forms features such as river terraces, and *deltas.

alp a gentle slope above the steep sides of a glaciated valley, often used for summer grazing. See *transhumance.

alpha share on the stock market, a share in any of the companies most commonly traded – that is, the larger companies.

alternative energy see *energy, alternative.

altimetry process of measuring altitude, or elevation. Satellite altimetry involves using an instrument – commonly a laser – to measure the distance between the satellite and the ground.

alumina or **corundum**; Al_2O_3, oxide of aluminium, widely distributed in clays, slates, and shales. It is formed by the decomposition of the feldspars in granite and used as an abrasive. Typically it is a white powder, soluble in most strong acids or caustic alkalis but not in water. Impure alumina is called 'emery'. Rubies, sapphires, and topaz are corundum gemstones. It is the chief component of bauxite.

aluminium ore raw material from which aluminium is extracted. The main ore is *bauxite, a mixture of minerals, found in economic quantities in Australia, Guinea, the West Indies, and several other countries.

Amazon Pact treaty signed 1978 by Bolivia, Brazil, Colombia, Ecuador, Guyana, Peru, Suriname, and Venezuela to protect and control the industrial or commercial development of the Amazon River.

amenity resources those resources which provide an opportunity for recreation and leisure pursuits. The national parks, forests and the coastline are good examples of these, as are areas of parkland and open space in cities.

amethyst variety of *quartz, SiO_2, coloured violet by the presence of small quantities of impurities such as manganese or iron; used as a semi-precious stone. Amethysts are found chiefly in the Ural Mountains, India, the USA, Uruguay, and Brazil.

Amoco Cadiz   US-owned oil tanker that ran aground off the French coast in 1978, releasing 225,000 tonnes of crude oil. The environmental impact was immense – 30,000 seabirds died, 130 beaches were immersed in a 30 cm layer of oil, and 230,000 tonnes of crustacea and fish perished. See *oil spill.

amphibole   any one of a large group of rock-forming silicate minerals with an internal structure based on double chains of silicon and oxygen, and with a general formula $X_2Y_5Si_8O_{22}(OH)_2$; closely related to *pyroxene. Amphiboles form orthorhombic, monoclinic, and triclinic crystals.

Amsterdam Treaty   amendment to the founding treaties of the *European Union (EU) adopted at the Amsterdam European Council in June 1997, signed on 2 October 1997, and in force from 1 May 1999. The treaty provided for the further protection and extension of the rights of citizens; new mechanisms to improve the coordination of the EU's common foreign and security policy; and the strengthening of the EU institutions (extending the co-decision procedure by which legislative proposals need the agreement of both the *Council of the European Union and the European Parliament).

anabatic wind   warm wind that blows uphill in steep-sided valleys in the early morning. As the sides of a valley warm up in the morning the air above is also warmed and rises up the valley to give a gentle breeze. By contrast, a *katabatic wind is cool and blows down a valley at night.

anabranch   (Greek *ana* 'again') stream that branches from a main river, then reunites with it.

For example, the Great Anabranch in New South Wales, Australia, leaves the Darling near Menindee, and joins the Murray below the Darling–Murray confluence.

anaerobic   not requiring oxygen for the release of energy from food molecules such as glucose. An organism is described as anaerobic if it does not require oxygen in order to survive. Instead, anaerobic organisms use anaerobic respiration to obtain energy from food. Most anaerobic organisms are micro-organisms such as bacteria, yeasts, and internal parasites that live in places where there is never much oxygen, such as in the mud at the bottom of a lake or pond, or in the alimentary canal. Anaerobic organisms release much less of the available energy from their food than do *aerobic organisms.

anarchism   (Greek *anarkhos* 'without ruler') political belief that society should have no government, laws, police, or other authority, but should be a free association of all its members. It does not mean 'without order'; most theories of anarchism imply an order of a very strict and symmetrical kind, but they believe that such order can be achieved by cooperation. Anarchism must not be confused with nihilism (a purely negative and destructive activity directed against society); anarchism is essentially a pacifist movement.

andalusite   aluminium silicate, Al_2SiO_5, a white to pinkish mineral crystallizing as square- or rhombus-based prisms. It is common in metamorphic rocks formed from clay sediments under low-pressure conditions. Andalusite, kyanite, and sillimanite are all polymorphs of Al_2SiO_5.

Andean Group   Spanish **Grupo Andino**, South American organization aimed at economic and social cooperation between member states. It was established under the Treaty of Cartagena (1969), by Bolivia, Chile, Colombia, Ecuador, and Peru. Venezuela joined in 1973, but Chile withdrew in 1976 and Peru suspended membership and became an observer in 1992. The Andean Pact was established in 1992 to create a free trade area. The organization is based in Lima, Peru.

anemometer an instrument for measuring the velocity of the wind. An anemometer should be fixed on a post at least 5 m above ground level. The wind blows the cups round and the speed is read off the dial in km/hr (or knots).

andesite volcanic igneous rock, intermediate in silica content between rhyolite and basalt. It is characterized by a large quantity of feldspar *minerals, giving it a light colour. Andesite erupts from volcanoes at destructive plate margins (where one plate of the Earth's surface moves beneath another; see *plate tectonics), including the Andes, from which it gets its name.

aneroid barometer kind of *barometer.

angle of declination angle at a particular point on the Earth's surface between the direction of the true or geographic North Pole and the magnetic north pole. The angle of declination has varied over time because of the slow drift in the position of the magnetic north pole.

angle of dip or **angle of inclination**, angle at a particular point on the Earth's surface between the direction of the Earth's magnetic field and the horizontal; see *dip, magnetic.

animal rights the extension of the concept of human rights to animals on the grounds that animals may not be able to reason but can suffer and are easily exploited by humans. The **animal-rights movement** is a general description for a wide range of organizations, both national and local, that take a more radical approach than the traditional **welfare** societies.

More radical still is the concept of **animal liberation**, that animals should not be used or exploited in any way at all.

annual rings or **growth rings**, concentric rings visible on the wood of a cut tree trunk or other woody stem. Each ring represents a period of growth when new xylem is laid down to replace tissue being converted into wood (secondary xylem). The wood formed from xylem produced in the spring and early summer has larger and more numerous vessels than the wood formed from xylem produced in autumn when growth is slowing down. The result is a clear boundary between the pale spring wood and the denser, darker autumn wood. Annual rings may be used to estimate the age of the plant (see *dendrochronology), although occasionally more than one growth ring is produced in a given year.

annual variation in earth science, regular change in the strength and direction of the Earth's magnetic field, which reaches a maximum during years of high sunspot activity. It must be allowed for by navigators using a compass.

Antarctic Circle imaginary line that encircles the South Pole at latitude 66° 32′ S. The line encompasses the continent of Antarctica and the Antarctic Ocean.

The region south of this line experiences at least one night in the southern summer during which the Sun never sets, and at least one day in the southern winter during which the Sun never rises.

Antarctic Treaty international agreement between 13 nations aiming to promote scientific research and keep Antarctica free from conflict, dating from 1961. In 1991 a 50-year ban on mining activity was secured. An environmental protection protocol, addressing the issues of wildlife conservation, mineral exploitation, and marine pollution, came into effect in January 1998 after it was ratified by Japan. Antarctica is now a designated 'natural reserve devoted to peace and science'.

The agreement was signed 1959 between 13 nations with an interest in Antarctica (including the USA and Britain), and came into force 1961 for a 30-year period. A total of 39 countries are party to it (as of 1996). Its provisions (covering the area south of latitude 60° S) neither accepted nor rejected any nation's territorial claims, but barred any new ones; imposed a ban on military operations and large-scale mineral extraction; and allowed for free exchange of scientific data from bases. In 1980 the treaty was extended to conserve marine resources within the larger area bordered by the Antarctic Convergence; and in 1991 an agreement was signed extending the Antarctic Treaty and imposing a 50-year ban on mining activity.

antecedent drainage   an established river drainage system able to maintain its original course by downcutting at the same rate as the surrounding land surface is uplifted and folded by earth movements. Examples of antecedent drainage include the Indus and Brahmaputra in the Himalayas and the Colorado in the USA.

anthracite   (Greek *anthrax* 'coal') hard, dense, shiny variety of *coal, containing over 90% carbon and a low percentage of ash and impurities, which causes it to burn without flame, smoke, or smell. Because of its purity, anthracite gives off relatively little sulphur dioxide when burnt.

anthropology   (Greek *anthropos* 'man', *logos* 'discourse') the study of humankind. It investigates the cultural, social, and physical diversity of the human species, both past and present. It is divided into two broad categories: biological or physical anthropology, which attempts to explain human biological variation from an evolutionary perspective; and the larger field of social or cultural anthropology, which attempts to explain the variety of human cultures. This differs from sociology in that anthropologists are concerned with cultures and societies other than their own.

anticline   in geology, rock layers or beds folded to form a convex arch (seldom preserved intact) in which older rocks comprise the core. Where relative ages of the rock layers, or stratigraphic ages, are not known, convex upward folded rocks are referred to as antiforms.

The fold of an anticline may be undulating or steeply curved. A steplike bend in otherwise gently dipping or horizontal beds is a **monocline**. The opposite of an anticline is a *syncline.

anticyclone   area of high atmospheric pressure caused by descending air, which becomes warm and dry. Winds radiate from a calm centre, taking a clockwise direction in the northern hemisphere and an anticlockwise direction in the southern hemisphere. Anticyclones are characterized by clear weather and the absence of rain and violent winds. In summer they bring hot, sunny days and in winter they bring fine, frosty spells, although fog and low cloud are not uncommon in the UK. Blocking anticyclones, which prevent the normal air circulation of an area, can cause summer droughts and severe winters.

For example, the summer drought in Britain in 1976, and the severe winters of 1947 and 1963, were caused by blocking anticyclones.

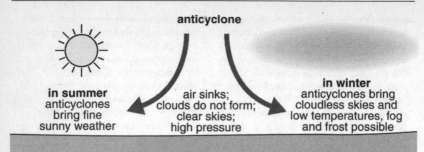

anticyclone

in summer
anticyclones
bring fine
sunny weather

air sinks;
clouds do not form;
clear skies;
high pressure

in winter
anticyclones bring
cloudless skies and
low temperatures, fog
and frost possible

anticyclone Anticyclones are formed when areas of descending air result in high pressure. They bring more stable conditions and can stay in place for days or even months. Summer anticyclones bring hot sunny days. Winter anticyclones bring cooler weather with a possibility of fog and freezing conditions.

antipodes (Greek 'opposite feet') places at opposite points on the globe.
  In the UK, Australia and New Zealand are called the Antipodes.

antitrust laws in economics, regulations preventing or restraining trusts, monopolies, or any business practice considered to be unfair or uncompetitive. In the US, antitrust laws prevent mergers and acquisitions that might create a monopoly situation or ones in which restrictive practices might be stimulated.

apartheid (Afrikaans 'apartness') racial-segregation policy of the government of South Africa from 1948 to 1994. Under the apartheid system, non-whites – classified as Bantu (black), coloured (mixed), or Indian – did not share full rights of citizenship with the white minority. For example, black people could not vote in parliamentary elections, and until 1990 many public facilities and institutions were restricted to the use of one race only. The establishment of Black National States was another manifestation of apartheid. In 1991, after years of internal dissent and violence and the boycott of South Africa, including the imposition of international trade sanctions by the United Nations (UN) and other

organizations, President F W de Klerk repealed the key elements of apartheid legislation and by 1994 apartheid had ceased to exist.
  The term apartheid has also been loosely applied to similar movements and other forms of racial separation, for example social or educational, in other parts of the world.

apatite common calcium phosphate mineral, $Ca_5(PO_4)_3(F,OH,Cl)$. Apatite has a hexagonal structure and occurs widely in igneous rocks, such as pegmatite, and in contact metamorphic rocks, such as marbles. It is used in the manufacture of fertilizer and as a source of phosphorus. Carbonate hydroxylapatite, $Ca_5(PO_4,CO_3)_3(OH)_2$, is the chief constituent of tooth enamel and, together with other related phosphate minerals, is the inorganic constituent of bone. Apatite ranks 5 on the *Mohs scale of hardness.

Appleton layer or **F-layer**, band containing ionized gases in the Earth's upper atmosphere, at a height of 150–1000 km, above the E-layer (formerly the Kennelly-Heaviside layer). It acts as a dependable reflector of radio signals as it is not affected by atmospheric conditions, although its ionic composition varies with the sunspot cycle.

appropriate technology simple or small-scale machinery and tools that, because they are cheap and easy to produce and maintain, may be of most use in the developing world; for example, hand ploughs and simple looms. This equipment may be used to supplement local crafts and traditional skills to encourage small-scale industrialization.

aquaculture the management and use of water environments for the raising and harvesting of plant and animal food products. Whereas fishing and whaling remain essentially hunting activities, aquaculture represents a more intensive utilization of water resources more akin to ranching and agriculture in which fish and shellfish are reared in enclosed ponds, tanks and cages, or on protected beds.

Aquaculture is almost exclusively confined to inland waters and estuarine or other near-shore coastal waters. Developments further out on the continental shelf remain at the experimental stage. The use of ponds to breed and raise fish such as carp dates back several thousand years in parts of China and Southeast Asia, but the extension of aquaculture to marine areas is a recent development. The raising of freshwater or marine fish for commercial purposes is often separately classed as *fish farming. Marine aquaculture enterprises have concentrated on raising shellfish, especially molluscs such as oysters, mussels and clams which are relatively immobile and command high market prices.

aquamarine blue variety of the mineral *beryl. A semi-precious gemstone, it is used in jewellery.

aquifer a body of rock through which appreciable amounts of water can flow. The rock of an aquifer must be porous and permeable (full of interconnected holes) so that it can conduct water.

Aquifers are an important source of fresh water, for example for drinking and irrigation, in many arid areas of the world, and are exploited by the use of *artesian wells.

An aquifer may be underlain, overlain, or sandwiched between less permeable layers, called **aquicludes** or **aquitards**, which impede water movement. Sandstones and porous limestones make the best aquifers.

Arab Common Market organization providing for the abolition of customs duties on agricultural products, and reductions on other items, between the member states: Egypt, Iraq, Jordan, Libya, Mauritania, Syria, and Yemen. It was founded in 1964.

Arab Fund for Economic and Social Development AFESD, association established 1968 by the *League of Arab States to finance economic and social developments in Arab states. Its members include the 21 Arab states that make up the League plus the Palestine Liberation Organization. Its headquarters are in Kuwait.

Arab League or **League of Arab States**, organization of Arab states established in Cairo in 1945 to promote Arab unity, primarily in opposition to Israel. The original members were Egypt, Syria, Iraq, Lebanon, Transjordan (Jordan 1949), Saudi Arabia, and Yemen. They were later joined by Algeria, Bahrain, Comoros, Djibouti, Kuwait, Libya, Mauritania, Morocco, Oman, Palestine, Qatar, Somalia, Sudan, Tunisia, and the United Arab Emirates. In 1979 Egypt was suspended and the league's headquarters transferred to Tunis in protest against the Egypt–Israeli peace, but Egypt was readmitted as a full member in May 1989, and in March 1990 its headquarters returned to Cairo. Despite the strains imposed on it by the 1990–91 Gulf War, the alliance survived.

arable farming the practice of cultivating land to produce cereal and root crops – as opposed to the keeping of livestock. Arable land is usually cultivated annually or is at least fit for ploughing, and may be distinguished from pastures and land devoted to permanent crops of trees and bushes. In mixed cropping regimes, however, arable farming of annual crops may be undertaken between the rows of tree crops.

The geographic boundaries of arable farming are determined by locational deficiencies, specifically in terms of climatic conditions, soil types and relief. The margins of cultivation are reached at about 70° N where the cropping of spring barley and potatoes becomes limited by the shortness of the growing season. In parts of the Andes, the potato is grown at heights of 4300 m, while in the Sahel arable farming of millet survives on barely 250 mm of annual rainfall. In any farming system dependent at least partly on cash crops a further limiting factor on the distribution of arable farming is the distance of the farm from market.

Arab Maghreb Union AMU, association formed in 1989 by Algeria, Libya, Mauritania, Morocco, and Tunisia to formulate common policies on military, economic, international, and cultural issues.

Arab Monetary Fund AMF, money reserve established in 1976 by many Arab states plus the Palestine Liberation Organization (PLO) to provide a mechanism for promoting greater stability in exchange rates and to coordinate Arab economic and monetary policies. It operates mainly by regulating petrodollars within the Arab community to make member countries less dependent on the West for the handling of their surplus money. The fund's headquarters are in Abu Dhabi in the United Arab Emirates.

aragonite white, yellowish or grey mineral, calcium carbonate, $CaCO_3$, a denser, harder polymorph (substance having the same chemical composition but different structure) of calcite.

Secreted by corals, molluscs and green algae, it is an important constituent of shallow marine muds and pearl.

arch in geomorphology (the study of landforms), any natural bridge-like land feature formed by *erosion. Most sea arches are formed from the wave erosion of a *headland where the backs of two caves have met and broken through. The roof of the arch eventually collapses to leave part of the headland isolated in the sea as a *stack. A natural bridge is formed on land by wind or water erosion and spans a valley or ravine.

In some cases, as at Stair Hole, Dorset, England, an arch is formed when the sea has battered a hole through a cliff of hard rock (such as limestone) and has removed soft rock (such as clay) behind.

Archaean or **Archaeozoic**, widely used term for the earliest era of geological time; the first part of the Precambrian **Eon**, spanning the interval from the formation of Earth to about 2500 million years ago.

archipelago group of islands, or an area of sea containing a group of islands. The islands of an archipelago are usually volcanic in origin, and they sometimes represent the tops of peaks in areas around continental margins flooded by the sea.

Volcanic islands are formed either when a hot spot within the Earth's mantle produces a chain of volcanoes on the surface, such as the Hawaiian Archipelago, or at a destructive plate margin (see *plate tectonics) where the subduction of one plate beneath another produces an arc-shaped island group called an 'island arc', such as the Aleutian Archipelago. Novaya Zemlya in the Arctic Ocean, the northern extension of the Ural Mountains, resulted from continental flooding.

Arctic Circle imaginary line that encircles the North Pole at latitude 66° 30′ N. Within this line there is at least one day in the summer during which the Sun never sets, and at least one day in the winter during which the Sun never rises.

The length of the periods of continuous daylight in summer and darkness in winter varies with the nearness to the North Pole; the nearer the Pole the longer the period during which the sun is continually above or below the horizon.

arctic climate Antarctica, all of Greenland, the north of Alaska, Canada, and Russia are areas with continuous *permafrost. Winters are severe and the sea freezes; summers have continuous periods of daylight but the monthly temperatures struggle to rise above freezing point. Nearer each pole the climate is constant frost. Precipitation is light and falls mainly as snow.

arcuate delta see *delta.

arête German **grat**; North American **combe-ridge**, sharp narrow ridge separating two *glacial troughs (U-shaped valleys), or *corries. They are formed by intense *freeze–thaw weathering on the sides of mountains. The typical U-shaped cross-sections of glacial troughs give arêtes very steep sides. Arêtes are common in glaciated mountain regions such as the Rockies, the Himalayas, and the Alps. There are also several in the UK, for example Striding Edge and Swirral Edge in the English Lake District.

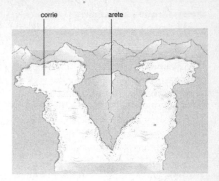

arête An arête is a steep-sided ridge between two glacially-formed valleys, or corries. If three corries all erode towards the same point, a triangular arête may be formed.

arid region in earth science, a region that is very dry and has little vegetation. Aridity depends on temperature, rainfall, and evaporation, and so is difficult to quantify, but an arid area is usually defined as one that receives less than 250 mm of rainfall each year. (By comparison, New York City receives 1120 mm per year.) There are arid regions in North Africa, Pakistan, Australia, the USA, and elsewhere. Very arid regions are *deserts.

The scarcity of fresh water is a problem for the inhabitants of arid zones, and constant research goes into discovering cheap methods of distilling sea water and artificially recharging natural groundwater reservoirs. Another problem is the eradication of salt in irrigation supplies from underground sources or where a surface deposit forms in poorly drained areas.

arkose in earth science, type of *sandstone composed of grains of quartz and feldspar, derived from crystalline rocks such as gneiss or granite. It may be grey, pink, or red, often coloured by iron oxide impurities.

artesian basin a shallow syncline with a layer of permeable rock, e.g. chalk, sandwiched between two layers of impermeable rock, e.g. clay. Where the permeable rock is exposed at the surface, rainwater will enter the rock and the rock will become saturated. This is known as an *aquifer*. Boreholes can be sunk into the structure to tap the water in the aquifer. If there is sufficient pressure of water within the aquifer, the water will rise freely to the surface (*artesian well). The London Basin consists of a shallow syncline formed of a layer of chalk between two layers of clay. London's *water table dropped considerably during the period of significant industrialization (due to demand by the water companies and industry). However, industry has now relocated elsewhere and the water table is rising, causing problems to buildings with deep foundations.

artesian well well that is supplied with water rising naturally from an underground water-saturated rock layer (*aquifer). The water rises from the aquifer under its own pressure. Such a well may be drilled into an aquifer that is confined by impermeable rocks both above and below. If the water table (the top of the region of water saturation) in that aquifer is above the level of the well head, hydrostatic pressure will force the water to the surface.

Artesian wells are often over-exploited because their water is fresh and easily available, and they eventually become unreliable. There is also some concern that pollutants such as pesticides or nitrates can seep into the aquifers.

Much use is made of artesian wells in eastern Australia, where aquifers filled by water in the Great Dividing Range run beneath the arid surface of the Simpson Desert. The artesian well is named after Artois, a French province, where the phenomenon was first observed.

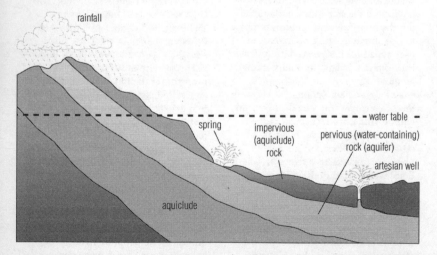

artesian well In an artesian well, water rises from an underground water-containing rock layer under its own pressure. Rain falls at one end of the water-bearing layer, or aquifer, and percolates through the layer. The layer fills with water up to the level of the water table. Water will flow from a well under its own pressure if the well head is below the level of the water table.

artificial fertilizer   a chemical product containing one or all of the following: nitrogen, potash, phosphates and trace elements, which are derived from petrochemicals, natural phosphate deposits and other industrial sources. In contrast *organic fertilizers include animal dung, rotted vegetation (compost) and animal derivatives such as bone meal. Artificial fertilizers, if leached into streams and rivers, can cause problems of *eutrophication.

artificial fibres   textile materials, e.g. nylon, rayon, viscose, and polyester, made from hydro-carbons, derived from such materials as oil, *coal and wood fibres, in contrast with natural fibres such as wool and cotton.

asbestos   any of several related minerals of fibrous structure that offer great heat resistance because of their nonflammability and poor conductivity. Commercial asbestos is generally either made from serpentine ('white' asbestos) or from sodium iron silicate ('blue' asbestos). The fibres are woven together or bound by an inert material. Over time the fibres can work loose and, because they are small enough to float freely in the air or be inhaled, asbestos usage is now strictly controlled; exposure to its dust can cause cancer.

ASEAN   acronym for *Association of South East Asian Nations.

Asian Development Bank   ADB, bank founded in 1966 to stimulate growth in Asia and the Far East by administering direct loans and technical assistance. Members include 30 countries within the region and 14 countries of Western Europe and North America. The headquarters are in Manila, Philippines.

Asian geography   Asia is the largest continent, occupying the northern portion of the eastern hemisphere, extending beyond the Arctic Circle, and nearly reaching the Equator. It contains about one-third of the whole of the dry land, and one-twelfth of the whole surface of the globe.

Geographically speaking, Europe is an appendix to Asia, and exact delimitation between the two is impossible. The Ural River and mountains are the common conventional boundaries with Europe north of the Caspian, while the Manych depression is used as the limit of Asia between the Black and the Caspian seas, and the Bering Strait, 60 km wide, separates Asia from North America. Asia's northern boundary is the Arctic Ocean, the extreme north point being Cape Sievero-Vostochny. The southern boundary is impossible to fix exactly, but the volcanic chain of islands that can be traced through the Molucca and Sundra Islands may be taken as the limit. The southern coastline is much more irregular, and broken by the three great peninsulas of Arabia, India, and Indochina. The Mediterranean and Black seas form natural western limits to the continent, as does the Red Sea lying between Asia and Africa.

relief Asia can be divided into five broad zones: northern, highland, arid, tropical, and insular. It is a continent of contrasts, containing both the highest point on the Earth's surface (Mount Everest) and the lowest (the Dead Sea).

the northern region This comprises the Arctic wastes of Siberia, a region of plains, plateaux, and folded mountain ranges, where the major influences on relief are frost weathering and *permafrost. The far north has been subjected to several periods of glaciation and here erosive influences dominate, while in the south of the region, sedimentary features become more prevalent. South–north-flowing rivers such as the Ob, Lena, and Yenisey form the major transport arteries of the region, despite being frozen over for most of the year.

highland Asia This region includes not only the Himalayas but also the great mountain ranges that radiate from the high Pamirs to the northwest of the Indian subcontinent: the Tian Shan to the northeast, the Kunlun Shan to the east, the Karakoram and Himalayas to the southeast, and the Hindu Kush to the southwest. Between these ranges are the high plateaux of Tibet (the most extensive in the world) and Mongolia. This central mass of mountains and plateaux forms an effective east–west and north–south barrier, enclosing regions of inland drainage such as the Tarim Basin, yet giving rise to many of the major rivers of Southeast Asia, namely the Indus, Ganges, Brahmaputra, Salween, Mekong, Chang Jiang, and Huang He. From this central mass, ridges like fingers point east and southeast into China and Southeast Asia, forming the mountainous backbone of Indochina and Peninsular Malaysia. The offshore islands are also very mountainous, forming a broken arc from the Kamchatka peninsula to Java and Sumatra, and are related formations.

the arid zone The isolating barrier effect of the mountain mass appears clearly in the area lying to the north: the Central Asian arid zone, comprising the Takla Makan, Gobi, and Ordos deserts. These are areas of inland drainage covered with wind-borne *loess and silt deposited by rivers.

tropical Asia This comprises South and Southeast Asia, the latter including the Indonesian and Philippine archipelagos. South Asia is usually limited to peninsular India, the island of Sri Lanka, and the Indo-Gangetic Plain. The peninsula is an uplifted and terraced plateau with lava intrusions (the Deccan. The Indo-Gangetic Plain is formed from the combined flood-plains of the Indus,

Ganges, and Brahmaputra, and is a deep marginal depression running west–east parallel to the Himalayas. Mainland Southeast Asia is a complex region of folded mountain chains, decreasing in height southwards, interspersed with fertile river valleys.

the insular or island arc This consists of the archipelagos that border the southeast margin of Asia. These islands are a product of volcanism and coral reef building. They are bordered by deep ocean trenches and are characteristically unstable and mountainous.

climate Differences between the climatic conditions of the various regions are determined to a considerable degree by topography. Extremes are again evident; Siberia is the world's coldest place, the wettest is Cherrapunji in Assam, and Asia also has the world's driest regions. The climates range from Arctic in the far north to equatorial in the south. To the south and east of the central mountain mass the dominant feature is the monsoon, where changes of pressure and temperature in the interior bring cool, dry, outward-blowing winds in winter and hot, moisture-laden, inward-blowing winds in summer. Vegetation and soil types correspond roughly to the climate regimes that prevail; a continental climate prevails over a large part of Asia. The Arctic air mass from the north and the tropical air mass from the south converge over the central mountain mass of the Himalayas, but it is the alternate heating and cooling of the land mass of Asia and the reversal of the trade winds causing the winter and summer monsoons that always dominate any discussion of Asia's climate.

Asian geology the continent of Asia is made up of a large number of separate geological units, of varying age and origin. The Himalayan belt separates the main mass of central and northern Asia from the two separate stable blocks of peninsular India and Arabia. These southerly blocks were originally part of the ancient southern continent of *Gondwanaland, and became joined to the rest of Asia only in late Phanerozoic times, after the splitting up of Gondwanaland. The northern part of Asia, part of the ancient continent of *Laurasia, is made up of ancient cratons separated by mobile belts active in early Phanerozoic times. The Himalayan belt runs east–west from the Mediterranean through the Middle East to the Himalayas, then turns south through Burma, Malaysia, and Indonesia. It is a very large and long-lived mobile belt, active intermittently throughout the Phanerozoic, and still active in the present day.

Asiatic Laurasia This vast continental area to the north of the Himalayas is at present a stable massif, containing a number of ancient shield areas of Precambrian age.

the Siberian platform The largest of the shield areas, it occupies the northern part of the continent and is composed of Precambrian rocks which emerge as massifs from a thick cover of more recent sediments. In the **Anabar massif**, in the central part of the platform, high-grade metamorphic rocks of early to middle Precambrian age are surrounded by a flat-lying succession of late Precambrian age.

The **Aldan massif**, in the southeast of the Siberian platform, contains very old early Precambrian high-grade metamorphic rocks, together with middle Precambrian metasediments and granites. Gold occurs as lodes and placers in these rocks. In the southeast and southwest parts of the platform are large areas of late Precambrian rocks, representing mobile belts that were active 1600–800 million years ago.

the Taimyr Peninsula To the north of the Siberian platform, the Taimyr Peninsula contains a massif of metamorphic Precambrian rocks that has been largely incorporated in a younger Palaeozoic fold belt. There are also a number of stable blocks of Precambrian age in eastern Asia; these have been preserved in later Phanerozoic fold belts in eastern China, Manchuria, and Korea.

The younger rocks of Asiatic Laurasia occur in a number of widely separated areas. To the west of the Siberian platform a thick sequence of Mesozoic and Tertiary sediments underlie the West Siberian lowlands; this area is bounded to the west by the Upper Palaeozoic mobile belt of the Ural Mountains. To the southwest, the Siberian platform is cut by the Lake Baikal Rift Valley System. To the south of the platform is a complex area of Palaeozoic mobile belts, while to the east in China lies a complex of Mesozoic fold belts. At the eastern margin of the continent are the island arcs of the still active circum-Pacific mobile belt.

the Baikal Rift Valley System An arcuate rift system of late Tertiary age, the Baikal Rift Valley System runs for almost 2000 km along the southern margin of the Siberian craton. A series of fault troughs contain thick accumulations of sedimentary and volcanic rocks. Lake Baikal lies in the central part of the rift some 3 km below the blocks on each side of the rift valley.

Central Asia The Palaeozoic mobile belts cover the mountainous area of Mongolia and Central Asia to the south of the Siberian platform; they contain sediments of late Precambrian to Palaeozoic age and have been involved in phases of orogenic activity (see *orogeny) in the late Precambrian, and in Caledonian and Hercynean times, with associated granite intrusions. Mesozoic and Tertiary sediments of continental type occur within and alongside the fold belts, and are generally undisturbed. The whole area suffered massive differential uplift, and thick conglomerates were eroded from the uplifted blocks and deposited in a number of basins.

northeast Asia Mesozoic fold belts occur in northeast Asia, flanking the northeast margin of the Siberian platform. Thick sequences of detrital sediments of Permian to Jurassic age adjoin the platform; these are folded and intruded by Mesozoic granites containing gold, tin, and tungsten ores. Further east, towards the present continental margin, is a mobile belt active since early Phanerozoic times; this belt suffered its major orogenic activity in Alpine times. Tertiary granites and Tertiary to Quaternary andesite, dacite, and basalt lavas are widespread. To the east again this belt is fringed by the island arcs of the present active oceanic margin Kamchatka, the Kurils, Japan, and the Philippines.

The Precambrian Siberian platform itself is largely covered by shallow-water shelf-sea deposits of evaporites and non-marine sediments, ranging in age from Cambrian to Cretaceous and containing coals at a number of horizons. Thick plateau lavas, with associated basic sills and dykes, were products of Permian and Triassic times. These are known as the Siberian Traps. Diamond-bearing kimberlite pipes were intruded in and near the plateau basalt province in Triassic times.

The Indian Craton The triangular southern part of India and Sri Lanka is geologically quite distinct from the Himalayan mobile belt to the north; the southern part was originally joined to Africa as part of Gondwanaland,

and has drifted north to join the continent of Asia only in geologically recent times. The Indian craton is composed of an ancient Precambrian crystalline basement, containing rocks formed during several episodes of Precambrian mobility. The oldest Precambrian rocks are of early to middle Precambrian age and are found in the Dharwar belt of southwest India, the Aravalli region between Mumbai (formerly Bombay) and Delhi, and the iron-ore belt of the Singbhum area.

The Dharwar belt is made up of an assemblage of gneisses, migmatites, and granites containing linear greenstone belts in which volcanics and sedimentary rocks, metamorphosed to a greater or lesser degree, are preserved. Some of the sediments contain rich haematite ores and manganese ores. Gold also occurs associated with basic volcanic rocks or intrusions in the greenstone belts. The whole of the Dharwar belt is intruded by swarms of basic dykes, intruded towards the end of the main metamorphic episode, 2600–2300 million years ago. In the Aravalli region, banded gneisses of the same general age occur underlying later cover rocks.

The iron-ore belt of the Singbhum area contains an ancient complex of metamorphic rocks, formed in an early Precambrian cycle of activity ending some 3200 million years ago. Above these older metamorphic rocks is the iron-ore group, a folded but unmetamorphosed succession of shales, sandstones, lavas, and banded haematite-jasper, of great economic importance as one of India's main sources of iron ore. This cycle of deposition ended with the intrusion of the Singbhum granite 2700 million years ago. A further cycle of activity resulted in the reactivation of part of this belt.

All these early massifs are truncated by mobile belts of late Precambrian age; these are the **Eastern Ghats belt**, lying along the east coast; the **Aravalli-Delhi belt**, running southwest from Delhi, and the **Satpura belt**, extending inland from Kolkata (formerly Calcutta). The Eastern Ghats belt is a belt of high-grade metamorphic rocks, gneisses, and charnockites, formed 2000–1200 million years ago. To the west of this belt late tectonic sediments and volcanics of the Eastern Ghats cycle are preserved relatively unaltered in the Cuddapah basin, northwest of Chennai (formerly Madras).

The Aravalli-Delhi belt runs north–northeast for some 800 km through Rajasthan to Delhi. The Aravalli system at the base of the succession is a highly metamorphosed and granitised complex containing older Precambrian basement rocks. Above this are thick sedimentary accumulations, metamorphosed during further periods of orogenic activity and invaded by post-orogenic granites 1650–1000 million years ago.

The Satpura belt runs west–southwest from the Ganges delta for about 800 km before disappearing under younger rocks. In the western part of the belt two formations occur: the Sakoli series, sediments metamorphosed to a relatively low grade, and the Sausar group, highly metamorphosed and migmatised rocks formed about 1000 million years ago. The metamorphosed sediments contain rich deposits of manganese ore. In the eastern part, the Satpura belt is involved in activity of the Singbhum cycle, which also involves older rocks of the iron-ore belt. The margin of the older iron-ore belt is the thrust zone of the copper belt, containing important copper and uranium deposits. During this cycle, thick sediments and volcanics accumulated in the mobile belt to the north of the copper belt, and were later metamorphosed and folded.

Between the Aravalli-Delhi and Satpura belts and to the southwest is a huge area covered by rocks of the Vindhyan formation, shallow-water quartzites and shales laid down at various times on a stable platform after the stabilization of the main Indian craton around 900 million years ago. The Vindhyan sandstones are extensively quarried as building stones.

A mobile belt of late Precambrian to early Palaeozoic age (the Indian Ocean belt) crosses the southeast tip of India and runs through Sri Lanka and up the east coast as far as Chennai. In Sri Lanka a complex series of high-grade metamorphic rocks were formed during a series of metamorphic episodes, the final one occurring 500–600 million years ago. In late Palaeozoic times India was still part of Gondwanaland and formed a shield area, bounded to the north by the mobile belt of the Himalayas. On the platform adjoining the mobile belt sediments of Cambrian to Mesozoic age were laid down.

glaciation Towards the end of the Palaeozoic a widespread glaciation affected Gondwanaland and thick glacial deposits accumulated. In India the Gondwana system, preserved in the northeast of the peninsula, begins with glacial tillites which pass up into coal measures, sandstones, and shales. The coal measures contain thick seams that provide most of India's coal. The upper part of the Gondwana succession consists of continental sandstones and shales of early Triassic to mid-Cretaceous age.

In late Jurassic to Cretaceous times marine sediments began to accumulate in new basins on the eastern and western margins of the craton, and in late Cretaceous to early Tertiary times thick plateau lavas, the Deccan Traps, were poured out over much of northern peninsular India. They are up to 2 km thick, and were associated with the disruption of Gondwanaland, which began in Cretaceous times. From this time on, India moved north on a crustal plate, and in late Tertiary and Quaternary times this resulted in the collision of the continent with the ancient Laurasian massif, and the rise of the new Himalayan mountain ranges on the site of the mobile belt at the old continental margin. Thick Tertiary and post-Tertiary sediments were eroded from the rising mountain chains and deposited on the north India plains.

The Himalayan Mobile Belt This belt has a long history of activity, beginning in the earliest Palaeozoic and continuing up to the present day. It began as part of a long belt encircling the ancient northern continent of Laurasia, and was the later site of the collision between the fragments of Gondwanaland – Africa, Arabia, and peninsular India and Laurasia – as they moved northwards after the disruption of the supercontinent. During Palaeozoic and Mesozoic times sediments were deposited without break in the ancient *Tethys Sea, along the margin of the Laurasian continent. These are preserved in the more northerly parts of the Himalayas; further south, the main part of the belt is made up of slices and fragments of the Indian subcontinent, sliced off as the continent collided with and underthrust the massif to the north. The main orogenic period began in late Mesozoic times. Fold belts generated during this cycle formed the Pamirs, the Karakoram, and the Hindu Kush ranges, linking up westwards through Afghanistan with the Alpine orogenic belt of the Middle East and Europe.

The southeast part of mainland Asia is occupied by mountain ranges running southeast from the eastern Himalayas; although the structures in these ranges are of similar Cretaceous to Quaternary age as those in the Himalayas, they do not seem actually to join the Himalayan ranges. The Malayan peninsula contains important granite intrusions from which valuable tin ores are derived.

The Pacific Island Arcs These form part of the active circum-Pacific mobile belt. They run from the Kurils in the north through Japan, the Philippines, the Mariana Islands, and on to the Tonga Isles and New Zealand. The arcs are flanked on their ocean (convex) side by deep ocean trenches, the sites where ocean floor is being thrust below the continental mass. This is therefore a region of strong seismic activity, suffering very deep earthquakes. The history of the island arcs stretches back to Palaeozoic times, and throughout their history they have been the site of frequent volcanic activity and metamorphism.

Asia-Pacific Economic Cooperation Conference APEC, trade group comprising 21 Pacific Asian countries, formed in November 1989 to promote multilateral trade and economic cooperation between member states. Its members are the USA, Canada, Japan, Australia, New Zealand, South Korea, Brunei, Indonesia, Malaysia, the Philippines, Singapore, Thailand, China, Hong Kong, Taiwan, Papua New Guinea, Chile, Mexico, Russia, Vietnam, and Peru.

APEC members account for more than 50% of the world's economic production and 41% of its overall trade value in 1995.

aspect in earth sciences, the direction in which a slope faces. In the northern hemisphere a slope with a southerly aspect receives more sunshine than other slopes and is therefore better suited for growing crops that require many hours of sunshine in order to ripen successfully. Vineyards in northern Europe are usually situated on south-facing slopes.

asphalt mineral mixture containing semisolid brown or black *bitumen, used in the construction industry. Asphalt is mixed with rock chips to form paving material, and the purer varieties are used for insulating material and for waterproofing masonry. It can be produced artificially by the distillation of *petroleum.

assembly industry a firm which assembles outputs from other industries into a finished product. For example, the motor-car industry assembles parts such as engines, bodies, wheels, windscreens and electrical components into a final product. The individual components are made by a number of other firms. Assembly industry is distinguished from a manufacturing industry such as steel-making, which uses primary products such as *coal, *limestone and iron ore.

asset in accounting, anything owned by or owed to the company that is either cash or can be turned into cash. The term covers physical assets such as land or property of a company or individual, as well as financial assets such as cash, payments due from bills, and investments. Assets are divided into *fixed assets – assets that are expected to be used in the business for some time such as land, plant, machinery, buildings – and current assets – assets which are frequently turnover in the course of business, such as stock. On a company's balance sheet, total assets must be equal to total liabilities (money and services owed).

assisted area region that is receiving some help from the central government, usually in the form of extra funding,

as part of a regional policy. Most policies concentrate on identifying and then assisting 'backward' or 'problem' areas so that economic activity may be more equally shared within the country.

Association of Caribbean States ACS, association of 25 states in the Caribbean region, formed in 1994 in Colombia to promote social, political, and economic cooperation and eventual integration. Its members include the states of the Caribbean and Central America plus Colombia, Suriname, and Venezuela. Associate membership has been adopted by 12 dependent territories in the region. Its creation was seen largely as a reaction to the *North American Free Trade Agreement between the USA, Canada, and Mexico, although its far smaller market raised doubts about its vitality.

Association of South East Asian Nations ASEAN, regional alliance formed in Bangkok in 1967; it took over the non-military role of the Southeast Asia Treaty Organization in 1975. Its members are Indonesia, Malaysia, the Philippines, Singapore, Thailand, (from 1984) Brunei, (from 1995) Vietnam, (from 1997) Laos and Myanmar, and (from 1999) Cambodia; its headquarters are in Jakarta, Indonesia. North Korea took part in the organization for the first time at the 2000 annual meeting of foreign ministers.

asthenosphere layer within Earth's *mantle lying beneath the *lithosphere, typically beginning at a depth of approximately 100 km and extending to depths of approximately 260 km. Sometimes referred to as the 'weak sphere', it is characterized by being weaker and more elastic than the surrounding mantle.

The asthenosphere's elastic behaviour and low viscosity allow the overlying, more rigid plates of lithosphere to move laterally in a process known as *plate tectonics. Its elasticity and viscosity also allow overlying crust and mantle to move vertically in response to gravity to achieve isostatic equilibrium (see *isostasy).

asymmetrical fold folded *strata where the two limbs are at different angles to the horizontal.

asymmetric valley valley with one steep side and one shallow side. They usually form in *periglacial environments – regions that are close to glaciers but which thaw out in summer, and where one side of the valley is largely in the sun and the other in the shade. The area affected by the sun is actively cut down by *freeze–thaw weathering and develops a shallow, gentle gradient. The other side is largely unaffected and remains steep.

atlas book of maps. The first modern atlas was the *Theatrum orbis terrarum* (1570); the first English atlas was a collection of the counties of England and Wales by Christopher Saxton (1579). Mercator began work on the first great world atlas in 1585; it was completed by his son in 1594. Early atlases had a frontispiece showing Atlas (a figure in Greek mythology) supporting the globe.

atmosphere mixture of gases surrounding a planet. Planetary atmospheres are prevented from escaping by the pull of gravity. On Earth, atmospheric pressure decreases with altitude. In its lowest layer, the atmosphere consists of nitrogen (78%) and oxygen (21%), both in molecular form (two atoms bonded together) and argon (1%). Small quantities of other gases are important to the chemistry and physics of the Earth's atmosphere, including water, carbon dioxide, and traces of other noble gases (rare gases),

as well as ozone. The atmosphere plays a major part in the various cycles of nature (the *water cycle, the carbon cycle, and the *nitrogen cycle). It is the principal industrial source of nitrogen, oxygen, and argon, which are obtained by the fractional distillation of liquid air.

The Earth's atmosphere is divided into four regions of atmosphere classified by temperature.

troposphere This is the lowest level of the atmosphere (altitudes from 0–10 km) and it is heated to an average temperature of 15°C by the Earth, which in turn is warmed by infrared and visible radiation from the Sun. Warm air cools as it rises in the troposphere and this rising of warm air causes rain and most other weather phenomena. The temperature at the top of the troposphere is approximately –60°C.

stratosphere Temperature increases with altitude in this next layer (from 10–50 km), from –60°C to near 0°C.

mesosphere Temperature again decreases with altitude through the mesosphere (50–80 km), from 0°C to below –100°C.

thermosphere In the highest layer (80 km to about 700 km), temperature rises with altitude to extreme values of thousands of degrees. The meaning of these extreme temperatures can be misleading. High thermosphere temperatures represent little heat because they are defined by motions among so few atoms and molecules spaced widely apart from one another.

The thermal structure of the Earth's atmosphere is the result of a complex interaction between the electromagnetic radiation from the Sun, radiation reflected from the Earth's surface, and molecules and atoms in the atmosphere. High in the thermosphere temperatures are high because of collisions between ultraviolet (UV) photons and atoms

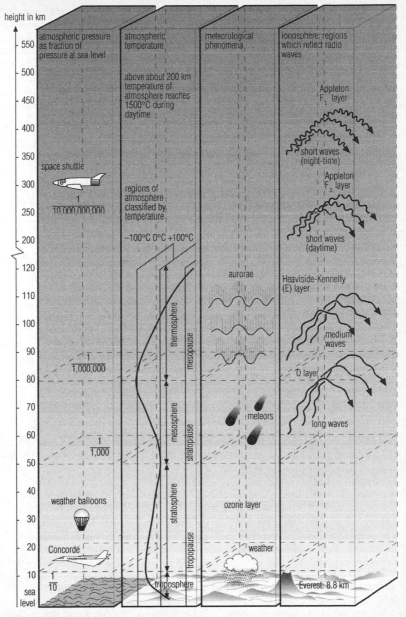

atmosphere All but 1% of the Earth's atmosphere lies in a layer 30 km above the ground. At a height of 5500 m, air pressure is half that at sea level. The temperature of the atmosphere varies greatly with height; this produces a series of layers, called the troposphere, stratosphere, mesosphere, and thermosphere.

of the atmosphere. Temperature decreases at lower levels because there are fewer UV photons available, having been absorbed by collisions higher up. The thermal minimum that results at the base of the thermosphere is called the **mesopause**. The temperature maximum near the top of the stratosphere is called the **stratopause**. Here, temperatures rise as UV photons are absorbed by heavier molecules to form new gases. An important example is the production of *ozone molecules (oxygen atom triplets, O_3) from oxygen molecules. Ozone is a better absorber of ultraviolet radiation than ordinary (two-atom) oxygen, and it is the ozone layer within the stratosphere that prevents lethal amounts of ultraviolet from reaching the Earth's surface. The temperature minimum between the stratosphere and troposphere marks the influence of the Earth's warming effects and is called the **tropopause**.

ionosphere At altitudes above the ozone layer and above the base of the mesosphere (50 km), ultraviolet photons collide with atoms, knocking out electrons to create a plasma of electrons and positively charged ions. The resulting **ionosphere** acts as a reflector of radio waves, enabling radio transmissions to 'hop' between widely separated points on the Earth's surface.

solar activity Far above the atmosphere lie the **Van Allen radiation belts**. These are regions in which high-energy charged particles travelling outwards from the Sun (the solar wind) have been captured by the Earth's magnetic field. The outer belt (about 1600 km) contains mainly protons, the inner belt (about 2000 km) contains mainly electrons. Sometimes electrons spiral down towards the Earth, noticeably at polar latitudes, where the magnetic field is strongest. When such particles collide with atoms

and ions in the thermosphere, light is emitted. This is the origin of the glows visible in the sky as the aurora borealis (northern lights) and the aurora australis (southern lights).

During periods of intense solar activity, the atmosphere swells outwards; there is a 10–20% variation in atmosphere density. One result is to increase drag on satellites. This effect makes it impossible to predict exactly the time of re-entry of satellites.

atmospheric chemistry The chemistry of atmospheres is related to the geology of the planets they envelop. Unlike Earth, Venus's dense atmosphere is dominantly carbon dioxide (CO_2). The carbon dioxide-rich atmosphere of Venus absorbs infrared radiation eminating from the planet's surface, causing the very high surface temperatures capable of melting lead (see *greenhouse effect). If all of the carbon dioxide that has gone to form carbonate rock (see *limestone) on Earth were liberated into the troposphere, our atmosphere would be similar to that of Venus. It is the existence of liquid water, which enables carbonate rock to form on Earth, that has caused the Earth's atmosphere to differ substantially from the Venusian atmosphere.

Other atmospheric ingredients are found in particular localities: gaseous compounds of sulphur and nitrogen in towns, salt over the oceans; and everywhere dust composed of inorganic particles, decaying organic matter, tiny seeds and pollen from plants, and bacteria. Of particular importance are the human-made *chlorofluorocarbons (CFCs) that destroy stratospheric ozone.

development of the atmosphere It is thought that the Earth was formed about 4600 million years ago. The inside of the Earth was very hot and full of chemical activity, causing hot gases to burst out through volcanoes on the Earth's surface. Slowly, over millions of years, the Earth's atmosphere developed from these gases. The most common gases produced by volcanoes are water vapour, nitrogen, and carbon dioxide. The water vapour cooled and condensed to form the seas and oceans. Much of the carbon dioxide dissolved in rainwater and *sea water, leaving nitrogen as the major component of the atmosphere. Following the evolution of green plants 2200 million years ago and the start of photosynthesis, more carbon dioxide was used up and oxygen was produced. The result was the development of the Earth's unique atmosphere, mainly oxygen and nitrogen with small amounts of carbon dioxide. The composition of gases in dry air has been about the same for 200 million years.

atmosphere symbol atm; or **standard atmosphere**, in physics, a unit of pressure equal to 760 mmHg, 1013.25 millibars, or 1.01325×10^5 pascals, or newtons per square metre. The actual pressure exerted by the atmosphere fluctuates around this value, which is assumed to be standard at sea level and 0°C, and is used when dealing with very high pressures.

atmospheric cell a thermally driven element of the atmospheric circulation system which operates to eliminate the horizontal heat gradient between the equator and the poles. It is thought that three main atmospheric cells are located in each hemisphere.

A *Hadley cell operates at low latitudes in both hemispheres whilst a high-latitude cell functions between the polar anti-cyclone and the polar front. There is an equatorward low-level air flow within the high-latitude cell, with a compensating high-level poleward flow. The mid-latitude cell

is less strongly developed and is probably largely maintained by the Hadley and high-latitude cells. This mid-latitude cell is located between the subtropical anticyclone belt and the polar front. Air flow at low levels is poleward in the mid-latitude cell, with a compensating high-level return flow. The presence of jet streams and the passage of a depression or a migratory anticyclone can modify the operation of atmospheric cells.

atmospheric circulation large-scale movement of air within the lower *atmosphere. Warm air at the *Equator rises, creating a zone of low pressure. This air moves towards the poles, losing energy and becoming cooler and denser, until it sinks back to the surface at around 30° latitude, creating an area of high pressure. At the surface, air moves from this high pressure zone back towards the low pressure zone at the Equator, completing a circulatory movement.

atmospheric pollution contamination of the atmosphere with the harmful by-products of human activity; see *air pollution.

atmospheric pressure pressure at any point on the Earth's surface that is due to the weight of the column of air above it; it therefore decreases as altitude increases, because there is less air above. Particles in the air exert a force (pressure) against surfaces; when large numbers of particles press against a surface, the overall effect is known as air pressure. At sea level the average pressure is 101 kilopascals (1013 millibars, or 760 mm Hg, or 1 atmosphere). Changes in atmospheric pressure, measured with a barometer, are used in weather forecasting. Areas of relatively high pressure are called *anticyclones; areas of low pressure are called *depressions.

For every square metre of a surface, a force of 10 tonnes of air pressure is exerted. This air pressure does not crush objects because they exert an equal amount of force to balance the air pressure. At higher altitudes the air is thinner and the air pressure is lower. Here, water boils at a temperature less than that at sea level. In space there are no air particles and so no air pressure is exerted on an astronaut's body. Astronauts wear spacesuits that supply an air pressure against their bodies.

A barometer is an instrument used to measure air pressure. If a glass tube is filled with mercury and turned upside down with its end in a bowl of mercury, then the height of the column of mercury is held by the air particles pressing on the mercury in the bowl. This measures the air pressure. A standard measurement for atmospheric pressure at sea level is a column of mercury 760 mm high. At higher altitudes the height of the mercury column would be less, as the air pressure is lower.

atoll   continuous or broken circle of *coral reef and low coral islands surrounding a lagoon.

attrition   in earth science, the process by which particles of rock are rounded and gradually reduced in size by hitting one another as they are transported by rivers, wind, or the sea.

The rounding of particles is a good indication of how far they have been transported. This is particularly true for particles carried by rivers, which become more rounded and smaller as the distance downstream increases. Thus the load of a river is often large and jagged in its upper course, and smaller and rounder in its lower course.

aurora   spectacular array of light in the night sky, caused by charged particles from the Sun hitting the Earth's upper atmosphere. The **aurora borealis** is seen in the north of the northern hemisphere; the **aurora australis** in the south of the southern.

Australasian and Oceanian geology
the continental mass of Australasia is composed of five geologically distinct units; the oldest of these is the **Precambrian craton** of central and western Australia. The southeastern margin of the craton adjoins the younger **Adelaidean mobile belt**, of late Precambrian and early Palaeozoic age. The eastern margin of the Australian land mass is occupied by the **Tasman mobile belt**, a long-lived belt active from Palaeozoic to Cretaceous times. Within the craton are a number of large Phanerozoic sedimentary basins, such as the **Great Artesian Basin**. The youngest province contains the still-active mobile belts of the **Indonesian Islands** and **New Zealand**.

the **Precambrian Craton** The ancient shield area of Western Australia has a cover of younger rocks over much of its area; Precambrian rocks are only exposed in a number of large isolated blocks. The Yilgam and Pilbara blocks are areas of middle Precambrian rocks, composed of a highly metamorphosed granitic basement complex containing greenstone belts. The greenstone belts contain metamorphosed volcanic and sedimentary rocks, and the whole complex was intruded by a swarm of basic dykes about 2400 million years ago. The greenstone belts also contain layered basic intrusions associated with the volcanics, and these contain important nickel ores. The gold ores of this area are also found associated with mineralization of the basic rocks in the greenstone belts.

The **Musgrave and Arunta blocks** in central Australia are composed of high-grade metamorphic Precambrian complexes. The **Gawler block** in South Australia is composed of granitic basement with remnants of a greenstone-belt succession, containing high-grade iron ores in the Middleback Ranges.

In late Precambrian times a number of mobile belts became established; a new sedimentary basin was developed in the Hamersley belt, and cover sediments were laid down on the established craton.

The mobile belts are the **Halls Creek** belt, **East Kimberleys**, and the **West Kimberleys** in Western Australia, the **Pine Creek** belt in the Northern Territories, and the **Willyama Complex** in South Australia. The latter belt is of economic importance as it contains the Broken Hill mining district, well known for its silver-lead-zinc mineralization.

The **Hamersley belt** of Western Australia is a large area containing a remarkably well-preserved and thick accumulation of late Precambrian sediments. The **Mount Bruce** supergroup at the base is a largely unmodified sequence, 10 km thick, containing banded iron formations. These were gently folded and then a further thick sequence of late Precambrian sandstones, lavas, and dolomites were deposited. At the same time thick sedimentary successions were laid down on the southwestern side of the Gulf of Carpenteria, the Kimberley basin, and on the Gawler block of South Australia.

the Adelaidean mobile belt
This belt stretches over 1100 km following the southeastern margin of the Precambrian craton, and marking the site of an extensive late Precambrian and early Palaeozoic mobile belt. It contains a 16-km thick geosynclinal sequence of mainly shallow-water sediments; there are well-preserved, thick glacial deposits, and these are also found extensively outside the geosynclinal area resting directly on the craton. In the uppermost Precambrian rocks is an important group of fossils, the Ediacara Fauna, a remarkably well-preserved collection of soft-bodied organisms. The youngest sediments in the belt are of Cambrian age, and *orogeny began in the Cambrian, resulting in low-grade regional metamorphism and folding of the rocks.

the Tasman mobile belt This 3000-km belt occupies the whole east coast of Australia. It initiated in the Lower Palaeozoic and continued for some 400 million years. Thick sequences of sediments and volcanics accumulated in a number of north–south troughs, and suffered a number of orogenic disturbances culminating in major orogeny in Silurian times. Sedimentation continued in some areas until Jurassic times. The belt is notable for the amount and variety of the igneous rocks produced over its long history. Large volumes of granite and acid volcanics were produced over a period of some 200 million years, from Devonian to Permian times. Mineralization is associated with some of the granites and volcanics; gold, lead-zinc, copper, and tin are all important economic deposits. Late movements were associated with widespread post-orogenic volcanism in Tertiary times, continuing into the Plio-Pleistocene.

the intra-cratonic basins
The **Carnarvon**, **Canning**, and **Bonaparte Gulf** basins near the west and northwest borders of the craton came into existence in mid-Palaeozoic times, and contain marine sandstones, shales, and limestones mainly of Devonian and Lower Carboniferous age. These were followed by glacial deposits in Permian times, evidence of the widespread Permian glaciation of *Gondwanaland, of which Australia was still a part. Many of these glacial deposits spread outside the basins and onto the old craton. These were followed by thick Jurassic and Cretaceous sequences. The **Great**

Artesian Basin came into existence in Jurassic times, and contains a thin continental succession of Mesozoic rocks.

the Indonesian island arcs The islands of Southeast Asia form a number of seismically and volcanically active island arcs. This area is one where three crustal plates meet, with complex geological results. The area has been active since the late Palaeozoic, and activity continues to the present day.

the New Zealand island arc system The earliest-formed rocks in New Zealand are those of the Palaeozoic metamorphic complex in western New Zealand. These consist of metamorphosed Palaeozoic volcanics and greywackes. Much of the rest of the country is occupied by the New Zealand geosyncline, which began forming in Permian times and contains thick sediments of Permian to Cretaceous age. These were deformed during the Rangitata orogeny in the Cretaceous, and suffered low-grade regional metamorphism. The Alpine fault of New Zealand is an important fault which runs northeast through the geosyncline, and has been active since late Mesozoic times up to the present day.

The present land mass of New Zealand came into being with the uplift of new mountain ranges during the Rangitata orogeny. Volcanic activity continued through the Tertiary along the east side of South Island. In North Island acid volcanics erupted in Neogene and recent times along the still active central volcanic belt.

authoritarianism rule of a country by a dominant elite who repress opponents and the press to maintain their own wealth and power. They are frequently indifferent to activities not affecting their security, and rival power centres, such as trade unions and political parties, are often allowed to exist, although under tight control. An extreme form is *totalitarianism.

autochthon (Greek 'from the earth') the first inhabitants of a country or area of land. In Greek mythology, the autochthons were supposed to have sprung from the rocks and trees.

autocracy form of government in which one person holds absolute power. The autocrat has uncontrolled and undisputed authority. Russian government under the tsars was an autocracy extending from the mid-16th century to the early 20th century.

automobile emissions the collection of car exhaust gases comprising the NO_x group (nitrogen oxides), carbon monoxide, lead, unburnt hydrocarbons, water vapour, carbon dioxide and aldehydes. The first three pollutants in this list can now be found in major concentration in most urban areas throughout the world. Traffic congestion can produce atmospheric carbon monoxide levels of up to 100 parts per million (ppm) compared to clean air concentrations of 0.1 ppm.

In many of the wealthiest world cities, such as Los Angeles, automobile emissions, rubber from car tyres and tarmacadam particles from the road surfaces combine in the presence of high sunshine levels to produce a highly toxic form of pollutant mist called photochemical smog. Most industrial nations have now applied legislation to reduce the level of automobile emissions. Unleaded petrol, exhaust gas cleaners (after-burners or catalytic converters) and lean-burn engines are the main methods so far used to combat high pollution emission levels.

autonomy in politics, a term used to describe political self-government of a state or, more commonly, a subdivision of a state. Autonomy may be based upon cultural or ethnic differences and often leads eventually to independence.

autotroph   any self-feeding, organic green cell or plant capable of combining energy from the Sun with simple inorganic salts from the soil to form complex, manufactured sugars via the process of photosynthesis. Autotrophs form the first trophic level in a food chain, and as such are of fundamental importance to all heterotrophs.

avalanche   (French *avaler* 'to swallow') fall or flow of a mass of snow and ice down a steep slope under the force of gravity. Avalanches occur because of the unstable nature of snow masses in mountain areas.

Changes of temperature, sudden sound, or earth-borne vibrations may trigger an avalanche, particularly on slopes of more than 35°. The snow compacts into ice as it moves, and rocks may be carried along, adding to the damage caused.

Avalanches leave slide tracks, long gouges down the mountainside that can be up to 1 km long and 100 m wide. These slides have a similar beneficial effect on *biodiversity as do forest fires, clearing the land of snow and mature mountain forest and so enabling plants and shrubs that cannot grow in shade, to recolonize and creating wildlife corridors.

Avalanches are particularly hazardous to people in ski resort areas such as the French Alps. In 1991 a massive avalanche considerably altered the shape of Mount Cook in New Zealand.

azonal soil   any immature soil displaying little, if any, differentiation between *soil horizons. This results from the processes of soil formation having insufficient time to develop the characteristics found in zonal soils. Typically, azonal soils lack a defined B horizon with the A horizon lying directly above the parent material of the C horizon.

B

backwash   the retreat of a wave that has broken on a *beach. When a wave breaks, water rushes up the beach as *swash and is then drawn back towards the sea as backwash.

bacteria   a large group of diverse, unicellular microorganisms which exist singly, in chains or in clusters. They form the smallest of the living organisms (1–10 microns in size) and together with fungi, form the decomposer group of organisms. They can be subdivided into the non-photosynthetic filamentous gliding forms, such as myxobacteria, and the true bacteria, aubacteria. The former group are common in decaying plant materials and are the active decomposers. The aubacteria contain bacteriochlorophylls and carry out photosynthesis anaerobically.

Bacteria occur in soil, water and air, as symbiants, parasites or pathogens of humans, animals, plants and other microorganisms. Saprophytic species are important in the major biospheric movements of matter (for example, the nitrogen and sulphur cycles). Certain species form symbiotic relationships with higher plants. While some bacteria are vital to the continuation of human life, other forms are responsible for highly dangerous human diseases such as anthrax, tetanus and tuberculosis.

badlands   barren landscape cut by erosion into a maze of ravines, pinnacles, gullies, and sharp-edged ridges. Areas in South Dakota and Nebraska, USA, are examples.

Badlands, which can be created by overgrazing, are so called because of their total lack of value for agriculture and their inaccessibility.

balance of nature   in ecology, the idea that there is an inherent equilibrium in most *ecosystems, with plants and animals interacting so as to produce a stable, continuing system of life on Earth. The activities of human beings can, and frequently do, disrupt the balance of nature.

Organisms in the ecosystem are adapted to each other – for example, waste products produced by one species are used by another and resources used by some are replenished by others; the oxygen needed by animals is produced by plants while the waste product of animal respiration, carbon dioxide, is used by plants as a raw material in photosynthesis. The *nitrogen cycle, the *water cycle, and the control of animal populations by natural predators are other examples.

The idea of a balance of nature is also expressed in the *Gaia hypothesis, which likens the Earth to a living organism, constantly adjusting itself to circumstances so as to increase its chances of survival.

balance of payments   in economics, an account of a country's debit and credit transactions with other countries. Items are divided into the current account, which includes both visible trade (imports and exports of goods) and invisible trade (services such as transport, tourism, interest, and dividends), and the capital account, which includes investment in and out of the country, international grants, and loans. Deficits or surpluses on these accounts are brought into balance by buying and selling reserves of foreign currencies.

balance of power   in politics, the theory that the best way of ensuring international order is to have power so distributed among states that no single state is able to achieve a dominant position. The term, which may also refer more simply to the actual distribution of power, is one of the most enduring concepts in international relations. Since the development of nuclear weapons, it has been asserted that the balance of power has been replaced by a 'balance of terror'.

balance of trade   the balance of trade transactions of a country recorded in its current account; it forms one component of the country's *balance of payments.

bar   in earth sciences, deposit of sand or silt formed in a river channel, or a long ridge of sand or pebbles running parallel to a coastline (see *coastal erosion). Coastal bars can extend across estuaries to form **bay bars** and are formed in one of two ways. *Longshore drift can transport material across a bay and deposit it, thereby closing off the bay. Alternatively, an offshore bar (formed where waves touch the seabed and disturb the sediments, causing a small ridge to be formed) may be pushed towards the land as the sea level rises. These bars are greatly affected by the beach cycle. The high tides and high waves of winter erode the beach and deposit the sand as offshore bars. These are known as barrier beaches in the USA.

bar   in physics, unit of pressure equal to 10^5 pascals or 10^6 dynes/cm^2, approximately 750 mmHg or 0.987 atm. Its diminutive, the **millibar** (one-thousandth of a bar), is commonly used by meteorologists.

barchan   a type of crescent-shaped sand *dune formed in desert regions where the wind direction is very constant. Wind blowing round the edges of the dune causes the crescent shape, while the dune may advance in a downwind direction as particles are blown over the crest.

bar graph   a graph on which the values of a certain variable are shown by the length of shaded columns, which are numbered in sequence. Compare *histogram.

barograph   an aneroid *barometer connected to an arm and inked pen which records pressure changes continuously on a rotating drum. The drum usually takes a week to make one rotation.

barometer   instrument that measures atmospheric pressure as an indication of weather. Most often used are the mercury barometer and the aneroid barometer.

In a mercury barometer a column of mercury in a glass tube, roughly 0.75 m high (closed at one end, curved upwards at the other), is balanced by the pressure of the atmosphere on the open end; any change in the height of the column reflects a change in pressure. In an aneroid barometer, a shallow cylindrical metal box containing a partial vacuum expands or contracts in response to changes in pressure. See diagram, p. 42.

barrage   in geography, a structure built across a river or estuary in order to manage the water. The Thames barrier is an example of a flood-control barrage. A barrage may be used to regulate the water supply by controlling floods or storing water for *irrigation, or to generate power by, for example, harnessing tidal energy in estuaries.

barrier beach   a long narrow beach that extends across a bay and which can lead to the formation of a *lagoon on the landward side.

barrier island   long island of sand, lying offshore and parallel to the coast.

Some are over 100 km in length. Most barrier islands are derived from marine sands piled up by shallow longshore currents that sweep sand parallel to the seashore. Others are derived from former spits, connected to land and built up by drifted sand, that were later severed from the mainland.

Often several islands lie in a continuous row offshore. Coney Island and Jones Beach near New York City are well-known examples, as is Padre Island, Texas. The Frisian Islands are barrier islands along the coast of the Netherlands.

barrier reef *coral reef that lies offshore, separated from the mainland by a shallow lagoon.

barrier to entry factor that makes it difficult or impossible for a company to enter an industry and compete with existing producers. Barriers to entry to an industry are high when, for example, it is very costly to buy the physical *capital needed to set up in an industry, or when there are very strong brands in the market, or when the law grants privileges such as patent rights to existing firms.

barrier to exit obstacle deterring a company from leaving a market. Examples would be companies located in an industry-specific cluster, or companies employing a workforce that would be difficult to retrain.

barter exchange of goods or services without the use of money. Exchanging ships for oil would be an example of barter. Children swapping cards is another example. On an international level, there are many instances of barter today.

For example, a deal between Saudi Arabia and British Aerospace involved British Aerospace exchanging fighter planes for Saudi oil.

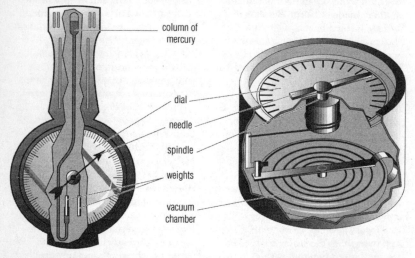

column of mercury

dial

needle

spindle

weights

vacuum chamber

mercury barometer · · · · · · · · · · · · aneroid barometer

barometer The mercury barometer (left) and the aneroid barometer (right). In the mercury barometer, the weight of the column of mercury is balanced by the pressure of the atmosphere on the lower end. A change in height of the column indicates a change in atmospheric pressure. In the aneroid barometer, any change of atmospheric pressure causes the metal box which contains the vacuum to be squeezed or to expand slightly. The movements of the box sides are transferred to a pointer and scale via a chain of levers.

baryte barium sulphate, $BaSO_4$, the most common mineral of barium. It is white or light-coloured, and has a comparatively high density (specific gravity 4.6); the latter property makes it useful in the production of high-density drilling muds (muds used to cool and lubricate drilling equipment). Baryte occurs mainly in ore veins, where it is often found with calcite and with lead and zinc minerals. It crystallizes in the orthorhombic system and can form tabular crystals or radiating fibrous masses.

basalt commonest volcanic *igneous rock in the Solar System. Basalt is an *extrusive rock, created by the outpouring of volcanic magma. The magma cools quickly, allowing only small crystals to form. Much of the surfaces of the terrestrial planets Mercury, Venus, Earth, and Mars, as well as the Moon, are composed of basalt. Earth's ocean floor is virtually entirely made of basalt. Basalt is mafic, that is, it contains relatively little *silica: about 50% by weight. It is usually dark grey but can also be green, brown, or black. Its essential constituent minerals are calcium-rich *feldspar, and calcium- and magnesium-rich *pyroxene.

baseflow or **groundwater flow**, movement of water from land to river through rock. It is the slowest form of such water movement, and accounts for the constant flow of water in rivers during times of low rainfall, and makes up the river's base line on a *hydrograph.

base level level, or altitude, at which a river reaches the sea or a lake. The river erodes down to this level. If base level falls (due to uplift or a drop in sea level), *rejuvenation takes place.

basic economic problem in economics, the problem posed by the fact that human wants are infinite but resources are scarce. Resources therefore have to be allocated, which then involves an *opportunity cost.

basic rock igneous rock with relatively low silica contents of 45–52% by weight, such as gabbro and basalt. Along with the terms **acid rock** and **ultrabasic rock** it is part of an outdated classification system based on the erroneous belief that silicon in rocks is in the form of silicic acid. Geologists today are more likely to use the descriptive term *mafic rock or report the amount of SiO_2 in percentage by weight.

basin of internal drainage in certain desert regions there are depressions (sometimes below sea level) from which there is no natural outlet. Intermittent streams drain to the centre of the basin, from which *drainage occurs by evaporation. Lakes occupying such basins fluctuate in depth and area according to the balance of inflow and evaporation. Extensive salt deposits (e.g. halite, gypsum) may be left behind as lakes evaporate, e.g. Dead Sea, Makgadikgadi Pan (Botswana).

basti or **bustee**, (Urdu 'settlement') Indian name for an area of makeshift housing; see *shanty town.

batholith large, irregular, deep-seated mass of intrusive *igneous rock, usually granite, with an exposed surface of more than 100 sq km. The mass forms by the intrusion or upswelling of magma (molten rock) through the surrounding rock. Batholiths form the core of some large mountain ranges like the Sierra Nevada of western North America.

According to *plate tectonic theory, magma rises in subduction zones along continental margins where one plate sinks beneath another. The solidified magma becomes the central axis of a rising mountain range, resulting in the deformation (folding and overthrusting) of rocks on either side. Gravity measurements indicate that the downward extent or thickness of many batholiths is some 10–15 km.

In the UK, a batholith underlies SW England and is exposed in places to form areas of high ground such as Dartmoor and Land's End.

bathyal zone upper part of the ocean, which lies on the continental shelf at a depth of between 200 m and 2000 m.

Bathyal zones (both temperate and tropical) have greater biodiversity than coral reefs, according to a 1995 study by the Natural History Museum in London. Maximum biodiversity occurs between 1000 m and 3000 m.

bauxite principal ore of aluminium, consisting of a mixture of hydrated aluminium oxides and hydroxides, generally contaminated with compounds of iron, which give it a red colour. It is formed by the *chemical weathering of rocks in tropical climates. Chief producers of bauxite are Australia, Guinea, Jamaica, Russia, Kazakhstan, Suriname, and Brazil.

To extract aluminium from bauxite, high temperatures (about 800°C) are needed to make the ore molten. Electric currents are then passed through the molten ore. The process is economical only if cheap electricity is readily available, usually from a hydroelectric plant.

bay in earth science, a wide-curving indentation of the sea into the land. Bays are often eroded in beds of rocks that are weaker than the adjacent *headlands. Deposition is most intense in bays due to *wave refraction.

A bay is larger than a cove, but smaller than a gulf. A **bay-head beach** is a small beach of sand or shingle found between two peninsulas or headlands as seen along the coastlines of Cornwall, England, and south Wales.

bay bar a bank of sand or shingle extending as a barrier almost or

totally across a *bay, caused by *longshore drift.

bayou (corruption of French *boyau* 'gut') in the southern USA, an *oxbow lake or marshy offshoot of a river.

Bayous may be formed, as in the lower Mississippi, by a river flowing in wide curves or meanders in flat country, and then cutting a straight course across them in times of flood, leaving loops of isolated water behind.

beach strip of land bordering the sea, normally consisting of boulders and pebbles on exposed coasts, or *sand on sheltered coasts. Beaches lie between the high- and low-water marks (high and low tides). A *berm, a ridge of sand and pebbles, may be found at the farthest point that the water reaches, generally at high tide.

The material of the beach consists of a rocky debris eroded from exposed rocks and headlands by the processes of *coastal erosion, or material carried in by rivers. The material is transported to the beach, and along the beach, by *longshore drift.

When the energy of the waves decreases, more sand is deposited than is transported, building depositional features such as *spits, *bars, and **tombolos**.

Concern for the condition of bathing beaches led in the 1980s to a directive from the European Union on water quality. In the UK, beaches free of industrial pollution, litter, and sewage, and with water of the highest quality, have the right (since 1988) to fly a blue flag.

In 1991 the English Tourist Board and Tidy Britain Group decided jointly to award a Golden Starfish prize to smaller beaches that were clean and had safe access, but whose local authorities had not the funds to finance the public telephones, beach guards, and daily cleaning stipulated for the blue flag award.

artificial barriers In some places attempts are made to artificially halt longshore drift and increase deposition on a beach by placing barriers (*groynes) at right angles to the beach. These barriers cause sand to build up on their upstream side but remove the beach on the downstream side, causing beach erosion. The finer sand can also be moved about by the wind, forming sand *dunes.

the beach cycle Beach erosion also occurs due to the natural seasonal **beach cycle**. Spring high tides and the high waves of winter storms tend to carry sand away from the beach and deposit it offshore (as an offshore bar). In the summer, calmer waves and neap (low) tides cause increased deposition of sand on the beach.

commercial threats to beaches Apart from the natural process of longshore drift, a beach may be threatened by the commercial use of sand and pebbles by the mineral industry, and by pollution (for example, by oil spilled or dumped at sea).

beach replenishment Although it is expensive, the high value of tourism, industry, and residential property can make beach replenishment a feasible solution. Miami Beach, Florida, is an excellent example. Between 1976 and 1982 an 18 km long, 200 m wide beach was constructed using 18 million cubic metres of material dredged from a zone offshore. It replicated a natural beach as far as possible.

beach deposits beach deposits may be varied, comprising sand and shingle deposited by the sea, banks of pebbles deposited in storm tides, and boulders which have been weathered out of the *coastal cliffs behind. Sand is the final product of marine *erosion, formed largely by *attrition of the marine *load.

beach nourishment the adding of extra sand to the foreshore of a beach to act as a buffer against the sea and reduce erosion.

bearing direction of a fixed point, or the path of a moving object, from a point of observation. Bearings are angles measured in degrees (°) from the north line in a clockwise direction. A bearing must always have three figures. For instance, north is 000°, northeast is 045°, south is 180°, and southwest is 225°.

True north differs slightly from magnetic north (the direction in which a compass needle points), hence northeast may be denoted as 045M or 045T, depending on whether the reference line is magnetic (M) or true (T) north. True north also differs slightly from grid north since it is impossible to show a sphere on a flat map.

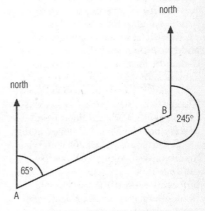

the bearing of B from A is 065°
the backbearing, or bearing of A from B, is 245°

bearing A bearing is the direction of a fixed point, or the path of a moving object, from a point of observation on the Earth's surface, expressed as an angle from the north. In the diagram, the bearing of a point A from an observer at B is the angle between the line BA and the north line through B, measured in a clockwise direction from the north line.

Beaufort scale system of recording wind velocity, devised by Sir Francis Beaufort in 1806. It is a numerical scale ranging from 0–17, calm being indicated by 0 and a hurricane by 12; 13–17 indicate degrees of hurricane force.

In 1874 the scale received international recognition; it was modified in 1926. Measurements are made at 10 m above ground level.

bed in geology, a single *sedimentary rock unit with a distinct set of physical characteristics or contained fossils, readily distinguishable from those of beds above and below. Well-defined partings called **bedding planes** separate successive beds or strata.

The depth of a bed can vary from a fraction of a centimetre to several metres, and can extend over any area. The term is also used to indicate the floor beneath a body of water (lake bed), and a layer formed by a fall of particles (ash bed). Beds of Portland and Purbeck limestone are clearly shown at Stair Hole, next to Lulworth Cove, Dorset, England.

bedding plain see *bed.

bedload material rolled or bounced (by *saltation) along a river bed. The particles carried as bedload are much larger than those carried in suspension in the water. During a flood many heavy boulders may be moved in this way. Such boulders can be seen lying on the river bed during times of normal flow.

Before Present (BP) a timescale devised by geologists, archeologists and palaeoecologists to allow unambiguous dating of events from the Pleistocene epoch to the present. The year chosen to represent the beginning of the present has been set at 1950. The term is most frequently used in palaeo-environmental reconstruction studies, for example, pollen analysis, glacial geomorphology and palaeosoil studies.

ben (Scottish Gaelic *beinn* 'summit') mountain peak, first element in the names of numerous Scottish mountains such as Ben Lomond, Ben More, and Ben Nevis. The related Brittonic form of *ban* (plural *bannau*) is found in some Welsh place names, including Bannau Brycheiniog (the Brecon Beacons).

Benelux acronym for BElgium, the NEtherlands, and LUXembourg, customs union of Belgium, the Netherlands, and Luxembourg, an agreement for which was signed in London by the three governments in exile in 1944, and ratified in 1947. It came into force in 1948 and was further extended and strengthened by the Benelux Economic Union Treaty in 1958. The full economic union between the three countries came into operation in 1960. The three Benelux countries were founder-members of the European Economic Community (now the *European Union), for which the Benelux union was an important stimulus.

Benguela current cold ocean current in the South Atlantic Ocean, moving northwards along the west coast of southern Africa and merging with the south equatorial current at a latitude of 15° S. Its rich plankton supports large, commercially exploited fish populations.

Benioff zone seismically active zone inclined from a deep sea trench. Earthquakes along Benioff zones apparently define a lithospheric plate that descends into the mantle beneath another, overlying plate. The zone is named after Hugo Benioff, a US seismologist who first described this feature.

benthic describing the environment on the sea floor supporting bottom-dwelling plants (such as seaweeds) and animals (including corals, anemones, sponges, and shellfish).

bentonite soft porous rock consisting mainly of clay minerals, such as montmorillonite, and resembling *fuller's earth, which swells when wet. It is formed by the chemical alteration of glassy volcanic material, such as tuff. Bentonite is used in papermaking, moulding sands, drilling muds for oil wells, and as a decolorant in food processing.

bergschrund a large *crevasse located at the rear of a *corrie icefield in a glaciated region, formed by the weight of the ice in the corrie dragging away from the rear wall as the *glacier moves downslope.

Beringia or **Bering Land Bridge**, former land bridge 1600 km wide between Asia and North America; it existed during the ice ages that occurred before 35,000 BC and during the period 24,000–9000 BC. As the climate warmed and the ice sheets melted, Beringia flooded. It is now covered by the Bering Strait and Chukchi Sea.

berm on a beach, a ridge of sand or pebbles running parallel to the water's edge, formed by the action of the waves on beach material. Sand and pebbles are deposited at the farthest extent of swash (advance of water) to form a berm. Berms can also be formed well up a beach following a storm, when they are known as **storm berms**.

beryl in full **beryllium aluminium silicate**, $3BeO.Al_2O_3.6SiO_2$, mineral that forms crystals chiefly in granite. It is the chief ore of beryllium. Two of its gem forms are aquamarine (light-blue crystals) and emerald (dark-green crystals).

beta share on the stock exchange, a share traded less actively than an *alpha share.

B horizon see *soil.

bid-rent theory assumption that land value and rent decrease as distance from the central business district increases. Shops and offices have greater need for central, accessible locations than other users (such as those requiring land for residential purposes) and can pay higher prices. They therefore tend to be located within the expensive central area.

biennial a plant which extends its life cycle over two years.

bight coastal indentation, crescent-shaped or gently curving, such as the Bight of Biafra in West Africa and the Great Australian Bight.

bill of lading document giving proof of particular goods having been loaded on a ship. The person to whom the goods are being sent normally needs to show the bill of lading in order to obtain the release of the goods. For air freight, there is an **airway bill**.

biodegradable capable of being broken down by living organisms, principally bacteria and fungi. In biodegradable substances, such as food and sewage, the natural processes of decay lead to compaction and liquefaction, and to the release of nutrients that are then recycled by the ecosystem.

This process can have some disadvantageous side effects, such as the release of methane, an explosive greenhouse gas. However, the technology now exists for waste tips to collect methane in underground pipes, drawing it off and using it as a cheap source of energy. Nonbiodegradable substances, such as glass, heavy metals, and most types of plastic, present serious problems of disposal.

biodiversity contraction of biological diversity, measure of the variety of the Earth's animal, plant, and microbial species, of genetic differences within species, and of the *ecosystems that support those species. High biodiversity means there are lots of different species in an area. The maintenance of biodiversity is important for ecological stability and as a resource for research into, for example, new drugs and crops.

Estimates of the number of species vary widely because many species-rich ecosystems, such as tropical forests, contain unexplored and unstudied habitats. Among small organisms in particular many are unknown. For example, it is thought that less than 1% of the world's bacterial species have been identified.

The most significant threat to biodiversity comes from the destruction of rainforests and other habitats. It is estimated that 7% of the Earth's surface hosts 50–75% of the world's biological diversity. Costa Rica, for example, has an area less than 10% of the size of France but possesses three times as many vertebrate species.

loss of biodiversity Since the start of the 20th century there has been the most severe and rapid loss of biodiversity in the history of the planet. We are in a period of mass extinction. In the geological past, several periods of extinction have occurred and they appear to be associated with catastrophic events, such as a large meteorite hitting the Earth or massive volcanic activity. It is believed that the current loss of species is mainly due to the loss of *habitats. Although the most significant threat to biodiversity comes from the destruction of rainforests, even in Britain the loss of species continues at a worrying rate. On average each county in Britain loses a species every ten years.

the Rio Summit In 1992 an international convention for the preservation of biodiversity was signed by over 100 world leaders (the Rio Summit). The convention called on industrialized countries to give financial and technological help to developing countries in order to allow them to protect and manage

their natural resources, and to profit from the commercial demand for genes and chemicals from wild species. However, the convention was weakened when the USA refused to sign because of fears it would undermine the patents and licences of biotechnology companies.

marine protection areas In 1995, 1306 sites worldwide were designated as marine protection areas (MPAs). Of these, 155 areas were singled out for further protection, including the Bering Strait and Kachemak Bay, both in Alaska. The MPAs were selected on the basis of genetic diversity, biological productivity, and the extent to which they provided habitats for endangered species.

bioeconomics theory put forward in 1979 by Chicago economist Gary Becker that the concepts of sociobiology apply also in economics. The competitiveness and self-interest built into human genes are said to make capitalism an effective economic system, whereas the selflessness and collectivism proclaimed as the socialist ideal are held to be contrary to human genetic make-up and to produce an ineffective system.

biofeedback in biology, modification or control of a biological system by its results or effects. For example, a change in the position or *trophic level of one species affects all levels above it.

Many biological systems are controlled by negative feedback. When enough of the hormone thyroxine has been released into the blood, the hormone adjusts its own level by 'switching off' the gland that produces it. In ecology, as the numbers in a species rise, the food supply available to each individual is reduced. This acts to reduce the population to a sustainable level.

biofouling build-up of barnacles, mussels, seaweed, and other organisms on underwater surfaces, such as ships' hulls. Marine industries worldwide spend at least £1.4 billion controlling biofouling by scraping affected surfaces and painting with antifouling paint.

Antifoulants can contribute to marine pollution as toxic components leach from the paint into surrounding water, for example *tributyl tin. Research is underway to develop natural antifoulants using compounds produced by immobile marine plants and creatures, themselves at risk from biofouling.

biofuel   any solid, liquid, or gaseous fuel produced from organic (once living) matter, either directly from plants or indirectly from industrial, commercial, domestic, or agricultural wastes. There are three main methods for the development of biofuels: the burning of dry organic wastes (such as household refuse, industrial and agricultural wastes, straw, wood, and peat); the fermentation of wet wastes (such as animal dung) in the absence of oxygen to produce biogas (containing up to 60% methane), or the fermentation of sugar cane or maize to produce alcohol and esters; and energy forestry (producing fast-growing wood for fuel).

Fermentation produces two main types of biofuels: alcohols and esters. These could theoretically be used in place of fossil fuels but, because major alterations to engines would be required, biofuels are usually mixed with fossil fuels. The EU allows 5% ethanol, derived from wheat, beet, potatoes, or maize, to be added to fossil fuels. A quarter of Brazil's transportation fuel in 1994 was ethanol.

biogas   the production of methane and carbon dioxide, which can be obtained from plant or crop waste. Biogas is an example of a renewable source of energy (see *renewable resources, *non-renewable resources).

biogeochemistry   emerging branch of *geochemistry involving the study of how chemical elements and their isotopes move between living organisms and geological materials. For example, the analysis of carbon in bone gives biogeochemists information on how the animal lived, its diet, and its environment.

biological control   control of pests such as insects and fungi through biological means, rather than the use of chemicals. This can include breeding resistant crop strains; inducing sterility in the pest; infecting the pest species with disease organisms; or introducing the pest's natural predator. Biological control tends to be naturally self-regulating, but as ecosystems are so complex, it is difficult to predict all the consequences of introducing a biological controlling agent.

biological oxygen demand (BOD) the amount of dissolved oxygen consumed by micoorganisms as they decompose organic material in polluted water. Measurement of the rate of oxygen take-up is used as a standard test to detect the polluting capacity of effluent; the greater the BOD value (and hence the greater presence of oxygen-consuming microorganisms) the greater the volume of pollutant present. The test measures the mass (in grams) of dissolved oxygen consumed per litre of water when a sample is incubated in a dark chamber at 25°C for five days. Examples of BOD values expressed in g/m^3 are as follows:

Domestic sewage	350
Brewing	550
Distilling	7000
Paper pulp mill	25000

biological weathering   or **organic weathering**, form of *weathering caused by the activities of living organisms – for example, the growth of roots or the burrowing of animals.

Tree roots are probably the most significant agents of biological weathering as they are capable of prising apart rocks by growing into cracks and joints. Plants also give off organic acids that help to break down rocks chemically (see *chemical weathering).

biomass total mass of living organisms present in a given area. It may be used to describe the mass of a particular species (such as earthworm biomass), for a general category (such as herbivore biomass – animals that eat plants), or for everything in a *habitat. Estimates also exist for the entire global plant biomass. Biomass can be the mass of the organisms as they are – wet biomass – or the mass of the organisms after they have been dried to remove all the water – dry biomass. Measurements of biomass can be used to study interactions between organisms, the stability of those interactions, and variations in population numbers. Growth results in an increase in biomass, so biomass is a good measure of the extent to which organisms thrive in particular habitats. For a plant, biomass increase occurs as a result of the process of photosynthesis. For a herbivore, biomass increase depends on the availability of plant food. Studying biomass in a habitat is a useful way to see how food is passed from organism to organism along *food chains and through food webs.

Some two-thirds of the world's population cooks and heats water by burning biomass, usually wood. Plant biomass can be a renewable source of energy as replacement supplies can be grown relatively quickly. Fossil fuels, however, originally formed from biomass, accumulate so slowly that they cannot be considered renewable. The burning of biomass (defined either as natural areas of the ecosystem or as forest, grasslands, and fuel woods)

produces 3.5 million tonnes of carbon in the form of carbon dioxide each year, accounting for up to 40% of the world's annual carbon dioxide production.

Plant biomass can be changed into liquid or gaseous fuels to generate electricity or heat, or to fuel internal combustion engines. Fuel from biomass is burned in a reactor to generate heat energy, which is then converted into mechanical energy to turn turbine blades in a generator to produce electricity. Biomass power stations of 80 megawatts can produce electricity to power approximately 42,000 homes.

biome broad natural assemblage of plants and animals shaped by common patterns of vegetation and climate. Examples include the *tundra biome, the *rainforest biome, and the *desert biome.

bioreactor sealed vessel in which microbial reactions can take place. The simplest bioreactors involve the slow decay of vegetable or animal waste, with the emission of methane that can be used as fuel. Laboratory bioreactors control pH, acidity, and oxygen content and are used in advanced biotechnological operations, such as the production of antibiotics by genetically engineered bacteria.

biosphere narrow zone that supports life on our planet. It is limited to the waters of the Earth, a fraction of its crust, and the lower regions of the atmosphere. The biosphere is made up of all the Earth's *ecosystems. It is affected by external forces such as the Sun's rays, which provide energy, the gravitational effects of the Sun and Moon, and cosmic radiations.

BioSphere 2 BS2, ecological test project, a 'planet in a bottle', in Arizona, USA. Under a sealed glass and metal dome, different habitats are recreated, with representatives of nearly 4000 species, to test the effects that various environmental factors have on ecosystems. Simulated ecosystems, or 'mesocosms', include savannah, desert, rainforest, marsh and Caribbean reef. The response of such systems to elevated atmospheric concentrations of carbon dioxide gas (CO_2) are among the priorities of Biosphere 2 researchers.

biotechnology industrial use of living organisms. Examples of its uses include fermentation, genetic engineering (gene technology), and the manipulation of reproduction. The brewing and baking industries have long relied on the yeast micro-organism for fermentation purposes, while the dairy industry employs a range of bacteria and fungi to convert milk into cheeses and yoghurts. Enzymes, whether extracted from cells or produced artificially, are central to most biotechnological applications. Recent advances include genetic engineering, in which single-celled organisms with modified DNA are used to produce insulin and other drugs.

There are many medical and industrial applications of the use of micro-organisms, such as drug production. One important area is the production of antibiotics such as penicillin.

It is thought that biotechnology may be helpful in reducing world food shortages. Micro-organisms grow very quickly in suitable conditions and they often take substances that humans cannot eat and use them to produce foods that we can eat.

biotic factor organic variable affecting an ecosystem – for example, the changing population of elephants and its effect on the African savannah.

biotic potential the total theoretical reproductive capacity of an individual organism or species under ideal environmental conditions. The biotic potential of many small organisms

such as bacteria, annual plants, and small mammals is very high but rarely reached, as other elements of the ecosystem such as predators and nutrient availability keep the population growth in check.

biotite dark mica, $K(Mg, Fe)_3Al Si_3O_{10}(OH, F)_2$, a common silicate mineral. It is brown to black with shiny surfaces, and like all micas, it splits into very thin flakes along its one perfect cleavage. Biotite is a mineral found in igneous rocks, such as granites, and metamorphic rocks such as schists and gneisses.

bird's foot delta see ˄delta.

birth control the use of contraceptives prevent pregnancy. It is part of the general practice of family planning.

birth rate the number of live births per 1000 of the population over a period of time, usually a year (sometimes it is also expressed as a percentage). For example, a birth rate of 20/1000 (or 2%) would mean that 20 babies were being born per 1000 of the population. It is sometimes called **crude birth rate** because it takes in the whole population, including men and women who are too old to bear children.

The UK's birth rate fell from 28/1000 at the beginning of the 20th century to 13.6/1000 in 1993, owing to increased use of contraception, better living standards, and falling infant mortality rate. It was estimated that by the year 2000 the UK birth rate would be under 12. The birth rate remains high in poorer countries – for example, in Nigeria it stands at 57.4/1000 (1993), in Bangladesh 35.4/1000 (1993), and in Uzbekistan 30.6/1000 (1993).

bise cold dry northerly wind experienced in southern France and Switzerland.

bitumen involatile, tarry material, containing a mixture of hydrocarbons (mainly alkanes), that is the residue from the fractional distillation of crude oil (unrefined *petroleum). Sometimes the term is restricted to a soft kind of pitch resembling asphalt.

bituminous coal sometimes called house coal – a medium-quality *coal with some impurities; the typical domestic coal. It is also the major fuel source for *thermal power stations.

black earth exceedingly fertile soil that covers a belt of land in northeastern North America, Europe, and Asia. See *chernozem.

black economy unofficial economy of a country, which includes undeclared earnings from a second job ('moonlighting'), and enjoyment of undervalued goods and services (such as company 'perks'), designed for tax evasion purposes. In industrialized countries, it has been estimated to equal about 10% of *gross domestic product.

black ice see *glazed frost.

blizzard severe winter weather condition characterized by strong winds, blowing snow, and low temperatures.

bloc (French) group, generally used to describe politically allied countries, as in the former 'Soviet bloc'.

blockade cutting-off of a place by hostile forces by land, sea, or air so as to prevent any movement to or fro, in order to compel a surrender without attack or to achieve some other political aim (for example, the Berlin blockade (1948) and Union blockade of Confederate ports during the American Civil War). Economic sanctions are sometimes used in an attempt to achieve the same effect.

block faulting the dissection of a region by an extensive system of vertical or semi-vertical *faults. The faults divide the landscape into a series of blocks which may be relatively uplifted or depressed, to produce a series of *block mountains (*horsts*) and *rift valleys.

blocking anticyclone see *anticyclone.

block mountain or horst. A section of the Earth's *crust uplifted by faulting. Mt. Ruwenzori in the East African Rift System is an example of a block mountain.

blowhole a crevice, *joint or *fault in coastal rocks, enlarged by marine *erosion. A blowhole often leads from the rear of a cave (formed by wave action at the foot of a *coastal cliff) up to the cliff top. As waves break in the cave they erode the roof at the point of weakness and eventually a hole is formed. Air and sometimes spray are forced up the blowhole to erupt at the surface. At times wave action can be so strong that bursts of seawater are sent through the blowhole, enlarging it further.

blow-out hollow or depression of bare sand in an area of *dunes on which vegetation grows. Blow-outs are common in coastal dune complexes and are formed by wind erosion, which can be triggered by the destruction of small areas of vegetation by people or animals; lack of sand supply from a beach; and localized dryness.

blue-collar worker a worker who is either a manual worker or who works in potentially dirty conditions. The term 'blue-collar' derives from the wearing of dark-coloured overalls which show less dirt than light-coloured clothing. Compare *white-collar worker.

blue ice heavily compressed ice at the bottom of a *corrie icefield or *glacier, the expulsion of air caused by the compression making the ice appear blue. Contrast this with *white ice.

Blueprint for Survival environmental manifesto published in 1972 in the UK by the editors of the *Ecologist* magazine. The statement of support it attracted from a wide range of scientists helped draw attention to the magnitude of environmental problems.

bluff alternative name for a *river cliff.

boat people illegal emigrants travelling by sea, especially those Vietnamese who left their country after the takeover of South Vietnam in 1975 by North Vietnam. In 1979, almost 69,000 boat people landed in Hong Kong.

bog type of wetland where decomposition is slowed down and dead plant matter accumulates as *peat. Bogs develop under conditions of low temperature, high acidity, low nutrient supply, stagnant water, and oxygen deficiency. Typical bog plants are sphagnum moss, rushes, and cotton grass; insectivorous plants such as sundews and bladderworts are common in bogs (insect prey make up for the lack of nutrients).

Large bogs are found in Ireland and northern Scotland. They are dominated by heather, cotton grass, and over 30 species of sphagnum mosses which make up the general bog matrix. The rolling blanket bogs of Scotland are a southern outcrop of the arctic tundra ecosystem, and have taken thousands of years to develop.

According to government figures November 1994, Scotland's peat bogs contain three-quarters of all carbon locked up in organic material in Britain.

boiling water reactor see *water-cooled reactor.

bolson basin without an outlet, found in desert regions. Bolsons often contain temporary lakes, called playa lakes, and become filled with sediment from inflowing intermittent streams.

borax hydrous sodium borate, $Na_2B_4O_7.10H_2O$, found as soft, whitish crystals or encrustations on the shores of hot springs and in the dry beds of salt lakes in arid regions, where it occurs with other borates, halite, and *gypsum. It is used in bleaches and washing powders.

A large industrial source is Borax Lake, California. Borax is also used in glazing pottery, in soldering, as a mild antiseptic, and as a metallurgical flux.

bore surge of tidal water up an estuary or a river, caused by the funnelling of the rising tide by a narrowing river mouth. A very high tide, possibly fanned by wind, may build up when it is held back by a river current in the river mouth. The result is a broken wave, a metre or a few feet high, that rushes upstream.

Famous bores are found in the rivers Severn (England), Seine (France), Hooghly (India), and Chang Jiang (China), where bores of over 4 m have been reported.

boreal forest see *taiga.

borehole see *artesian basin.

borough (Old English burg, 'a walled or fortified place') urban-based unit of local government in the UK and USA. It existed in the UK from the 8th century until 1974, when it continued as an honorary status granted by royal charter to a district council, entitling its leader to the title of mayor. In England in 1998 there were 32 London borough councils and 36 metropolitan borough councils. The name is sometimes encountered in the USA: New York City has five administrative boroughs, Alaska has local government boroughs, and in other states some smaller towns use the name.

Bouguer anomaly anomaly in the local gravitational force that is due to the density of rocks rather than local topography, elevation, or latitude. A positive anomaly, for instance, is generally indicative of denser and therefore more massive rocks at or below the surface. A negative anomaly indicates less massive materials. Calculations of Bouguer anomalies are used for mineral prospecting and for understanding the structure beneath the Earth's surface. The Bouguer anomaly is named after its discoverer, the French mathematician Pierre Bouguer, who first observed it in 1735.

boulder clay   another name for *till, a type of glacial deposit.

bourne   a spring with a fluctuating origin on the *dip slope of a chalk escarpment. In wet seasons, the spring may emerge higher up the dip slope, i.e. when the *water table is higher. Drier weather will cause a drop in the level of the water table and the spring will then emerge further down the dip slope.

The term 'bourne' is commonly applied to streams in the chalk regions of southern England.

BP   *abbreviation for* Before Present, a term used in geological time scales to denote any time before the present.

braiding   the subdivision of a river into several channels caused by deposition of sediment as islets in the channel. Braided channels are common in meltwater streams.

Brandt Commission   international committee (1977–83) set up to study global development issues. It produced two reports, stressing the interdependence of the countries of the wealthy, industrialized North and the poor South (or developing world), and made detailed recommendations for accelerating the development of poorer countries (involving the transfer of resources to the latter from the richer countries).

breakeven analysis   management accounting technique to determine the *breakeven point by analysing fixed and variable costs in relation to the total revenue. This technique is used to determine the levels of activity required to generate a certain level of profit, and shows how fixed and variable costs affect profit levels. A *breakeven chart is often used in breakeven analysis.

breakeven chart   line graph showing fixed costs, total revenue, and total costs for a given activity over the entire range of possible output levels. The point at which the total revenue curve meets the total cost curve is the

*breakeven point. As well as providing the *breakeven point, a breakeven chart also allows potential profit or loss to be determined for any level of output.

breakeven point   level of output where costs equal revenue and neither profit nor loss is made.

break-of-bulk point   place where goods are transferred from one form of transport to another. This frequently involves the goods being repackaged into smaller quantities ready for individual users.

Break of bulk often occurs at ports, where cargo carried on inland waterways is transferred to ocean-going vessels; Rotterdam is an example. Break-of-bulk points may be important industrial locations.

breakwater   see *groyne.

breccia   coarse-grained *sedimentary rock, made up of broken fragments (clasts) of pre-existing rocks held together in a fine-grained matrix. It is similar to *conglomerate but the fragments in breccia are jagged in shape.

British Council   semi-official organization set up in 1934 (royal charter 1940) to promote a wider knowledge of the UK, excluding politics and commerce, and to develop cultural relations with other countries. It employs more than 6000 people and is represented in 109 countries, running libraries, English-teaching operations, and resource centres.

brown earth   **brown forest soil, sol brun, cambisol or inceptisol**, a medium-textured, brown soil with high organic content commonly found beneath deciduous forests in humid temperate zones. Characteristic features of this soil type include a B horizon rich in iron compounds brought about by leaching from the free-draining A horizon which has a high organic content. However, zonation may be difficult to detect from simple visual inspection.

Because of the rich humus content of their upper horizons, brown earth soils are very fertile and have been in continuous agricultural use in Europe since the middle ages. They have proved to be of immense value, providing a base for intensive arable agriculture with few management problems.

brownfield site   site that has been developed in the past for industry or for some human activity other than agriculture. Most brownfield sites are found in urban areas, where redevelopment is encouraged. However, many sites are contaminated by industrial pollutants, and it is costly to make the land safe for future development.

Brundtland Report   the findings of the World Commission on Environment and Development, published in 1987 as *Our Common Future*. It stressed the necessity of environmental protection and popularized the phrase 'sustainable development'. The commission was chaired by the Norwegian prime minister Gro Harlem Brundtland.

Brussels, Treaty of   pact of an economic, political, cultural, and military alliance established in 17 March 1948, for 50 years, by the UK, France, and the Benelux countries, joined by West Germany and Italy in 1955. It was the forerunner of the North Atlantic Treaty Organization and the European Community (now the European Union).

budget   estimate of income and expenditure for some future period, used in financial planning. National budgets set out estimates of government income and expenditure and generally include projected changes in taxation and growth.

bulk decreasing good   good that decreases in size and/or weight during the manufacturing process. The resulting reduction in transportation costs means that bulk decreasing goods tend to be manufactured close to the raw materials they use and transported to market from there.

bulk increasing good   good that increases in size and/or weight during the manufacturing process. The resulting increase in transportation costs means that bulk increasing goods tend to be manufactured close to the markets they are sold in.

bureaucracy   organization whose structure and operations are governed to a high degree by written rules and a hierarchy of offices; in its broadest sense, all forms of administration, and in its narrowest, rule by officials.

Burgess model   see *concentric-ring theory.

burgh   or **burh** or **borough**, archaic form of *borough.

burgh   former unit of Scottish local government, referring to a town enjoying a degree of self-government. Burghs were abolished in 1975; the terms **burgh** and **royal burgh** once gave mercantile privilege but are now only an honorary distinction.

bush   any uncultivated or sparsely settled land, especially forested land in varying conditions of naturalness, as found in Africa, Australia and New Zealand. Bush may vary from open shrubby country to dense rainforest.

bush fallowing   or **shifting cultivation**, a system of *agriculture in which there are no permanent fields. For example in the tropical *rainforest, remote societies cultivate forest clearings for one year and then move on. The system functions successfully when forest *regeneration occurs over a sufficiently long period to allow the soil to regain its fertility.

business cycle   or **trade cycle**, period of time that includes a peak and trough of economic activity, as measured by a country's national income. The economy passes through phases of boom and recession, causing changes in the levels of output, unemployment, and inflation.

business park   low-density office development of a type often established by private companies on *greenfield sites. The sites are often landscaped to create a pleasant working environment.

Business parks tend to be located near motorway junctions and may have a high proportion of high-tech firms.

bustee   alternative spelling of **basti**, an Indian name for a *shanty town.

butte   steep-sided, flat-topped hill, formed in horizontally layered sedimentary rocks, largely in arid areas. A large butte with a pronounced tablelike profile is a *mesa.

Buttes and mesas are characteristic of semi-arid areas where remnants of resistant rock layers protect softer rock underneath, as in the plateau regions of Colorado, Utah, and Arizona, USA.

bycatch   or **bykill**, in commercial fishing, that part of the catch that is unwanted. Bycatch constitutes approximately 25% of global catch, and consists of a variety of marine life, including fish too small to sell or otherwise without commercial value, seals, dolphins, sharks, turtles, and even seabirds.

C

cabinet government alternative term for *parliamentary government.

Cainzoic Era see *Cenozoic Era.

calcification the accumulation of calcium carbonate in the B horizon of a *soil. Calcification usually occurs in the low-rainfall continental interiors such as the New South Wales grasslands, the North American prairies, and the Soviet steppes. Limited leaching results in the downward movement of calcium carbonate but only as far as the B horizon, where a *duricrust may form; this calcium carbonate duricrust is known as *calcrete* or *caliche*. Calcification is greatly accelerated by the upward movement of calcium rich groundwater through capillarity; high air and ground temperatures cause evaporation and the deposition of calcium salts from the groundwater.

calcimorphic soil a type of soil found above calcium-rich parent material. Typical examples include rendzina and terra rosa soils.

calcite colourless, white, or light-coloured common rock-forming mineral, calcium carbonate, $CaCO_3$. It is the main constituent of *limestone and marble and forms many types of invertebrate shell.

Calcite often forms *stalactites and stalagmites in caves and is also found deposited in veins through many rocks because of the ease with which it is dissolved and transported by groundwater; *oolite is a rock consisting of spheroidal calcite grains. It rates 3 on the *Mohs scale of hardness. Large crystals up to 1 m have been found in Oklahoma and Missouri, USA. *Iceland spar is a transparent form of calcite used in the optical industry; as limestone it is used in the building industry.

calcrete see *calcification.

caldera in geology, a very large basin-shaped *crater. Calderas are found at the tops of volcanoes, where the original peak has collapsed into an empty chamber beneath. The basin, many times larger than the original volcanic vent, may be flooded, producing a crater lake, or the flat floor may contain a number of small volcanic cones, produced by volcanic activity after the collapse.

Typical calderas are Kilauea, Hawaii; Crater Lake, Oregon, USA; and the summit of Olympus Mons, on Mars. Some calderas are wrongly referred to as craters, such as Ngorongoro, Tanzania.

caliche see *calcification.

California current cold ocean *current in the eastern Pacific Ocean flowing southwards down the west coast of North America. It is part of the North Pacific *gyre (a vast, circular movement of ocean water).

calima (Spanish 'haze') dust cloud in Europe, coming from the Sahara Desert, which sometimes causes heatwaves and eye irritation.

call centre office that deals with customers' telephone enquiries. Call centres often employ several hundred staff in a tightly regulated environment, and they are often described as the developed world's equivalent of the sweatshop, or Taylorism (see *scientific management) in the 21st century. Call centres are now being located in countries with skilled but low-cost labour, for example India.

cambisol the term used in the FAO soil classification system for a brown earth soil.

Cambrian period period of geological time roughly 570–510 million years ago; the first period of the Palaeozoic Era. All invertebrate animal life appeared, and marine algae were widespread. The **Cambrian Explosion** 530–520 million years ago saw the first appearance in the fossil record of modern animal phyla; the earliest fossils with hard shells, such as trilobites, date from this period.

The name comes from Cambria, the medieval Latin name for Wales, where Cambrian rocks are typically exposed and were first described.

Canaries current cold ocean current in the North Atlantic Ocean flowing southwest from Spain along the northwest coast of Africa. It meets the northern equatorial current at a latitude of 20° N.

Cancún resort in Yucatan, Mexico, created on a barrier island 1974 by the Mexican government to boost tourism; population around 30,000 (almost all involved in the tourist industry). It is Mexico's most popular tourist destination, with beaches, a coral reef, and a lagoon. The hotel zone is on the sandbar, and the workers live in Cancún city on the mainland.

canopy continuous layer of dense treetop foliage in woodland or forest. The leaves filter the sunlight, reaching the lower layers of foliage, so the nearer to the ground plants grow, the less sunlight they receive. The canopy provides a *habitat for diverse specially adapted species, for example those found in the *rainforest canopy include various monkey species, bats, numerous insects, and birds.

canton in France, an administrative district, a subdivision of the *arrondissement*; in Switzerland, one of the 23 subdivisions forming the Confederation.

canyon (Spanish *cañon* 'tube') deep, narrow valley or gorge running

canyon

canyon Cross section of a canyon. Canyons are formed in dry regions where rivers maintain a constant flow of water over long periods of time. The Grand Canyon, for example, was first cut around 26 million years ago, in Miocene times.

through mountains. Canyons are formed by stream down-cutting, usually in arid areas, where the rate of down-cutting is greater than the rate of weathering, and where the stream or river receives water from outside the area.

There are many canyons in the western USA and in Mexico, for example the Grand Canyon of the Colorado River in Arizona, the canyon in Yellowstone National Park, and the Black Canyon in Colorado.

capacity in economics, the maximum amount that can be produced when all the resources in an economy, industry, or firm are employed as fully as possible. Capacity constraints can be caused by lack of investment and skills shortages, and spare capacity can be caused by lack of demand.

capillarity the ability of a soil to retain a film of water around individual soil particles and in pores through the action of surface tension. This is achieved despite the action of gravity which serves to move water molecules downwards through the soil profile. In some soils in semi-arid regions, capillary water is drawn upwards and dissolved chemicals are precipitated in the upper layers to form duricrusts.

capillary water see soil water.

capital in a country, the city where the government headquarters are. The capital is usually the most important and largest city in a country; for example, London. This is not the case in the USA and Canada. Some countries have moved the seat of government to reduce strain on the largest city's infrastructure; for example, Brasília is the specially built capital of Brazil.

capital in economics, the stock of goods used in the production of other goods. Classical economics regards capital as a factor of production, distinguishing between **financial capital** and **physical capital**. Financial capital is accumulated or inherited wealth held in the form of assets, such as stocks and shares, property, and bank deposits, while physical capital is wealth in the form of physical assets such as machinery and plant. The term is also used to describe investment in a company as either share capital or debt (called loan capital).

Fixed capital is durable, examples being factories, offices, plant, and machinery. **Circulating capital** is capital that is used up quickly, such as raw materials, components, and stocks of finished goods waiting for sale. **Private capital** is usually owned by individuals and private business organizations. **Social capital** is usually owned by the state and is the *infrastructure of the economy, such as roads, bridges, schools, and hospitals. Investment is the process of adding to the capital stock of a nation or business.

capital employed in business, the total assets of a company (excluding intangibles, such as goodwill) less its current liabilities (including overdrafts, short-term loans, trade and other creditors).

capital expenditure spending on fixed assets such as plant and equipment, trade investments, or the purchase of other businesses.

capital-intensive production form of production where a relatively high amount of capital is used in relation to labour or other factors of production. Production in primary and secondary *industrial sectors is becoming increasingly capital-intensive. As more and more capital is used, labour productivity increases but at the same time employment in these industries falls.

capitalism economic system in which the principal means of production, distribution, and exchange are in private (individual or corporate) hands and competitively operated for profit. A **mixed economy** combines the private enterprise of capitalism and a degree of state monopoly, as in nationalized industries and welfare services.

capitalization in business, total market value of a company's issued share capital. This is calculated by multiplying the number of shares issued by a company by the current price of the shares.

The term is also used to describe the capital structure of a company, that is, the proportions of debt or equity, and the proportions of the equity that are ordinary or preference shares.

cap rock 1. (also called fall maker) A stratum of resistant *rock at the lip of a *waterfall.
2. hard rock protecting the top of a *mesa.

carbonaceous composed of or containing carbon.

carbonation in earth science, a form of *chemical weathering caused by rainwater that has absorbed carbon dioxide from the atmosphere and formed a weak carbonic acid. The slightly acidic rainwater is then capable of dissolving certain minerals in rocks. Water can also pick up acids when it passes through soil. This water – enriched with organic acid – is also capable of dissolving rock. *Limestone is particularly vulnerable to this form of weathering.

carbon cycle   the natural circulation of carbon in the biosphere. Separate but interconnected cycles exist for the circulation of carbon on land and in the sea. The cycles are connected at the junction of the ocean and the atmosphere. The carbon cycle begins with the fixation of atmospheric carbon dioxide by the process of photosynthesis, conducted by plants and certain microorganisms. Thereafter, the cycle extends to the animal kingdom, the animals eating the plants thereby respiring and releasing carbon dioxide back into the atmosphere. Carbon is also released through the decomposition of dead plant and animal matter, and the burning of *fossil fuels such as coal and oil, which produce carbon dioxide that is released into the atmosphere.

Until 1860, the amount of carbon in circulation was approximately stable. However, following the industrial revolution and the dramatic increase in the burning of fossil fuels, the amount of carbon in the atmosphere has rapidly increased. The carbon cycle is in danger of being disrupted by the increased burning of fossil fuels, and the destruction of large areas of tropical forests. The rising levels of carbon dioxide in the atmosphere are probably increasing the temperature on Earth (enhanced *greenhouse effect). It is thought that by limiting the production of carbon dioxide through human activities we can slow the rate at which temperatures on Earth will rise

carbon dioxide CO_2, colourless, odourless gas, slightly soluble in water, and denser than air. Discovered by Scottish scientist Joseph Black (1727–1799) in 1754. It is formed by the complete oxidation of carbon.

Carbon dioxide is produced by living things during the processes of respiration and the decomposition of organic matter, and it is used up during photosynthesis. It therefore plays a vital role in the *carbon cycle.

Its increasing quantity in the atmosphere contributes to the *greenhouse effect and *global warming. Britain has 1% of the world's population, yet it produces 3% of CO_2 emissions; the USA has 5% of the world's population and produces 25% of CO_2 emissions.

Carboniferous period   period of geological time roughly 362.5–290 million years ago, the fifth period of the Palaeozoic Era. In the USA it is divided into two periods: the Mississippian (lower) and the Pennsylvanian (upper).

Typical of the lower-Carboniferous rocks are shallow-water *limestones, while upper-Carboniferous rocks have *delta deposits with *coal (hence the name). Amphibians were abundant, and reptiles evolved during this period.

carbon sequestration   disposal of carbon dioxide waste in solid or liquid form. From 1993 energy conglomerates such as Shell, Exxon, and British Coal have been researching ways to reduce their carbon dioxide emissions by developing efficient technologies to trap the gas and store it securely – for example, by burying it or dumping it in the oceans. See also *greenhouse effect.

carcinogen   any substance capable of causing cancers in animal tissues. The mechanism by which carcinogens cause cancers in living tissue is not fully understood; nevertheless, much is known about the substances which predispose to the disease. The first carcinogenic material to be discovered was arsenic, but since the 1930s beryllium, cadmium, cobalt, chromium and asbestos have all been added to the list of carcinogenic materials. The range of suspected or proven carcinogens include tyre rubber, tar macadam and the synthetic PCBs, of which there are now over 200 different types used in everyday items such as plastics, electrical components and hydraulic fluids in motor vehicles. Many of the synthetic compounds used in industrial processes have been found to be carcinogenic, for example, many of the epoxy resins and adhesives.

Caribbean Community and Common Market CARICOM, organization for economic and foreign policy coordination in the Caribbean region, established by the Treaty of Chaguaramas in 1973 to replace the former Caribbean Free Trade Association. Its members are Antigua and Barbuda, Bahamas, Barbados, Belize, Dominica, Grenada, Guyana, Haiti, Jamaica, Montserrat, St Kitts and Nevis, St Lucia, St Vincent and the Grenadines, and Trinidad and Tobago. The Bahamas is a member of the Community but not of the Caribbean Single Market and Economy (CSME).

The British Virgin Islands and the Turks and Caicos Islands are associate members, and Anguilla, the Dominican Republic, Mexico, Haiti, Puerto Rico, and Venezuela are observers. CARICOM headquarters are in Georgetown, Guyana.

CARICOM acronym for *Caribbean Community and Common Market.

carnelian semi-precious gemstone variety of *chalcedony consisting of quartz (silica) with iron impurities, which give it a translucent red colour. It is found mainly in Brazil, India, and Japan.

carnivore any animal which gains the majority of its food supply by consuming other animals. The term is often restricted to any mammal of the order carnivora, which includes cats, dogs, bears, raccoons, hyenas and weasels. The feeding habits of mammalian carnivores has resulted in the development of a distinct tooth design, featuring strong, pointed teeth (the canines) for tearing and ripping meat. Carnivores form the third trophic level of the food chain; they are the secondary consumers.

carnotite potassium uranium vanadate, $K_2(UO_2)_2(VO_4)_2.3H_2O$, a radioactive ore of vanadium and uranium with traces of radium.

A yellow powdery mineral, it is mined chiefly in the Colorado Plateau, USA; Radium Hill, Australia; and Shaba, the Democratic Republic of Congo (formerly Zaire).

carrying capacity in ecology, the maximum number of animals of a given species that a particular *habitat can support. If the carrying capacity of an ecosystem is exceeded by overpopulation, there will be insufficient resources and one or more species will decline until an equilibrium, or *balance of nature, is restored. Similarly, if the number of species in an environment is less than the carrying capacity, the population will tend to increase until it balances the available resources. Human interference frequently causes disruption to the carrying capacity of an area, for instance by the establishment of too many grazing animals on grassland, the over-culling of a species, or the introduction of a non-indigenous species into an area.

cartel (German *Kartell* 'a group') agreement among national or international firms not to compete with one another. Cartels can be formed to fix prices by maintaining the price of a product at an artificially low level, to deter new competitors, or to restrict production of a commodity in order to maintain prices at an artificially high level to boost profits. The members of a cartel may also agree on which member should win a contract, known as bid rigging, or which customers they will supply. Cartels therefore represent a form of *oligopoly. *OPEC, for example, is an example of a transnational cartel restricting the output of a commodity, in this case oil. In many countries, including the USA and the UK, companies operating a cartel may be breaching legislation designed to abolish anticompetitive practices.

cartography art and practice of drawing *maps, originally with pens and drawing boards, but now mostly with computer-aided drafting programs.

cash flow input of cash required to cover all expenses of a business, whether revenue or capital. Alternatively, the actual or prospective balance between the various outgoing and incoming movements which are designated in total. Cash flow is positive if receipts are greater than payments; negative if payments are greater than receipts. Money may be received through cash sales of products or assets, and receipts of debts. Money may flow out through purchase of raw materials, the settlement of debts, and the payment of salaries.

cassiterite or **tinstone**, mineral consisting of reddish-brown to black stannic oxide (SnO_2), usually found in granite rocks. It is the chief ore of tin. When fresh it has a bright ('adamantine') lustre. It was formerly extensively mined in Cornwall, England; today Malaysia is the world's main supplier. Other sources of cassiterite are Africa, Indonesia, and South America.

caste (Portuguese *casta* 'race') a system of stratifying a society into ranked groups defined by marriage, descent, and occupation. Most common in South Asia, caste systems are also found in other societies. such as in Mali and Rwanda. In the past, such systems could be found in Japan, in South Africa under apartheid, and among the Natchez – an American Indian people.

The caste system in Hindu society dates from ancient times. Traditional society is loosely ranked into four varnas (social classes): Brahmin (priests), Kshatriyas (nobles and warriors), Vaisyas (traders and farmers), and Sudras (servants), plus a fifth group, Harijan (untouchables). Their subdivisions, jati, number over 3000, each with its own occupation. A Hindu's dharma, or holy path in life, depends not only on the stage of life (ashrama) that he or she is currently in, but also on caste; it is a duty to follow the caste into which one is born by the laws of rebirth. Traditionally, Hindus would only mix with and marry people of their own caste.

cataclastic rock metamorphic rock, such as a breccia, containing angular fragments of preexisting rock produced by the grinding and crushing action (cataclasis) of *faults.

catalytic converter device fitted to the exhaust system of a motor vehicle in order to reduce toxic emissions from the engine. It converts the harmful exhaust products that cause *air pollution to relatively harmless ones.

It does this by passing them over a mixture of catalysts coated on a metal or ceramic honeycomb (a structure that increases the surface area and therefore the amount of active catalyst with which the exhaust gases will come into contact). **Oxidation catalysts** (small amounts of precious palladium and platinum metals) convert hydrocarbons (unburnt fuel) and carbon monoxide into carbon dioxide and water, but do not affect nitrogen oxide emissions. **Three-way catalysts** (platinum and rhodium metals) also convert nitrogen oxide gases into nitrogen and oxygen.

catastrophism   theory that the geological features of the Earth were formed by a series of sudden, violent 'catastrophes' beyond the ordinary workings of nature. The theory was largely the work of Georges Cuvier. It was later replaced by the concepts of *uniformitarianism and evolution.

catch crop   a minor crop (in terms of output) that is planted in the same year as, but immediately after, the main crop to 'catch' remaining soil moisture and other nutrients. Catch cropping is a form of multiple cropping. In temperate zones leafy crops may be sown in late summer or autumn to be grazed by sheep later in the year or the following spring. In parts of the tropics, cassava is planted as a catch crop to be harvested when required up to four years after planting, often after the plot has otherwise been abandoned in the course of shifting cultivation.

catchment area   **1.** (social sciences) the area surrounding, or related to, a particular service point, for example, a shopping centre or school, in which a defined proportion of the population makes frequent use of that service. In densely populated regions, catchment areas are likely to overlap. **2.** (earth sciences) also called drainage basin, the area of land bounded by watersheds draining into a river, basin or reservoir.

catena   or **hydrologic sequence** a sequence of soils in a particular locality which are derived from the same parent material but have differing characteristics due to variations in topography and drainage. A catena may be apparent from soil profiles which traverse a hillslope.

cave   passage or tunnel – or series of tunnels – formed underground by water or by waves on a coast. Caves of the former type commonly occur in areas underlain by limestone, such as Kentucky, USA, and many Balkan regions, where the rocks are soluble in water. A **pothole** is a vertical hole in rock caused by water descending a crack; it is thus open to the sky.

Coastal caves are formed where rocks with lines of weakness, like *basalt at tide level, are exposed to severe wave action. The erosion (*corrasion and *corrosion) of the rock layers is increased by subsidence, and the hollow in the cliff face grows still larger because of air compression in the chamber (*hydraulic action). Where the roof of a cave has fallen in, the vent up to the land surface is called a *blowhole. If this grows, finally destroying the cave form, the outside pillars of the cave are known as stacks or columns. The Old Man of Hoy (137 m high), in the Orkney Islands, is a fine example of a stack.

Most inland caves are found in *karst (limestone) regions, because *limestone is soluble when exposed to acid water. As the water makes its way along the main joints, fissures, and bedding planes, they are constantly enlarged into potential cave passages, which ultimately join to form a complex network. *Stalactites and stalagmites form due to water that is rich in calcium carbonate dripping from the roof of the cave. The collapse of the roof of a cave produces features such as **natural arches** and **steep-sided gorges**.

cavitation   in hydraulics, *erosion of rocks caused by the forcing of air into cracks. Cavitation results from the pounding of waves on the coast and the swirling of turbulent river currents, and exerts great pressure, eventually causing rocks to break apart.

The process is particularly common at waterfalls, where the turbulent falling water contains many air bubbles, which burst and send shock waves into the rocks of the river bed and banks. In addition, as water is forced into cracks in the rock, air within the crack is compressed and literally explodes, helping to break down the rock.

CBD *abbreviation for* *central business district.

celestine or **celestite**, mineral consisting of strontium sulphate, $SrSO_4$, occurring as white or light blue crystals. Celestine occurs in cavity linings associated with calcite, dolomite, or fluorite. It is the principal source of strontium.

cell-based manufacturing manufacturing system where common production processes among different products within an organization have been identified and combined. This is done to reduce the production costs for the organization.

cell production in business, the division of continuous flow production into several separate cells. Each cell carries out a specific part of the production process. This causes the workers to be more committed to their contribution and cell team than if they were a small part of a huge production line.

Celsius scale of temperature, previously called centigrade, in which the range from freezing to boiling of water is divided into 100 degrees, freezing point being 0 degrees and boiling point 100 degrees.

The degree centigrade (°C) was officially renamed Celsius in 1948 to avoid confusion with the angular measure known as the centigrade (one hundredth of a grade). The Celsius scale is named after the Swedish astronomer Anders Celsius, who devised it in 1742 but in reverse (freezing point was 100°; boiling point 0°).

cement any bonding agent used to unite particles in a single mass or to cause one surface to adhere to another.

Portland cement is a powder which when mixed with water and sand or gravel turns into mortar or concrete.

In geology, cement refers to a chemically precipitated material such as carbonate that occupies the interstices of clastic rocks.

The term 'cement' covers a variety of materials, such as fluxes and pastes, and also bituminous products obtained from tar. In 1824 English bricklayer Joseph Aspdin (1779–1855) created and patented the first Portland cement, so named because its colour in the hardened state resembled that of Portland stone, a limestone used in building.

Cement is made by heating limestone (calcium carbonate) with clay (which contains a variety of silicates along with aluminium). This produces a grey powdery mixture of calcium and aluminium silicates. On addition of water, a complex series of reactions occurs and calcium hydroxide is produced. Cement sets by losing water.

cementation the fusion of rock *sediments or the remains of microscopic marine organisms into *sedimentary rocks. As the material builds up it is compacted together. Cementation occurs as chemical precipitation deposits micro crystals in the spaces between the mineral fragments and grains, holding them together. The 'cement' formed may be *silica, iron oxide, or carbonate. Cementation contributes to the processes involved in the *rock cycle, the formation, change, and reformation of rocks.

Clastic sedimentary rocks, such as *sandstone, siltstone, and claystone, are made from the sediments of other rocks. Non-clastic sedimentary rocks, such as *limestone and *chalk, are of organic origin. **Silica cementation** takes place below the water table surface; its source is groundwater.

Carbonate cementation, also sourced from groundwater, takes place either at above or below ground level. **Iron-oxide cementation** is less frequent.

Cenozoic Era or **Caenozoic**, era of geological time that began 65 million years ago and continues to the present day. It is divided into the Tertiary and Quaternary periods. The Cenozoic marks the emergence of mammals as a dominant group, and the rearrangement of continental masses towards their present positions.

census official count of the population of a country, originally for military call-up and taxation, later for assessment of social trends as other information regarding age, sex, and occupation of each individual was included. The data collected are used by government departments in planning for the future in such areas as health, education, transport, and housing.

Most countries have a census of some sort. In the UK a census has been conducted every ten years since 1801. Although the information about individual households remains secret for 100 years, data is available on groups of households down to about 200 (an enumeration district), showing such characteristics as age and sex structure, employment, housing types, car ownership, and qualifications held. The larger-scale information on population numbers, movements, and origins is published as a series of reports by the Office of Population Censuses and Surveys. The most recent census took place on 29 April 2001. The first US census was taken in 1790.

centigrade former name for the *Celsius temperature scale.

Central American Common Market CACM; Spanish **Mercado Común Centroamericana (MCCA)**, economic alliance established in 1961 by El Salvador, Guatemala, Honduras (seceded in 1970), and Nicaragua;

Costa Rica joined in 1962. Formed to encourage economic development and cooperation between the smaller Central American nations and to attract industrial capital, CACM failed to live up to early expectations: nationalist interests remained strong and by the mid-1980s political instability in the region and border conflicts between members were hindering its activities. Its offices are in Guatemala City, Guatemala.

central bank the bank responsible for issuing currency in a country. Often it is also responsible for foreign-exchange dealings on behalf of the government and for supervising the banking system in the country (it holds the commercial reserves of the nation's clearing banks).

Although typically independent of central government, a central bank will work closely with it, especially in implementing monetary policy. The earliest bank to take on the role of central bank was the Bank of England. In the USA the Federal Reserve System was established 1913.

central business district CBD, area of a town or city where most of the commercial activity is found. This area is dominated by shops, offices, entertainment venues, and local-government buildings. Usually the CBD is characterized by high rents and rates, tall buildings, and chain stores, and is readily accessible to pedestrians. It may also occupy the historic centre of the city and is often located where transport links meet.

Central European Free Trade Agreement CEFTA, international trade agreement signed on 21 December 1992 by the former Czechoslovakia, Hungary, and Poland, and in force from March 1993 (by which time Czechoslovakia had become the Czech and Slovak republics). The main objective of the agreement was gradually to reduce and eliminate tariffs between CEFTA countries and to establish a free-trade zone by 1 January 2001, as a first step towards integrating these countries into Western Europe. Slovenia joined as a full member in 1996, Romania in 1997, and Bulgaria in 1999.

central government in the UK, that part of the public sector controlled by the nationally-elected government at Westminster, as opposed to *local government, which is controlled by local councillors in counties, boroughs, parishes, and so on.

centralization in business, a form of organization where decisionmaking for the whole business is taken by individuals or groups of people at the centre of the business. This compares with 'decentralization', where decisionmaking is devolved throughout the whole business.

central place or **service centre**, place to which people travel from the surrounding area (*hinterland) to obtain various goods or services. Central places will often be towns or areas within towns; for example, a shopping arcade that serves people in the immediate neighbourhood with low-order goods. Central places vary in importance; higher-order goods and services are provided where the *threshold population is reached.

According to **central-place theory**, if each settlement of a particular order acts as a central place for certain levels of goods or services, there should be a regular pattern and distribution of settlements within an area. This theory was first put forward by German geographer Walter Christaller in 1903. He suggested that each settlement would be surrounded by a hexagonal sphere of influence (hexagonal rather than circular because circles cannot fit together exactly). The size of these hexagons depends on the order of the

central place – village, town, or city. Each order would have a market area three times that of the settlement below. Settlements of each order would therefore be spaced at regular intervals in a spatial hierarchy. Christaller took a number of factors for granted; for example, he assumed that transport was equally possible in all directions. In the real world this is not the case; however, the theory provides a starting point for explaining settlement distribution.

central planning alternative name for *command economy.

cereal farming see *arable farming.

CFC *abbreviation for* *chlorofluorocarbon.

chain of command in business, the path down which orders and decisions are communicated, from the board of directors of a company at the top of the hierarchy down to shop-floor workers at the bottom. The shorter the chain of command, the faster communication is likely to be. There is also less likely to be misinterpretation of communication. A short chain of command also tends to motivate workers because they are able to interact with those in positions of authority and see their decisions being implemented by workers below them.

chain of production the different production stages through which a good or service passes before being sold to the buyer. For example, a loaf of bread passes from agriculture through milling to baking before finally being sold in a retail shop.

chalcedony form of the mineral quartz, SiO_2, in which the crystals are so fine-grained that they are impossible to distinguish with a microscope (cryptocrystalline). Agate, onyx, and carnelian are *gem varieties of chalcedony.

chalcopyrite copper iron sulphide mineral, $CuFeS_2$, the most common ore of copper. It is brassy yellow in colour and may have an iridescent surface tarnish. It occurs in many different types of mineral vein, in rocks ranging from basalt to limestone.

chalk soft, fine-grained, whitish sedimentary rock composed of calcium carbonate, $CaCO_3$, extensively quarried for use in cement, lime, and mortar, and in the manufacture of cosmetics and toothpaste. **Blackboard chalk** in fact consists of *gypsum (calcium sulphate, $CaSO_4.2H_2O$).

Chalk was once thought to derive from the remains of microscopic animals or foraminifera. In 1953, however, it was seen under the electron microscope to be composed chiefly of *coccolithophores, unicellular lime-secreting algae, and hence primarily of plant origin. It is formed from deposits of deep-sea sediments called oozes.

channel efficiency measure of the ability of a river channel to move water and sediment. Channel efficiency can be measured by calculating the channel's *hydraulic radius (cross-sectional area/length of bed and bank). The most efficient channels are generally semicircular in cross-section, and it is this shape that water engineers try to create when altering a river channel to reduce the risk of *flooding.

Channel, English see *English Channel.

channel of distribution path taken to distribute goods from the manufacturer to the retailer, often through wholesalers, down to the final consumer. In the case of mail order, the manufacturer can distribute directly to consumers, although many large mail-order catalogue companies act as a retailer.

chaos theory or **chaology** or **complexity theory**, branch of mathematics that attempts to describe irregular, unpredictable systems–that

is, systems whose behaviour is difficult to predict because there are so many variable or unknown factors. Weather is an example of a chaotic system.

chaparral thick scrub country of the southwestern USA. Thorny bushes have replaced what was largely evergreen oak trees.

chargé d'affaires (French 'entrusted with affairs') diplomatic agent, ranking next below a 'minister resident', and holding his or her credentials from the head of the Foreign Office. *Chargés d'affaires* may either act as their country's representative to a minor state or be empowered to take the place of an ambassador.

charters, town royal grants of certain privileges, rights, or immunities made to towns from early times.

chemical oxygen demand COD, measure of water and effluent quality, expressed as the amount of oxygen (in parts per million) required to oxidize the reducing substances present.

Under controlled conditions of time and temperature, a chemical oxidizing agent (potassium permanganate or dichromate) is added to the sample of water or effluent under consideration, and the amount needed to oxidize the reducing materials present is measured. From this the chemically equivalent amount of oxygen can be calculated. Since the reducing substances typically include remains of living organisms, COD may be regarded as reflecting the extent to which the sample is polluted. Compare biological oxygen demand.

chemical weathering form of *weathering brought about by chemical attack on rocks, usually in the presence of water. Chemical weathering involves the breakdown of the original minerals within a rock to produce new minerals (such as *clay minerals, *bauxite, and *calcite). The breakdown of rocks occurs because of chemical reactions between the minerals in the rocks and substances in the environment, such as water, oxygen, and weakly acidic rainwater. Some chemicals are dissolved and carried away from the weathering source, while others are brought in.

Material worn away from rocks by weathering, either as fragments of rock or dissolved material, may be transported and deposited as sediments, which eventually become compacted to form *sedimentary rocks. Thus chemical weathering plays a part in the *rock cycle.

Chemical processes involved in weathering include *carbonation (breakdown by weakly acidic rainwater), *hydrolysis (breakdown by water), *hydration (breakdown by the absorption of water), and *oxidation (breakdown by the oxygen in air and water).

The reaction of carbon dioxide gas in the atmosphere with *silicate minerals in rocks to produce carbonate minerals is called the 'Urey reaction' after the chemist who proposed it, Harold Urey. The Urey reaction is an important link between Earth's climate and the geology of the planet. It has been proposed that chemical weathering of large mountain ranges like the Himalayas can remove carbon dioxide from the atmosphere by the Urey reaction (or other more complicated reactions like it), leading to a cooler climate as the *greenhouse effects of the lost carbon dioxide are diminished.

Chernobyl town in northern Ukraine, 100 km north of Kiev; site of a former nuclear power station. The town is now abandoned. On 26 April 1986, two huge explosions occurred at the plant, destroying a central reactor and breaching its 1000-tonne roof. In the immediate vicinity of Chernobyl, 31 people died (all firemen or workers at the plant) and 135,000 were permanently evacuated. It has been estimated that there will be an additional 20,000–40,000 deaths from cancer over 60 years; 600,000 people are officially classified as at risk. According to World Health Organization (WHO) figures from 1995, the incidence of thyroid cancer in children increased 200-fold in Belarus as a result of fallout from the disaster. The last remaining nuclear reactor at Chernobyl was shut down in December 2000.

The Chernobyl disaster occurred as the result of an unauthorized test being conducted, in which the reactor was run while its cooling system was inoperative. The resulting clouds of radioactive isotopes spread all over Europe, from Ireland to Greece. A total of 9 tonnes of radioactive material were released into the atmosphere, 90 times the amount produced by the Hiroshima A-bomb. In all, 5 million people are thought to have been exposed to radioactivity following the blast. In Ukraine, Belarus, and Russia more than 500,000 people were displaced from affected towns and villages and thousands of square miles of land were contaminated.

chernozem a deep, rich *soil of the plains of southern Russia. The upper horizons are rich in lime and other plant nutrients; in the dry *climate the predominant movement of *soil moisture is upwards (contrast with *leaching*), and lime and other chemical nutrients therefore accumulate in the upper part of the *soil profile.

Child, Convention on the Rights of the United Nations document designed to make the well-being of children an international obligation. It was adopted in 1989 and covers children from birth up to 18. It laid down international standards for: **provision** of a name, nationality, health care, education, rest, and play; **protection**

from commercial or sexual exploitation, physical or mental abuse, and engagement in warfare; **participation** in decisions affecting a child's own future.

child labour   work done by children less than 15 years old. The 'International Labour Organization (ILO) estimates (1996) that there are at least 250 million child labourers aged 5–14, 150 million working on a full-time basis; in some countries 20% of child workers in rural areas are aged under 10, 5% in urban areas. Asia (excluding Japan) has an estimated 153 million child workers, Africa 80 million, and Latin America (including the Caribbean) some 17.5 million (1996). However, the estimated incidence of child labour is largest in Africa, at approximately 41% of all 5–14 year olds (two out of five children), compared with 21% in Asia (one in five), and 17% in Latin America (one in six). Child labour in Asia is declining as a result of growth in income per head, the spread of basic education, and a reduction in the size of families. In Africa and Latin America, however, it is increasing owing to rapid population growth and lowering standards of living. Child labour also exists in richer industrialized countries in seasonal activities, street trades, farming, or small workshops, and in those regions moving towards a market economy, such as Eastern Europe. More than two-thirds of all child labour is found in the agricultural sector. In the UK, the Trade Union Congress (TUC) estimated that some 500,000 schoolchildren were working illegally in 2001. In the USA the growth of the service sector, the rapid increase in the supply of part-time jobs, and the search for a more flexible workforce have contributed to the expansion of the child-labour market. In the 1990s

there was a major international effort to end child labour.

china clay   commercial name for *kaolin.

chinook   (American Indian 'snow-eater') warm dry wind that blows downhill on the east side of the Rocky Mountains of North America. It often occurs in winter and spring when it produces a rapid thaw, and so is important to the agriculture of the area.

The chinook is similar to the *föhn in the valleys of the European Alps.

Chipko Movement   Indian grass-roots villagers' movement campaigning against the destruction of their forests. Its broad principles are nonviolent direct action, a commitment to the links between village life and an unplundered environment, and a respect for all living things.

chlordane   organochlorine *pesticide, used especially against ants and termites.

chlorofluorocarbon   CFC, a class of **synthetic chemicals** that are odourless, nontoxic, nonflammable, and chemically inert. The first CFC was synthesized in 1892, but no use was found for it until the 1920s. Their stability and apparently harmless properties made CFCs popular as propellants in *aerosol cans, as refrigerants in refrigerators and air conditioners, as degreasing agents, and in the manufacture of foam packaging. They are now known to be partly responsible for the destruction of the *ozone layer. In June 1990 representatives of 93 nations, including the UK and the USA, agreed to phase out production of CFCs and various other ozone-depleting chemicals by the end of the 20th century.

When CFCs are released into the atmosphere, they drift up slowly into the stratosphere, where, under the influence of ultraviolet radiation from the Sun, they react with ozone (O_3) to form free chlorine (Cl) atoms and molecular oxygen (O_2), thereby destroying the ozone layer that protects Earth's surface from the Sun's harmful ultraviolet rays. The chlorine liberated during ozone breakdown can react with still more ozone, making the CFCs particularly dangerous to the environment. CFCs can remain in the atmosphere for more than 100 years. Replacements for CFCs are being developed, and research into safe methods for destroying existing CFCs is being carried out.

C horizon see *soil.

choropleth map map on which the average numerical value of some aspect of an area (for example, unemployment by county) is indicated by a scale of colours or *isoline shadings. An increase in average value is normally shown by a darker or more intense colour or shading. Choropleth maps are visually impressive but may mislead by suggesting sudden changes between areas.

Christaller, Walter see *central place.

chromite $FeCr_2O_4$, iron chromium oxide, the main chromium ore. It is one of the *spinel group of minerals, and crystallizes in dark-coloured octahedra of the cubic system. Chromite is usually found in association with ultrabasic and basic rocks; in Cyprus, for example, it occurs with *serpentine, and in South Africa it forms continuous layers in a layered *intrusion.

chromium ore essentially the mineral *chromite, $FeCr_2O_4$, from which chromium is extracted. South Africa and Zimbabwe are major producers.

chrysotile mineral in the *serpentine group, $Mg_3Si_2O_5(OH)_4$. A soft, fibrous, silky mineral, the primary source of asbestos.

c.i.f. or **CIF**, *abbreviation for* cost, insurance, and freight, or charged in full, way to value a commodity. Many

countries value their imports on this basis, whereas exports are usually valued *f.o.b. (**free-on-board**). For balance of payments purposes, figures are usually adjusted to include the freight and insurance costs.

cinnabar mercuric sulphide mineral, HgS, the only commercially useful ore of mercury. It is deposited in veins and impregnations near recent volcanic rocks and hot springs. The mineral itself is used as a red pigment, commonly known as **vermilion**. Cinnabar is found in the USA (California), Spain (Almadén), Peru, Italy, and Slovenia.

cirque French name for a *corrie, a steep-sided armchair-shaped hollow in a mountainside.

cirrocumulus a high-altitude cloud consisting of thin layers of white globular masses. Cirrocumulus clouds usually occur in groups which form a rippled pattern known as a 'mackerel sky'.

cirrostratus a thin, whitish veil of high-altitude cloud that produces solar or lunar halo phenomena. It is associated with the approach of the warm front of a depression.

cirrus high, wispy or strand-like, thin *cloud associated with the advance of a *depression.

CIS *abbreviation for* Commonwealth of Independent States, established in 1992 by 11 former Soviet republics.

CITES acronym for Convention on International Trade in Endangered Species, international agreement under the auspices of the *World Conservation Union with the aim of regulating trade in *endangered species of animals and plants. The agreement came into force in 1975 and by 1997 had been signed by 138 states. It prohibits any trade in a category of 8000 highly endangered species and controls trade in a further 30,000 species.

Animals and plants listed in Appendix 1 of CITES are classified endangered, and all trade in that species is banned; those listed in Appendix 2 are classified vulnerable, and trade in the species is controlled without a complete ban; those listed in Appendix 3 are subject to domestic controls while national governments request help in controlling international trade.

city (French *cité* from Latin *civitas*) generally, a large and important town. In the Middle East and historic Europe, and in the ancient civilizations of Mexico and Peru, cities were states in themselves. In the early Middle Ages, European cities were usually those towns that were episcopal sees (seats of bishops).

Cities cover only 2% of the Earth's surface but use 75% of all resources. In April 1996, the World Resources Report predicted that two-thirds of the world's population will live in cities by 2025. Cities with more than 10 million inhabitants are sometimes referred to as **megacities**. In 1997 there were 18 megacities, 13 of them in developing nations.

In the UK, a city is a town or *borough, traditionally a cathedral town, that has acquired the title of 'city' by long custom or which is so designated in its charter, or which is specially created as a city by order in council. There have been, however, many variations to this definition. For example, Dorchester and Sherborne in Dorset were once episcopal sees, but have never been called cities, not even when they had corporations. In the Domesday Book, Gloucester and Leicester are called both *civitas* and *burgum*. In 1889 Birmingham, though not an Anglican episcopal see, was raised to the rank of a city on account of its industrial importance. Since then the title has been conferred on many other incorporated towns in the UK.

UK cities occupy 7.7% of land and contain 89% of the nation's population.

origins The Romans used the word *civitas* to denote the whole state or body politic, *urbs* and *municipium* being applied to towns. This meaning of the word has been totally lost in modern times, but the large cities of the modern world do somewhat resemble the cities of ancient Greece in their local self-government.

The Greek *polis* (city state) represented a collection of families, gathered together within a certain space, who administered their own foreign and domestic affairs, and had their own religion. These cities were only bound by affection to the *metropolis* (mother city), of which they were, in a sense, colonies.

The indeterminate use of the word 'city' probably began at a very early date. Du Cange in his glossary of medieval Latin words defines the word *civitas* as *urbs episcopalis*, and says that towns were called *oppida* or *castra*.

modern cities In 1950, 83 cities worldwide had a population of 1 million or more; by 1996 there were 280, and it is estimated that by 2015 there will be more than 500.

In the 1980s the term **edge city** was coined to denote the growth of business sites, supermarkets, and other retail sites around the edge of both urban and suburban regions. Typically they are used by day but have no residential population.

clan (Scottish Gaelic *clann* 'children') social grouping based on *kinship. Some traditional societies are organized by clans, which are either matrilineal or patrilineal, and whose members must marry into another clan in order to avoid in-breeding.

Familiar examples are the Highland clans of Scotland. Theoretically each clan is descended from a single ancestor from whom the name is derived – for example, clan MacGregor ('son of Gregor').

class in sociology, the main grouping of social stratification in industrial societies, based primarily on economic and occupational factors, but also referring to people's style of living or sense of group identity.

Within the social sciences, class has been used both as a descriptive category and as the basis of theories about industrial society. Theories of class may see such social divisions either as a source of social stability (Emile Durkheim) or social conflict (Karl Marx).

The most widely used descriptive classification in the UK divides the population into five main classes, with the main division between manual and nonmanual occupations. Such classifications have been widely criticized, however, on several grounds: they reflect a middle-class bias that brain is superior to brawn; they classify women according to their husband's occupation rather than their own; they ignore the upper class, the owners of land and industry.

clastic rock in earth science, type of sedimentary rock consisting mainly of broken fragments of a parent rock deposited by some transport mechanism. Examples include *conglomerate, *sandstone, and *shale.

clay very fine-grained *sedimentary deposit that has undergone a greater or lesser degree of consolidation. When moistened it is plastic, and it hardens on heating, which renders it impermeable. It may be white, grey, red, yellow, blue, or black, depending on its composition. Clay minerals consist largely of hydrous silicates of aluminium and magnesium together with iron, potassium, sodium, and organic substances. The crystals of clay minerals have a layered structure, capable of holding water, and are responsible for its plastic properties. According to international classification, in mechanical analysis of soil, clay has a grain size of less than 0.002 mm.

Types of clay include adobe, alluvial clay, building clay, brick, cement, china clay (or kaolinite), ferruginous clay, fireclay, fusible clay, puddle clay, refractory clay, and vitrifiable clay. Clays have a variety of uses, some of which, such as pottery and bricks, date back to prehistoric times.

clay–humus complex the chemically active part of the soil formed by the intimate association between finely weathered clay mineral particles and the decomposed remains of plants and animals (in particular, *mull humus). Each clay–humus particle or **micelle** acts as a weakly charged **anion** (a negatively charged ion). Surrounding the micelle and attracted to it by the negative charge are numerous bases or **cations** (positively charged ions) such as calcium and magnesium. These adsorbed cations are capable of being exchanged between micelles, and between micelles and plant rootlets. The total amount of exchangeable ions is known as the **base exchange capacity** or **total exchange capacity** of the soil.

Cations can be easily replaced by the accumulation of hydrogen ions (which are them selves aggressive cations). After rainfall, the hydrogen ions in the percolating soil water cause the leaching of cations, leading to a temporary increase in acidity of the micelles. As the water drains from the soil so some of the hydrogen ions are removed, allowing a reduction in the soil acidity.

clay mineral one of a group of hydrous silicate minerals that form most of the fine-grained particles in clays. Clay minerals are normally formed by weathering or alteration of other silicate minerals. Virtually all have sheet silicate structures similar to the *micas. They exhibit the following useful properties: loss of water on heating; swelling and shrinking in different conditions; cation exchange with other media; and plasticity when wet. Examples are kaolinite, illite, and montmorillonite.

Kaolinite $Al_2Si_2O_5(OH)_4$ is a common white clay mineral derived from alteration of aluminium silicates, especially feldspars. Illite contains the same constituents as kaolinite, plus potassium, and is the main mineral of clay sediments, mudstones, and shales; it is a weathering product of feldspars and other silicates. Montmorillonite contains the constituents of kaolinite plus sodium and magnesium; along with related magnesium- and iron-bearing clay minerals, it is derived from alteration and weathering of mafic igneous rocks. Kaolinite (the mineral name for kaolin or china clay) is economically important in the ceramic and paper industries. Illite, along with other clay minerals, may also be used in ceramics. Montmorillonite is the chief constituent of fuller's earth, and is also used in drilling muds (muds used to cool and lubricate drilling equipment). Vermiculite (similar to montmorillonite) will expand on heating to produce a material used in insulation.

claypan   see *hardpan.

Clean Air Act   legislation designed to improve the quality of air by enforcing pollution controls on industry and households. The first Clean Air Act in the UK was passed in 1956 after the London Smog killed 4000 people. The USA enacted a Clean Air Act in 1970, further amended in 1990.

In Europe, national legislation is supplemented by EU and UN agreements and conventions such as EU regulations on levels of ozone, nitrogen oxide, and sulphur dioxide. In addition the UK has signed protocols drawn up by the UN Economic Commission for Europe to reduce transboundary pollution.

cleavage   in geology and mineralogy, the tendency of a rock or mineral to split along defined, parallel planes related to its internal structure; the clean splitting of slate is an example. It is a useful distinguishing feature in rock and mineral identification. Cleavage occurs as a result of realignment of component minerals during deformation or metamorphism. It takes place where bonding between atoms is weakest, and cleavages may be perfect, good, or poor, depending on the bond strengths; a given rock or mineral may possess one, two, three, or more orientations along which it will cleave.

Some minerals have no cleavage, for example, quartz will fracture to give curved surfaces similar to those of broken glass. Some other minerals, such as apatite, have very poor cleavage that is sometimes known as a parting. Micas have one perfect cleavage and therefore split easily into very thin flakes. Pyroxenes have two good cleavages and break (less perfectly) into long prisms. Galena has three perfect cleavages parallel to the cube edges, and readily breaks into smaller and smaller cubes. Baryte has one perfect cleavage plus good cleavages in other orientations.

cliff   see *coastal cliff.

climate   combination of weather conditions at a particular place over a period of time – usually a minimum of 30 years. A *climate classification encompasses the averages, extremes, and frequencies of all meteorological elements such as temperature, atmospheric pressure, precipitation, wind, humidity, and sunshine, together with the factors that influence them.

The primary factors that influence the climate of an area are: latitude (as a result of the Earth's rotation and orbit); ocean currents; large-scale movements of wind belts and air masses over the Earth's surface; temperature differences between land and sea surfaces; topography; continent positions; and vegetation. In the long term, changes in the Earth's orbit and the angle of its axis inclination also affect climate. Climatologists are especially concerned with the influences of human activity on climate change, among the most important of which, at both local and global levels, are those currently linked with *ozone depleters, the *greenhouse effect, and *global warming.

prevailing winds Regions are affected by different wind systems, which result from the rotation of the Earth and the uneven heating of surface air. As air is heated by radiation from the Sun, it expands and rises, and cooler air flows in to take its place. This movement of air produces belts of prevailing winds. Because of the rotation of the Earth, these are deflected to the right in the northern hemisphere and to the left in the southern hemisphere. This effect, which is greater in the higher latitudes, is known as the *Coriolis effect. Because winds transport heat and moisture, they affect the temperature, humidity, precipitation, and cloudiness of an area. As a result, regions with different prevailing wind directions have different climates.

temperature variations The amount of heat received by the Earth from the Sun varies with latitude and season. In equatorial regions, there is no large seasonal variation in the mean daily temperature of the air near the ground, while in the polar regions, temperatures in the long winters, when there is little incoming solar radiation, fall far below summer temperatures. The temperature of the sea, and of the air above it, varies little in the course of day or night, whereas the surface of the land is rapidly cooled by lack of solar radiation. This is due to the specific heat capacity – the amount of energy required to raise the temperature of land or sea. It takes about two and a half times as much energy to raise the temperature of the sea by 1°C as it does to raise the temperature of the land. This is because the sea is transparent, and energy passes down through the sea. Also, ocean currents spread the energy over a wide area. Similarly, the annual change of temperature is relatively small over the sea but much greater over the land. Thus, continental areas are colder than maritime regions in winter, but warmer in summer. This results in winds blowing from the sea which, relative to the land, are warm in winter and cool in summer, while winds originating from the central parts of continents are hot in summer and cold in winter. On average, air temperature drops with increasing land height at a rate of 1°C per 90 m – so even in equatorial regions, the tops of mountains can be snow-covered throughout the year.

vegetation-based climates Rainfall is produced by the condensation of water vapour in air. When winds blow against a range of mountains the air is forced to ascend, resulting in *precipitation (rain or snow), the amount depending on the height of the ground and the humidity of the air. Centred on the Equator is a belt of tropical *rainforest, which may be either constantly wet or monsoonal (having wet and dry seasons in each year). On either side of this is a belt of savannah, with lighter seasonal rainfall and less dense vegetation, largely in the form of grasses. Then there is usually a transition through *steppe (semi-arid) to *desert (arid), with a further transition through steppe to what is termed *Mediterranean climate with dry summers. Beyond this is the moist temperate climate of middle latitudes, and then a zone of cold climate with moist winters. Where the desert extends into middle latitudes, however, the zones of Mediterranean and moist temperate climates are missing, and the transition is from desert to a cold climate with moist winters. In the extreme east of Asia a cold climate with dry winters extends from about 70° N–35° N. The polar caps have *tundra and glacial climates, with little or no precipitation.

climate change   change in the *climate of an area or of the whole world over an appreciable period of time. That is, a single winter that is colder than average does not indicate climate change. It is the change in average weather conditions from one period of time (30–50 years) to the next. Changes in climate can occur naturally or as a result of human activity. Natural variations can be caused by fluctuations in the amount of solar radiation reaching the Earth – for example, sunspot activity is thought to produce changes in the Earth's climate. Variations in the Earth's orbit around the Sun, known as the *Milankovitch hypothesis, is also thought to bring about climatic changes. Natural events on the surface of the Earth, such the eruption of *volcanoes and the effects of *El Niño, can result in temporary climate changes on a worldwide scale, sometimes extending over several months or even years.

There is indisputable evidence for natural changes in the world's climate:
(a) *fossil records*: fossils of species such as the mammoth have been found in what are now temperate regions;

(b) *topographical evidence*: in North Africa, now largely desert, it can be seen where rivers once ran, indicating that the area's rainfall was once much higher;
(c) *historical records*: the freezing over of the River Thames was widely reported in the 17th and 18th centuries (see *Ice Age, Little);
(d) *meteorological records*: records of the weather which exist for many areas of the world and some of which date from early times; comparisons can be made with present-day climate;
(e) *geological evidence*: the landscape of, for example, much of Britain has been shaped by the action of glaciers (see *Ice Age). The study of past climates (paleoclimatogy) involves the investigation of climate changes from the ice ages to the beginning of instrumental recording in the 19th century.

There is increasing evidence to suggest that human industrial activity affects the global climate, particularly in terms of *global warming.

climate classification   description of different types of *climate, taking into account the averages, extremes, and frequencies of all meteorological elements such as temperature, atmospheric pressure, precipitation, wind, humidity, and sunshine, together with the factors that influence them.

The different types of climate can be described as tropical (hot), warm temperate (or subtropical), cool temperate, cold (or polar), and arctic.

tropical (hot) climates There are three distinctive tropical climates: equatorial, tropical continental, and hot deserts. The mean monthly temperature in these climate areas never falls below 21°C.

Areas with an **equatorial** climate are located within 5° north and south of the Equator. These areas include the Amazon and Congo basins and the coastal lands of Ecuador and West

Africa. Temperatures are high and
constant throughout the year because
the Sun is always high in the sky.
Each day has approximately 12 hours
of daylight and 12 hours of darkness.
Annual rainfall totals usually exceed
2000 mm; most afternoons there are
heavy showers. There is high daytime
humidity and winds are light and
variable.

Areas with a **tropical continental**
climate are mainly located between
latitudes 5° and 15° north and south
of the Equator, and within central parts
of continents. These areas include the
Campos (Brazilian highlands), most of
Central Africa surrounding the Congo
basin, and parts of northern Australia.
Temperatures are high throughout the
year but there is a short season, slightly
cooler than the equatorial climate,
when the sun is overhead at either the
Tropic of Cancer or the Tropic of
Capricorn. The annual temperature
range is slightly greater than that of
the equatorial climate due to the sun
being at a slightly lower angle in the
sky for part of the year. The higher
temperature is also due to tropical
continental areas being at a greater
distance from the sea and having less
cover from cloud and vegetation. The
main feature of this climate is the
alternate wet and dry seasons.

Hot deserts are usually found on
the west coast of continents between
15° and 30° north or south of the
Equator and in the *trade wind belt.
The exception is the extensive Sahara-
Arabian-Thar desert which owes its
existence to the size of the Afro-Asian
continent. Desert temperatures are
characterized by their extremes. The
annual range is often 20–30°C and the
diurnal (daytime) range over 50°C.
Due to the lack of cloud cover and the
bare rocks or sand surface, daytimes
receive intense insolation (exposure)
from the overhead sun. In contrast,

nights may be extremely cold with
temperatures likely to fall below 0°C.
No deserts are truly dry even though
they suffer from extreme water
shortages. The amount of precipitation
is extremely unreliable; some desert
areas may receive rain only once every
two or three years. Rain, when it does
fall, is heavy.

**Warm temperate (subtropical)
climates** can be classified into two
areas: Mediterranean climate and the
Eastern margin climate.

The *Mediterranean climate is
found on the west coasts of continents
between 30° and 40° north and
south of the Equator – that is, in
Mediterranean Europe, California,
parts of southern Australia, Cape
Province (South Africa), and central
Chile. The characteristics of the climate
are hot, dry summers and warm, wet
winters. Summers in southern Europe
are hot with little cloud cover, and the
winters are mild. Other Mediterranean
climate areas are less warm in summer.

Eastern margin climates, which are
found in southeast and eastern Asia,
are dominated by the *monsoon.
Temperature figures and rainfall
distributions are similar to those of
tropical continental areas although
annual amounts of rain are much
higher.

Cool temperate climates can be
classified into two main areas
*continental climate and western
margins. Periods of one to five months
are below 6°C in these climate areas.

Continental climates are to be found
in the centre of continents between
approximately 40° and 60° north of
the Equator. The annual range of
temperature is high as there is no
moderating influence from the sea.
Maximum mean monthly summer
temperatures are around 20°C.
Precipitation decreases rapidly as
distance from the sea increases.

Annual rainfall averages at around 500 mm and there is a threat of drought. The ground is snow-covered for several winter months.

Western margin climates are often referred to as 'northwest European', and are found on west coasts between approximately 40° and 60° north and south of the equator. Other areas with similar climatic characteristics are northwest USA and British Columbia, southern Chile, New Zealand's South Island, and Tasmania. Summers are cool with the warmest month at a temperature of 15–20°C. This is due to the low angle of the sun in the sky combined with frequent cloud cover and the cooling influence of the sea. Winters are mild in comparison. Mean monthly temperatures remain a few degrees above freezing due to the warming effect of the sea. Autumn is usually warmer than spring; seasonal temperature variations depend on prevailing air masses. Precipitation often exceeds 2000 mm annually and falls throughout the year, the highest amount falling during winter when depressions are more frequent and intense. Snow is common in the mountains.

Cold or polar climates can be found in the subarctic regions 60° north of North America, Europe, and Asia. This type of climate also occurs at higher altitudes in more temperate latitudes and in southern Chile. For over six months of the year the mean temperature remains below 6°C. There is a period each year in the area north of the Arctic Circle during which time the Sun never rises. Winters are long and cold; the minimum mean monthly temperatures may be as low as –30°C. The wind-chill factor is high with strong winds. Summers are short but the long hours of daylight and clear skies mean they are relatively warm. Precipitation is light as the cold

air can only hold a limited amount of moisture, and the small amount of winter snowfall is frequently blown about in blizzards.

Arctic climates are found in Antarctica, all of Greenland, the north of Alaska, Canada, and Russia where is continuous *permafrost. Winters are severe and the sea freezes; summers have continuous periods of daylight but the monthly temperatures struggle to rise above freezing point. Nearer each *pole the climate is constant frost. Precipitation is light and falls mainly as snow.

climate model computer simulation, based on physical and mathematical data, of a climate system, usually the global (rather than local) climate. It is used by researchers to study such topics as the possible long-term disruptive effects of the greenhouse gases, or of variations in the amount of radiation given off by the Sun.

climatic region geographical area which experiences certain *weather conditions. No two localities on Earth may be said to have exactly the same *climate, but widely separated areas of the world can possess similar climates. These climatic regions have some comparable physical and environmental features as well as having similar weather patterns. Climatic regions are differentiated by weather conditions (including temperature, humidity, precipitation type and amount, wind speed and direction, atmospheric pressure, sunshine, cloud types, and cloud coverage), and by weather phenomena (such as thunderstorms, *fog, and *frost) that have prevailed there over a long period of time, usually 30 years.

climatology study of climate, its global variations and causes.

Climatologists record mean daily, monthly, and annual temperatures and monthly and annual rainfall totals, as well as maximum and minimum values. Other data collected relate to pressure, humidity, sunshine, cold cover, and the frequency of days of frost, snow, hail, thunderstorms, and gales. The main facts are summarized in tables and climatological atlases published by nearly all the national meteorological services of the world. Climatologists also study climates of the past (paleoclimates) by gathering information from such things as tree rings, deep sea sediments, and ice cores – all of which record various climate factors as they form.

climax community group of plants and animals that is best able to exploit the environment in which it exists. It is brought about by *succession (a change in the species present) and represents the point at which succession ceases to occur.

In temperate or tropical conditions, a typical climax community comprises woodland or forest and its associated fauna (for example, an oak wood in the UK). In essence, most land management is a series of interferences with the process of succession.

climax vegetation the plants in a *climax community.

climograph diagram that shows both the average monthly temperature and *precipitation of a place.

clint one of a number of flat-topped limestone blocks that make up a *limestone pavement. Clints are separated from each other by enlarged joints called grykes.

cloud water vapour condensed into minute water particles that float in masses in the atmosphere. Clouds, like fogs or mists, that occur at lower levels, are formed by the cooling of air containing water vapour, which generally condenses around tiny dust particles.

Clouds are classified according to the height at which they occur, and by their shape. **Cirrus** and **cirrostratus**

clouds occur at around 10 km. The former, sometimes called mares'-tails, consist of minute specks of ice and appear as feathery white wisps, while cirrostratus clouds stretch across the sky as a thin white sheet. Three types of cloud are found at 3–7 km: cirrocumulus, altocumulus, and altostratus. **Cirrocumulus** clouds occur in small or large rounded tufts, sometimes arranged in the pattern called mackerel sky. **Altocumulus** clouds are similar, but larger, white clouds, also arranged in lines. **Altostratus** clouds are like heavy cirrostratus clouds and may stretch across the sky as a grey sheet. **Stratocumulus** clouds are generally lower, occurring at 2–6 km. They are dull grey clouds that give rise to a leaden sky that may not yield rain. Two types of clouds, **cumulus** and **cumulonimbus**, are placed in a special category because they are produced by daily ascending air currents, which take moisture into the cooler regions of the atmosphere. Cumulus clouds have a flat base generally at 1.4 km where condensation begins, while the upper part is dome-shaped and extends to about 1.8 km. Cumulonimbus clouds have their base at much the same level, but extend much higher, often up to over 6 km. Short heavy showers and sometimes thunder may accompany them. **Stratus** clouds, occurring below 1–2.5 km, have the appearance of sheets parallel to the horizon and are like high-level fogs.

In addition to their essential role in the water cycle, clouds are important in the regulation of radiation in the Earth's atmosphere. They reflect short-wave radiation from the Sun, and absorb and re-emit long-wave radiation from the Earth's surface.

Club of Rome informal international organization that aims to promote greater understanding of the

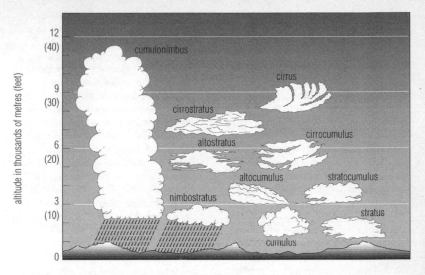

cloud Standard types of cloud. The height and nature of a cloud can be deduced from its name. Cirrus clouds are at high levels and have a wispy appearance. Stratus clouds form at low level and are layered. Middle-level clouds have names beginning with 'alto'. Cumulus clouds, ball or cottonwool clouds, occur over a range of height.

interdependence of global economic, political, natural, and social systems. Members include industrialists, economists, and research scientists. Membership is limited to 100 people. It was established in 1968.

cluster in business, a collection of businesses operating in the same industry gathered in a single geographical location. For example, Internet clusters include Silicon Valley, California, in the USA, and Silicon Fen around Cambridge in England.

cluster sample small selected sample chosen by market researchers to represent a particular target market. Clusters are often chosen on a geographical basis because this cuts costs. A cluster sample from the Scottish Highlands, for example, might be asked to test midge repellent.

coal black or blackish mineral substance formed from the compaction of ancient plant matter in tropical swamp conditions. It is used as a fuel and in the chemical industry. Coal is classified according to the proportion of carbon it contains. The main types are *anthracite (shiny, with about 90% carbon), **bituminous coal** (shiny and dull patches, about 75% carbon), and **lignite** (woody, grading into peat, about 50% carbon). Coal can be burned to produce heat energy, for example in power stations to produce electricity. Coal burning is one of the main causes of *acid rain, which damages buildings and can be detrimental to aquatic and plant life.

By 1700 Britain was the world's largest coal producer, and over 50% of the country's energy needs were met by coal. From about 1800, coal was carbonized commercially to produce coal gas for gas lighting and coke for smelting iron ore. By the second half of the 19th century, study of the by-products (coal tar, pitch, and ammonia) formed the basis of organic chemistry, which eventually led to the

COALIFICATION

80

development of the plastics industry in the 20th century. The York, Derby, Notts coalfield is Britain's chief reserve, extending north of Selby. Under the Coal Industry Nationalization Act 1946, Britain's mines were administered by the National Coal Board, now known as British Coal.

coalification the lithification of organically rich sediments to form coal.

coal measures a sequence of rocks from the carboniferous period which often, but not always, includes coal seams of commercially exploitable thickness.

coal mining extraction of coal from the Earth's crust. Coal mines may be opencast, adit, or deepcast. The least expensive is opencast but this may result in scars on the landscape.

history In Britain, coal was mined on a small scale from Roman times, but production expanded rapidly between 1550 and 1700. Coal was the main source of energy for the Industrial Revolution, and many industries were located near coalfields to cut transport costs. Competition from oil as a fuel, cheaper coal from overseas (USA, Australia), the decline of traditional users (town gas, railways), and the exhaustion of many underground workings resulted in the closure of mines (850 in 1955, 54 in 1992), but rises in the price of oil, greater productivity, and the discovery of new, deep coal seams suitable for mechanized extraction (for example, at Selby in Yorkshire) improved the

vegetation (mostly from the Carboniferous) decays to form peat beds

peat is sandwiched between layers of sediment and compressed to form lignite

bituminous coal forms after further compression

anthracite

coal The formation of coal. Coal forms where vegetable matter accumulates but is prevented from complete decay and forms peat beds. Over time it becomes buried and compressed, forming lignite. Increased pressure and temperature produces bituminous coal with a higher carbon content. At great depths, high temperatures reduce methane and anthracite is formed with a very high carbon concentration.

position of the British coal industry 1973–90. It remains very dependent on the use of coal in electricity generation, however, and is now threatened by a trend towards using natural gas from the North Sea and Irish Sea gas fields for this purpose. The percentage of electricity generated from coal dropped from 74% 1992 to just over 50% in 1995.

coal tar   a dark viscid substance obtained from the destructive distillation of bituminous coal and used in the manufacture of plastics, pesticides, drugs and dyes.

coast   meeting place of land and sea. The coast is changed by the physical processes of *coastal erosion and *coastal deposition, which include the action of waves, rain, wind, and frost.

coastal cliff   steep slope rising from the sea to a considerable height, and consisting of either bare rock or an accumulation of deposited material such as clay. The nature of the cliff depends on a number of factors such as the nature of the rock (its strength, weaknesses in the rock and bedding planes), its exposure to waves, and whether any *coastal protection measures have been used.

Limestone has a well-developed jointing and bedding structure and forms geometric cliff profiles with steep, angular cliff faces, and flat tops. *Coastal erosion causes complete blocks to fall away resulting in angular overhangs such as those at Tresilian Bay, Llantwit Major, Wales. The rate of cliff retreat is much more rapid in chalk, and steep cliffs are created by erosion and landslides. Cliffs can also be found in areas of softer rock such as the clay cliffs of Norfolk and Holderness, England.

The dip of rock bedding planes also affects the cliff profile. If beds dip vertically, a sheer cliff face will be formed, as at the Old Red Sandstone cliffs in Dyfed, Wales. If they dip steeply seaward then steep, shelving cliffs will occur, as at the cliffs at Tenby, Wales.

coastal deposition   the laying down of sediment (*deposition) in a low-energy environment with constructive *waves. Coastal deposition occurs where there is a large supply of material from cliffs, rivers, or beaches, *longshore drift, and an irregular coastline. Geographical features include the *spit, *bar, *beach, foreland, and tombolo, such as at Chesil Beach, England.

Most beaches display a number of features of coastal deposition. These include **cusps**, semi-circular scalloped embankments found in the shingle or shingle / sand junction; **ripples** formed by wave action or tidal currents; **storm beaches**, noticeable ridges found at the level of the highest spring tides; and small-scale beach ridges known as **berms**, which are built up by successive levels of tides or storms.

coastal erosion   the erosion of the land by the constant battering of the sea, primarily by the processes of hydraulic action, corrasion, attrition, and corrosion. *Hydraulic action occurs when the force of the waves compresses air pockets in coastal rocks and cliffs. The air expands explosively, breaking the rocks apart. It is also the force of the water on the cliff. During severe gales this can be as high as 6 tonnes / cm^2 – the force of a bulldozer. Rocks and pebbles flung by waves against the cliff face wear it away by the process of *corrasion, or abrasion as it is also known. Chalk and limestone coasts are often broken down by *solution (also called *corrosion). *Attrition is the process by which the eroded rock particles themselves are worn down, becoming smaller and more rounded.

Frost shattering (or *freeze–thaw), caused by the expansion of frozen water in cracks, and *biological weathering, caused by the burrowing

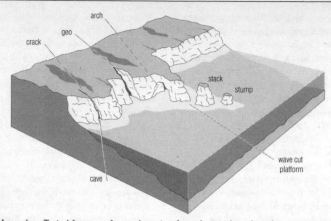

coastal erosion Typical features of coastal erosion: from the initial cracks in less resistant rock through to arches, stacks, and stumps that can occur as erosion progresses.

of rock-boring molluscs and plants, also lead to the breakdown of rock.

Where resistant rocks form *headlands, the sea erodes the coast in successive stages. First it exploits weaknesses such as faults and cracks to form caves. Then it gradually wears away the interior of the caves and enlarges them. In some cases the roofs may be broken through to form blowholes. In other cases the caves at either side of a headland may unite to form a natural arch. When the roof of the arch collapses, a *stack is formed. This may be worn down further to produce a *stump. There are good examples of stacks at The Needles, Isle of Wight, England.

Some areas may be eroding at a rate of over 100 m per century. Some experts believe that it is better to surrender the land to the sea, rather than build costly sea defences in rural areas (see *coastal protection). Between 1926 and 1996, 29 villages disappeared from the Yorkshire coast as a result of tidal battering.

In Britain, the southern half of the coastline is slowly sinking (on the east coast, at the rate of 0.5 cm per year) whilst the northern half is rising (due to

the removal of ice after the last ice age). This intensifies the loss of land by erosion.

Beach erosion Beach erosion occurs when more sand is eroded and carried away from the beach than is deposited by *longshore drift. Beach erosion can occur due to the construction of artificial barriers, such as *groynes, or due to the natural high tides and high waves during winter storms, which tend to carry sand away from the beach and deposit it offshore in the form of *bars. During the calmer summer season some of this sand is redeposited on the beach.

coastal marsh a marsh formed by the growth of a *spit across a *bay, and the gradual silting up of the resulting *lagoon. A well known example is Romney Marsh in southeast England.

coastal protection measures taken to prevent *coastal erosion. Many stretches of coastline are so severely affected by erosion that beaches are swept away, threatening the livelihood of seaside resorts, and buildings become unsafe.

To reduce erosion, several different forms of coastal protection are used. Structures such as sea walls attempt to prevent waves reaching the cliffs by deflecting them back to sea. Such structures are expensive and of limited success. Adding sediment (beach nourishment) to make a beach wider causes waves to break early so that they have less power when they reach the cliffs. Wooden or concrete barriers called groynes may also be constructed at right angles to the beach in order to block the movement of sand along the beach (*longshore drift). However, this has the effect of 'starving' beaches downshore: 'protection' of one area usually means destruction of another.

Rock armour refers to large blocks of stone dumped on a beach or at the base of a cliff to reduce erosion. Hard engineering refers to constructed/built devices, while soft engineering refers to natural features such as salt marshes or sand dunes, which may help to protect against erosion.

Coastal protection may also refer to the process of simply leaving the coast to the elements but removing the harmful factor of human population and development.

cobalt ore   cobalt is extracted from a number of minerals, the main ones being smaltite, $(CoNi)As_3$; linnaeite, Co_3S_4; cobaltite, $CoAsS$; and glaucodot, $(CoFe)AsS$.

All commercial cobalt is obtained as a by-product of other metals, usually associated with other ores, such as copper. The Democratic Republic of Congo (formerly Zaire) is the largest producer of cobalt, and it is obtained there as a by-product of the copper industry. Other producers include Canada and Morocco. Cobalt is also found in the manganese nodules that occur on the ocean floor, and was successfully refined in 1988 from the Pacific Ocean nodules, although this process has yet to prove economic.

coccolithophorid   microscopic, planktonic marine alga, which secretes a calcite shell. The shells (coccoliths)

of coccolithophores are a major component of deep sea ooze. Coccolithophores were particularly abundant during the late *Cretaceous period and their remains form the northern European chalk deposits, such as the white cliffs of Dover.

coke   a solid-fuel product comprising approximately 80% carbon produced by the distillation of coal to drive off its volatile impurities. Coke is used in the manufacture of steel.

col   pass or saddle, a depression or gap in a line of hills or mountains. Cols are formed in several ways including river capture and glacial erosion. Lines of communication often utilize cols as they allow relatively easy access through otherwise difficult terrain.

cold climate   or polar climate, the climate of the subarctic regions 60° north of North America, Europe, and Asia. This type of climate also occurs at higher altitudes in more temperate latitudes and in southern Chile. For over six months of the year the mean temperature remains below 6°C. There is a period each year in the area north of the Arctic Circle during which time the Sun never rises. Winters are long and cold; the minimum mean monthly temperatures may be as low as −30°C. The *wind-chill factor is high with strong winds. Summers are short but the long hours of daylight and clear skies mean they are relatively warm. Precipitation is light as the cold air can only hold a limited amount of moisture, and the small amount of winter snowfall is frequently blown about in blizzards.

cold front   see *depression.

collective security   system for achieving international stability by an agreement among all states to unite against any aggressor. Such a commitment was embodied in the post-World War I League of Nations and also in the *United Nations (UN), although the League was not able to live up to the ideals of its founders, nor has the UN been able to do so.

collectivism in politics, a position in which the collective (such as the state) has priority over its individual members. It is the opposite of *individualism, which is itself a variant of anarchy.

Collectivism, in a pure form impossible to attain, would transfer all social and economic activities to the state, which would assume total responsibility for them. In practice, it is possible to view collectivism as a matter of degree and argue that the political system of one state is more or less collectivist than that of another; for example, in the provision of state-controlled housing.

collision zone in plate tectonics, a region in which two pieces of continental crust collide after an intervening ocean has disappeared. It may be an area of mountain building and volcanic and seismic activity.

Colombo Plan plan for cooperative economic and social development in Asia and the Pacific, established 1950. The 26 member countries are Afghanistan, Australia, Bangladesh, Bhutan, Cambodia, Canada, Fiji Islands, India, Indonesia, Iran, Japan, South Korea, Laos, Malaysia, Maldives, Myanmar (Burma), Nepal, New Zealand, Pakistan, Papua New Guinea, Philippines, Singapore, Sri Lanka, Thailand, UK, and USA. They meet annually to discuss economic and development plans such as irrigation, hydroelectric schemes, and technical training.

colonial influence the consequences of, for example, British colonial activity in many parts of the developing world. There are two broad areas where colonialism has had long-term and continuing influence: in government and administration, and in economic systems. Many former colonies have governments and bureaucracies modelled on institutions of the former colonial power, and this may not always be to the best advantage of the nation concerned. Economically, the majority of former colonies continue to be suppliers of *primary industry products to the industrial world, a situation which may well perpetuate underdevelopment.

colonialism another name for *imperialism.

colonial trusteeship idea that colonies were held by the colonial power in trust for the indigenous population. This became an important and widely accepted principle of colonial policy in Britain in the 19th century. It was implicitly and explicitly invoked in support of Britain's efforts to eradicate the slave trade and slavery and, in the pursuit of political, social, and economic policies, would, it was argued, prepare indigenous peoples for self-government.

colonization in ecology, the spread of species into a new habitat, such as a freshly cleared field, a new motorway verge, or a recently flooded valley. The first species to move in are called **pioneers**, and may establish conditions that allow other animals and plants to move in (for example, by improving the condition of the soil or by providing shade). Over time a range of species arrives and the habitat matures; early colonizers will probably be replaced, so that the variety of animal and plant life present changes. This is known as *succession.

colony (Latin *colonia* from *colere* 'to till') country under the control of immigrants who remain subject to the jurisdiction of the parent state. Historically, the acquisition of colonies occurred for a variety of reasons. In general, rivalry with other powers and the expansion of commercial interests were the major factors, but a variety of factors influenced the establishment of particular colonies. The need to protect trade routes and the desire to control the sources of various products were important in colonial development in Africa and Asia, including India.

combe or **coombe**, steep-sided valley found on the scarp slope of a chalk *escarpment. The inclusion of 'combe' in a placename usually indicates that the underlying rock is chalk.

Comecon acronym for COuncil for Mutual ECONomic Assistance; or **CMEA**, economic organization from 1949 to 1991, linking the USSR with Bulgaria, Czechoslovakia, Hungary, Poland, Romania, East Germany (1950–90), Mongolia (from 1962), Cuba (from 1972), and Vietnam (from 1978), with Yugoslavia as an associated member. Albania also belonged between 1949 and 1961. Its establishment was prompted by the Marshall Plan. Comecon was formally disbanded in June 1991.

command economy or **planned economy**, economy planned and directed by government, where resources are allocated to factories by the state through central planning. This system is unresponsive to the needs and whims of consumers and to sudden changes in conditions (for example, crop failure or fluctuations in the world price of raw materials).

commercial agriculture *capital intensive production of crops for sale and profit, although the farmers and their families may use a small amount of what they produce. Profits may be reinvested to improve the farm. Large-scale commercial farming is *agribusiness; see also *plantation. The opposite of commercial farming is subsistence farming where no food is produced for sale.

commodity something produced for sale. Commodities may be consumer goods, such as radios, or producer goods, such as copper bars.

Commodity markets deal in raw or semi-raw materials that are amenable to grading and that can be stored for considerable periods without deterioration.

Commodity markets developed to their present form in the 19th century, when industrial growth facilitated trading in large, standardized quantities of raw materials. Most markets encompass trading in **commodity futures** – that is, trading for delivery several months ahead. Major commodity markets exist in Chicago, Tokyo, London, and elsewhere. Although specialized markets exist, such as that for silkworm cocoons in Tokyo, most trade relates to cereals and metals. **Softs** is a term used for most materials other than metals.

Common Agricultural Policy CAP, system of financial support for farmers in *European Union (EU) countries, a central aspect of which is the guarantee of minimum prices for part of what they produce. The objectives of the CAP were outlined in the Treaties of Rome (1957): to increase agricultural productivity, to provide a fair standard of living for farmers and their employees, to stabilize markets, and to assure the availability of supply at a price that was reasonable to the consumer. The CAP has been criticized for its role in creating overproduction, and consequent environmental damage, and for the high price of food subsidies.

common land unenclosed wasteland, forest, and pasture used in common by the community at large. Poor people have throughout history gathered fruit, nuts, wood, reeds, roots, game, and so on from common land; in dry regions of India, for example, the landless derive 20% of their annual income in this way, together with much of their food and fuel. Codes of conduct evolved to ensure that common resources were not depleted. But in the 20th century, in the developing world as elsewhere, much common land has been privatized or appropriated by the state, and what remains is overburdened by those who depend upon it.

Common Market popular name for the **European Economic Community**; see *European Union.

common market organization of autonomous countries formed to promote trade; see *customs union.

Common Market for Eastern and Southern Africa COMESA, integrated trading bloc formed in December 1994 by 20 states of eastern and southern Africa (with a combined population of 385 million) to replace the former *Preferential Trade Area for Eastern and Southern Africa, which had existed since 1981. COMESA's member states have 'agreed to cooperate in developing their natural and human resources for the good of all their people', to promote peace and security in the region, and establish a free trade area (FTA). An FTA was launched by nine of the member states – Djibouti, Egypt, Kenya, Madagascar, Malawi, Mauritius, Sudan, Zambia, and Zimbabwe – in October 2000. A customs union, with a common external tariff, is planned.

Commonwealth, the (British) voluntary association of 54 sovereign (self-ruling) countries and their dependencies, the majority of which once formed part of the British Empire and are now independent sovereign states. They are all regarded as 'full members of the Commonwealth'; the newest member being Mozambique, which was admitted in November 1995. Additionally, there are 13 territories that are not completely sovereign and remain dependencies of the UK or one of the other fully sovereign members, and are regarded

as 'Commonwealth countries'. Heads of government meet every two years, apart from those of Nauru and Tuvalu; however, Nauru and Tuvalu have the right to take part in all functional activities. The Commonwealth, which was founded in 1931, has no charter or constitution, and is founded more on tradition and sentiment than on political or economic factors. However, it can make political statements by withdrawing membership: Nigeria was suspended from the Commonwealth between November 1995 and May 1999 because of human-rights abuses. Fiji was readmitted in October 1997, ten years after its membership had been suspended as a result of discrimination against its ethnic Indian community.

commune group of people or families living together, sharing resources and responsibilities. There have been various kinds of commune through the ages, including a body of burghers or burgesses in medieval times, a religious community in America, and a communal division in communist China.

communications the contacts and linkages in an *environment.

For example, roads and railways are communications, as are telephone systems, newspapers, and radio and television.

communism (French *commun* 'common, general') revolutionary socialism based on the theories of the political philosophers Karl Marx and Friedrich Engels, emphasizing common ownership of the means of production and a planned, or *command economy. The principle held is that each should work according to his or her capacity and receive according to his or her needs. Politically, it seeks the overthrow of capitalism through a proletarian (working-class) revolution. The first communist state was the Union of Soviet Socialist Republics (USSR) after the revolution of 1917. Revolutionary socialist parties and groups united to form communist parties in other countries during the inter-war years. After World War II, communism was enforced in those countries that came under Soviet occupation. Communism as the ideology of a nation state survives in only a few countries in the 21st century, notably China, Cuba, North Korea, Laos, and Vietnam, where market forces are being encouraged in the economic sphere.

community 1. in ecology, an assemblage (group) of plants, animals, and other organisms living within a defined area. Communities are usually named by reference to a dominant feature, such as characteristic plant species (for example, a beech-wood community), or a prominent physical feature (for example, a freshwater-pond community).
2. in the social sciences, the sense of identity, purpose, and companionship that comes from belonging to a particular place, organization, or social group. The idea dominated sociological thinking in the first half of the 20th century, and inspired academic courses in **community studies**.

community charge in the UK, a charge (commonly known as the *poll tax) levied by local authorities from 1989 in Scotland and 1990 in England and Wales; it was replaced in 1993 by a council tax.

community council in Scotland and Wales, name for a *parish council.

commuter person who travels into a large town or city for work. For example, each working day more than 1.2 million people commute into London. A **commuter belt** is the area around a town in which commuters live (see also *dormitory town). Commuter settlements are generally more affluent than others, reflecting the wages of their residents, who prefer to live in a more attractive rural environment and can afford the daily costs of travelling to work.

commuter village any village, usually on the fringe of a metropolitan region, in which a high proportion of the workforce commutes substantial distances to jobs elsewhere in the region

comparative advantage law of international trade first elaborated by English economist David Ricardo showing that trade becomes worthwhile if the cost of production of particular items differs between one country and another.

For example, if France can produce cheese at a cost of 100 units and milk at a cost of 300 units whereas Spain can produce cheese at 200 units and milk at 400 units, then France has an absolute advantage in the production of both cheese and milk because it can produce both more cheaply in absolute cost terms. However, it will still be advantageous for France to trade with Spain because in France milk is more expensive relative to cheese (milk costs three times more to produce than cheese) than in Spain (where milk costs only twice as much). So France would specialize in the production of cheese and Spain in the production of milk and they would trade.

comparative method in sociology, the comparison of different societies or social groups as a means of elucidating their differences and/or similarities. It was originally used by philologists to analyse the common characteristics of different languages in order to trace their common origins.

comparison goods expensive goods, such as hi-fi systems and furniture, that the shopper will buy only after making a comparison between various models. A high *threshold population is needed to sustain a shop selling comparison goods, and people are prepared to travel some distance (*range) to obtain them. Shops selling comparison goods often cluster in the central business district to share customers and increase trade.

competition   in ecology, the interaction between two or more organisms, or groups of organisms, that use a common resource in short supply. There can be competition between members of the same species and competition between members of different species. Competition invariably results in a reduction in the numbers of one or both competitors, and in evolution contributes both to the decline of certain species and to the evolution of adaptations.

The resources in short supply for which organisms compete may be obvious things, such as mineral salts for animals and plants, or light for plants. However, there are less obvious resources. For example, competition for suitable nesting sites is important in some species of birds. Competition results in a reduction in breeding success for one or other organism(s). Because of this it is one of the most important aspects of natural selection, which may result in evolutionary change if the *environment is changing. Competition also results in the distribution of organisms we see in *habitats. It is believed that organisms tend to occur where the pressures of competition are not as great as in other areas. In agriculture cultivation methods are designed to reduce competition. For example, a crop of wheat is sown at a density that minimizes competition within the same species. The plants are grown far enough apart to reduce competition between the roots of neighbouring wheat plants for soil *mineral nutrients. The spraying of

the ground to kill weeds reduces competition between the wheat and weed plants. Some weeds would grow taller than the wheat and deprive it of light.

competition, perfect   in commerce, see *perfect competition.

competition policy   government policy on competition in markets. Competition policy is usually aimed at increasing the level of competition in the market, for example by breaking up *monopolies and making restrictive trade practices illegal.

competitiveness   the extent to which a producer is able to sell products in a market where other producers are selling similar products. For example, it can be argued that the UK manufacturing industry has lost competitiveness to Japanese and German rivals because UK prices have been too high and the products are not of such good quality. Companies can compete on many different levels, including price, quality, service, product features, exclusivity, and the degree to which a product is fashionable.

composite volcano   steep-sided conical *volcano formed above a *subduction zone at a destructive *plate margin. It is made up of alternate layers of ash and lava. The *magma (molten rock) associated with composite volcanoes is very thick and often clogs up the vent. This can cause a tremendous build-up of pressure, which, once released, causes a very violent eruption. Examples of composite volcanoes are Mount St Helens in the USA, and Stromboli and Vesuvius in Italy.

Composite volcanoes are usually found in association with island arcs and coastal mountain chains. The magma is mostly derived from plate material, and is rich in silica. This makes a very viscous lava, such as andesite, which solidifies rapidly to form a high, steep-sided volcanic mountain. This magma often clogs the volcanic vent. After the eruption, the crater may collapse to form a caldera.

Vesuvius lies at a destructive plate margin where the African plate is subducting beneath the Eurasian plate. Its lava is andesite in composition.

compost organic material decomposed by bacteria under controlled conditions to make a nutrient-rich natural fertilizer for use in gardening or farming. A well-made compost heap reaches a high temperature during the composting process, killing most weed seeds that might be present.

comprehensive redevelopment a policy of clearing substandard housing and completely rebuilding the area to create a new environment. Comprehensive redevelopment usually takes place in *inner city areas where conditions of overcrowding may be severe, and where houses are often mingled with industry and business. Comprehensive redevelopment aims to provide houses with good amenities, with areas of open space and parks, and which are away from industry and business. Some schemes have been successful but many schemes in the 1960s, where high-rise blocks of flats were included, have come in for much criticism.

concealed coalfield a coalfield in which *coal measures are located at some depth beneath the overlying *strata. In contrast is the *exposed coalfield* where coal measures outcrop at or near the surface.

concentric-ring theory hypothetical pattern of land use within an urban area, where different activities occur at different distances from the urban centre. The result is a sequence of rings. The theory was first suggested by the US sociologist E W Burgess in 1925. He said that towns expand outwards evenly from an original core

so that each zone grows by gradual colonization into the next outer ring. He based his model on Chicago, proposing five major concentric zones of urban land-use. The central business district was at the centre of this model. It was surrounded by the transition zone, or factory zone. This was surrounded by a zone of housing for working people, which was surrounded by a zone of single family dwellings. The outermost zone was the commuter zone.

In addition, the cost of land may decrease with increased distance from the city centre as demand for it falls (see *bid-rent theory). This means that commercial activity that can afford high land values will be concentrated in the city centre.

concordant coastline a coastline that is parallel to mountain ranges immediately inland. A rise in sea level or a sinking of the land will cause the valleys to be occupied by the sea and the mountains to become a line of islands. This has occurred along the coast of Croatia in the Adriatic sea. Compare *discordant coastline.

condensation conversion of a vapour to a liquid. This is frequently achieved by letting the vapour come into contact with a cold surface. It is the process by which water vapour turns into fine water droplets to form a *cloud.

Condensation in the atmosphere occurs when the air becomes completely saturated and is unable to hold any more water vapour. As air rises it cools and contracts – the cooler it becomes the less water it can hold. Rain is frequently associated with warm weather fronts because the air rises and cools, allowing the water vapour to condense as rain. The temperature at which the air becomes saturated is known as the **dew point**. Water vapour will not condense in air if there are not enough condensation nuclei (particles of dust, smoke, or salt) for the droplets to form on. It is then said to be supersaturated. Condensation is an important part of the *water cycle.

conduction the direct transfer of heat from one medium to another. Much of the heat from the Earth's surface is lost to the atmosphere through the conduction of outgoing long wave radiation despite air being a poor conductor.

confluence point at which two rivers join, for example the River Thames and River Cherwell at Oxford, England, or the White Nile and Blue Nile at Omdurman, Sudan.

congestion in traffic, the over-crowding of a route, leading to slow and inefficient flow. Congestion on the roads is a result of the large increase in car ownership. It may lead to traffic jams and long delays as well as pollution. Congestion within urban areas may also restrict *accessibility.

conglomerate company that has a number of subsidiaries in a number of nonrelated markets. For example, British American Tobacco (BAT) is a conglomerate owning a tobacco company and a Californian insurance company. A conglomerate merger is a merger between two companies which produce unrelated products.

conglomerate in geology, a type of *sedimentary rock composed of rounded fragments ranging in size from pebbles to boulders, cemented together in a finer sand or clay material.

A *bed (layer) of conglomerate is often associated with a break in a sequence of rock beds (known as an unconformity), where it marks the advance in the past of the sea over an older eroded landscape.

coniferous forest a forest of *evergreen trees such as pine, spruce and fir. Natural coniferous forests occur considerably further north than forests of broad-leaved *deciduous species, as coniferous trees are able to withstand harsher climatic conditions. The *taiga areas of the northern hemisphere consist of coniferous forests. Not all coniferous forests are indigenous to an area as some have been planted for commercial reasons (see *afforestation). It is anomalous that the larch, a species common to many coniferous forests, is a deciduous conifer.

connate water or **fossil water,** the water trapped in sedimentary rocks during their formation. Connate water is an important source of groundwater and is tapped by wells for domestic, industrial and agricultural purposes, especially in arid regions.

conservation in the life sciences, action taken to protect and preserve the natural world, usually from pollution, overexploitation, and other harmful features of human activity. The late 1980s saw a great increase in public concern for the environment, with membership of conservation groups, such as *Friends of the Earth, *Greenpeace, and the US *Sierra Club, rising sharply and making the *green movement an increasingly-powerful political force. Globally the most important issues include the depletion of atmospheric ozone by the action of *chlorofluorocarbons (CFCs), the build-up of carbon dioxide in the atmosphere (thought to contribute to the *greenhouse effect), and *deforestation.

Conservation may be necessary to prevent an endangered species from dying out in an area or even becoming extinct. But conservation of particular *habitats may be as important, if not more important. Habitat loss is believed to be the main cause of the great reduction of biodiversity and the rate of extinction occurring on Earth. There is concern about loss of species and *biodiversity, because living organisms contribute to human health, wealth, and happiness in several ways. Humans often enjoy being in natural environments, especially those who spend much of their lives in towns and cities. Human cultures may be dependent on the natural environment to sustain them and maintain a stable society. This is particularly true of societies in the developing world. The spread of desert into arid areas around the Sahara has probably contributed to the unstable societies of some areas there. There is also an economic argument for conservation. It is believed that many undiscovered useful chemicals may exist within organisms on Earth that could be developed into important drugs – but when a plant or animal becomes extinct, the chemicals it contains are also lost.

UK conservation groups The first conservation groups in Britain include the Commons Preservation Society (1865) in London, which fought successfully against the enclosure of Hampstead Heath and Epping Forest; the National Footpaths Preservation Society (1844); and the National Trust (1895). Government bodies with a role in conservation include English Heritage (1983), English Nature (1991), formerly the Nature Conservancy Council, and the Countryside Commission.

In the UK the conservation debate has centred on water quality, road-building schemes, the safety of nuclear power, and the ethical treatment of animals. Twelve coastal sites in Great Britain, including five Special Areas of Conservation, have been designated by the European Union (EU) to be part of a network of Natura 2000 sites. The EU will provide funds to help preserve these sites from development, overfishing, and pollution, and to monitor rare plants. These sites include the north Northumberland coast, with

its sea caves, its breeding population of grey seals in the Farne Islands, and Arctic species such as the wolf fish; the Wash and north Norfolk coast, with its population of common seals, waders, and wildfowl, and its extensive salt marshes; and Plymouth Sound and estuaries, with their submerged sandbanks.

'Turning the Tide' is a £10 million project, launched in 1997 by the Millennium Commission, to protect and restore Britain's only magnesian limestone cliffs, between Hartlepool and Sunderland. The area is rich in wild flowers, with grassland and denes (steep, wooded valleys). Intensive farming and the use of fertilizers have damaged the flora and fauna of the area. The beaches are polluted as a result of more than two centuries of coal-mining along the Durham coast. Waste from the mines was dumped into the sea and onto the beaches, leaving heaps of spoil 3.7–4.6 m high. The restoration project aims to remove spoil from the beaches and return the cliffs to their natural grassland. Action by governments has been prompted and supplemented by private agencies, such as the World Wide Fund for Nature. In attempts to save particular species or habitats, a distinction is often made between preservation – that is, maintaining the pristine state of nature exactly as it was or might have been – and conservation, the management of natural resources in such a way as to integrate the requirements of the local human population with those of the animals, plants, or the habitat being conserved.

conservative margin in *plate tectonics, see *transform margin

constructive margin in *plate tectonics, a boundary between two lithospheric plates, along which new crust is being created. The term usually refers to *divergent margins, where

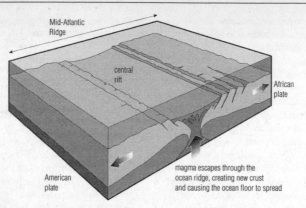

Mid-Atlantic
Ridge

central
rift

African
plate

American
plate

magma escapes through the
ocean ridge, creating new crust
and causing the ocean floor to spread

constructive margin A rift in the crustal plates where new material is being formed causes the plates to be pushed apart. This usually occurs as a result of volcanic action.

two oceanic plates are moving away from each other. As they diverge, magma (molten rock) wells up to fill the open space, and solidifies, forming new oceanic crust. Similar processes occur where a continent plate is beginning to split apart.

*Subduction zones could also be classified as constructive. As the oceanic plate subducts, it releases water to the overlying mantle. This decreases the melting point of the mantle rocks, turning them into magma. This magma then migrates upwards and erupts, forming chains of volcanoes and adding to the crust above the subduction zone.

consumable good that is consumed within a short period after purchase. Consumables include foodstuffs, beverages, detergents, and petrol.

consumer person who purchases goods and services. Consumers demand goods which businesses provide, and they need to be protected by law from unfair traders; hence the need for consumer protection acts and the work of consumer bodies such as the Consumers' Association.

consumption in economics, the purchase of goods and services for final use, as opposed to spending by firms on capital goods, known as capital formation.

container habitat in ecology, small self-contained ecosystem, such as a water pool accumulating in a hole in a tree. Some ecologists believe that much can be learned about larger ecosystems through studying the dynamics of container habitats, which can contain numerous leaf-litter feeders and their predators.

container port any large area of land, strategically located on, or adjacent to, a railway network, or alongside deep-water harbour frontage, where freight containers can be assembled into groups for loading onto trains or ships. At the terminal, large travelling cranes manoeuvre containers either directly on to their onward transport vehicle or into temporary storage clusters to await despatch. Access to a local road system is an important feature of their location.

contaminated land land that is considered to pose a health risk to humans because of pollution; usually land that has been the site of industrial activity.

continent any one of the seven large land masses of the Earth, as distinct from the oceans. They are Asia, Africa, North America, South America, Europe, Australia, and Antarctica. Continents

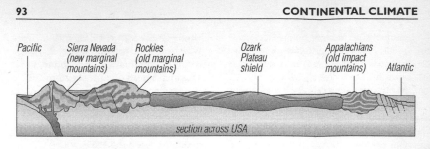

| Pacific | Sierra Nevada (new marginal mountains) | Rockies (old marginal mountains) | Ozark Plateau shield | Appalachians (old impact mountains) | Atlantic |

section across USA

continent The North American continent is growing in the west as a result of collision with the Pacific plate. On the east of the wide area of the Ozark Plateau shield lie the Appalachian Mountains, showing where the continent once collided with another continent. The eastern coastal rifting formed when the continents broke apart. On the western edge, new impact mountains have formed.

are constantly moving and evolving (see *plate tectonics). A continent does not end at the coastline; its boundary is the edge of the shallow continental shelf, which may extend several hundred kilometres out to sea.

Continental crust, as opposed to the crust that underlies the deep oceans, is composed of a wide variety of igneous, sedimentary, and metamorphic rocks. The rocks vary in age from recent (currently forming) to almost 4000 million years old. Unlike the ocean crust, the continents are not only high standing, but extend to depths as great at 70 km under high mountain ranges. Continents, as high, dry masses of rock, are present on Earth because of the density contrast between them and the rock that underlies the oceans. Continental crust is both thick and light, whereas ocean crust is thin and dense. If the crust were the same thickness and density everywhere, the entire Earth would be covered in water.

At the centre of each continental mass lies a shield or *craton, a deformed mass of old *metamorphic rocks dating from Precambrian times. The shield is thick, compact, and solid (the Canadian Shield is an example), and is usually worn flat. Around the shield is a concentric pattern of fold mountains, with older ranges, such as the Rockies, closest to the shield, and

younger ranges, such as the coastal ranges of North America, farther away. This general concentric pattern is modified when two continental masses have drifted together and they become welded with a great mountain range along the join, the way Europe and northern Asia are joined along the Urals. If a continent is torn apart, the new continental edges have no mountains; for instance, South America has mountains (the Andes) along its western flank, but none along the east where it tore away from Africa 200 million years ago.

continental climate type of climate typical of a large, mid-latitude land mass. This type of climate is to be found in the centre of continents between approximately 40° and 60° north of the Equator. The two main areas are the North American prairies and the Russian steppes (both grasslands). Areas with a continental climate are a long way from the oceans, which affects their climate in two ways: they have very low annual rainfall because little atmospheric moisture is available; and their temperature range is very large over a year, because the temperature-moderating effect of the sea has been lost. Maximum mean monthly summer temperatures for continental climates are around 20°C. During the winter there are several months when the temperature remains below freezing point. Precipitation decreases rapidly towards the east in Russia as distance from the sea increases, whereas in North America totals are lowest to the west in the rain shadow (area sheltered from rain by hills) of the Rockies. In both areas, annual rainfall averages 500 mm, and there is a threat of drought. The ground is snow-covered for several months between October and April.

continental crust see *crust.

continental drift in geology, the theory that, about 250–200 million years ago, the Earth consisted of a

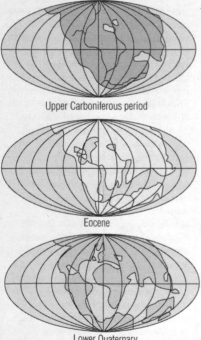

Upper Carboniferous period

Eocene

Lower Quaternary

continental drift The continents are slowly shifting their positions, driven by fluid motion beneath the Earth's crust. Over 200 million years ago, there was a single large continent called Pangaea. By 200 million years ago, the continents had started to move apart. By 50 million years ago, the continents were approaching their present positions.

single large continent (*Pangaea), which subsequently broke apart to form the continents known today. The theory was first proposed in 1912 by German meteorologist Alfred Wegener, but such vast continental movements could not be satisfactorily explained or even accepted by geologists until the 1960s.

The theory of continental drift gave way to the theory of *plate tectonics. Whereas Wegener proposed that continents pushed their way through underlying mantle and ocean floor, plate tectonics states that continents are just part of larger lithospheric plates (which include ocean crust as well) that move laterally over the Earth's surface.

continental rise portion of the ocean floor rising gently from the abyssal plain toward the steeper continental slope. The continental rise is a depositional feature formed from sediments transported down the slope mainly by turbidity currents. Much of the continental rise consists of coalescing submarine alluvial fans bordering the continental slope.

continental shelf submerged edge of a continent, a gently sloping plain that extends into the ocean. It typically has a gradient of less than 1°. When the angle of the sea bed increases to 1–5° (usually several hundred kilometres away from land), it becomes known as the *continental slope.

continental slope sloping, submarine portion of a continent. It extends downward from the edge of the continental shelf. In some places, such as south of the Aleutian Islands of Alaska, continental slopes extend directly to the ocean deeps or abyssal plain. In others, such as the east coast of North America, they grade into the gentler continental rises that in turn grade into the abyssal plains.

contour on a map, a line drawn to join points of equal height. Contours are drawn at regular height intervals; for example, every 10 m. The closer together the lines are, the steeper the slope. Contour patterns can be used to interpret the relief of an area and to identify land forms.

contour ploughing a method of soil *conservation whereby ploughing is undertaken along *contours rather than with the slope. The effect of this strategy is to reduce the rate of runoff and thus to retain *soil that would otherwise be washed off downslope. The risk of sheet erosion and gully erosion (see *soil erosion) is thus reduced.

contract farming 1. a system in which individual farmers contract with food-processing firms to produce a specific crop. The firm will often provide seed and other inputs at the start of the season, and will purchase the entire crop at harvest time. Such contract farming is common in East Anglia. 2. a system where farmers use contractors to farm their land. See also *agribusiness.

conurbation or **metropolitan area**, large continuous built-up area formed by the joining together of several urban settlements. Conurbations are often formed as a result of *urban sprawl. Typically, they have populations in excess of 1 million and some are many times that size; for example, the Osaka–Kobe conurbation in Japan, which contains over 16 million people.

convection transfer of heat energy that involves the movement of a fluid (gas or liquid). Fluid in contact with the source of heat expands and tends to rise within the bulk of the fluid. Cooler fluid sinks to take its place, setting up a convection current.

convectional rainfall rainfall associated with hot climates, resulting from the uprising of convection currents of warm air. Air that has been warmed by the extreme heating of the ground surface rises to great heights and is abruptly cooled. The water vapour carried by the air condenses and rain falls heavily. Convectional rainfall is often associated with a *thunderstorm.

convection current current caused by the expansion of a liquid, solid, or gas as its temperature rises. The expanded material, being less dense, rises, while colder, denser material sinks. Material of neutral buoyancy moves laterally. Convection currents arise in the atmosphere above warm land masses or seas, giving rise to *sea breezes and land breezes, respectively. In some heating systems, convection currents are used to carry hot water upwards in pipes.

Convection currents in the hot, solid rock of the Earth's mantle help to drive the movement of the rigid plates making up the Earth's surface (see *plate tectonics).

convenience good low-order product that is purchased frequently, such as stamps or bread. A *comparison good is a seldom purchased high-order product.

convergent evolution or **convergence**, in biology, the independent evolution of similar structures in species (or other taxonomic groups) that are not closely related, as a result of living in a similar way. Thus, birds and bees have wings, not because they are descended from a common winged ancestor, but because their respective ancestors independently evolved flight.

convergent margin or **convergent boundary**, in *plate tectonics, the boundary or active zone between two lithospheric plates that are moving towards one another. Convergent margins are characterized by *folds, reverse faulting, destructive high-magnitude earthquakes, and in some cases volcanic activity.

There are three types of convergent margins: ocean–ocean, ocean–continent, and continent–continent (referring to the types of plates converging). Both ocean–ocean and ocean–continent convergences result in *subduction zones. Examples include the Lesser Antilles (ocean–ocean) and the Andes (ocean–continent). Most ocean–ocean and ocean–continent convergence occurs around the edge of the Pacific Ocean. Because continental crust is too buoyant to sink into the mantle, continent–continent convergences cause large-scale folding, resulting in the formation of mountain ranges such as the Himalayas.

cool temperate climate   two main cool temperate areas, classified as *continental climate and western margins. Periods of one to five months are below 6°C in these climate areas.

Western margin climates are often referred to as 'northwest European', and are found on west coasts between approximately 40° and 60° north and south of the equator. Other areas with similar climatic characteristics are northwest USA and British Columbia, southern Chile, New Zealand's South Island, and Tasmania. Summers are cool with the warmest month at a temperature of 15–20°C. This is due to the low angle of the sun in the sky combined with frequent cloud cover and the cooling influence of the sea. Winters are mild in comparison. Mean monthly temperatures remain a few degrees above freezing due to the warming effect of the sea. Autumn is usually warmer than spring; seasonal temperature variations depend on prevailing air masses. Precipitation often exceeds 2000 mm annually and falls throughout the year, the highest amount falling during winter when depressions are more frequent and

intense. Snow is common in the mountains.

cooperative   business organization with limited liability where each shareholder has only one vote however many shares they own and individuals pool their *resources in order to optimize individual gains.

In a worker cooperative, it is the workers who are the shareholders and own the company. The workers decide on how the company is to be run. In a consumer cooperative, consumers control the company.

A farming cooperative would enable farmers to have the use of machinery which could not normally be afforded by individuals. Cooperative systems exist in many developing countries, for example there is the Ujamaa system in Tanzania.

Cooperatives may also facilitate small-scale manufacturing projects in developing countries, such as the production of handicrafts for sale to tourists and for export to the developed world.

cooperative movement   the banding together of groups of people for mutual assistance in trade, manufacture, the supply of credit, housing, or other services. The original principles of the cooperative movement were laid down in 1844 by the Rochdale Pioneers, under the influence of Robert Owen, and by Charles Fourier in France.

co-opetition   combination of competitive and cooperative approaches to doing business. The competitive approach to business views business as similar to warfare while cooperative approaches embrace the ideas of teams and partnerships. The concept of co-opetition was introduced by Adam Brandenburger from Harvard Business School and Barry Nalebuff from Yale School of Management.

copper ore any mineral from which copper is extracted, including native copper, Cu; chalcocite, Cu_2S; chalcopyrite, $CuFeS_2$; bornite, Cu_5FeS_4; azurite, $Cu_3(CO_3)_2(OH)_2$; malachite, $Cu_2CO_3(OH)_2$; and chrysocolla, $CuSiO_3.2H_2O$.

Native copper and the copper sulphides are usually found in veins associated with igneous intrusions. Chrysocolla and the carbonates are products of the weathering of copper-bearing rocks. Copper was one of the first metals to be worked, because it occurred in native form and needed little refining. Today the main producers are the USA, Russia, Kazakhstan, Georgia, Uzbekistan, Armenia, Zambia, Chile, Peru, Canada, and the Democratic Republic of Congo (formerly Zaire).

coppicing woodland management practice of severe pruning where trees are cut down to near ground level at regular intervals, typically every 3–20 years, to promote the growth of numerous shoots from the base.

This form of *forestry was once commonly practised in Europe, principally on hazel and chestnut, to produce large quantities of thin branches for firewood, fencing, and so on; alder, eucalyptus, maple, poplar, and willow were also coppiced. The resulting thicket was known as a coppice or copse. See also *pollarding.

Some forests in the UK, such as Epping Forest near London, have coppice stretching back to the Middle Ages.

coral marine invertebrate of the class Anthozoa in the phylum Cnidaria, which also includes sea anemones and jellyfish. It has a skeleton of lime (calcium carbonate) extracted from the surrounding water. Corals exist in warm seas, at moderate depths with sufficient light. Some coral is valued for decoration or jewellery,

for example, Mediterranean red coral *Corallum rubrum*.

Corals live in a symbiotic relationship with microscopic algae (zooxanthellae), which are incorporated into the soft tissue. The algae obtain carbon dioxide from the coral polyps, and the polyps receive nutrients from the algae. Corals also have a relationship to the fish that rest or take refuge within their branches, and which excrete nutrients that make the corals grow faster. The majority of corals form large colonies although there are species that live singly. Their accumulated skeletons make up large coral reefs and atolls. The Great Barrier Reef, to the northeast of Australia, is about 1600 km long, has a total area of 20,000 sq km, and adds 50 million tonnes of calcium to the reef each year. The world's reefs cover an estimated 620,000 sq km.

Coral reefs provide a habitat for a diversity of living organisms. In 1997 some 93,000 species were identified. One third of the world's marine fishes live in reefs. The world's first global survey of coral reefs, carried out in 1997, found around 95% of reefs had experienced some damage from overfishing, pollution, dynamiting, poisoning, and the dragging of ships' anchors. A 1998 research showed that nearly two-thirds of the world's coral reefs were at risk, including 80% of the reefs in the Philippines and Indonesia, 66% of those in the Caribbean, and over 50% of those in the Indian Ocean, the Red Sea, and the Gulf of Arabia.

diseases Since the 1990s, coral reefs have been destroyed by previously unknown diseases. The **white plague** attacked 17 species of coral in the Florida Keys, USA, in 1995. The **rapid wasting disease**, discovered in 1997, affects coral reefs from Mexico to Trinidad. In the Caribbean, the fungus

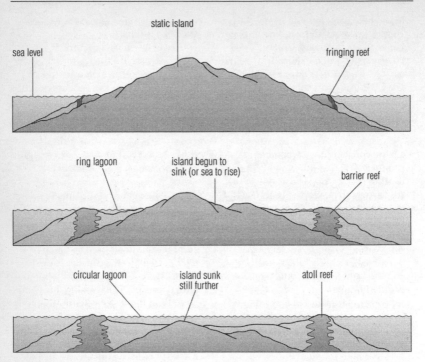

coral atoll The formation of a coral atoll by the gradual sinking of a volcanic island. The reefs fringing the island build up as the island sinks, eventually producing a ring of coral around the spot where the island sank.

Aspergillus attacks sea fans, a soft coral. It was estimated in 1997 that around 90% of the coral around the Galapagos islands had been destroyed as a result of 'bleaching', a whitening of coral reefs which occurs when the coloured algae evacuate the coral. This happens either because the corals produce toxins that are harmful to the algae or because they do not produce sufficient nutrients. Without the algae, the coral crumbles and dies away. Bleaching is widespread all over the Caribbean and the Indo-Pacific. A report published in November 1998 by Reef Check, an international organization, claimed that coral reefs are dying at a record rate throughout the world, and reefs which had existed for hundreds of years had suddenly died in 1998.

Worldwide warming has caused bleaching of coral reefs in 50 or more countries in that year.

cordierite silicate mineral, $(Mg,Fe)_2Al_4Si_5O_{18}$, blue to purplish in colour. It is characteristic of metamorphic rocks formed from clay sediments under conditions of low pressure but moderate temperature; it is the mineral that forms the spots in spotted slate and spotted hornfels.

cordillera group of mountain ranges and their valleys, all running in a specific direction, formed by the continued convergence of two tectonic plates (see *plate tectonics) along a line. The term is applied especially to the principal mountain belt of a continent. The Andes of South America are an example.

core 1. in earth science, the innermost part of the Earth. It is divided into an outer core, which begins at a depth of 2900 km, and an inner core, which begins at a depth of 4980 km. Both parts are thought to consist of iron-nickel alloy. The outer core is liquid and the inner core is solid.

The fact that *seismic shear waves disappear at the mantle–outer core boundary indicates that the outer core is molten, since shear waves cannot travel through fluid. Scientists infer the iron-nickel rich composition of the core from the Earth's density and its moment of inertia, and the composition of iron meteorites, which are thought to be pieces of cores of small planets. The temperature of the core, as estimated from the melting point of iron at high pressure, is thought to be at least 4000°C, but remains controversial. The Earth's magnetic field is believed to be the result of the movement of liquid metal in the outer core.

2. in *human geography, a *central place or central region, usually the centre of economic and political activity in a region or nation, see *core/periphery model.

core market the most important market for an individual producer. A company may have one or more core markets. For example, the core market of British Airways is air travel services; for British American Tobacco the core markets are cigarette manufacture and insurance. In the 1960s and 1970s it became fashionable for companies in the UK to diversify and enter peripheral markets. However, low profitability subsequently led many firms to sell off peripheral activities and concentrate on what they did best, providing products for their core markets.

core/periphery model this model, developed by John Friedman, identifies relationships between growing and stagnant regions during the process of economic development as follows:

(a) *Preindustrial phase:* independent villages with local *spheres of influence.

(b) *Industrializing phase:* a core region emerges, based on a growing industrial urban centre – the rest of the nation remains an undeveloped *periphery, supplying labour, food and other *resources to the core.

(c) *Fully developed phase:* regional urban centres develop to spread the benefits of economic and social progress to the former peripheries.

This is a highly simplified version of a complex model of regional development.

Coriolis effect effect of the Earth's rotation on the atmosphere, oceans, and theoretically all objects moving over the Earth's surface. In the northern hemisphere it causes moving objects and currents to be deflected to the right; in the southern hemisphere it causes deflection to the left. The effect is named after its discoverer, French mathematician Gaspard de Coriolis (1792–1843).

corn circle see *crop circle.

corporatism belief that the state in capitalist democracies should intervene to a large extent in the economy to ensure social harmony. In Austria, for example, corporatism results in political decisions often being taken after discussions between chambers of commerce, trade unions, and the government.

corporative state state in which the members are organized and represented not on a local basis as citizens, but as producers working in a particular trade, industry, or profession. Originating with the syndicalist workers' movement, the idea was superficially adopted by the fascists during the 1920s and 1930s.

Catholic social theory, as expounded in some papal encyclicals, also favours the corporative state as a means of eliminating class conflict.

corrasion or **abrasion**, grinding away of solid rock surfaces by particles carried by water, ice, and wind. It is generally held to be the most significant form of *erosion in rivers. As the eroding particles are carried along they are eroded themselves (becoming rounder and smaller) due to the process of *attrition.

corridor, wildlife route linking areas of similar habitat, or between sanctuaries. For example there is a corridor linking the Masai Mara reserve in Kenya and the Serengeti in Tanzania. On a smaller scale, disused railways provide corridors into urban areas for foxes.

Corridors are anything from a few metres to several kilometres in width. The Grider Creek corridor connecting two areas of forest wilderness between the extreme north of California and southern Oregon is 5.5 km wide and 26 km long. The wider the corridor is the greater its conservation value as it will provide additional habitat rather than just a travel route. Corridors have been criticized as pathways for diseases and new predators.

corrie Welsh **cwm**; French, North American **cirque**, Scottish term for a steep-sided armchair-shaped hollow in the mountainside of a glaciated area. The weight and movement of the ice has ground out the bottom and worn back the sides. A corrie is open at the front, and its sides and back are formed of *arêtes. There may be a small lake in the bottom, called a tarn.

A corrie is formed as follows:
(1) snow accumulates in a hillside hollow (enlarging the hollow by *nivation), and turns to ice;
(2) the hollow is deepened by *abrasion and *plucking;

(3) the ice in the corrie moves under the influence of gravity, deepening the hollow still further;
(4) since the ice is at the foot of the hollow and moves more slowly, a rock lip (a ridge of solid rock) forms;
(5) when the ice melts, a lake or tarn may be formed in the corrie. The steep back wall may be severely weathered by *freeze–thaw weathering, providing material for further abrasion.

corrosion in earth science, an alternative name for *solution, the process by which water dissolves rocks such as limestone.

corundum Al_2O_3, native aluminium oxide and the hardest naturally occurring mineral known, apart from diamond (corundum rates 9 on the Mohs scale of hardness); lack of *cleavage also increases its durability. Its crystals are barrel-shaped prisms of the trigonal system. Varieties of gem-quality corundum are **ruby** (red) and **sapphire** (any colour other than red, usually blue). Poorer-quality and synthetic corundum is used in industry, for example as an *abrasive.

Corundum forms in silica-poor igneous and metamorphic rocks. It is a constituent of emery, which is metamorphosed bauxite.

cost for a business, the amount of money spent in order to meet a specific aim, such as producing goods and services for sale. It is also the term used to describe the amount spent acquiring a particular good or service.

In business studies, **direct costs** are costs that vary directly with output, such as raw material inputs. **Indirect costs** are costs that change as output changes but not in direct proportion. **Overhead costs** are the costs of running the business which do not change as output changes. In economics, these three cost concepts are called variable, semi-variable, and fixed costs. **Total cost** is the sum of all costs incurred in producing a given level of output. **Average cost** can be found by dividing total cost by total output. **Marginal cost** is the cost of producing an extra unit of output. The **opportunity cost** of production is what has to be given up because a particular choice has been made. It is the benefit foregone of the next best alternative. The **private cost** of production is the cost to the individual or business that created the cost. **Social costs** are all the costs incurred by society through production. Private cost may be less than social cost because, for example, a producer may not have to pay for polluting the environment.

cost–benefit analysis process whereby a project is assessed for its social and welfare benefits as well as the financial return on investment. Examples of factors considered might be the environmental impact of an industrial plant, or the convenience for users of a new railway. A major difficulty is finding a method of quantifying net social costs and benefits.

cost centre part of a business to which a cost is allocated for the purposes of strategic planning. Cost centres can be large divisions of a business, departments, small teams, or even individuals.

cost, insurance, and freight see *c.i.f.

cost of living cost of goods and services needed for an average standard of living.

In Britain the cost-of-living index was introduced in 1914 and based on the expenditure of a working-class family of a man, woman, and three children; the standard is 100. Known from 1947 as the Retail Price Index (RPI), it is revised to allow for inflation. Supplementary to the RPI are the Consumer's Expenditure Deflator (formerly Consumer Price Index) and the Tax and Price Index (TPI), introduced in 1979. Comprehensive indexation has been advocated as a means of controlling inflation by linking all forms of income (such as wages and investment), contractual debts, and tax scales to the RPI. Index-linked savings schemes were introduced in the UK in 1975. In the USA a consumer price index, based on the expenditure of families in the iron, steel, and related industries, was introduced in 1890. The present index is based on the expenditure of the urban wage-earner and clerical-worker families in 46 large, medium, and small cities, the standard being 100. Increases in social security benefits are linked to it, as are many wage settlements.

cost of sales in business, the cost incurred directly in making sales. This may include the cost of raw materials or goods bought for resale and labour costs incurred in producing goods. It does not include overhead costs.

cottage industry manufacture undertaken by employees in their homes and often using their own equipment. Cottage industries frequently utilize a traditional craft such as weaving or pottery, but may also use high technology.

council in local government in England and Wales, a popularly elected local assembly charged with the government of the area within its boundaries. Under the Local Government Act 1972, they comprise three types: *county councils, *district councils, and *parish councils. Many city councils exist in the USA.

Council for Mutual Economic Assistance CMEA, full name for *Comecon, an organization of Eastern bloc countries 1949–91.

Council for the Protection of Rural England countryside conservation group founded in 1926 by Patrick Abercrombie with interests ranging from planning controls to energy policy.

A central organization campaigns on national issues and 42 local groups lobby on regional matters.

The **Campaign for the Preservation of Rural Wales** is the Welsh equivalent.

Council of Europe   body constituted in 1949 to achieve greater unity between European countries, to help with their economic and social progress, and to uphold the principles of parliamentary democracy and respect for human rights. It has a **Committee** of foreign ministers, a **Parliamentary Assembly** (with members from national parliaments), and a **European Commission on Human Rights**, established by the 1950 *European Convention on Human Rights.

Council of the Entente   CE; **Conseil de l'Entente**, organization of West African states for strengthening economic links and promoting industrial development. It was set up 1959 by Benin, Burkina Faso, Côte d'Ivoire, and Niger; Togo joined 1966 when a Mutual Aid and Loan Guarantee Fund was established. The headquarters of the CE are in Abidjan, Côte d'Ivoire.

Council of the European Union formerly **Council of Ministers**, main decision-making and legislative body of the *European Union (EU). Member states are represented at council meetings by the ministers appropriate to the subject under discussion (for example, ministers of agriculture, environment, education, and so on). The presidency of the Council changes every six months and rotates in turn among the 15 EU member countries. The Council sets the EU's objectives, coordinates the national policies of the member states, resolves differences with the *European Commission and the European Parliament, and concludes international agreements on behalf of the EU. European

Community law is adopted by the Council, or by the Council and the European Parliament through the co-decision procedure, and may take the form of binding 'regulations', 'directives', and 'decisions'.

council tax   method of raising revenue (income) for local government in Britain. It replaced the community charge, or *poll tax, from April 1993. The tax is based on property values at April 1991, but takes some account of the number of people occupying each property.

counter-urbanization   the movement of people and places of employment from large cities to places outside the cities – these may be small towns, villages, or rural areas. Inner cities lose population as a result. In most areas, the movement is to small towns rather than to truly rural areas.

country code   set of ten 'rules' produced by the Countryside Commission, giving advice about behaviour in rural areas. The code aims to protect wildlife and farmland.

The ten points in the code are: guard against all risk of fire; fasten all gates; keep dogs under control; keep to the paths across farmland; avoid damaging fences, hedges, and walls; leave no litter; safeguard water supplies; protect wildlife, wild plants, and trees; go carefully on country roads; respect the life of the countryside.

country park   pleasure ground or park, often located near an urban area, providing facilities for the public enjoyment of the countryside. Country parks were introduced in the UK following the 1968 Countryside Act and are the responsibility of local authorities with assistance from the Countryside Commission. They cater for a range of recreational activities such as walking, boating, and horse-riding.

Countryside Commission official conservation body created for England and Wales under the Countryside Act 1968. It replaced the National Parks Commission, had by 1980 created over 160 country parks, and designates Areas of Outstanding Natural Beauty.

Countryside Council for Wales Welsh nature conservation body formed in 1991 by the fusion of the former *Nature Conservancy Council and the Welsh Countryside Commission. It is government-funded and administers conservation and land-use policies within Wales.

county (Latin *comitatus* through French *comté*) administrative unit of a country or state. It was the name given by the Normans to Anglo-Saxon 'shires', and the boundaries of many present-day English counties date back to Saxon times. There are currently 34 English administrative non-metropolitan counties and 6 metropolitan counties, in addition to 34 unitary authorities. Welsh and Scottish counties were abolished in 1996 in a reorganization of local government throughout the UK, and replaced by 22 and 33 unitary authorities respectively. Northern Ireland has 6 geographical counties, although administration is through 26 district councils. In the USA a county is a subdivision of a state; the power of counties differs widely among states.

county council in England, a unit of local government whose responsibilities include broad planning policy, highways, education, personal social services, and libraries; police, fire, and traffic control; and refuse disposal. The tier below the county council has traditionally been the district council, but with local government reorganization from 1996, there has been a shift towards unitary authorities (based on a unit smaller than the county) replacing both. By 1998 there were 34 two-tier non-metropolitan county councils under which there were 274 district councils. See also * local authority and *local government.

crag in previously glaciated areas, a large lump of rock that a glacier has been unable to wear away. As the glacier passed up and over the crag, weaker rock on the far side was largely protected from erosion and formed a tapering ridge, or **tail**, of debris.

An example of a crag-and-tail feature is found in Edinburgh in Scotland; Edinburgh Castle was built on the crag (Castle Rock), which dominates the city beneath.

crater bowl-shaped depression in the ground, usually round and with steep sides. Craters are formed by explosive events such as the eruption of a volcano or the impact of a meteorite.

The Moon has more than 300,000 craters over 1 km in diameter, mostly formed by asteroid and meteorite bombardment; similar craters on Earth have mostly been worn away by erosion. Craters are found on all of the other rocky bodies in the Solar System.

Craters produced by impact or by volcanic activity have distinctive shapes, enabling geologists to distinguish likely methods of crater formation on planets in the Solar System. Unlike volcanic craters, impact craters have raised rims and central peaks and are circular, unless the meteorite has an extremely low angle of incidence or the crater has been affected by some later process.

crater lake a lake occupying a *caldera of a dormant or extinct *volcano.

craton or **continental shield**, the relatively stable core of a continent that is not currently affected by tectonics along plate boundaries. Cratons generally consist of highly deformed *metamorphic rock that formed during ancient orogenic explosions.

Cratons exist in the hearts of all the continents, a typical example being the Canadian Shield.

crawling peg or **sliding peg**, or **sliding parity**, or **moving parity**, in economics, a method of achieving a desired adjustment in a currency exchange rate (up or down) by small percentages over a given period, rather than by major revaluation or devaluation.

Some countries use a formula that triggers a change when certain conditions are met. Others change values frequently to discourage speculations.

credit in economics, a means by which goods or services are obtained without immediate payment, usually by agreeing to pay interest. The three main forms are **consumer credit** (usually given to individuals by retailers), **bank credit** (such as overdrafts or personal loans), and **trade credit** (common in the commercial world both within countries and internationally).

creep the slow, almost imperceptible downslope movement of soil and rock debris under the influence of gravity. Creep is most effective in upper soil layers and is usually absent below depths of 90 cm. Rates of movement rarely exceed 2.5 cm per year. Creep is initiated by several processes including raindrop impact, the wetting and drying of soil, freeze–thaw action, the heating and cooling of soil and rock, root growth and the activity of animals on and below a slope. The economic cost of creep can be considerable due to the repairs required for such things as tilting telephone poles, bulging and broken walls and stress fractures in road surfaces.

Cretaceous period of geological time approximately 143–65 million years ago. It is the last period of the Mesozoic era, during which angiosperm (seed-bearing) plants evolved, and dinosaurs reached a peak. The end of the Cretaceous period is marked by a mass extinction of many lifeforms, most notably the dinosaurs. The north European chalk, which forms the white cliffs of Dover, was deposited during the latter half of the Cretaceous, hence the name Cretaceous, which comes from the Latin *creta*, 'chalk'.

crevasse a crack or fissure in a *glacier resulting from the stressing and fracturing of ice at a change in *gradient or valley shape. Crevasses range from small surface cracks to major fractures many metres in depth, and may occur at any angle although two main orientations are recognized: transverse and longitudinal. Transverse crevasses occur where there is a change of gradient in the valley floor; longitudinal where the valley becomes wider and the ice mass stretches to occupy the broader space.

critical path analysis procedure used in the management of complex projects to minimize the amount of time taken. The analysis shows which subprojects can run in parallel with each other, and which have to be completed before other subprojects can follow on.

crop circle circular area of flattened grain found in fields especially in southeast England, with increasing frequency every summer since 1980. More than 1000 such formations were reported in the UK in 1991. The cause is unknown, but they are thought to be made by people.

cross section a drawing of a vertical section of a line of ground, deduced from a map. It depicts the *topography of a system of *contours.

crowding out in economics, a situation in which an increase in government expenditure results in a fall in private-sector investment, either because it causes inflation or a rise in interest rates (as a result of increased government borrowing) or because it reduces the efficiency of production as a result of government intervention. Crowding out has been used in recent years as a justification of *supply-side economics such as the privatization of state-owned industries and services.

Crown colony any British colony that is under the direct legislative control of the Crown and does not possess its own system of representative government. Crown colonies are administered by a crown-appointed governor or by elected or nominated legislative and executive councils with an official majority. Usually the Crown retains rights of veto and of direct legislation by orders in council.

crust rocky outer layer of the Earth, consisting of two distinct parts – the oceanic crust and the continental crust. The **oceanic** crust is on average about 10 km thick and consists mostly of basaltic rock overlain by muddy sediments. By contrast, the **continental** crust is largely of granitic composition and has a more complex structure. Because it is continually recycled back into the mantle by the process of subduction, the oceanic crust is in no place older than about 200 million years. However, parts of the continental crust are over 3.5 billion years old.

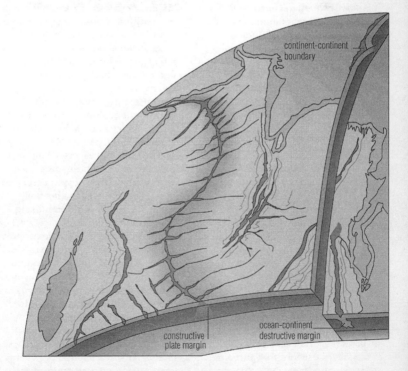

crust The crust of the Earth is made up of plates with different kinds of margins. In mid-ocean, there are constructive plate margins, where magma wells up from the Earth's interior, forming new crust. On continent–continent margins, mountain ranges are flung up by the collision of two continents. At an ocean–continent destructive margin, ocean crust is forced under the denser continental crust, forming an area of volcanic instability.

Beneath a layer of surface sediment, the oceanic crust is made up of a layer of *basalt, followed by a layer of gabbro. The continental crust varies in thickness from about 40–70 km, being deepest beneath mountain ranges, and thinnest above continental rift valleys. Whereas the oceanic crust is composed almost exclusively of basaltic igneous rocks and sediments, the continental crust is made of a wide variety of *sedimentary rocks, *igneous rocks, and *metamorphic rocks.

cryolite rare granular crystalline mineral (sodium aluminium fluoride), Na_3AlF_6, used in the electrolytic reduction of *bauxite to aluminium. It is chiefly found in Greenland.

In the extraction of aluminium from the ore bauxite (impure aluminium oxide), the alumina (pure aluminium oxide) has to be molten for electrolysis to take place. Alumina has a very high melting point (2047°C), which would require a lot of energy to keep it as a liquid, so the alumina is dissolved in molten cryolite instead. Far less energy is required and a temperature of 950°C can be used.

crystal regular-shaped solid that reflects light. Examples include diamonds, grains of salt, and sugar. Particles forming a crystal are packed in an exact and ordered pattern. When this pattern is repeated many millions of times, the crystal is formed. Such an arrangement of particles, that is regular and repeating, is called a giant molecular structure.

crystal system any of the seven crystal systems defined by symmetry, into which all known crystalline susbtances crystallize. The elements of symmetry used for this purpose are: (1) planes of **mirror symmetry**, across which a mirror image is seen, and (2) axes of **rotational symmetry**, about which, in a 360° rotation of the crystal, equivalent faces are seen two, three,

four, or six times.

To be assigned to a particular crystal system, a mineral must possess a certain minimum symmetry, but it may also possess additional symmetry elements. Since crystal symmetry is related to internal structure, a given mineral will always crystallize in the same system, although the crystals may not always grow into precisely the same shape. In cases where two minerals have the same chemical composition but different internal structures (for example graphite and diamond, or quartz and cristobalite), they will generally have different crystal systems.

CSCE *abbreviation for* **Conference on Security and Cooperation in Europe**, known after December 1994 as the *Organization on Security and Cooperation in Europe (OSCE).

cuesta alternative name for *escarpment.

cultural anthropology or **social anthropology**, subdiscipline of anthropology that analyses human culture and society, the nonbiological and behavioural aspects of humanity. Two principal branches are ethnography (the study at first hand of living cultures) and ethnology (the comparison of cultures using ethnographic evidence).

culture in sociology and anthropology, the way of life of a particular society or group of people, including patterns of thought, beliefs, behaviour, customs, traditions, rituals, dress, and language, as well as art, music, and literature. Archaeologists use the word to mean the surviving objects or artefacts that provide evidence of a social grouping.

culvert an artificial drainage channel for transporting water quickly from place to place. Culverting schemes are important in areas prone to flooding. Concrete-lined culverts, with less frictional resistance than a river channel, allow water to flow more freely, thus lessening the risk of flooding during periods of high rainfall.

cumecs *abbreviation for* 'cubic metres per second'. See *discharge.

cumulonimbus a heavy dark *cloud of great vertical height. It is the typical thunderstorm cloud, producing heavy showers of rain, snow or hail. Such clouds form where intense solar radiation causes vigorous convection.

cumulus a large *cloud (smaller than a *cumulonimbus) with a 'cauliflower' head and almost horizontal base. It is indicative of fair or, at worst, showery *weather in generally sunny conditions.

currency the type of money in use in a country; for example, the US dollar, the Australian dollar, the UK pound sterling, and the Japanese yen.

current in earth science, flow of a body of water or air, or of heat, moving in a definite direction. Ocean currents are fast-flowing bodies of seawater moved by the wind or by variations in water density between two areas. They are partly responsible for transferring heat from the Equator to the poles and thereby evening out the global heat imbalance. There are three basic types of ocean current: **drift currents** are broad and slow-moving; **stream currents** are narrow and swift-moving; and **upwelling currents** bring cold, nutrient-rich water from the ocean bottom.

cuspate delta see *delta.

customer person who purchases a product or service from an organization. The would-be purchaser remains a potential customer only until they make a purchase, or provide a firm commitment to buy.

customs duty tax imposed on goods coming into the country from abroad. In the UK, it is the responsibility of the Customs and Excise to collect these duties. The taxes collected are paid directly to the European Union (EU) as part of the UK's contribution to the EU budget.

customs union organization of autonomous countries where trade between member states is free of restrictions, but where a tariff or other restriction is placed on products entering the customs union from nonmember states. Examples include the *European Union (EU), the Caribbean Community (CARICOM), the Central American Common Market, and the Central African Economic Community.

In the 19th century, the establishment of the Zollverein between German states was a significant factor contributing to eventual political union in Germany.

cut-off see *oxbow lake.

cwm Welsh name for a *corrie.

cyclone alternative name for a *depression, an area of low atmospheric pressure with winds blowing in a anticlockwise direction in the northern hemisphere and in a clockwise direction in the southern hemisphere. A severe cyclone that forms in the tropics is called a tropical cyclone or *hurricane.

D

dairying a *pastoral farming system in which dairy cows produce milk that is used by itself or used to produce dairy products such as cheese, butter, cream and yoghurt.

DALR see *dry adiabatic lapse rate.

daminozide trade name **Alar**, chemical formerly used by fruit growers to make apples redder and crisper. In 1989 a report published in the USA found the consumption of daminozide to be linked with cancer, and the US Environment Protection Agency (EPA) called for an end to its use. The makers have now withdrawn it worldwide.

Danube Commission organization that ensures the freedom of navigation on the River Danube, from Ulm in Germany to the Black Sea, to people, shipping, and merchandise of all states, in conformity with the Danube Convention 1948. The commission comprises representatives of all the states through which the Danube flows: Germany, Austria, Slovak Republic, Hungary, Bulgaria, Russia, Ukraine, Serbia and Montenegro, and Romania. Its headquarters are in Budapest, Hungary.

dating in geology, the process of determining the age of minerals, rocks, fossils, and geological formations. There are two types of dating: relative and absolute. **Relative dating** involves determining the relative ages of materials, that is determining the chronological order of formation of particular rocks, fossils, or formations, by means of careful field work. **Absolute dating** is the process of determining the absolute age (that is the age in years) of a mineral, rock, or

fossil. Absolute dating is accomplished using methods such as radiometric dating (measuring the abundances of particular isotopes in a mineral), fission track dating, and even counting annual layers of sediment.

death rate number of deaths per 1000 of the population of an area over the period of a year. Death rate is a factor in *demographic transition.

Death rate is linked to a number of social and economic factors such as standard of living, diet, and access to clean water and medical services. The death rate is therefore lower in wealthier countries; for example, in the USA it is 9/1000; in Nigeria 18/1000.

debt something that is owed by a person, organization, or country, usually money, goods, or services. Debt usually occurs as a result of borrowing *credit. **Debt servicing** is the payment of interest on a debt. The **national debt** of a country is the total money owed by the national government to private individuals, banks, and so on; **international debt**, the money owed by one country to another, began on a large scale with the investment in foreign countries by *newly industrialized countries in the late 19th to early 20th centuries. By the end of the 20th century, the two main types of debt in developing countries were **multilateral debt** (owed to international financial institutions such as the *World Bank) and **bilateral debt** owed to governments, either for aid loans or export credit guarantee department (ECGD) loans (made to underwrite exports). International debt became a global problem as a result of the oil crisis of the 1970s. Debtor countries paid an ever-increasing share of their national output in **debt servicing** (paying off the interest on a debt, rather than paying off the debt itself). In 1996 the World Bank and International Monetary Fund (IMF) introduced the Heavily Indebted Poor Countries (HIPC) debt-relief initiative, a debt-relief programme. The Cologne Debt Initiative (or HIPC2), launched by the Group of Eight (G8) industrialized nations in 1999, sought to speed up this process and release funding for poverty reduction.

debt crisis any situation in which an individual, company, or country owes more to others than it can repay or pay interest on; more specifically, the massive indebtedness of many developing countries that became acute from the 1980s, threatening the stability of the international banking system as many debtor countries became unable to service their debts.

debt-for-nature swap agreement under which a proportion of a country's debts are written off in exchange for a commitment by the debtor country to undertake projects for environmental protection. Debt-for-nature swaps were set up by environment groups in the 1980s in an attempt to reduce the debt problem of poor countries, while simultaneously promoting conservation.

Most debt-for-nature swaps have concentrated on setting aside areas of land, especially tropical rainforest, for protection and have involved private conservation foundations. The first swap took place in 1987, when a US conservation group bought $650,000 of Bolivia's national debt from a bank for $100,000, and persuaded the Bolivian government to set aside a large area of rainforest as a nature reserve in exchange for never having to pay back the money owed. Other countries participating in debt-for-

nature swaps are the Philippines, Costa Rica, Ecuador, and Poland. However, the debtor country is expected to ensure that the area of land remains adequately protected, and in practice this does not always happen. The practice has also produced complaints of neocolonialism.

decentralization 1. of a business or organization, reorganizing into smaller units, often on separate sites. For many businesses, decentralization involves decision-making by individuals or groups throughout an organization rather than at the centre or headquarters. Decision-making is therefore 'devolved' throughout the business, dispersing authority away from the centre of the organization. Decentralization provides certain advantages, such as quick decision-making, empowerment of line managers, and corporate flexibility at a local level.

Since the 1970s, decentralization has been an increasingly popular organizational model. Globalization and the rise of the corporate brand may reverse this trend. 2. the dispersion of a population or industry away from a central point. A common form is *counter-urbanization (in developed countries, the movement of industries and people away from cities). Examples in the UK include the move of the then Department of Social Security to Newcastle and the DVLA to Swansea.

deciduous woodland trees which are generally of broad-leaved rather than *coniferous habit, and which shed their leaves during the cold season. The larch is a deciduous conifer and is thus the exception to the *evergreen norm in such trees.

decision-making choosing between two or more alternative courses of action. Decision-making can be subjective or objective. In objective decision-making, decision models are used in an attempt to eliminate bias or hunch, and to ensure a decision meets the objectives of the organization. There are several different decision models, including decision trees, Discounted Cash Flow, and *critical path analysis.

Subjective decision-making involves choosing an action that produces the best possible outcome based on the individual's preferences, prejudices, and other subjective factors. For example, in a business an individual might decide to produce a new product in red because they like the colour. An objective decision would involve the statistical analysis of relevant market research and other investigations as to the relative cost of different colours.

declining industry an industry whose output is declining over time. In the UK, the shipbuilding and textile industries declined in the 20th century. A declining industry is also associated with a fall in employment in the industry.

decolonization gradual achievement of independence by former colonies of the European imperial powers, which began after World War I. The process of decolonization accelerated after World War II with 43 states achieving independence between 1956 and 1960, 51 between 1961 and 1980, and 23 from 1981. The movement affected every continent: India and Pakistan gained independence from Britain in 1947; Algeria gained independence from France in 1962, the 'Soviet empire' broke up 1989–91.

decomposer in biology, any organism that breaks down dead matter. Decomposers play a vital role in the *ecosystem by freeing important chemical substances, such as nitrogen compounds, locked up in dead organisms or excrement. They feed on some of the released organic matter,

but leave the rest to filter back into the soil as dissolved nutrients, or pass in gas form into the atmosphere, for example as nitrogen and carbon dioxide.

The principal decomposers are bacteria and fungi, but earthworms and many other invertebrates are often included in this group. The *nitrogen cycle relies on the actions of decomposers.

Deep-Sea Drilling Project research project initiated by the USA in 1968 to sample the rocks of the ocean *crust. In 1985 it became known as the *Ocean Drilling Program (ODP).

deep-sea trench another term for *ocean trench.

deflation **1.** in economics, a reduction in the level of economic activity, usually caused by an increase in interest rates and reduction in the money supply, increased taxation, or a decline in government expenditure. **2.** the removal of loose sand by wind *erosion in desert regions. It often exposes a bare rock surface beneath.

deforestation destruction of forest for timber, fuel, charcoal burning, and clearing for agriculture and extractive industries, such as mining, without planting new trees to replace those lost (reforestation) or working on a cycle that allows the natural forest to regenerate.

Deforestation causes fertile soil to be blown away or washed into rivers, leading to *soil erosion, drought, flooding, and loss of wildlife, and affecting the *biodiversity (biological variety) of ecosystems. It may also increase the carbon dioxide content of the atmosphere and intensify the *greenhouse effect, because there are fewer trees absorbing carbon dioxide from the air for photosynthesis.

Many people are concerned about the rate of deforestation as great damage is being done to the *habitats of plants and animals. Deforestation ultimately leads to famine, and is thought to be partially responsible for the flooding of lowland areas – as for example in Bangladesh – because trees help to slow down water movement.

deglaciation the retreat and erosion of an ice sheet or glacier. During the late pleistocene epoch, deglaciation led to the disappearance of several continental ice sheets, including the Scandinavian Ice Sheet that centred on the Baltic Sea, and the Laurentide Ice Sheet that centred on the Hudson Bay in Canada.

degradation in geography, the lowering and flattening of land through *erosion, especially by *rivers. Degradation is also used to describe a reduction in the quality of usefulness of environmental resources such as vegetation or soil.

deindustrialization decline in the share of manufacturing industries in a country's economy. Typically, industrial plants are closed down and not replaced, and service industries increase.

delta river sediments deposited when a river flows into a standing body of water with no strong currents, such as a lake, lagoon, sea, or ocean. A delta is the result of fluvial and marine processes. *Deposition is enhanced when water is saline because salty water causes small clay particles to adhere together. Other factors influencing deposition include the type of sediment, local geology, sea-level changes, plant growth, and human impact. Some examples of large deltas are those of the Mississippi, Ganges and Brahmaputra, Rhône, Po, Danube,

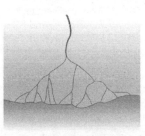

arcuate delta

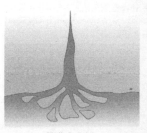

bird's foot delta

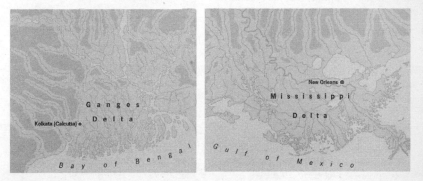

delta An arcuate delta and a bird's foot delta. The Mississippi delta is an example of a bird's foot delta and the Ganges delta is an example of an arcuate delta.

and Nile rivers. The shape of the Nile delta is like the Greek letter *delta* or D, and gave rise to the name.

The **arcuate (arc-shaped) delta** of the Nile, found in areas where *longshore drift keeps the seaward edge of the delta trimmed and relatively smooth, is only one form of delta. Others include **birdfoot (projecting) deltas**, where the river brings down enormous amounts of fine silt, as at the Mississippi delta; and **tidal (linear) deltas**, like that of the Mekong, China, where most of the material is swept to one side by sea currents. **Cuspate deltas** are pointed, and are shaped by regular opposing, gentle water movement, as seen at the Ebro and Tiber deltas in Italy.

The material deposited as a delta can be divided into three types. **bottomset beds** The lower parts of the delta are built outwards along the sea floor by turbidity currents, which are loaded with material. These bottomset beds are composed of very fine material. **foreset beds** Inclined, sloping layers of coarse material are deposited over the bottomset beds. Each bed is deposited above and in front of the previous one, the material moving by rolling and *saltation. The delta thus builds up in a seaward direction. **topset beds** These are made up of fine material and are a continuation of the river's flood plain. They are extended and built up by the work of numerous distributaries formed when the main river splits into several smaller channels.

demand in economics, the quantity of a product or service that customers want to buy at any given price. Also, the desire for a commodity, together with ability to pay for it.

demand curve in economics, a curve on a graph that shows the relationship between the quantity demanded for a good and its price. It is typically downward-sloping, showing that as

the price of the good goes down, the quantity demanded goes up. The demand curve will shift if there is a change in a variable which affects demand other than the price of the good.

democracy (Greek *demos* 'the community', *kratos* 'sovereign power') government by the people, usually through elected representatives, such as local councillors or members of a parliamentary government. In the modern world, democracy has developed from the American and French revolutions.

demographic transition any change in birth and death rates; over time, these generally shift from a situation where both are high to a situation where both are low. This may be caused by a variety of social factors (among them education and the changing role of women) and economic factors (such as higher standard of living and improved diet). The **demographic transition model** suggests that it happens in four or five stages:

(a) high birth rate, fluctuating but high death rate;

(b) birth rate stays high, death rate starts to fall, giving maximum population growth;

(c) birth rate starts to fall, death rate continues falling;

(d) birth rate is low, death rate is low.

A fifth stage is thought to happen in developed countries where the population is aging, death rates exceed the birth rates, and the population declines.

demography study of the size, structure, dispersement, and development of human *populations to establish reliable statistics on such factors as birth and death rates, marriages and divorces, life expectancy, and migration. Demography is used to calculate life tables, which give the life expectancy of members of the population by sex and age.

Demography is significant in the social sciences as the basis for industry and for government planning in such areas as education, housing, welfare, transport, and taxation. Demographic changes are important for many businesses. For example, the fall in the number of people in the UK aged 10–20 since the 1980s has led to many school closures, a shrinkage in the potential market for teenage clothes, and a fall in the number of young people available for recruitment into jobs by employers. Equally, the forecast rise in the number of people aged 75+ over the next 20 years will lead to an expansion of demand for accommodation for the elderly.

dendritic drainage   see *drainage pattern.

dendrochronology   the science of reconstructing and dating past bioclimatic events by means of studying the annual growth rings in tree trunks. Temperate forest species produce distinctive summer and winter accumulations of conducting tissue. It is possible to establish visual and statistical relationships between the rate of tissue growth and the prevailing climate. The technique requires that a living tree be cut down and the pattern of tissue growth to be carefully plotted. The sequence of growth over the age span of the tree is then compared and correlated with climatic data (average temperature, wetness, dryness, wind patterns, etc.) and a statistical model established between growth rate and climate. This model can then be tested with other trees of the same species growing in the same area. If proved valid, the model can be applied to dead trees of the same species, or to timber beams in old houses and the climatic conditions inferred from the relationship already established.

Many tree species have been successfully used in dendrochronology work, in particular those which live to great age, such as the oak and the yew. Without doubt the most useful and spectacular species has been the bristlecone pine (*Pinus aristata*) which grows at high elevation in the Sierra Nevada range of North America. This species can live for up to 4600 years while dead trunks have been dated back 8200 years. Correlation of past climatic events using the bristlecone pine and radio-carbon dating of materials has shown that the incorporation of ^{14}C in living tissue has not been at a constant rate as had previously been thought. A correction factor of 1000 years at 6000 years *before present must be applied.

denudation   natural loss of soil and rock debris, blown away by wind or washed away by running water, laying bare the rock below. Over millions of years, denudation causes a general lowering of the landscape.

dependent territories   term used as a means of referring collectively to colonies, protectorates, protected states, and trust territories for which Britain remains responsible. The term 'dependencies' is normally used to refer to territories placed under the authority of another; for example Ascension Island and Tristan da Cunha are dependencies of St Helena.

depopulation   the decline of the population of a given area, usually caused by people moving to other areas for economic reasons, rather than an increase in the *death rate or decrease in *birth rate.

deposition   in earth science, the dumping of the load carried by a river, glacier, or the sea. Deposition occurs when the river, glacier, or sea is no longer able to carry its load for some reason, for example a shallowing of gradient, decreasing speed, decreasing energy, decrease in the volume of water in the channel, or an increase in the friction between water and channel. *Glacial deposition occurs when ice melts.

Many types of deposition are found along the course of a river.

alluvial fans These are found in semi-arid areas where mountain streams enter a main valley or plain at the foot of the mountains. The sudden decrease in velocity causes the stream to deposit its load. Smaller fans are common in glaciated areas at the edge of major glacial troughs, particularly at the base of a *hanging valley.

riffles These are small ridges of material deposited where the river velocity is reduced midstream. If there are many riffles the river is said to be **braided**.

levees and flood plain deposits These are formed, over a long period of time, in places where a river regularly bursts its banks. Water loses velocity quickly leading to the rapid deposition of coarse material near the river channel edge to form embankments, called levees. Finer material is carried further away and deposited on the flood plain.

depreciation in economics, the decline of a currency's value in relation to other currencies. Depreciation is also an accounting procedure applied to tangible assets. It describes the decrease in value of the asset (such as factory machinery) resulting from usage, obsolescence, or time. Amortization is used for intangible assets and depletion for wasting assets. Depreciation is applied to assets yearly, each time reducing the net book value of the asset. It is an important factor in assessing company profits and tax liabilities.

depressed area region with substandard economic performance, perhaps as a result of a change in

industrial structure, such as a decline in manufacturing industry. An example in the UK is Clydeside, where traditional heavy industries have closed because of reduced demand and exhaustion of raw materials (such as coal and iron ore). Depressed areas may be characterized by high unemployment, low-quality housing, and poor educational standards. Government aid may be needed to reverse such decline (see *assisted area).

depression or **cyclone** or **low**, in meteorology, a region of relatively low atmospheric pressure. In mid-latitudes a depression forms as warm, moist air from the tropics mixes with cold, dry polar air, producing warm and cold boundaries (*fronts) and unstable weather – low cloud and drizzle, showers, or fierce storms. The warm air, being less dense, rises above the cold air to produce the area of low pressure on the ground. Air spirals in towards the centre of the depression in an anticlockwise direction in the northern hemisphere, clockwise in the southern hemisphere, generating winds up to gale force. Depressions tend to travel eastwards and can remain active for several days.

A deep depression is one in which the pressure at the centre is very much lower than that around it so that it produces very strong winds, as opposed to a shallow depression, in which the winds are comparatively light. A severe depression in the tropics is called a *hurricane, tropical cyclone, or typhoon, and is a great danger to shipping; a *tornado is a very intense, rapidly swirling depression, with a diameter of only a few hundred metres or so.

deprivation, area of an area, usually within a city containing a concentration of people experiencing social, economic and environmental problems. Such areas are generally

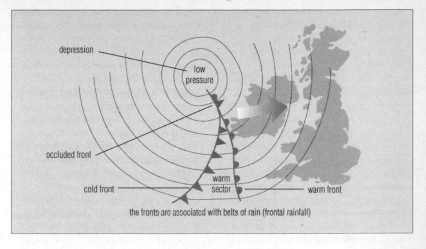

depression A typical depression showing low pressure at the centre.

defined so that remedial measures aimed at improving the living conditions of the deprived population can be more precisely targeted. In the UK, deprivation areas were first designated in the late 1960s by education authorities; **educational priority areas** were defined around schools which were deemed to have a concentration of pupils disadvantaged in their school work by reason of family poverty, ill-health or a poor educational inheritance, and thus requiring a special input of resources. Town planners adopted the concept of areas of deprivation from the early 1970s onwards, based on wider criteria which included high rates of unemployment, infant mortality, delinquency, family poverty, low standards of education, chronic sickness, together with inaccessibility to services such as shops, health and social support services. In addition, housing quality, overcrowding and the lack of standard amenities is often taken into consideration, as is the proportion of elderly and one-parent families.

Such areas have become foci for the selective input of extra resources within many large cities, through policies of **positive discrimination:** priority is accorded to such areas under city-wide programmes for community regeneration and renewal, at the expense of other areas whose population enjoys a better quality of life.

deregulation action to abolish or reduce government controls and supervision of private economic activities, with the aim of improving competitiveness. In Britain, the major changes in the City of London in 1986 (the Big Bang) were in part deregulation. Another UK example was the Building Societies Act 1985 that enabled building societies to compete in many areas with banks.

derelict land any land left standing without a present use as a result of a past activity which has physically despoiled it or aesthetically disfigured it. Physical despoilation may be the outcome of past mineral workings, especially opencast workings, while past chemical manufacture may leave soils polluted or poisoned so that natural vegetation is slow to regenerate and premature re-use poses a health hazard to potential occupiers of the site. Sometimes the dereliction is a characteristic of the ruins of buildings, rather than of the physical land surface itself. Clearance and rehabilitation of derelict land is a major objective of urban renewal in an effort to promote new industrial and residential development.

desalination removal of salt, usually from sea water, to produce fresh water for irrigation or drinking. Distillation has usually been the method adopted, but in the 1970s a cheaper process, using certain polymer materials that filter the molecules of salt from the water by reverse osmosis, was developed.

Desalination plants have been built along the shores of Middle Eastern countries where fresh water is in short supply.

desert arid area with sparse vegetation (or, in rare cases, almost no vegetation). Soils are poor, and many deserts include areas of shifting sands. Deserts can be either hot or cold. Almost 33% of the Earth's land surface is desert, and this proportion is increasing. Arid land is defined as receiving less than 250 mm rain per year.

The **tropical desert** belts of latitudes from 5–30° are caused by the descent of air that is heated over the warm land and therefore has lost its moisture. Other natural desert types are the **continental deserts**, such as the Gobi, that are too far from the sea to receive any moisture; **rain-shadow deserts**, such as California's Death Valley, that lie in the lee of mountain ranges, where the ascending air drops its rain only on the windward slopes; and **coastal deserts**, such as the Namib, where cold ocean currents cause local dry air masses to descend. Desert surfaces are usually rocky or gravelly, with only a small proportion being covered with sand (about 3%). Deserts can be created by changes in climate, or by the human-aided process of desertification.

Characteristics common to all deserts include irregular rainfall of less than 250 mm per year, very high evaporation rates of often 20 times the annual precipitation, and low relative humidity and cloud cover. Temperatures are more variable; tropical deserts have a big diurnal temperature range and very high daytime temperatures (58°C has been recorded at Azizia in Libya), whereas mid-latitude deserts have a wide annual range and much lower winter temperatures (in the Mongolian desert the mean temperature is below freezing point for half the year).

Desert soils are infertile, lacking in *humus and generally grey or red in colour. The few plants capable of surviving such conditions are widely spaced, scrubby and often thorny. Long-rooted plants (phreatophytes) such as the date palm and musquite commonly grow along dry stream channels. Salt-loving plants (halophytes) such as saltbushes grow in areas of highly saline soils and near the edges of *playas (dry saline lakes). Xerophytes are drought-resistant and survive by remaining leafless during the dry season or by reducing water losses with small waxy leaves. They frequently have shallow and widely branching root systems and store water during the wet season (for example, succulents and cacti with pulpy stems).

desertification spread of deserts by changes in climate, or by human-aided processes. Desertification can sometimes be reversed by special planting (marram grass, trees) and by the use of water-absorbent plastic grains, which, added to the soil, enable crops to be grown. About 30% of land worldwide is affected by desertification (1998), including 1 million hectares in Africa and 1.4 million hectares in Asia. The most rapid desertification is in developed countries such as the USA, Australia, and Spain.

Natural causes of desertification include decreased rainfall, increased temperatures, lowering of the *water table, and *soil erosion.

The human-aided processes leading to desertification include overgrazing, destruction of forest belts, and exhaustion of the soil by intensive cultivation without restoration of fertility – all of which may be prompted by the pressures of an expanding population or by concentration in land ownership. About 135 million people are directly affected by desertification, mainly in Africa, the Indian subcontinent, and South America. The *Sahel region in Africa is one example.

deskilling removing the need for individual skill as part of an operation. Technology has continually deskilled operations, from the advent of the production line, through to computerization of the service industries. Experienced insurance underwriters, for example, were once required to analyse risks and decide on premiums for all forms of insurance. Today, in many classes of insurance, computers calculate premiums on the basis of input data.

despotism (Greek **despotes** 'master') arbitrary and oppressive rule of a despot or autocrat, whose decisions are not controlled by law or political institutions; another term for tyranny.

destructive margin in *plate tectonics, the boundary between two lithospheric plates, along which crust is being destroyed. The term refers to a *subduction zone, a type of *convergent margin in which an oceanic plate subducts (dives) into the mantle beneath a continental plate or another oceanic plate.

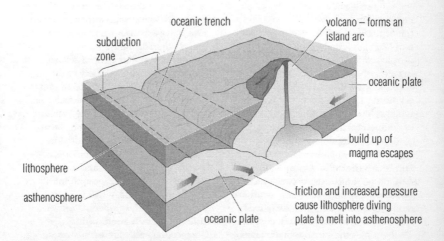

destructive margin When crustal plates meet, and one plate is denser than the other, the denser plate is forced under the other plate (at the subduction zone) and melts to form magma. If both plates are of equal density they collide and crumple up against each other forming mountains.

detritus in biology, the organic debris produced during the decomposition of animals and plants.

devaluation in economics, the lowering of the official value of a currency against other currencies, so that exports become cheaper and imports more expensive. Used when a country is badly in deficit in its balance of trade, it results in the goods the country produces being cheaper abroad, so that the economy is stimulated by increased foreign demand.

developing world or **Third World** or **the South**, those countries that are less developed than the industrialized free-market countries of the West and the industrialized former communist countries. Countries of the developing world are the poorest, as measured by their income per head of population, and are concentrated in Asia, Africa, and Latin America. The early 1970s saw the beginnings of attempts by countries in the developing world to act together in confronting the powerful industrialized countries over such matters as the level of prices of primary products, with the nations regarding themselves as a group that had been exploited in the past by the developed nations and that had a right to catch up with them. Countries that adopted a position of political neutrality towards the major powers, whether poor or wealthy, are known as *non-aligned movement.

Many development studies refer to developing countries as 'the South', and to developed and industrialized nations as 'the North', because most developing nations are in the southern hemisphere and most industrialized nations are in the northern hemisphere. Developing countries are themselves divided into low income, or least-developed countries (LDCs), such as Angola, Sudan, Bangladesh, and

Myanmar; middle-income countries, such as Nigeria, Indonesia, and Bolivia; and upper-middle-income countries, such as Brazil, Algeria, and Malaysia. The developing world has 75% of the world's population but consumes only 20% of its resources. In 1990 the average income per head of population in the northern hemisphere was $12,500, 18 times higher than that in the southern hemisphere, and developing countries accounted for 10% of world exports of manufactured goods. In the 1990s the developing world increased its global share of merchandise exports by 17%, most of the share being in office and electronic equipment, particularly from Mexico, China, and East Asia. However, the exports of the majority of least-developed countries were still confined to primary commodities (cash crops and unprocessed minerals), and growth here remained slow and unpredictable over the period, declining in some years. More than a third of low-income countries saw exports decline in 2000. At the beginning of the 21st century, 1.2 billion people were still existing on less than $1 a day, with another 1.6 billion living on less than $2 a day.

development in the social sciences, the acquisition by a society of industrial techniques and technology; hence the use of the term 'developed' to refer to the nations of the Western capitalist countries and the Eastern communist countries, and the term 'underdeveloped' or '*developing world' to refer to poorer, non-aligned nations. The terms 'more economically-developed countries' (MEDC) and 'less economically-developed countries' (LEDC) are now used.

Since the 1960s there has been growing awareness of the damaging effects of human activities on the natural environment, and the assumption that industrial development is good has been increasingly questioned. Many universities now have academic departments of **development studies** that address the theoretical questions involved in proposed practical solutions to problems in the developing world. These nations face a number of conflicts between their need to develop economically and the environmental consequences of that development. Most developing countries have an increasing population, chiefly because death rates are decreasing and birth rates remain high, and this puts new pressures on already scarce resources. Developing nations are being encouraged to work their way out of their problems through sustainable development, using technology appropriate to local needs and resources. The World Bank and other international organizations provide funds for development.

development aid   see *aid.

development area   in Britain, a region designated by government as in need of special assistance for economic reconstruction. See *depressed area.

development process   the sequence of events by which a nation moves from a predominantly subsistence agricultural economy to one based on *commercial agriculture, industry and a highly urbanized society.

devil wind   minor form of *tornado, usually occurring in fine weather; formed from rising thermals of warm air (as is a *cyclone). A fire creates a similar updraught.

    A **fire devil** or **firestorm** may occur in oil-refinery fires, or in the firebombings of cities, for example Dresden, Germany, in World War II.

devolution   delegation of authority and duties; in the later 20th century, the movement to decentralize governmental power.

Devonian period   period of geological time 408–360 million years ago, the fourth period of the Palaeozoic era. Many desert sandstones from North America and Europe date from this time. The first land plants flourished in the Devonian period, corals were abundant in the seas, amphibians evolved from air-breathing fish, and insects developed on land.

    The name comes from the county of Devon in southwest England, where Devonian rocks were first studied.

dew   precipitation in the form of moisture that collects on plants and on the ground. It forms after the temperature of the ground has fallen below the *dew point of the air in contact with it. As the temperature falls during the night, the air and its water vapour become chilled, and condensation takes place on the cooled surfaces.

dew point   temperature at which the air becomes saturated with water vapour. At temperatures below the dew point, the water vapour condenses out of the air as droplets. If the droplets are large they become deposited on plants and the ground as dew; if small they remain in suspension in the air and form mist or fog.

dewpond   drinking pond for farm animals on arid hilltops such as chalk downs. In the UK, dewponds were excavated in the 19th century and lined with mud and clay. It is uncertain where the water comes from but it may be partly rain, partly sea mist, and only a small part dew.

diabase   alternative name for *dolerite (a form of basalt that contains very little silica), especially dolerite that has metamorphosed.

diagenesis   in geology, the physical, chemical, and biological processes by which a sediment becomes a *sedimentary rock. The main processes involved include compaction of the grains, and the cementing of the grains

sand or other sediment, grains separate

after compaction, grains crushed together and interlocked

after cementation, mineral crystals cemented grains together

diagenesis The formation of sedimentary rock by diagenesis. Sand and other sediment grains are compacted and cemented together.

together by the growth of new minerals deposited by percolating groundwater. As a whole, diagenesis is actually a poorly understood process.

diamond generally colourless, transparent mineral, an allotrope of carbon. It is regarded as a precious gemstone, and is the hardest substance known (10 on the *Mohs scale). Industrial diamonds, which may be natural or synthetic, are used for cutting, grinding, and polishing.

diamond-anvil cell device composed of two opposing cone-shaped diamonds that when squeezed together by a lever-arm exert extreme pressures. The high pressures result from applying force to the small areas of the opposing diamond faces. The device is used to determine the properties of materials at pressures corresponding to those of planetary interiors. One discovery made with the diamond-anvil cell is $MgSiO_3$-perovskite,

thought to be the predominant mineral of Earth's lower *mantle.

diapirism geological process in which a particularly light rock, such as rock salt, punches upwards through the heavier layers above. The resulting structure is called a salt dome, and oil is often trapped in the curled-up rocks at each side.

dictatorship term or office of an absolute ruler, overriding the constitution. (In ancient Rome a dictator was a magistrate invested with emergency powers for six months.)

differential erosion the unequal *erosion of interbedded hard and soft rocks, the softer *strata being worn away more quickly. See, for example, *bay.

dilution reduction in the value of earnings and asset values to shareholders caused by the issue of more shares. A rights issue, for example, results in an increase in the total number of shares issued, with a corresponding dilution of the value of each share. A dilution of earnings, where the increase in issued shares is not accompanied by a commensurate increase in profits, may also arise when a less profitable company buys a more profitable one.

diminishing returns, law of in economics, the principle that additional application of one factor of production, such as an extra machine or employee, at first results in rapidly increasing output but eventually yields declining returns, unless other factors are modified to sustain the increase.

diorite igneous rock intermediate in composition between mafic (consisting primarily of dark-coloured minerals) and felsic (consisting primarily of light-coloured minerals) – the coarse-grained plutonic equivalent of *andesite. Constituent minerals include *feldspar and *amphibole or pyroxene with only minor amounts of *quartz.

dioxin   any of a family of over 200 organic chemicals, all of which are heterocyclic hydrocarbons. The term is commonly applied, however, to only one member of the family, 2,3,7,8-tetrachlorodibenzo-*p*-dioxin (2,3,7,8-TCDD), a highly toxic chemical that occurs, for example, as an impurity in the defoliant Agent Orange, used in the Vietnam War, and sometimes in the weedkiller 2,4,5-T. It has been associated with chloracne (a disfiguring skin complaint), birth defects, miscarriages, and cancer.

dip   in earth science, angle at which a structural surface, such as a fault or a bedding plane, is inclined from horizontal. Measured at right angles to the *strike of that surface, it is used together with strike to describe the orientation of geological features in the field. Rocks that are dipping have usually been affected by *folding.

dip, magnetic   angle at a particular point on the Earth's surface between the direction of the Earth's magnetic field and the horizontal. It is measured using a **dip circle**, which has a magnetized needle suspended so that it can turn freely in the vertical plane of the magnetic field. In the northern hemisphere the needle dips below the horizontal, pointing along the line of the magnetic field towards its north pole. At the magnetic north and south poles, the needle dips vertically and the angle of dip is 90°. See also *angle of declination.

dip slope   the gentler of the two slopes on either side of an escarpment crest; the dip slope inclines in the direction of the dipping *strata; the steep slope in front of the crest is the scarp slope.

direct cost   or **variable cost** cost of production materials, fuel, and so on which varies directly with the volume of output. For example, steel is a direct cost for a car manufacturer because if twice the number of cars are produced, twice the amount of steel will be used in the production process.

discharge   in a river, the volume of water passing a certain point per unit of time. It is usually expressed in cubic metres per second (cumecs). The discharge of a particular river channel may be calculated by multiplying the channel's cross-sectional area (in square metres) by the velocity of the water (in metres per second).

discordant coastline   a coastline that is at right angles to the mountains and valleys immediately inland. A rise in sea level or a sinking of the land will cause the valleys to be flooded. A flooded river valley is known as a *ria, whilst a flooded glaciated valley is known as a *fiord. Compare *concordant coastline.

diseconomies of scale   increase in the average cost of production as output increases in the long run. Diseconomies of scale are said to arise because larger businesses find it more difficult to manage their resources, particularly their workers, than small firms.

For example, a large firm may employ too many managers, which pushes up their average costs.

disinvestment   withdrawal of investments in a country for political reasons. The term is also used in economics to describe non-replacement of stock as it wears out.

distributary   river that has branched away from a main river. Distributaries are most commonly found on a *delta, where the very gentle gradient and large amounts of silt deposited encourage channels to split.

Channels are said to be **braided** if they branch away from a river and rejoin it at a later point.

distribution   in commerce, the process by which goods are sent from manufacturers through to the consumer. **Channels of distribution** usually involve both wholesalers and retailers. Distribution or **place** is one of the four elements (or four Ps) of the marketing mix. It is important for a business to get its product distributed in the right outlets if it is to generate sales at the right price.

distribution theory   the study of how the national income of a country is allocated between different individuals and groups.

district council   lower unit of local government in England. In 1998 there were 274 district councils under 34 (two-tier) non-metropolitan county councils, and 36 single-tier metropolitan district councils. Their responsibilities cover housing, local planning and development, roads (excluding trunk and classified), bus services, environmental health (refuse collection, clean air, food safety and hygiene, and enforcement of the Offices, Shops and Railway Premises Act 1963), council tax, museums and art galleries, parks and playing fields, swimming baths, cemeteries, and so on.

divergent margin   in *plate tectonics, the boundary, or active zone between two lithospheric plates that are moving apart. Divergent margins are characterized by extensive normal faulting, low-magnitude earthquakes, and, in most cases, volcanic activity.

The most common type of divergent margin is the **oceanic spreading centre**, where two oceanic plates move apart. As they diverge, magma (molten rock) wells up to fill the open space, and solidifies, forming new oceanic crust. These boundaries are characterized by long, segmented volcanic ridges known as *mid-ocean ridges. Mid-ocean ridges, which include the *Mid-Atlantic Ridge and the East Pacific Rise, are found in all the world's oceans.

The second type of divergent margin occurs on the continents, in regions of incipient continental break-up. As the

continent stretches, the crust breaks (faults) and thins. The most famous example is the Great Rift Valley in East Africa. Like their oceanic counterparts, continental rifts are characterized by normal faulting and volcanism. If rifting continues, the continent is broken in two, oceanic crust forms between the new continents, and the margin becomes a mid-ocean ridge.

diversification   a broadening of an agricultural or industrial product range, in order to reduce dependence on a single, perhaps vulnerable, product. Diversification in *agriculture is ecologically sound since a more varied *ecosystem will have a healthier pest-predator complex – uniform wheatfields, for example, are susceptible to explosions of plant-specific pests which require expensive (and perhaps polluting) artificial control. Such *monoculture also progressively drains the *soil of nutrients. Diversification is economically sound as an insurance against falling markets for a single product.

division of labour   system of work where a task is split into several parts and done by different workers. For example, on a car assembly line, one worker will fit doors, another will make the engine block, and another will work in the paint shop. The division of labour is an example of *specialization.

doldrums   area of low atmospheric pressure along the Equator, in the *intertropical convergence zone where the northeast and southeast trade winds converge. The doldrums are characterized by calm or very light winds, during which there may be sudden squalls and stormy weather. For this reason the areas are avoided as far as possible by sailing ships.

dolerite   igneous rock formed below the Earth's surface, a form of basalt, containing relatively little silica (mafic in composition).

123 DRAINAGE BASIN

Dolerite is a medium-grained (hypabyssal) basalt and forms in shallow intrusions, such as dykes, which cut across the rock strata, and sills, which push between beds of sedimentary rock. When exposed at the surface, dolerite weathers into spherical lumps.

doline a steep-sided circular depression in karst landscapes formed by solution weathering or from the collapse of underground caverns. Ancient dolines may be totally dry while dolines of more recent origin may be the sites at which overland flow disappears underground.

dolomite in mineralogy, white mineral with a rhombohedral structure, calcium magnesium carbonate ($CaMg(CO_3)_2$). Dolomites are common in geological successions of all ages and are often formed when *limestone is changed by the replacement of the mineral calcite with the mineral dolomite.

dolomite in sedimentology, type of limestone rock where the calcite content is replaced by the mineral *dolomite. Dolomite rock may be white, grey, brown, or reddish in colour, commonly crystalline. It is used as a building material. The region of the Alps known as the Dolomites is a fine example of dolomite formation.

dome a geologic feature that is the reverse of a basin. It consists of anticlinally folded rocks that dip in all directions from a central high point, resembling an inverted but usually irregular cup.

Such structural domes are the result of pressure acting upward from below to produce an uplifted portion of the crust. Domes are often formed by the upwelling of plastic materials such as salt or magma. The salt domes along the North American Gulf Coast were produced by upwelling ancient sea salt deposits, while the Black Hills of South Dakota are the result of structural domes pushed up by intruding igneous masses.

dormitory town rural settlement that has a high proportion of *commuters in its population. The original population may have been displaced by these commuters and the settlements enlarged by housing estates. Dormitory towns have increased in the UK since 1960 as a result of *counter-urbanization.

downhole measurement in geology, any one of a number of experiments performed by instruments lowered down a borehole. Such instruments may be detectors that study vibrations passing through the rock from a generator, or may measure the electrical resistivity of the rock or the natural radiation.

downsizing restructuring of an organization, usually involving a significant reduction in workforce. A popular practice during the late 1980s and early 1990s, downsizing was seen as a way to deliver better shareholder value by reducing costs, and was associated with the practice of delayering.

drainage the removal of water from the land surface by processes such as streamflow and infiltration. Drainage can be hastened artificially by the laying of pipes and *culverts.

drainage basin or **catchment area**, area of land drained by a river system (a river and its tributaries). It includes the surface run-off in the *water cycle, as well as the *water table. Drainage basins are separated by *watersheds. A drainage basin is an example of an **open system** because it is open to inputs from outside, such as precipitation, and is responsible for outputs out of the system, such as output of water into the sea and evaporation of water into the atmosphere.

drainage pattern the arrangement of a river and its tributaries within a catchment area. The drainage pattern is influenced by the nature of the land surface, the type and structure of underlying bedrock and the climatic regime of the area. There are three main types:
(a) *dendritic drainage* which is characterized by the irregular branching of tributaries. It forms in areas where geological structure exerts little influence on the course of a river;
(b) *trellis drainage* which consists of a river network where stream junctions are at approximate right angles, and result from the marked geological control of drainage;
(c) *radial drainage* which occurs when streams radiate outwards from an elevated dome-shaped structure such as a volcanic cone.

Antecedent drainage and superimposed drainage are recognized as being inherited drainage patterns.

dreikanter in earth science, eroded stone with three edges, formed by the action of wind-blown sand.

drift material transported and deposited by glacial action on the Earth's surface. See also *boulder clay.

drift mine a system of mining in which an inclined plane gives access to the ore. In areas of the Yorkshire coalfield inclined roadways lead to the coalface, allowing free access for plant and machinery. Compare *shaft mine.

drift net long straight net suspended from the water surface and used by commercial fishermen. They are controversial as they are indiscriminate in what they catch. Dolphins, sharks, turtles, and other marine animals can drown as a consequence of becoming entangled.

drought long, continuous period of dry weather, leading to a shortage of water.

drumlin long streamlined hill created in formerly glaciated areas. Debris

(*till) is transported by the glacial icesheet and moulded to form an egg-shaped mound, 8–60 m in height and 0.5–1 km in length. Drumlins commonly occur in groups on the floor of *glacial troughs, producing what is called a 'basket-of-eggs' landscape.

Drumlins are important indicators of the direction of ice flow, as their blunt ends point upstream, and their gentler slopes trail off downstream.

dry adiabatic lapse rate (DALR) the rate at which heat is lost from an unsaturated air mass as it rises through the atmosphere. The DALR has been calculated at 1°C decline per 100 m rise, although some minor variations can occur in this value.

dry farming a system of extensive agriculture in which grain crops are grown in semi-arid areas without irrigation. In areas where annual rainfall may be less than 500 mm, the conservation of soil moisture is critical. This may be achieved by the introduction of a bare fallow every second year as bare ground loses less moisture through *evapotranspiration than do vegetated surfaces. Part of the rain falling during the fallow period is thus stored in the soil for the following year's crops. Exposing the soil surface to the elements increases the likelihood of soil erosion although this may be lessened by the application of a mulch, or, as in parts of the West African savanna and Sahel regions, by the practice of placing lines of stones along the contours of gentle slopes to impede run-off and allow moisture to infiltrate the soil. Frequent cultivation of the fallow land by cross-ploughing or pulverizing the soil into a fine tilth also aids water absorption and moisture conservation, although at some risk of wind erosion.

dry valley valley without a river at its bottom. Such valleys are common on the dip slopes of chalk *escarpments, and were probably formed by rivers. However, chalk is permeable (water passes through it) and so cannot retain surface water. Two popular theories have arisen to explain how this might have happened:

(1) During the last ice age the chalk might have frozen and been rendered impermeable. During the summer thaw, water would then have flowed over the land, unable to sink into it, and river valleys would have been formed. When, after the ice age, the chalk thawed and became permeable again, rivers could no longer flow along the valleys and so these became dry.

(2) At the end of the last ice age so much meltwater might have been created that the *water table would be far higher than it is today. This would have enabled water to flow over the chalk surface without being absorbed, and create valleys. As the water table fell with time, however, water passed through the chalk once more and the valleys became dry. Good examples include Devil's Dyke, Fulking, England, and the Vale of the White Horse, Oxfordshire, England.

dumping in international trade, the selling of goods by one country to another at below marginal cost or at a price below that in its own country. Countries dump in order to get rid of surplus produce or to improve their competitive position in the recipient country. The practice is deplored by *free trade advocates because of the artificial, unfair advantage it yields.

dune mound or ridge of wind-drifted sand, common on coasts and in deserts. Loose sand is blown and bounced along by the wind, up the windward side of a dune (*saltation). The sand particles then fall to rest on the lee side, while more are blown up from the windward side. In this way a dune moves gradually downwind.

Dunes occur in areas where there is a large supply of sand, strong winds, low rainfall, and some vegetation or obstructions to trap the sand.

In sandy deserts, the typical crescent-shaped dune is called a **barchan**. **Seif dunes** are longitudinal and lie parallel to the wind direction, and **star-shaped dunes** are formed by irregular winds.

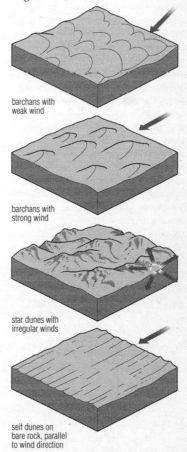

barchans with weak wind

barchans with strong wind

star dunes with irregular winds

seif dunes on bare rock, parallel to wind direction

dune The shape of a dune indicates the prevailing wind pattern. Crescent-shaped dunes form in sandy desert with winds from a constant direction. Seif dunes form on bare rocks, parallel to the wind direction. Irregular star dunes are formed by variable winds.

duricrust a hard, compact, cemented layer in the upper horizons of some *soils formed by the evaporation of mineral solutions; the concentration of the deposited minerals accounts for the hardness of the layer. These solutions are usually ferruginous, siliceous, calcareous, aluminous or magnesian in content; this high mineral content sometimes allows commercial exploitation, as occurs with certain laterites in Brazil. Duricrusts may be up to several metres thick and are most commonly found in semi-arid regions.

dust bowl area in the Great Plains region of North America (Texas to Kansas) that suffered extensive wind erosion as the result of drought and poor farming practice in once-fertile soil. Much of the topsoil was blown away in the droughts of the 1930s and the 1980s.

Similar dust bowls are being formed in many areas today, noticeably across Africa, because of overcropping and overgrazing.

dust devil small dust storm caused by intense local heating of the ground in desert areas, particularly the Sahara. The air swirls upwards, carrying fine particles of dust with it.

dust storm weather feature caused when large quantities of fine particles are raised into the atmosphere, reducing visibility to less than 1000 m. They happen most often in very dry places. Kuwait International Airport, for example, records an average of 27 dust storms per year.

duty a tax on a good. A customs duty is a tax on goods entering a country (a tax on imports). An excise duty is a type of indirect tax on goods consumed such as petrol, alcohol, or tobacco.

dyke 1. an artificial drainage channel. 2. an artificial bank built to protect low-lying land from flooding. 3. in earth science, a sheet of *igneous rock created by the intrusion of magma (molten rock) across layers of pre-existing rock. (By contrast, a sill is intruded *between* layers of rock.) It may form a ridge when exposed on the surface if it is more resistant than the rock into which it intruded.

dynamo theory theory of the origin of the magnetic fields of the Earth and other planets having magnetic fields in which the rotation of the planet as a whole sets up currents within the planet capable of producing a weak magnetic field.

E *abbreviation for* **east**.

Earth third planet from the Sun.
It is almost spherical, flattened slightly
at the poles, and is composed of five
concentric layers: inner *core, outer
core, *mantle, *crust, and atmosphere.
About 70% of the surface (including
the north and south polar ice caps)
is covered with water. The Earth is
surrounded by a life-supporting
atmosphere and is the only planet
on which life is known to exist.

mean distance from the Sun:
149,500,000 km
   equatorial diameter: 12,755 km
   circumference: 40,070 km
   rotation period: 23 hours 56 minutes
4.1 seconds
   year: (complete orbit, or sidereal
period) 365 days 5 hours 48 minutes
46 seconds.
   The Earth's average speed around
the Sun is 30 kps. The plane of its orbit
is inclined to its equatorial plane at an

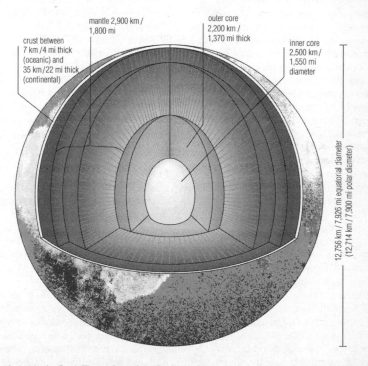

mantle 2,900 km /
1,800 mi

outer core
2,200 km /
1,370 mi thick

crust between
7 km /4 mi thick
(oceanic) and
35 km /22 mi thick
(continental)

inner core
2,500 km /
1,550 mi
diameter

12,756 km / 7,926 mi equatorial diameter
(12,714 km / 7,900 mi polar diameter)

Earth Inside the Earth. The surface of the Earth is a thin crust about 6 km thick under the sea and
40 km thick under the continents. Under the crust lies the mantle, about 2900 km thick and with
a temperature of 1500–3000°C. The outer core is about 2250 km thick, of molten iron and nickel.
The inner core is probably solid iron and nickel, at about 5000°C.

angle of 23.5°; this is the reason for the changing seasons

atmosphere: nitrogen 78.09%; oxygen 20.95%; argon 0.93%; carbon dioxide 0.03%; and less than 0.0001% neon, helium, krypton, hydrogen, xenon, ozone, and radon

surface: land surface 150,000,000 sq km (greatest height above sea level 8872 m Mount Everest); water surface 361,000,000 sq km (greatest depth 11,034 m *Mariana Trench in the Pacific). The interior is thought to be an inner core about 2600 km in diameter, of solid iron and nickel; an outer core about 2250 km thick, of molten iron and nickel; and a mantle of mostly solid rock about 2900 km thick. The crust and the uppermost layer of the mantle form about twelve major moving plates, some of which carry the continents. The plates are in constant, slow motion, called tectonic drift

satellite: the Moon

age: 4.6 billion years. The Earth was formed with the rest of the Solar System by consolidation of interstellar dust. Life began 3.5–4 billion years ago.

earthflow the rapid downslope movement of partially saturated soil and rock debris. Earthflows are more viscous than mudflows and often occur following spring thaws or heavy rainfalls. Earthflows may range in size from a few square metres up to several square kilometres.

Small earthflows may cause minor inconvenience by blocking roads and railways. They are often caused by the excavation of slope bases by river erosion and the artificial steepening of slopes that accompanies the construction of cuttings and embankments in association with the building of roads and railways.

Major earthflows are usually associated with areas of weak or impermeable underlying bedrock and may involve the movement of

millions of tonnes of soil and rock. These movements can cause considerable loss of life and damage to property, as happened with the Nicolet earthflow in Quebec, Canada in 1955 and the Aberfan disaster in South Wales in 1966.

In areas prone to earthflows, movements can be limited by remedial action such as the drainage of slopes and their stabilization through planting schemes and the erection of retaining walls.

earth pressure in civil engineering pressure exerted horizontally by earth behind a retaining wall.

earthquake abrupt motion of the Earth's surface. Earthquakes are caused by the sudden release in rocks of strain accumulated over time as a result of *plate tectonics. The study of earthquakes is called *seismology. Most earthquakes occur along *faults (fractures or breaks) and *Benioff zones. As two plates move past each other they can become jammed. When sufficient strain has accumulated, the rock breaks, releasing a series of elastic waves (*seismic waves) as the plates spring free. The force of earthquakes (magnitude) is measured on the *Richter scale, and their effect (intensity) on the *Mercalli scale. The point at which an earthquake originates is the **focus** or **hypocentre**; the point on the Earth's surface directly above this is the *epicentre.

The Alaskan (USA) earthquake of 27 March 1964 ranks as one of the greatest ever recorded, measuring 8.3–8.8 on the Richter scale. The 1906 San Francisco earthquake is among the most famous in history. Its magnitude was 8.3 on the Richter scale. The deadliest, most destructive earthquake in historical times is thought to have been in China in 1556. In 1987 a Californian earthquake was successfully predicted by measurement of underground pressure waves;

prediction attempts have also involved the study of such phenomena as the change in gases issuing from the *crust, the level of water in wells, slight deformation of the rock surface, a sequence of minor tremors, and the behaviour of animals. The possibility of earthquake prevention is remote. However, rock slippage might be slowed at movement points, or promoted at stoppage points, by the extraction or injection of large quantities of water underground, since water serves as a lubricant. This would ease overall pressure. Human activity can create earthquakes. Mining, water extraction, and oil extraction can cause subsidence that helps generate earthquakes, while the building of large dams and underground nuclear testing have also been linked with earthquakes.

Most earthquakes happen at sea and cause little damage. However, when severe earthquakes occur in highly populated areas they can cause great destruction and loss of life. A reliable form of earthquake prediction has yet to be developed, although the *seismic gap theory has had some success in identifying likely locations.

The San Andreas Fault in California, where the North American and Pacific plates move past each other, is a notorious site of many large earthquakes.

Earthquakes have been responsible for moving the North Pole towards Japan at a rate of about 6 cm every 100 years. This is because most major earthquakes occur along the Pacific Rim, and tend to tilt the pole towards the earthquake epicentres.

earthquake The propagation of seismic waves through the Earth. Waves vibrate outwards from the focus of the earthquake, deep within the Earth. As well as the movement of waves from the epicentre the wave frequency contracts and dilates.

earth science scientific study of the planet Earth as a whole. The mining and extraction of minerals and gems, the prediction of weather and earthquakes, the pollution of the atmosphere, and the forces that shape the physical world all fall within its scope of study. The emergence of the discipline reflects scientists' concern that an understanding of the global aspects of the Earth's structure and its past will hold the key to how humans affect its future, ensuring that its resources are used in a sustainable way. It is a synthesis of several traditional subjects such as *geology, *meteorology, *oceanography, *geophysics, *geochemistry, and *palaeontology.

Earth Summit or **United Nations Conference on Environment and Development**, international meetings aiming at drawing up measures towards environmental protection of the world. The first summit took place in Rio de Janeiro, Brazil, in June 1992. Treaties were made to combat global warming and protect wildlife ('biodiversity') (the latter was not signed by the USA). The second Earth Summit was held in New York in June 1997 to review progress on the environment. The meeting agreed to work towards a global forest convention in 2000 with the aim of halting the destruction of tropical and old-growth forests.

The Rio summit, which cost $23 million to stage (of which 60% was spent on security), was attended by 10,000 official delegates, 12,000 representatives of non-governmental organizations, and 7000 journalists.

In 1993, the Clinton administration overturned certain decisions made by George Bush at the Earth Summit. The USA, which had failed to ratify the Convention of Biological Diversity pact along with other nations, came under renewed pressure to sign it in April 1995 after India threatened to prevent US pharmaceutical and cosmetic companies from gaining access to its natural resources.

By 1996 most wealthy nations estimated that they would exceed their emissions targets, including Spain by 24%, Australia by 25%, and the USA by 3%. Britain and Germany were expected to meet their targets.

The second summit (1997) failed to agree a new deal to address the world's growing environmental crisis. Dramatic falls in aid to countries of the developing world, which the 1992 Earth Summit promised to increase, were at the heart of the breakdown. British prime minister Tony Blair condemned the USA, Japan, Canada, and Australia for failing to deliver on commitments to stabilize rising emissions of climate-changing greenhouse gases. The European Community as a whole was on target to meet its stabilization commitment because of cuts in emissions in Germany and the UK.

Deforestation was the main problem tackled at the second summit. The World Bank and the World Wide Fund for Nature signed an agreement aimed at protecting 250 million hectares of forest (10% of the world's forests). The importance of the issue was highlighted by the fact that deforestation progressed rapidly in developing countries since the first summit.

In June 2001, the US president George W Bush announced that the USA would not ratify the 1997 Kyoto Protocol, which committed the world's industrialized countries to cutting their annual emissions of harmful gases.

east one of the four cardinal points of the compass, indicating that part of the horizon where the Sun rises; when facing north, east is to the right.

East African Community EAC, economic alliance established by Kenya, Tanzania, and Uganda in November 2000. A Customs Union Protocol is to be implemented in July 2004, with no internal tariffs, common external tariffs, and free movement of goods, labour, and capital. The aim is to encourage foreign investment in the region. As Kenya had a more established industrial base, Tanzania and Uganda were given a period of adjustment.

easting the distance eastward of a point from a given meridian indicated by the first half of a map grid reference.

EC *abbreviation for* European Community, former name (to 1993) of the *European Union.

ecology (Greek *oikos* 'house') study of the relationship among organisms and the environments in which they live, including all living and nonliving components. The chief environmental factors governing the distribution of plants and animals are temperature, humidity, soil, light intensity, day length, food supply, and interaction with other organisms. The term ecology was coined by the biologist Ernst Haeckel in 1866.

Ecology may be concerned with individual organisms (for example, behavioural ecology, feeding strategies), with populations (for example, population dynamics), or with entire communities (for example, competition between species for access to resources in an ecosystem, or predator–prey relationships). Applied ecology is concerned with the management and conservation of habitats and the consequences and control of pollution.

econometrics application of mathematical and statistical analysis to the study of economic relationships, including testing economic theories and making quantitative predictions.

economic activity the production and distribution of goods and services to satisfy the wants and needs of consumers and other businesses. The many different activities that lead to the production of these goods and services are grouped into three broad categories – *primary industry (extraction of raw materials, such as mining), *secondary industry (manufacturing and construction), and tertiary industry (services). Recently, a fourth category has been added to these divisions, namely, 'quaternary industry', which includes the *high-tech industries using computers and microelectronics, particularly those processing information.

Economic and Monetary Union EMU, development of a unitary economy across the member states of *European Union (EU) with a single currency, single market, and harmonized interest and taxation rates. In June 1989, the *European Council decided to initiate moves towards EMU from 1 July 1990, on the basis of a report by Jacques Delors, the then president of the *European Commission. The *Maastricht Treaty then set out a timetable and criteria for the achievement of EMU and agreed to the establishment of a single European currency, the *euro.

economic community or **common market**, organization of autonomous countries formed to promote trade. Examples include the European Union, which was formed as the European Community in 1957, Caribbean Community (CARICOM) 1973, Latin American Economic System 1975, and Central African Economic Community 1985.

Economic Community of Central African States or **Communauté Economique des Etats de l'Afrique Centrale, CEEAC**, organization formed in 1983 to foster economic cooperation between member states, which include

Burundi, Cameroon, Central African Republic, Chad, the Republic of the Congo, Equatorial Guinea, Gabon, Rwanda, São Tomé and Principe, and the Democratic Republic of Congo (formerly Zaire).

Angola has observer status.

Economic Community of West African States ECOWAS; or **Communauté Economique des Etats de l'Afrique de l'Ouest**, organization promoting economic cooperation and development, established in 1975 by the Treaty of Lagos. Its members include Benin, Burkina Faso, Cape Verde, Gambia, Ghana, Guinea, Guinea-Bissau, Côte d'Ivoire, Liberia, Mali, Mauritania, Niger, Nigeria, Senegal, Sierra Leone, and Togo. Its headquarters are in Abuja, Nigeria.

Economic Cooperation Organization ECO, Islamic regional grouping formed in 1985 by Iran, Pakistan, and Turkey to reduce customs tariffs and promote commerce, with the aim of eventual customs union. In 1992 the newly independent republics of Azerbaijan, Kyrgyzstan, Tajikistan, Turkmenistan, and Uzbekistan were admitted into ECO.

economic growth rate of growth of output of all goods and services in an economy, usually measured as the percentage increase in gross domestic product or gross national product from one year to the next. It is regarded as an indicator of the rate of increase or decrease (if economic growth is negative) in the standard of living.

economic rent payment to a factor of production over and above its *transfer earnings, in other words, what it could earn in its next best use. For example, if a footballer is paid £50,000 a year but could only earn £20,000 a year in his next best job, then his economic rent would be £30,000.

economics (from Greek for 'household management') social science devoted to

studying the production, distribution, and consumption of wealth. It consists of the disciplines of *microeconomics (the study of individual producers, consumers, or markets), and *macroeconomics, (the study of whole economies or systems – in particular, areas such as taxation and public spending).

economy set of interconnected activities concerned with the production, distribution, and consumption of goods and services. The contemporary economy is very complex and includes transactions ranging from the distribution and spending of children's pocket money to global-scale financial deals being conducted by *multinational corporations.

economy of scale in economics, where the average cost of production, and therefore the unit cost, decreases as output increases. The high capital costs of machinery or a factory are spread across a greater number of units as more are produced. This may be a result of automation or mass production. If output increased by a factor of two, for example, the cost of production would increase by less than a factor of two. Economies of scale can be categorized as *external economies of scale or as *internal economies of scale.

ecosystem in ecology, a unit consisting of living organisms and the environment that they live in. A simple example of an ecosystem is a pond. The pond ecosystem includes all the pond plants and animals and also the water and other substances that make up the pond itself. Individual organisms interact with each other and with their environment in a variety of relationships, such as two organisms in competition, predator and prey, or as a food source for other organisms in a *food chain. These relationships are usually complex and finely balanced, and in natural ecosystems should be self-sustaining. However, major changes to an ecosystem, such as climate change, overpopulation, or the removal of a species, may threaten the system's sustainability and result in its eventual destruction. For instance, the removal of a major carnivore predator can result in the destruction of an ecosystem through overgrazing by herbivores. Ecosystems can be large, such as the global ecosystem (the ecosphere), or small, such as the pools that collect water in the branch of a tree, and they can contain smaller systems.

system diversity Ecosystems can be identified at different scales or levels, ranging from macrosystems (large scale) to microsystems (local scale). The global ecosystem (the ecosphere), for instance, consists of all the Earth's physical features – its land, oceans, and enveloping atmosphere (the geosphere) – together with all the biological organisms living on Earth (the biosphere); on a smaller scale, a freshwater-pond ecosystem includes the plants and animals living in the pond, the pond water and all the substances dissolved or suspended in that water, together with the rocks, mud, and decaying matter at the bottom of the pond. Thus ecosystems can contain smaller systems and be contained within larger ones.

equilibrium and succession The term 'ecosystem' was first coined in 1935 by a British ecologist, A G Tansley, to refer to a community of interdependent organisms with dynamic relationships between consumer levels, that can respond to change without altering the basic characteristics of the system. For example, cyclical changes in populations can sometimes result in large fluctuations in the numbers of a species, and are a fundamental part of

most ecosystems, but because of the interdependence of all the components, any change in one part of its nature will result in a reaction in other parts of the community. In most cases, these reactions work to restore the equilibrium or *balance of nature, but on occasions the overall change or disruption will be so great as to alter the system's balance irreversibly and result in the replacement of one type of ecosystem with another. Where this occurs as a natural process, as in the colonization of barren rock by living organisms, or the conversion of forest to grassland as a result of fires started by lightning, it is known as ecosystem development or ecological *succession. The maximum number of organisms that can be supported by a particular environment is termed its *carrying capacity.

the human threat The biosphere, or ecosphere, is an interactive layer incorporating elements of the atmosphere, hydrosphere and lithosphere, and involving natural cycles such as the carbon cycle, the *nitrogen cycle, and the *water cycle. Human interference in the Earth's natural systems, which began with the transition of human society from nomadic hunter-gatherer tribes into settled agriculture-based communities, gathered pace in the 18th and 19th centuries with the coming of the agrarian revolution and the Industrial Revolution. The technological revolution of the 20th and 21st centuries, with its programmes of industrialization and urbanization and intensive farming practices, has become a major threat, damaging the planet's ecosystems at all levels.

Gaia hypothesis The concept of the Earth as a single organism, or ecosystem, was formulated in the mid-1960s by the British scientist James Lovelock, while researching the possibility of life on Mars for NASA's space programme. The Gaia hypothesis, named after an Ancient Greek earth goddess, views the planet as a self-regulating system in which all the individual elements coexist in a symbiotic relationship. In developing this hypothesis, Lovelock realized that the damage effected by humans on many of the Earth's ecosystems was posing a threat to the viability of the planet itself. The effects of this disruption are now becoming apparent in the changing landscapes and climates of almost every region or *biome of the planet. They can be seen in the desertification of the Sahel, the shrinking of the Aral Sea in central Asia, the destruction of tropical rainforests, and the creation of the holes in the ozone layer over the Arctic and Antarctic because of the pollution of the atmosphere with greenhouse gases. These gases include carbon dioxide and sulphur dioxide emissions from the combustion of *fossil fuels, and the *chlorofluorocarbons (CFCs) widely used as propellants and refrigerants. The thinning of the protective ozone layer surrounding the planet, with its consequent threat of *global warming, affects the basic functioning of energy flow within every ecosystem of the planet, from micro-organisms to the ecosphere itself.

oldest ecosystem In 1999, palaeontologists discovered evidence in Western Australia, near the town of Marble Bar (1200 km) north of Perth, of what is believed to be the world's oldest ecosystem. The evidence consists of fossilized *stromatolites that have been dated at 3.46 billion years old.

ecotourism growing trend in tourism to visit sites that are of ecological interest, for example the Galapagos Islands, or Costa Rica. Ecotourism can bring about employment and income for local people, encouraging conservation, and is far less environmentally damaging than mass tourism.

One of the ideas behind ecotourism is that it is the practice of using money raised through tourism to pay for conservation and community projects, and putting this spending power in the hands of local people. However, if carried out unscrupulously it can lead to damage of environmentally-sensitive sites. Recently, ecotourism became increasingly popular, with many 'green' travel groups attempting to make entire holiday packages ecologically sound.

ECOWAS acronym for *Economic Community of West African States.

ECSC *abbreviation for* *European Coal and Steel Community.

ECU *abbreviation for* European Currency Unit, a unit of account the value of which depended on the underlying value of the constituent currencies of participating *European Union states. The ECU, which converted into the *euro on 1 January 1999, was not legal tender and was not represented by official banknotes and coins – unlike the euro which is a true currency in its own right.

EEC *abbreviation for* *European Economic Community.

efficiency, economic production at lowest cost. Efficiency also relates to how resources are allocated. Resources are said to be allocated efficiently if business organizations are producing the best-quality goods for the lowest price.

effluent liquid discharge of waste from an industrial process, usually into rivers or the sea. Effluent is often toxic but is difficult to control and hard to trace.

In some cases, as at *Minamata, Japan, where 43 people died of mercury poisoning, effluent can be lethal but usually its toxic effects remain unclear, because it quickly dilutes in the aquatic ecosystem.

EFTA acronym for *European Free Trade Association.

egalitarianism belief that all citizens in a state should have equal rights and privileges. Interpretations of this can vary, from the notion of equality of opportunity to equality in material welfare and political decision-making. Some states reject egalitarianism; most accept the concept of equal opportunities but recognize that people's abilities vary widely. Even those states which claim to be socialist find it necessary to have hierarchical structures in the political, social, and economic spheres. Egalitarianism was one of the principles of the French Revolution.

Ekman spiral effect in oceanography, theoretical description of a consequence of the *Coriolis effect on ocean currents, whereby currents flow at an angle to the winds that drive them. It derives its name from the Swedish oceanographer Vagn Ekman (1874–1954).

In the northern hemisphere, surface currents are deflected to the right of the wind direction. The surface current then drives the subsurface layer at an angle to its original deflection. Consequent subsurface layers are similarly affected, so that the effect decreases with increasing depth. The result is that most water is transported at about right-angles to the wind direction. Directions are reversed in the southern hemisphere.

elasticity in economics, the measure of response of one variable to changes in another. Such measures are used to test the effects of changes in prices and incomes on demand and supply.

elasticity of demand measurement of the change in demand in response to a

change in price, incomes, or other factor. Price elasticity, for example, is calculated by dividing the percentage change in price by the percentage change in demand.

elasticity of supply measurement of the change in supply in response to a change in price.

E-layer formerly **Kennelly-Heaviside layer**, lower regions (90–120 km) of the ionosphere, which reflect radio waves, allowing their reception around the surface of the Earth. The E-layer approaches the Earth by day and recedes from it at night. Its existence was proved by the British physicist Edward Victor Appleton (1892–1965).

electoral geography study of the geography of elections, including an analysis of the role of *demography and sociological factors on people's voting behaviour. It also includes the study of how constituency boundaries affect the outcomes of elections.

electromagnetic pollution the electric and magnetic fields set up by high-tension power cables, local electric sub-stations, and domestic items such as electric blankets. There have been claims that these electromagnetic fields are linked to increased levels of cancer, especially leukaemia, and to headaches, nausea, dizziness, and depression.

Although the issue has failed to receive official recognition in the UK, physicists there linked electromagnetic pollution with radon gas as a cause of cancer in 1996. The electric fields were found to attract the radioactive decay products of radon and cause them to vibrate, making them more likely to adhere to skin and mucous membranes. However, in December 1999 British researchers announced that the world's largest study into the safety of electromagnetic fields had not found a link between electromagnetic fields and childhood cancers.

electromagnetic radiation transfer of energy in the form of *electromagnetic waves.

electromagnetic spectrum complete range, over all wavelengths and frequencies, of *electromagnetic waves. These include (in order of decreasing wavelength) radio and television waves, microwaves, infrared radiation, visible light, ultraviolet light, X-rays, and gamma radiation.

The colour of sunlight is made up of a whole range of colours. A glass prism can be used to split white light into separate colours that are sensitive to the human eye, ranging from red (longer wavelength) to violet (shorter wavelength). The human eye cannot detect electromagnetic radiation outside this range. Some animals, such as bees, are able to detect ultraviolet light.

electromagnetic waves oscillating electric and magnetic fields travelling together through space at a speed of nearly 300,000 kps. Visible light is composed of electromagnetic waves. The **electromagnetic spectrum** is a family of waves that includes radio waves, infrared radiation, visible light, ultraviolet radiation, X-rays, and gamma rays. All electromagnetic waves are transverse waves. They can be reflected, refracted, diffracted, and polarized.

Radio and television waves lie at the **long wavelength–low frequency** end of the spectrum, with wavelengths longer than 10^{-4} m. Infrared radiation has wavelengths between 10^{-4} m and 7×10^{-7} m. Visible light has yet shorter wavelengths from 7×10^{-7} m to 4×10^{-7} m. Ultraviolet radiation is near the **short wavelength–high frequency** end of the spectrum, with wavelengths between 4×10^{-7} m and 10^{-8} m. X-rays have wavelengths from 10^{-8} m to 10^{-12} m. Gamma radiation has the shortest wavelengths (less than 10^{-10} m).

The different wavelengths and frequencies lend specific properties to electromagnetic waves. While visible light is diffracted by a diffraction grating, X-rays can only be diffracted by crystals. Radio waves are refracted by the atmosphere; visible light is refracted by glass or water.

electron microprobe instrument used to determine the relative and absolute abundances of elements at a particular point within an material (such as a mineral).

electrostatic precipitator device that removes dust or other particles from air and other gases by electrostatic means. An electric discharge is passed through the gas, giving the impurities a negative electric charge. Positively charged plates are then used to attract the charged particles and remove them from the gas flow. Such devices are attached to the chimneys of coal-burning power stations to remove ash particles.

elite a small group with power in a society, having privileges and status above others. An elite may be cultural, educational, religious, political (also called 'the establishment' or 'the governing circles'), or social. Sociological interest has centred on how such minorities get, use, and hold on to power, and on what distinguishes elites from the rest of society.

El Niño (Spanish 'the child') marked warming of the east Pacific Ocean that occurs when a warm current of water moves from the western Pacific, temporarily replacing the cold Peru Current along the west coast of South America. This results in a reduction in marine plankton, the main food source in the ocean, and fish numbers decline. The atmospheric circulation in the region is also seriously disturbed, and may result in unusual climatic events, for example floods in Peru, and drought in Australia. El Niño events occur at irregular intervals of between two and seven years.

El Niño is believed to be caused by the failure of trade winds and, consequently, of the ocean currents normally driven by these winds. Warm surface waters then flow in from the east. The phenomenon can disrupt the climate of the area disastrously, and has played a part in causing famine in Indonesia, drought and bush fires in the Galapagos Islands, rainstorms in California and South America, and the destruction of Peru's anchovy harvest and wildlife in 1982–83. El Niño contributed to algal blooms in Australia's drought-stricken rivers and an unprecedented number of typhoons in Japan in 1991. It is also thought to have caused the 1997 drought in Australia and contributed to certain ecological disasters such as bush fires in Indonesia.

El Niño usually lasts for about 18 months, but the 1990 occurrence lasted until June 1995; US climatologists estimated this duration to be the longest in 2000 years. The last prolonged El Niño of 1939–41 caused extensive drought and famine in Bengal. It is understood that there might be a link between El Niño and *global warming.

In a small way, El Niño affects the entire planet. The wind patterns of the the 1998 El Niño have slowed the Earth's rotation, adding 0.4 milliseconds to each day, an effect measured on the Very Long Baseline Interferometer (VLBI).

By examining animal fossil remains along the west coast of South America, US researchers estimated in 1996 that El Niño began 5000 years ago.

eluviation   the process by which fine soil particles, soluble salts and organic material are carried downward from the A horizon to the B horizon of a

*soil through the action of percolating rainwater. The subsequent deposition of this material in the B horizon is known as *illuviation. The occurrence and the rate of eluviation is usually determined by the texture and structure of the soil and upon the climatic conditions under which the soil has formed.

embargo   the legal prohibition by a government of trade with another country, forbidding foreign ships to leave or enter its ports.

Trade embargoes, as economic *sanctions, may be imposed on a country seen to be violating international laws.

emerald   clear, green gemstone variety of the mineral *beryl. It occurs naturally in Colombia, the Ural Mountains in Russia, Zimbabwe, and Australia. The green colour is caused by the presence of the element chromium in the beryl.

emery   black to greyish form of impure *corundum that also contains the minerals magnetite and haematite. It is used as an *abrasive.

emigration   **1.** in social sciences, the departure of persons from one country, usually their native land, to settle permanently in another. (See *immigration and emigration).
2. in ecology, movement of individuals away from a population, contributing to a reduction in its numbers.

employee   one who is employed under a contract of service to work for some form of payment. Payment may include salary, commission, and piece rates. Distinguishing between those who are employed and those who are self-employed is important, in some instances, for tax purposes. An employee/employer relationship is suggested in the following situations: there is a written agreement to that effect; the employer directs the method of work; the employer provides tools and equipment; and the employee is bound to the employer and cannot offer their services elsewhere.

employer person or business who makes a payment to another person in exchange for the services of that person.

employment structure the distribution of the workforce between the different *industrial sectors of the economy. Primary employment is in *agriculture, mining, forestry and fishing; secondary in manufacturing; tertiary in the retail, service and administration category; quaternary in information and expertise. One way of looking at the level of development of a nation is to examine its employment structure. Most employment in the *Third World is primary, while that in the more developed nations is tertiary, with an increasing number of people being employed in the quaternary sector.

EMU *abbreviation for* *Economic and Monetary Union, the proposed *European Union (EU) policy for a single currency and common economic policies.

enclave portion of a state within the boundaries of another. To the proprietor state the enclave is known as its **exclave**. Examples of enclaves are: the former Cabinda enclave of Angola; East Pakistan (now Bangladesh), formerly part of the Republic of Pakistan; East Prussia (now part of Poland), which, between 1919 and 1939, was part of Germany; and Nagorno-Karabakh, a predominantly Armenian region within Azerbaijan, which is disputed by the two nations.

enclosure 1. the process in England and Wales, whereby land was enclosed by the establishment of permanent hedges or walls for the purpose of improving the quality of agriculture. Many types of land were enclosed:

open field strips, waste and forest land, water meadows and, most controversial of all, common land. Enclosure of land prevented the indiscriminate wandering of domesticated farm animals, with their consequent trampling and eating of field crops. Grazing management of pasture also became possible. The enclosure movement was responsible for a major change in the visual appearance of the landscape. In place of small scale, disorganized farmsteads, a regular field pattern with larger collections of farm buildings appeared. The impact of enclosure on the native fauna and flora was catastrophic. Woodlands were cleared, wet areas drained and old grasslands ploughed. The loss of habitats along with organized hunting resulted in the extinction of a few species (wild boar, wolf) and an increasing rarity of many others.
2. any area of ground from which animals are excluded by means of a fence or cage. Enclosures are usually constructed during experiments to find the amount of vegetation eaten by herbivores. Enclosures can be temporary, as in the case of most experiments, or of a more permanent form whenever grazing pressures are to be controlled in an effort to permit for example, tree regeneration.

endangered species plant or animal species whose numbers are so few that it is at risk of becoming extinct. Officially designated endangered species are listed by the *World Conservation Union (or IUCN).

Endangered species are not a new phenomenon; extinction is an integral part of evolution. The replacement of one species by another usually involves the eradication of the less successful form, and ensures the continuance and diversification of life in all forms. However, extinctions induced by humans are thought to be destructive, causing evolutionary dead-ends that do not allow for succession by a more fit species. The great majority of recent extinctions have been directly or indirectly induced by humans; most often by the loss, modification, or pollution of the organism's habitat, but also by hunting for 'sport' or for commercial purposes.

According to a 1995 report to Congress by the US Fish and Wildlife Service, although seven of the 893 species listed as endangered under the US Endangered Species Act 1968–93 have become extinct, 40% are no longer declining in number. In February 1996 a private conservation group, Nature Conservancy, reported around 20,000 native US plant and animal species to be rare or imperilled.

According to the Red Data List of endangered species, published in 1996 by the World Conservation Union, 25% of all mammal species (including 46% of primates, 36% of insectivores, and 33% of pigs and antelopes), and 11% of all bird species are threatened with extinction.

energy   1. the capacity of a body or system to do work.
2. a measure of this capacity, expressed as the work that it does in changing to some specified reference state. It is measured in joules (SI units).

The planet Earth can be considered a single great energy system which receives solar energy as an input while it reflects light energy and radiates heat energy as an output. The flow of energy constitutes a major renewable resource. Within the system many transformations occur between the different types of energy. Over time the Earth neither gains nor loses energy; it exists in a state of energy balance or homeostasis.

Over the millennia humans have attempted to channel energy sources to suit their needs. The major source of energy, the Sun, cannot be controlled. Solar energy is used in agriculture but as a 'passive' energy source to stimulate photosynthesis. Instead, many other sources of energy have been developed, most of which are based upon the combustion of wood, coal, natural gas or oil. These fossil fuels are non-renewable resources and attempts are presently underway, albeit on a limited scale to find and utilize renewable, alternative energy sources.

energy, alternative   energy from sources that are renewable and ecologically safe, as opposed to sources that are non-renewable with toxic by-products, such as coal, oil, or gas (fossil fuels), and uranium (for nuclear power). The most important alternative energy source is flowing water, harnessed as hydroelectric power. Other sources include the oceans' tides and waves, wind power (harnessed by windmills and wind turbines), the Sun (solar energy), and the heat trapped in the Earth's crust (geothermal energy).

The Centre for Alternative Technology, near Machynlleth in mid-Wales, was established 1975 to research and demonstrate methods of harnessing wind, water, and solar energy.

energy conservation   methods of reducing energy use through insulation, increasing energy efficiency, and changes in patterns of use. Profligate energy use by industrialized countries contributes greatly to air pollution and the *greenhouse effect when it draws on non-renewable energy sources.

It has been calculated that increasing energy efficiency alone could reduce carbon dioxide emissions in several high-income countries by 1–2% a year. The average annual decrease in energy consumption in relation to gross national product 1973–87 was 1.2% in France, 2% in the UK, 2.1% in the USA, and 2.8% in Japan.

By applying existing conservation methods, UK electricity use could be reduced by 4 gigawatts by the year 2000 – the equivalent of four Sizewell nuclear power stations – according to a study by the Open University. This would also be cheaper than building new generating plants.

energy farm   any area of land or water in which plants are specifically grown for their ability to produce large amounts of biomass rapidly which can be converted into a range of biofuels such as methane or ethanol. Aquatic algae and trees are widely used for such purposes, while sugar cane, manioc (cassava), molasses, maize, wheat and sugar beet have all been used with varying degrees of success. Brazil has made extensive use of energy farms to produce ethanol and all petroleum used in Brazil has at least a 20% ethanol content. However, economic assessments made in the USA have suggested that the produce of energy farms cannot compete with traditional fossil fuels on a direct cost basis.

energy sources   There are two main sources of energy: the Sun, the ultimate source; and decay of radioactive elements in the Earth. Plants use the Sun's energy and convert it into food and oxygen. The remains of plants and animals that lived millions of years ago have been converted into *fossil fuels such as coal, oil, and natural gas.

energy resources So-called energy resources are stores of convertible energy. **Non-renewable resources** include the fossil fuels (coal, oil, and gas) and nuclear-fission 'fuels', for example, uranium-235. The term 'fuel' is used for any material from which energy can be obtained. We use up fuel reserves such as coal and oil, and

convert the energy they contain into other, useful forms. The chemical energy released by burning fuels can be used to do work. **Renewable resources**, such as wind, tidal, and geothermal power, have so far been less exploited. Hydroelectric projects are well established, and wind turbines and tidal systems are being developed.

English Channel   stretch of water between England and France, leading in the west to the Atlantic Ocean, and in the east via the Strait of Dover to the North Sea; it is also known as **La Manche** (French 'the sleeve') from its shape. The Channel Tunnel, opened in 1994, runs between Folkestone, Kent, England, and Sangatte, west of Calais, France.

English Heritage   leading conservation organization, responsible for the conservation of historic remains in England. Under the National Heritage Act 1983, its duties are to secure the preservation of ancient monuments and historic buildings; to promote the preservation and enhancement of conservation areas; and to promote the public's enjoyment and understanding of ancient monuments and historic buildings.

English Nature   agency created in 1991 from the division of the *Nature Conservancy Council into English, Scottish, and Welsh sections.

enterprise zone   former special zone introduced in the UK in 1980 and designated by government to encourage industrial and commercial activity, usually in economically-depressed areas such as the Isle of Dogs in London's Docklands area. Investment was attracted by means of tax reduction and other financial incentives. Enterprise zones no longer exist, but *assisted areas and intermediate areas survive.

entrainment   in earth science, picking up and carrying of one substance, such as rock particles, by another, such as water, air, or ice. The term is also used in *meteorology, when one body of air is incorporated into another, for instance, when clear air is entrained into a cloud.

entrenched meander   or **intrenched meander**, an *incised meander formed by rapid vertical fluvial erosion. River valleys containing entrenched meanders are usually steep-sided and symmetrical in cross-section. If the neck of an entrenched meander is breached then a small hill is produced, surrounded on three sides by the abandoned stream channel and on the fourth side by the present river channel. Such sites were often the location of fortified settlements in medieval Europe, such as the town of Verdun on the River Meuse in France.

entrepreneur   in business, a person who successfully manages and develops an enterprise through personal skill and initiative. Examples include US industrialist John D Rockefeller, US manufacturer Henry Ford, English businesswoman Anita Roddick, and English businessman Richard Branson.

environment   in ecology, the sum of conditions affecting a particular organism, including physical surroundings, climate, and influences of other living organisms. Areas affected by *environmental issues include the *biosphere and *habitat. In biology, the environment includes everything outside body cells and fluid surrounding the cells. This means that materials enclosed by part of the body surface that is 'folded in' are, in fact, part of the environment and not part of the organism. So the air spaces in human lungs and the contents of the stomach are all part of the environment and not the organism, using these terms correctly. Ecology is the study of the way organisms and

their environment interact with each other. Important processes in biology involve the transfer of material between an organism and its environment in exchanges of gases and food, for example during nutrition, photosynthesis, or respiration.

In common usage, 'the environment' often means the total global environment, without reference to any particular organism. In genetics, it is the external influences that affect an organism's development, and thus its phenotype.

Organisms usually show adaptations that help to explain why an organism lives where it does; in its habitat. Adaptations to cope with changing seasons, for example, can be quite different from one organism to another. Survival of the winter can be achieved in several different ways. Some plants die back. Some animals, such as swallows, may migrate; others, such as the dormouse, may hibernate.

Adaptations occur as a result of evolution. For example, a predator may evolve to have forward-facing eyes, acute vision and sense of smell, and have claws, talons, or a beak, for killing. The prey also adapts as a result of evolution.

*Competition is the interaction between two or more organisms when they need the same resource which is in short supply.

The organisms and environment with all their interactions make a working unit called an *ecosystem. Various environmental factors, such as the availability of nutrients, and competition, will help to determine the *population of a species in an area. However, in food chains we can predict that the populations of organisms further along a food chain generally get smaller, because the plants at the start of the chain make the energy-rich food, and some of this energy is 'lost' at each step of the chain and not available to organisms further along.

Some organisms may be small but very numerous, so population size may not be a good measure of how much of an organism there is in a habitat. *Biomass may be a more useful measure. This is the total mass of organisms in an area. Natural habitats support many different species of organism. *Biodiversity is a measure of this. Some habitats, such as rainforests, have very high biodiversity. This is partly explained by very complex food chains and webs for the organism in the habitat, but this cannot fully explain it. However, biodiversity is decreasing fast in the world as a result of human activity and this is a cause of concern. It is hoped that if humans use sustainable development in the future, the damage done to habitats and biodiversity may be lessened.

environmental audit   another name for *green audit, the inspection of a company to assess its environmental impact.

environmental impact assessment EIA, in the UK, a process by which the potential environmental impacts of human activities, such as the construction of a power station, dam, or major housing development, are evaluated. The results of an EIA are published and discussed by different levels of government, non-governmental organizations, and the general public before a decision is made on whether or not the project can proceed.

Some developments, notably those relating to national defence, are exempt from EIA. Increasingly studies include the impact not only on the physical environment, but also the socio-economic environment, such as the labour market and housing supply.

environmental impact statement EIS, statement predicting the consequences a project or development will have upon the environment. It is

intended to help manage environmental change by rating the environmental effects of development against economic factors.

environmental issues   matters relating to the damaging effects of human activity on the biosphere, their causes, and the search for possible solutions. The political movement that supports protection of the environment is the green movement. Since the Industrial Revolution, the demands made by both the industrialized and developing nations on the Earth's natural resources are increasingly affecting the balance of the Earth's resources. Over a period of time, some of these resources are renewable – trees can be replanted, soil nutrients can be replenished – but many resources, such as minerals and fossil fuels (coal, oil, and natural gas), are *non-renewable and in danger of eventual exhaustion. In addition, humans are creating many other problems that may endanger not only their own survival, but also that of other species. For instance, *deforestation and *air pollution are not only damaging and radically altering many natural environments, they are also affecting the Earth's climate by adding to the *greenhouse effect and *global warming, while *water pollution is seriously affecting aquatic life, including fish populations, as well as human health.

Environmental pollution is normally taken to mean harm done to the natural environment by human activity. In fact, some environmental pollution can have natural sources, for example volcanic activity, which can cause major air pollution or water pollution and destroy flora and fauna. In terms of environmental issues, however, environmental pollution relates to human actions, especially in connection with energy resources. The demands of the industrialized nations for energy to power machines, provide light, heat, and so on are constantly increasing. The most versatile form of energy is electricity, which can be produced from a wide variety of other energy sources, such as the fossil fuels and nuclear power (produced from uranium). These are all non-renewable resources and, in addition, their extraction, transportation, utilization, and waste products all give rise to pollutants of one form or another. The effects of these pollutants can have consequences not only for the local environment, but also at a global level.

widespread effects of pollution Many people think of air, water, and soil pollution as distinctly separate forms of pollution. However, each part of the global *ecosystem – air, water, and soil – depends upon the others, and upon the plants and animals living within the environment. Thus, pollution that might appear to affect only one part of the environment is also likely to affect other parts. For example, the emission of vehicle exhausts or acid gases from a power plant might appear to harm only the surrounding atmosphere. But once released into the air they are carried by the prevailing winds, often for several hundred kilometres, before being deposited as *acid rain. This can produce an enormous range of adverse effects across a very large area, for example: increased acidity levels in lakes and rivers are harmful to fish stocks and other aquatic life; physical damage to trees and other vegetation results in widespread destruction of forest areas; increased acidity of soils reduces the range of crops that can be grown, as well as decreasing production levels; rocks such as limestone, both in the natural landscape and in buildings, are eroded – the effect of acid rain on some of the world's most important architectural

structures is having disastrous consequences. In addition, acid rain in the form of aerosols or attached to smoke particles can cause respiratory problems in humans. Pollution of the Arctic atmosphere is creating Arctic haze – the result of aerosol emissions, such as dust, soot, and sulphate particles, originating in Europe.

desertification The destruction of fertile topsoil, and consequent soil erosion, as a result of human activity is becoming a worldwide problem. About 25% of the planet's land surface is now thought to be at risk owing to increased demand from expanding populations. This damage and destruction results not only from increased demand for food, but also as a result of changes in agricultural practices. Desertification of vast areas, such as in the Sahel in northern Africa, have resulted from the replacement of traditional farming methods in these marginal lands for the present-day cultivation of cash crops such as groundnuts and cotton. The consequence has been that the soil has lost its fertility and the land has become arid. Similarly, changes in agricultural practices produced the dust bowl in the USA in the 1930s and, more recently, the move from mixed farming to arable and the removal of hedges in order to enlarge fields for the use of modern agricultural machinery has resulted in the loss of topsoil in the large areas of the English Fenlands.

effects of tourism Environmental problems are developing not only from demands on natural resources in order to satisfy basic needs. The greater affluence and leisure time that people in the developed nations now enjoy is giving rise to the increasing demands of tourism. Not only are the more accessible areas of scenic beauty in their own countryside at risk from overuse and tramping feet, but in the tourists' search for more exotic locations, the landscapes, lifestyles, and wildlife of some of the world's more remote regions are now being brought within the reach of – and despoilment by – an ever-expanding tourist industry.

public awareness Concern for the environment is not just a late-20th century issue. In England, the first smoke abatement law dates from 1273, while in 1306 the burning of coal was prohibited in London because of fears of air pollution. However, the inspiration for the creation of the modern environmental movements came about from the publication in 1962 of Rachel Carson's book *Silent Spring*, in which she attacked the indiscriminate use of pesticides. This, combined with the increasing affluence of Western nations which allowed people to look beyond their everyday needs, triggered an awareness of environmental issues on a global scale and resulted in the formation of the Green movement. In the mid-1960s, the detection of CFCs in the atmosphere by British scientist James Lovelock led to a realization of the damaging effects of ozone depletion and added to public concern for the environment, as did his development of the Gaia hypothesis, which views the Earth as a single integrated and self-sustaining organism.

international measures In 1972, the United Nations Environment Program (UNEP) was formed to coordinate international measures for monitoring and protecting the environment, and in 1985 the Vienna Convention for the Protection of the Ozone Layer, which promised international cooperation in research, monitoring, and the exchange of information on the problem of ozone depletion, was signed by 22 nations. Discussions arising out of this convention led to the signing in 1987 of the *Montréal Protocol. In 1992, representatives of 178 nations met in

Rio de Janeiro for the United Nations Conference on Environment and Development. Known as the *'Earth Summit', this was one of the most important conferences ever held on environmental issues. UN members signed agreements on the prevention of global warming and the preservation of forests and endangered species, along with many other environmental issues.

The second Earth Summit, held in New York in 1997, tackled deforestation and agreed to work towards the preservation of the world's tropical and old-growth forests. In the same year, the United Nations Framework Convention on Climate Change (UNFCCC) adopted the *Kyoto Protocol, which committed the world's industrialized countries to cutting their annual emissions of harmful gases. By July 2001 84 parties had signed and 37 ratified or acceded to the Protocol. However, US President George W Bush announced that the USA would not be ratifying the Kyoto Protocol in June 2001.

Environmentally Sensitive Area
ESA, scheme introduced by the UK Ministry of Agriculture in 1984, as a result of EC legislation, to protect some of the most beautiful areas of the British countryside from the loss and damage caused by agricultural change. The first areas to be designated ESAs were in the Pennine Dales, the North Peak District, the Norfolk Broads, the Breckland, the Suffolk River Valleys, the Test Valley, the South Downs, the Somerset Levels and Moors, West Penwith, Cornwall, the Shropshire Borders, the Cambrian Mountains, and the Lleyn Peninsula.

The total area designated as ESAs was estimated in 1997 at 3,239,000 hectares. The scheme is voluntary, with farmers being encouraged to adapt their practices so as to enhance or maintain the natural features of the landscape and conserve wildlife habitat. A farmer who joins the scheme agrees to manage the land in this way for at least five years. In return for this agreement, the Ministry of Agriculture pays the farmer a sum that reflects the financial losses incurred as a result of reconciling conservation with commercial farming.

Environmental Protection Agency EPA, US agency set up in 1970 to control water and air quality, industrial and commercial wastes, pesticides, noise, and radiation. In its own words, it aims to protect 'the country from being degraded, and its health threatened, by a multitude of human activities initiated without regard to long-ranging effects upon the life-supporting properties, the economic uses, and the recreational value of air, land, and water'.

environmental lapse rate   rate at which temperature changes with altitude. Usually, because the atmosphere is heated from below, temperatures decrease as altitude increases at an average rate of –6.4°C per 1000 m. The rate varies from place to place and according to the time of day.

Under certain conditions, the lapse rate can be reversed, so that temperatures increase with altitude. This can be caused by large-scale atmospheric movements, for example where a warm stable air mass overlies a cool, less stable air mass. This is common in *trade wind areas of the tropics. Smaller scale temperature inversions develop at night where surface temperatures fall rapidly, cooling the air near the surface. It can also be caused by cold air flowing down a slope into valleys.

Eocene epoch   second epoch of the Tertiary period of geological time, roughly 56.5–35.5 million years ago. Originally considered the earliest

division of the Tertiary, the name means 'early recent', referring to the early forms of mammals evolving at the time, following the extinction of the dinosaurs.

eolith   naturally shaped or fractured stone found in Lower Pleistocene deposits and once believed by some scholars to be the oldest known artefact type, dating to the pre-Palaeolithic era. They are now recognized as not having been made by humans.

eon   or **aeon**, in earth science, large amount of geological time consisting of several eras. The term is also used to mean a thousand million years (10^9 y).

ephemeral stream   any stream whose flow is intermittent. Most ephemeral streams occur in moisture-deficient environments such as deserts or in regions with permeable underlying rocks such as limestone. After heavy rainstorms, ephemeral streams may become raging torrents and cause considerable soil erosion. As both the level of water and the speed of flow rapidly decrease then large quantities of debris are deposited haphazardly along the stream bed to await transport in the next flood.

epicentre   point on the Earth's surface immediately above the seismic focus of an *earthquake. Most building damage takes place at an earthquake's epicentre. The term is also sometimes used to refer to a point directly above or below a nuclear explosion ('at ground zero').

epiphyte   any herbaceous plant which grows upon another, usually a tree, but which does not feed upon its host; examples include the staghorn fern and many members of the orchid family. The epiphyte feeds on nutrients obtained in solution from rainwater and absorbed from the atmosphere. Epiphytes are most common in the humid rainforests of low latitudes.

epoch   subdivision of a geological period in the geological time scale. Epochs are sometimes given their own names (such as the Palaeocene, Eocene, Oligocene, Miocene, and Pliocene epochs comprising the Tertiary period), or they are referred to as the late, early, or middle portions of a given period (as the Late Cretaceous or the Middle Triassic epoch).

Geological time is broken up into **geochronological units** of which epoch is just one level of division. The hierarchy of geochronological divisions is eon, *era, period, epoch, age, and chron. Epochs are subdivisions of periods and ages are subdivisions of epochs. Rocks representing an epoch of geological time comprise a **series**.

Equator or **terrestrial equator**, *great circle whose plane is perpendicular to the Earth's axis (the line joining the poles). Its length is 40,092 km, divided into 360 degrees of longitude. The Equator encircles the broadest part of the Earth, and represents 0° latitude. It divides the Earth into two halves, called the northern and the southern hemispheres.

The **celestial equator** is the circle in which the plane of the Earth's Equator intersects the celestial sphere.

equatorial bulge in astronomy, increase in the diameter of a planet (or the Sun) at its equator, which results from its spinning on its axis. The diameter of the Earth, for example, is 12,756 km at the equator, whereas its polar diameter is only 12,714 km.

equatorial climate see *climate classification.

equinox time when the Sun is directly overhead at the Earth's *Equator and consequently day and night are of equal length at all latitudes. This happens twice a year: 21 March is the spring, or vernal, equinox and 23 September is the autumn equinox.

era any of the major divisions of geological time that includes several periods but is part of an eon. The eras of the current Phanerozoic in

chronological order are the Palaeozoic, Mesozoic, and Cenozoic. We are living in the Recent epoch of the Quaternary period of the Cenozoic era.

Geological time is broken up into **geochronological units** of which era is just one level of division. The hierarchy of geochronological divisions is eon, era, period, *epoch, age, and chron. Eras are subdivisions of eons and periods are subdivisions of eras. Rocks representing an era of geological time comprise an **erathem**.

erosion wearing away of the Earth's surface by a moving agent, caused by the breakdown and transport of particles of rock or soil. Agents of erosion include the sea, rivers, glaciers, and wind. By contrast, *weathering does not involve transportation.

The most powerful forms of erosion are water, consisting of sea waves and currents, rivers, and rain; ice, in the form of glaciers; and wind, hurling sand fragments against exposed rocks and moving dunes along. People also contribute to erosion by poor farming practices and the cutting down of forests, which can lead to increased overland water run-off.

There are several processes of river erosion including *hydraulic action, *corrasion, *attrition, and *solution.

erratic in geology, a displaced rock that has been transported by a glacier or some other natural force to a site of different geological composition.

escarpment or **cuesta**, large ridge created by the erosion of dipping sedimentary rocks. It has one steep side (scarp) and one gently sloping side (dip). Escarpments are common features of chalk landscapes, such as the Chiltern Hills and the North Downs in England. Certain features are associated with chalk escarpments, including dry valleys (formed on the dip slope), combes (steep-sided valleys on the scarp slope), and springs.

esker   narrow, steep-walled ridge, often meandering, found in formerly glaciated areas. It was originally formed beneath a glacier. It is made of sands and gravels, and represents the course of a river channel beneath the glacier. Eskers vary in height from 3–30 m and can be up to 160 km or so in length. Eskers are often used for roads as they are areas of high ground above marshy clay lowlands (glacial deposits).

estuary   river mouth widening into the sea, where fresh water mixes with salt water and tidal effects are felt.

Etesian wind   north-northwesterly wind that blows June–September in the eastern Mediterranean and Aegean seas.

ethical tourism   approach to *tourism which seeks to ensure that the local population benefits from tourist development and activities. Although there has been a rapid increase in the number of tourists visiting developing countries, these countries do not always benefit economically. For example, when large international companies build and run tourist resorts, a high percentage of the revenue usually leaves the country and the local economy suffers.

Ethical tourism promotes the idea that local people should own or be involved in the development of tourism and not simply provide a cheap labour force. It may also involve the boycotting of countries with politically repressive regimes.

ethnic cleansing   the forced expulsion of one ethnic group by another to create a homogenous population, for example, of more than 2 million Muslims by Serbs in Bosnia-Herzegovina 1992–95. The term has also been used to describe the killing of Hutus and Tutsis in Rwanda and Burundi in 1994, and for earlier mass exiles, as far back as the book of Exodus.

ethnic group   a group of people with a common identity such as culture, religion or skin colour. See *ethnicity.

ethnicity   (Greek *ethnos* 'a people') people's own sense of cultural identity; a social term that overlaps with such concepts as race, nation, class, and religion.

Social scientists use the term **ethnic group** to refer to groups or societies who feel a common sense of identity, often based on a traditional shared culture, language, religion, and customs. It may or may not include common territory, skin colour, or common descent. The USA, for example, is often described as a **multi-ethnic society** because many members would describe themselves as members of an ethnic group (Jewish, black, or Irish, for example) as well as their national one (American).

ethnobotany   study of the relationship between human beings and plants. It combines knowledge of botany, chemistry, and anthropology. Many pharmaceutical companies, universities, and government health agencies have contracted ethnobotanists to conduct research among indigenous peoples, especially in the Amazon, to discover the traditional use of medicinal plants which can lead to the development of new drugs.

ethnocentrism   viewing other peoples and cultures from the standard of one's own cultural assumptions, customs, and values. In anthropology, ethnocentrism is avoided in preference for a position of relativism.

ethnography   study of living cultures, using anthropological techniques like participant observation (where the anthropologist lives in the society being studied) and a reliance on informants. Ethnography has provided much data of use to archaeologists as analogies.

ethnology study of contemporary peoples, concentrating on their geography and culture, as distinct from their social systems. Ethnologists make a comparative analysis of data from different cultures to understand how cultures work and why they change, with a view to deriving general principles about human society.

ethnoscience method of analysing cultural systems. In vogue in the USA in the late 1950s and 1960s, it was based on the componential analysis of structural linguistics. The aim was to operate in a foreign culture and 'think like the indigenous people' by considering a cultural system as a language and discovering the 'grammatical rules' of the system.

ethnotourism tourism centred around an indigenous group of people and their culture. Ethnotourists seek the experience of other cultures, a major part of which is participating in another way of life and seeing people carrying out their daily routines. It is important that what is seen and experienced is authentic. Inevitably though, the influx of visitors destroys the very thing they seek. Cultures have come to accommodate and put on shows for tourists. The most popular locations are remote areas of the Amazon, Thailand, and Indonesia. Groups are small, usually 8–12 people, and are led by guides familiar with the tribes and their customs, taboos, and codes of conduct.

EU *abbreviation for* *European Union.

euro single currency of the *European Union (EU), which was officially launched on 1 January 1999 in 11 of the 15 EU member states (Austria, Belgium, Finland, France, Germany, Republic of Ireland, Italy, Luxembourg, the Netherlands, Portugal, and Spain). Greece adopted the euro on 1 January 2001. Euro notes and coins were introduced from 1 January 2002, circulating in parallel with national currencies for two months. Thereafter the national currencies were abolished.

European Atomic Energy Community Euratom, organization established by the second Treaty of Rome 1957, which seeks the cooperation of member states of the *European Union in nuclear research and the rapid and large-scale development of non-military nuclear energy.

European Bank for Reconstruction and Development EBRD, international bank established in 1991 to support the development of market economies in Central and Eastern Europe following the widespread collapse of communist regimes. It has 62 members (60 countries, the *European Community, and the European Investment Bank) and is based in London, England.

European Central Bank ECB, central bank of the *European Union (EU), formally constituted on 1 June 1998. It is an integral aspect of European *economic and monetary union. Together with the European Union's 15 national central banks, it is part of the **European System of Central Banks** (ESCB) whose main responsibility is ensuring price stability in those countries which have adopted the *euro. The ESCB defines and implements monetary policy for the eurozone, conducts foreign exchange operations, and manages the foreign reserves of the participating member states. Its precursor was the **European Monetary Institute**.

European Coal and Steel Community ECSC, organization established by the Treaty of Paris 1951 (ratified 1952) as a single authority for the coal and steel industries of France, West Germany, Italy, Belgium, Holland, and Luxembourg, eliminating tariffs and other restrictions; in 1967 it became part of the European Community (now the European Union).

European Commission executive body that proposes legislation on which the *Council of the European Union and the European Parliament decide, and implements the decisions made in the *European Union (EU). The European Commission is the biggest of the European institutions, and must work in close partnership with the governments of the member states and with the other European institutions. The aim of the Commission is to ensure the close union of EU member states, and to defend the interests of Europe's citizens. As well as having responsibility for policy and legislative proposals, the European Commission ensures that legislation passed by the EU is applied correctly; if it is not, the Commission can take action against the public or private sector. The Commission also manages policies and negotiates international trade and cooperation agreements.

European Community EC, collective term for the *European Economic Community (EEC), the *European Coal and Steel Community (ECSC), and the *European Atomic Energy Community (Euratom). The EC is now a separate legal entity with the *European Union (EU), which was established under the *Maastricht Treaty (1992) and includes intergovernmental cooperation on security and judicial affairs.

European Convention on Human Rights in full **European Convention for the Protection of Human Rights and Fundamental Freedoms**, human-rights convention signed in 1950 by all member countries of the *Council of Europe. The Convention was the first attempt to give legal content to human rights in an international agreement, and to combine this with the establishment of authorities for supervision and enforcement. It was signed in Rome on 4 November 1950 and came into force on 3 September 1953.

European Council name given to the meetings or summits between the heads of state and government of the *European Union (EU) member states and the president of the *European Commission. The council meets at least twice a year, usually towards the end of each country's rotating six-month presidency, and gives overall direction to the work of the EU. Foreign ministers and other ministers attend by invitation. The member state holding the presidency hosts the European Council.

European Economic Area EEA, zone of economic cooperation between member states of the *European Union (EU)) and the *European Free Trade Association (EFTA), which entered into force in 1994. In essence, the EEA extends the benefits of the *single European market to the three non-EU EFTA states of Norway, Iceland, and Liechtenstein. Switzerland, though a member of EFTA, rejected membership of the EEA in a referendum in 1992 (and also continued to reject membership of the EU in a referendum in March 2001).

European Economic Community EEC, organization established, together with the *European Atomic Energy Community (Euratom), under the terms of the 1957 Treaties of Rome. The treaties provided for the establishment by stages of a common market based on a customs union, the convergence of economic policies, and the promotion of growth in the nuclear industries for peaceful purposes. The EEC was also the popular name for the union of states that later became the *European Union (EU).

European Free Trade Association EFTA, organization established in 1960 and consisting of Iceland, Norway, Switzerland, and (from 1991) Liechtenstein, previously a non-voting associate member. There are no import duties between members. Of the original EFTA members, Britain and Denmark left in 1972 to join the *European Community (EC), as did Portugal in 1985; Austria, Finland, and Sweden joined the *European Union (EU) in 1995.

European geology Europe is divided by three mountain belts that run roughly northeast–southwest. In the far north, the highlands of Scandinavia and northwestern Britain formed about 400 million years ago above the lowland gneisses and granites of Karelia, Finland, and the Scandinavian peninsula. The broad highland belt that fans out westwards across central Europe from southern Poland to northern Spain and southern Ireland was formed from a deep-sea trough that was raised about 300 million years ago. Europe's most southerly mountains were formed when Africa moved northwards, enclosing the Mediterranean and raising the seabed to form the mountains of southern Spain, the Pyrenees, the Alps, the Carpathians, and the mountains of the Balkan region. The first ice age was 2–3 million years ago; Britain was cut off from mainland Europe by the rise in sea level when the ice melted at the end of the last ice age 8000–10,000 years ago.

The oldest rocks in Europe – Precambrian rocks more than 1000 million years old – are those of the Baltic Shield. This forms the northern border of the continent and passes southeast through Sweden, Finland, and Russia, below a cover of flat-lying Phanerozoic rocks forming the Russian platform. Precambrian rocks reappear to the south of the Russian platform in the Ukraine area. All these areas taken together form the European *craton, an area that has remained stable throughout Phanerozoic time.

Part of another ancient stable shield area is seen in northwestern Scotland; the Precambrian rocks here are a small

fragment of a much larger shield area disrupted by continental drift. The other, larger fragments are seen in central and western Greenland.

Another major component in the geology of Europe is the younger Caledonian system of orogenic belts (see *orogeny), developed over the period 900–400 million years ago. This system extends from western Ireland through Britain and Scandinavia to northern Norway; another branch of it runs through central Europe, but here it is largely obliterated by the younger Hercynean or Variscan system of orogenic belts, developed 400–280 million years ago. These run west–east through southern and central Europe. In its northern part this belt is in turn overlaid by younger sedimentary rocks, and in its southern part the more recent Alpine orogenic belt has deformed and metamorphosed the pre-existing rocks. The Alpine orogenic belt, formed between 300 million years ago and the present day, is itself an intricate system of fold-mountain belts running roughly west–east across southern Europe, the Mediterranean region, and North Africa.

The eastern limit of the European continent in terms of its geology can be taken as the Ural Mountains, the site of a north–south orogenic belt which became stabilized at the end of the Palaeozoic.

The Baltic Shield occupies most of Sweden and Finland, southern Norway, and Karelia, an area of some 500,000 sq km. The oldest rocks, gneisses more than 3000 million years old, are found in the Kola Peninsula and in the Ukraine. They occur as discrete masses within a larger area of rocks of middle Precambrian age in the northeastern part of the shield. The bulk of the central and western part of the shield is composed of rocks belonging to the Sveccofennid group;

these are metamorphosed sediments and volcanics, the remnants of a well-defined mobile belt active in late Precambrian times. This belt became stabilized around 1700 million years ago; from that time on the greater part of the shield remained as a stable area. A belt of younger rocks occurs in southern Norway and Sweden; these are the remains of the **Grenville mobile belts**, again of Precambrian age, consisting of gneisses and high-grade metamorphic rocks, together with granites and anorthosites.

The Caledonian mobile belts were initiated in Precambrian times and continued their activity well into the Phanerozoic. Beginning around 800 million years ago, thick accumulations of sediment were laid down in a number of basins of deposition extending from Ireland through Scotland, England, and Wales and across to Sweden and Norway. These sediments range in age from Precambrian to Silurian, reaching thicknesses of up to 15 km. To the east and west of the main sedimentary basins, thinner successions of Lower Palaeozoic rocks were deposited on the stable continental platform areas; the base of these deposits is marked by a widespread marine transgression at the base of the Cambrian. Episodes of orogenic folding and metamorphism began in the Late Precambrian in the earlier-formed basins of deposition, and continued over a long period until the Silurian. In the later-formed basins of deposition, the metamorphism and folding were generally less marked in character. The final stages of the orogenic cycle were marked in the Caledonian belts by regional metamorphism and uplift, forming a new mountain chain in Silurian times. During this final metamorphism, granites were emplaced at depth in the orogenic belt, while on the surface

the thick deposits of the **Old Red Sandstone** were laid down in a desert environment as the rapidly rising mountain chain was eroded.

The metamorphic rocks of the Scottish Highlands represent the deposits originally laid down in one of the early basins of deposition.

The Moine series (7 km thick) and the Dalradian series (8 km thick) form a thick pile of metamorphosed sandstones, turbidites, and volcanics. In the extreme northwest a different series of sediments were laid down on the stable foreland bordering the mobile belt; these are the shallow-water sandstones of the Torridonian series (about 4 km thick), and these are succeeded by Cambro-Ordovician quartzites, shales, and limestones.

The remains of other early Caledonian basins of deposition are seen as fragmented remnants further to the south, underlying the later Caledonian basins of southeastern Ireland, Wales, and the Welsh Borders. In Scandinavia, early basins of deposition occur immediately alongside the eastern foreland of the Baltic Shield. The succession on the foreland ranges from Cambrian to Silurian in age. Within the early basins, the lowest unit is the Sparagmite group, sandstones of Precambrian age, which are followed by Lower Palaeozoic sediments of considerable thickness in the centre of the mobile belt.

In a number of areas in and around the Caledonian belts there are deposits from an ancient glacial period in the late Precambrian. The later basins of deposition in the Caledonian belts are a number of relatively narrow basins which formed in or alongside the early basins. They include the Southern Uplands basin of Scotland and northeastern Ireland, the Mayo basin of western Ireland, the Lake District basin, and the Welsh basin. All the

Upper Palaeozoic rocks that accumulated in these basins contain thick turbidites and volcanics.

The Hercynean mobile belts represent parts of the earlier Caledonian system which remained active throughout Upper Palaeozoic times. In Western Europe the west–east Hercynean belt diverges sharply in trend from the northeast–southwest Caledonian belt, to run as a broad belt along the southern edge of a stable area composed of the Baltic Shield, the Russian platform, and the Old Caledonides forming the Old Red Sandstone continent.

In much of Europe the Hercynean rocks have been fragmented and disturbed either by the later Alpine activity or by later sedimentary cover. In the centre of the belt, however, pre-Hercynean crystalline rocks are preserved. To the south is a Mediterranean *facies of deep-water Palaeozoic rocks, while to the north lies the broad tract of Devonian and Carboniferous geosynclinal deposits in the Rhine and Ardennes. These are a largely detrital facies, often containing volcanics. The Hercynian cycle proper includes rocks ranging in age from Devonian to Permian. The base of the Devonian is marked by a marine transgression which initiated the geosynclinal phase of deposition.

In Lower Carboniferous times there was a further transgression of the sea onto the Old Red Sandstone continent, while to the south, sandy and shaly culm sediments were laid down in the geosyncline. At the end of the Lower Carboniferous there was a phase of deformation and metamorphism leading to the formation of a new mountain range known as the Hercynides, running east–west through central Europe. To the north, in Upper Carboniferous times, sediments accumulated on the foreland

and in a narrow *foredeep zone between the foreland and the Hercynides. In the last stages of the Carboniferous, coal was deposited in narrow basins in the foredeep and also within the new mountain range to the south. It is the thick deposits of the foredeep that provide the economically important coalfields of southern Wales, northern France, the Namur basin in Belgium, and the Ruhr in Westphalia, Germany. To the northwest lie further coal-bearing rocks laid down in basins on the foreland.

The final stage of the Hercynean activity is the deposition of continental Permian and Triassic sediments, eroded from the newly risen Hercynean mountain chain. These form the **New Red Sandstone**. Granites were again emplaced at the end of the Hercynean phase in late Carboniferous times. In Western Europe they are associated with an important suite of ore minerals, including tin, tungsten, silver, copper, lead, and zinc.

The Ural belt, which runs north–south and continues north to Novaya Zemlya, is similar to the Hercynean belt in that it was a long-lived mobile belt which underwent several orogenic episodes and finally became stabilized in early Mesozoic times. It separated the Russian and Siberian platforms, and represents the site of a wide Palaeozoic ocean.

A wide range of economically important deposits such as iron, chromium, and aluminium has made it an important industrial region.

The Alpine mobile belt is the youngest orogenic belt in Europe, and was initiated in the early Mesozoic. It overlies the earlier Palaeozoic and late Precambrian mobile belts. The Alpine belt proper stretches, in a series of linked arcs, from the west and east Alps through the Carpathians and Balkanides in the north, and through the Dinarides and Hellenides to the southwest.

The Alpine cycle began with the advance of the Triassic seas over the old Hercynean complexes, and new basins of deposition began to be formed. In the Alpine region, 2–3 km of Triassic quartzites, limestones, and dolomites were laid down; to the north and south of this central region nonmarine sediments and evaporites were deposited.

By the end of the Triassic, marine conditions predominated throughout the whole belt. Deposition of limestones, dolomites, and marls continued through the Jurassic until mid-Cretaceous times. At this point the first major orogenic events began: marine sedimentation ceased in the central part of the belt, and as compression and uplift of the central zone occurred, thick deposits of flysch (unsorted turbidite deposits) were laid down outside the central orogenic zone along with Upper Cretaceous and Palaeogene sediments.

By Miocene times folding, metamorphism, and uplift had brought much of the belt above sea level to form a new mountain chain. As the mountain chain rose, thick deposits of molasse (unsorted continental-type sandstones and breccias) were laid down. There was intense deformation and regional metamorphism within the mobile belt, with the formation of huge nappes, recumbent folds of rock which were often detached from their original root zones and moved for long distances. Late orogenic igneous intrusions are found in the Carpathians and the Hellenides. The Apeninnes in Italy are a still-active branch of the Alpine mobile belt, containing very late orogenic and still-active volcanoes such as Vesuvius, Stromboli, and Etna.

The Alpine belt passes east into the Caucasus, Cyprus, and the Zagros range of Iran. In these areas the main orogenic activity occurred in late Jurassic and early Cretaceous times, and was accompanied by the intrusion of granodiorites and related volcanicity. Widespread uplift occurred in Miocene and Pliocene times.

The Mesozoic and Tertiary cover rocks on the **European craton** to the north of the Alpine mobile belt are generally less than 1 km thick, except in basins such as the North Sea. They consist of continental sediments at the base – Permo-Triassic sediments derived from the old Hercynean landmass. These were followed by various clays, limestones, ironstones, and greensands in Jurassic and Lower Cretaceous times. A widespread marine transgression in Upper Cretaceous times led to the deposition of the chalk over southern England and the northern part of continental Europe. At the same time the initial rifting of the Atlantic was beginning to the northwest, with the development first of narrow basins of deposition which were rapidly widened by continental rifting and *seafloor spreading. There was extensive volcanic activity associated with this rifting and the remains of a vast Brito-Arctic volcanic province can be seen in the Tertiary volcanic rocks of western Scotland and Ireland.

This phase was followed by gradual uplift of the craton, associated with the Alpine mountain-building movements to the south, and sedimentation became restricted to a number of marine and nonmarine basins. During the Pleistocene much of Britain, all of Fennoscandia, and the Alps were covered by the advancing polar ice sheets, and the landforms we now see in these areas are a direct result of this glaciation.

European Union EU, political and economic grouping, comprising 25 countries (in 2004). The six original members – Belgium, France, (West) Germany, Italy, Luxembourg, and the Netherlands – were joined by the United Kingdom, Denmark, and the Republic of Ireland in 1973, Greece in 1981, Spain and Portugal in 1986, and Austria, Finland, and Sweden in 1995. East Germany was incorporated on German reunification in 1990. In 2004 the EU was joined by Cyprus, the Czech Republic, Estonia, Hungary, Latvia, Lithuania, Malta, Poland, Slovakia and Slovenia. The *European Community (EC) preceded the EU, and comprised the *European Coal and Steel Community (set up by the 1951 Treaty of Paris), the *European Economic Community, and the *European Atomic Energy Community (both set up by the 1957 Treaties of Rome). The EU superseded the EC in 1993, following intergovernmental arrangements for a common foreign and security policy and for increased cooperation on justice and home affairs policy issues set up by the *Maastricht Treaty (1992). Other important agreements have been the *Single European Act (1986), the *Amsterdam Treaty (1997), and the Treaty of Nice (2000). The basic aims of these treaties have been the expansion of trade, the abolition of restrictive economic practices, the encouragement of free movement of capital and labour, and establishment of a closer union among European peoples.

eurozone   collective term for those European countries which have adopted the single currency of the European Union, the *euro. In 2004 the 12 countries were Austria, Belgium, Finland, France, Germany, Greece, Republic of Ireland, Italy, Luxembourg, the Netherlands, Portugal, and Spain.

eustatic change   global rise or fall in sea level caused by a change in the amount of water in the oceans (by contrast, isostatic adjustment involves a rising or sinking of the land, which

causes a local change in sea level). During the last ice age, global sea level was lower than today because water became 'locked-up' in the form of ice and snow, and less water reached the oceans.

eutrophication   excessive enrichment of rivers, lakes, and shallow sea areas, primarily by municipal sewage, by sewage itself, and by the nitrates and phosphates from fertilizers used in agriculture. These encourage the growth of algae and bacteria which use up the oxygen in the water, making it uninhabitable for fish and other animal life. In this way eutrophication is responsible for a particular type of *water pollution.

The dissolved fertilizers cause the rapid growth of water plants, especially green algae, which can clog up waterways and prevent light reaching plants below the surface. As the algae die, aerobic bacteria bring about decay, using up oxygen in the water as they do so. Anaerobic bacteria take over and convert part of the dead matter into smelly decay products. Slowly a layer of dead plant material builds up on the bottom of the lake or river. Deprived of oxygen, fish and other life forms die and decay, leaving putrid, poisonous water.

evacuation   removal of civilian inhabitants from an area liable to aerial bombing or other hazards (such as the aftermath of an environmental disaster) to safer surroundings. The term is also applied to military evacuation. People who have been evacuated are known as evacuees.

evaporation   the process whereby a substance changes from a liquid to a vapour. Heat from the sun evaporates water from seas, lakes, rivers, etc., and this process produces water vapour in the *atmosphere.

evaporite   sedimentary deposit precipitated on evaporation of salt water.

With a progressive evaporation of seawater, the most common salts are deposited in a definite sequence: calcite (calcium carbonate), gypsum (hydrous calcium sulphate), halite (sodium chloride), and finally salts of potassium and magnesium.

evapotranspiration the return of water vapour to the *atmosphere by evaporation from land and water surfaces and the *transpiration of vegetation.

evergreen a vegetation type in which leaves are continuously present. Compare *deciduous woodland.

evolution slow gradual process of change from one form to another, as in the evolution of the universe from its formation to its present state, or in the evolution of life on Earth. In biology, it is the process by which life has developed by stages from single-celled organisms into the multiplicity of animal and plant life, extinct and existing, that inhabits the Earth. The development of the concept of evolution is usually associated with the English naturalist Charles Darwin (1809–1882) who attributed the main role in evolutionary change to *natural selection acting on randomly occurring variations. These variations in species are now known to be adaptations produced by spontaneous changes or mutations in the genetic material of organisms. In short, evolution is the change in the genetic makeup of a population of organisms from one generation to another. Evidence shows that many species of organisms do not stay the same over generations. The most dramatic evidence of this comes from fossils.

Evolution occurs via the following processes of natural selection: individual organisms within a particular species may show a wide range of variation because of differences in their genes; predation, disease, and competition cause individuals to die; individuals with characteristics most suited to the environment are more likely to survive and breed successfully; and the genes that have enabled these individuals to survive are then passed on to the next generation, and if the environment is changing, the result is that some genes are more abundant in the next generation and the organism has evolved.

Evolutionary change can be slow, as shown in part of the fossil record. However, it can be quite fast. If a population is reduced to a very small number, evolutionary changes can be seen over a few generations. Because micro-organisms have very short life cycles, evolutionary change in micro-organisms can be rapid. Micro-organisms can evolve resistance to a new antibiotic only a few years after the drug is first used. As a result of evolution from common ancestors, we are able to use classification of organisms to suggest evolutionary origins.

evolutionary toxicology study of the effects of pollution on evolution. A polluted habitat may cause organisms to select for certain traits, as in **industrial melanism** for example, where some insects, such as the peppered moth, evolve to be darker in polluted areas, and therefore better camouflaged against predation.

exchange rate price at which one currency is bought or sold in terms of other currencies, gold, or accounting units such as the special drawing right (SDR) of the *International Monetary Fund. Exchange rates may be fixed by international agreement or by government policy; or they may be wholly or partly allowed to 'float' (that is, find their own level) in world currency markets.

exchange rate policy policy of government towards the level of the *exchange rate of its currency. It may want to influence the exchange rate by using its gold and foreign currency reserves held by its central bank to buy and sell its currency. It can also use interest rates (monetary policy) to alter the value of the currency.

exfoliation or **onion-skin weathering**, type of *physical weathering in which the outer layers of a rock surface peel off in flakes and shells. It is caused by the rapid expansion and contraction of the rock surface when subjected to extreme changes of temperature. Exfoliation occurs particularly in hot dry desert climates, and on sheets of rock that are jointed parallel to the surface, such as *granite. The process is encouraged by the effects of *chemical weathering on the inner edges of the joints. Exfoliation contributes to the *rock cycle, the formation, change, and reformation of the Earth's outer layers.

The combination of exfoliation and chemical weathering produces rounded *inselbergs (prominent steep-sided hills) in arid regions, such as Ayers Rock in central Australia.

exfoliation in earth science, the successive separation of thin shells of rock from the surface of a massive rock. It is caused by day–night temperature variations or the decrease in compressional forces that accompanies exposure of previous covered rock.

exosphere uppermost layer of the *atmosphere. It is an ill-defined zone above the thermosphere, beginning at about 700 km and fading off into the vacuum of space. The gases are extremely thin, with hydrogen as the main constituent.

expanded town an existing town where growth is planned in order to accommodate newcomers. Funding will have come from local and national government. Expanded towns are favoured as being less costly than *new towns. Examples in England include Swindon and Basingstoke.

experimental petrology field of geology that involves creating or

altering rocks and minerals in the laboratory in order to understand rock-forming processes that are impossible to observe, either because they happen out of the physical range of human observation (for example 100 km deep), or because they happen too slowly.

exponential growth a rapid form of growth, for example where a population doubles in every unit of time: 2, 4, 8, 16, 32. This can be contrasted with **arithmetic growth**, for example 2, 4, 6, 8, 10 where the growth is the same in every unit of time.

export goods or service produced in one country and sold to another. Exports may be visible (goods such as cars physically exported) or invisible (services such as banking and tourism, that are provided in the exporting country but paid for by residents of another country).

export credit loan, finance, or guarantee provided by a government or a financial institution enabling companies to export goods and services in situations where payment for them may be delayed or subject to risk.

exposed coalfield see *concealed coalfield.

extensive farming a system of *agriculture in which relatively small amounts of capital or labour investment are applied to relatively large areas of land. For example, sheep ranching is an extensive form of farming, and yields per unit area are low. Extensive farming usually occurs at the *margin* of the agricultural system – at a great distance from the market, or on poor land of limited potential. See also *Von Thünen theory.

external economies of scale fall in the average cost of production for a firm arising from factors that affect an economy, geographical region, or industry as a whole, including general technological advances and the development of industrial clusters. For example, growth in an industry might lead to an increase in the number of component supply firms for that industry, which might lead to a fall in the price of components.

external processes landscape-forming processes such as *weathering and *erosion, in contrast to *internal processes.

externality the difference between the *social cost or benefit of an economic activity and its private cost or benefit. For example, if a company pollutes the atmosphere but pays nothing to the government or the community for this, then the pollution becomes an externality.

extinction in biology, the complete disappearance of a species from the planet. Extinctions occur when a species becomes unfit for survival in its natural habitat usually to be replaced by another, better-suited species. An organism becomes ill-suited for survival because its environment is changed or because its relationship to other organisms is altered. For example, a predator's fitness for survival depends upon the availability of its prey.

past extinctions Mass extinctions are episodes during which large numbers of species have become extinct virtually simultaneously in the distant past, the best known being that of the dinosaurs, other large reptiles, and various marine invertebrates about 65 million years ago between the end of the *Cretaceous period and the beginning of the Tertiary period, the latter known as the **K–T extinction**.

There have been several others in the more distant past. There is disagreement about the causes, but one of several major catastrophes have been blamed, including meteorite impact, volcanic eruption, massive lava flows, and significant global warming. Another mass extinction occurred about 10,000 years ago when many giant species of mammal died out. This is known as the 'Pleistocene overkill' because their disappearance was probably hastened by the hunting activities of prehistoric humans. The greatest mass extinction occurred about 250 million years ago, marking the Permian–Triassic boundary (see *geological time), when up to 96% of all living species became extinct.

extinctions in British Isles The last mouse-eared bat *Myotis myotis* in the UK died in 1990. This was the first mammal to have become extinct in the UK for 250 years, since the last wolf was exterminated.

In 2000, Chinese palaeontologists examined fossils of marine species from the mass extinction at the end of the Permian period, and determined that the extinction took place over 160,000 years. What caused the mass extinction remains undecided, though most earth scientists believe it probably came about through increased volcanic activity or a meteor strike. Mass extinctions apparently occur at periodic intervals of approximately 26 million years.

current extinctions In the past 100 years species have become extinct at a rate as high as any thought to have occurred in the past. Where it is known, the causes of extinction in recent times are generally linked to human activities. This is either due to direct killing or, more often, due to loss of *habitat. For example, it is believed that many unknown species are becoming extinct due to the loss of rainforests around the world. The number being lost is not known, but it may be large. The extinction of the flightless bird of the island of Mauritius, the dodo, was due to it being killed for food. It became extinct around 1681. The moas of New Zealand, and the passenger pigeon of North America were also exterminated by hunting. Australia has the worst record for extinction: 18 mammals have disappeared since Europeans settled there, and 40 more are threatened. *Endangered species are close to extinction.

extreme climate   a climate that is characterized by large ranges of temperature and sometimes of rainfall. In central Asia for example, hot summers (average 21°C) alternate with very cold winters (average –45°C). Thus the average temperature range is normally more than 60°C. Most rainfall occurs in summer. Compare *temperate climate, *maritime climate.

extrusive rock   or **volcanic rock**, *igneous rock formed on the surface of the Earth by volcanic activity (as opposed to intrusive, or plutonic, rocks that solidify below the Earth's surface). Magma (molten rock) erupted from volcanoes cools and solidifies quickly on the surface. The crystals that form do not have time to grow very large, so most extrusive rocks are finely grained. The term includes fine-grained crystalline or glassy rocks formed from hot lava quenched at or near Earth's surface, and those made of welded fragments of ash and glass ejected into the air during a volcanic eruption. The formation of extrusive igneous rock is part of the *rock cycle.

Large amounts of extrusive rock called *basalt form at the Earth's *ocean ridges from lava that fills the void formed when two tectonic plates spread apart. Explosive volcanoes that deposit pyroclastics generally occur where one tectonic plate descends beneath another. *Andesite is often formed by explosive volcanoes. Magmas that give rise to pyroclastic extrusive rocks are explosive because they are viscous. The island of Montserrat, West Indies, is an example of an explosive volcano that spews pyroclastics of andesite composition. Magmas that produce crystalline or glassy volcanic rocks upon cooling are less viscous. The low viscosity allows the extruding lava to flow easily. Fluid-like lavas that flow from the volcanoes of the Hawaiian Islands have low viscosity and cool to form basalt.

facies body of rock strata possessing unifying characteristics usually indicative of the environment in which the rocks were formed. The term is also used to describe the environment of formation itself or unifying features of the rocks that comprise the facies.

Features that define a facies can include collections of fossils, sequences of rock layers, or the occurrence of specific minerals. Sedimentary rocks deposited at the same time, but representing different facies belong to a single **chronostratigraphic unit** (see *stratigraphy). But these same rocks may belong to different **lithostratigraphic units**. For example, beach sand is deposited at the same time that mud is deposited further offshore. The beach sand eventually turns to *sandstone while the mud turns to *shale. The resulting sandstone and shale **strata** comprise two different facies, one representing the beach environment and the other the offshore environment, formed at the same time; the sandstone and shale belong to the same chronostratigraphic unit but distinct lithostratigraphic units.

factoring lending money to a company on the security of money owed to that company; this is often done on the basis of collecting those debts. The lender is known as the factor. Factoring may also describe acting as a commission agent for the sale of goods.

factory fishing the use of highly mechanized and technical commercial fishing fleets to exploit deep-water fish stocks beyond the reach of smaller coastal fishing fleets. Shoals of fish are detected by sonar and may be netted, trawled or sucked on board by means of a giant pump before being transferred to a factory ship with special refrigeration equipment to be gutted and processed at sea. So efficient is factory fishing that *overfishing, that is the exhaustion of fish stocks, has taken place in many of the world's traditional fishing grounds, for example, the North Sea and the Atlantic Ocean.

factory ship or **mother ship**, any large purpose-built vessel which collects and processes its own and/or the accumulated catches of fish from auxiliary fishing vessels. The factory ship was a response to the need for fishing fleets to move further from their home waters as coastal fish stocks became depleted in the 1970s. As well as receiving the catches from the rest of the fleet, the factory ship serves as a supply point for the other ships, providing reserves of fuel, maintenance facilities, food, etc. The fish are processed on board the factory ship, frozen and stored for the duration of the voyage, which may last several weeks, before unloading at a suitable port.

factory system the basis of manufacturing in the modern world. In the factory system workers are employed at a place where they carry out specific tasks, which together result in a product. This is called the division of labour. Usually these workers will perform their tasks with the aid of machinery. Such mechanization is another feature of the factory system, which leads to mass production.

Fahrenheit scale temperature scale invented in 1714 by Polish-born Dutch physicist Gabriel Fahrenheit (1686–1736) that was commonly used in English-speaking countries until the 1970s, after which the *Celsius scale was generally adopted, in line with the rest of the world. In the Fahrenheit scale, intervals are measured in degrees (°F); $°F = (°C \times 9/5) + 32$.

fair trade a way of conducting international trade so that all parties involved receive fair payments. Fair trade schemes are designed to counteract unjust international trade systems, which often exploit workers in developing countries. By ensuring that producers are paid a fair price for their skills and products, they are able to develop their businesses.

Fair trade is often involved in the sales of crafts and textiles, or food items such as tea, coffee, honey, and chocolate, the traditional products of developing countries. Fair trade also promotes sustainable development by improving market access for disadvantaged producers. The **International Federation for Alternative Trade**, a global network of fair trade organizations, and the **European Fair Trade Association**, an alliance of fair trade organizations in European countries, are two organizations which bring together people to promote worldwide fair trade.

fallow any agricultural land left uncropped for one or more seasons in order to restore soil fertility. In developed countries the land may be ploughed and harrowed several times to aerate the soil, to destroy perennial weeds by desiccation, and to quicken the decomposition of crop residues which beneficially increases the nitrogen content of the soil. In tropical farming systems, such as *shifting cultivation, the land is abandoned to recolonization by natural vegetation for a period of up to 30 years before being brought back into cultivation. During the agricultural revolution of the 18th and 19th centuries in western Europe the need for bare fallows was reduced by the introduction of crop rotations. Between 1960 and 1982 the practise of fallowing land in Europe was generally discontinued as greater use of chemical fertilizers permitted the continuous cropping of the land.

fallow system a farming practice in which a period of cropping is followed by a period when the land is fallow. Fallow systems are characteristic of tropical regions and are made necessary by the rapid leaching of nutrients from the soil as soon as cultivation occurs. Different farming systems can be compared on the basis of the ratio of fallow years to crop years. Thus, *shifting cultivation will have a fallow: cropland ratio of more than 10:1.

false colour image see *satellite image.

famine severe shortage of food affecting a large number of people. A report made by the United Nations (UN) Food and Agriculture Organization (FAO), published in October 1999, showed that although the number of people in the developing world without sufficient food declined by 40 million during the first half of the 1990s, there were still, in 1999, 790 million hungry people in poor countries and 34 million in richer ones. The food availability deficit (FAD) theory explains famines as being caused by insufficient food supplies. A more recent theory is that famines arise when one group in a society loses its opportunity to exchange its labour or possessions for food.

Most Western famine-relief agencies, such as the International Red Cross, set out to supply food or to increase its local production, rather than becoming involved in local politics. The FAD theory was challenged in the 1980s. Crop failures do not inevitably lead to famine; nor is it always the case that adequate food supplies are not available nearby. In 1990, for example, the Ethiopian air force bombed grain depots in a rebel-held area.

In sub-Saharan Africa, a third of the population remained undernourished in 1999. India and China were reported by the UN to have the largest amount of ill-fed people: 204 million and 164 million people respectively. However, from 1979–1997, India reduced the proportion of those undernourished from 38–22%, and China from 30–13%. Rates of undernourishment are highest in North Korea, Mongolia, and Central Africa, at nearly 50% of the population.

farming system in agriculture, a way of organizing resources depending on **inputs** such as nature of the land, labour, government subsidies, climate, knowledge and abilities of the farmer, and other factors; **processes** such as ploughing and harvesting; and **outputs** such as milk, eggs, and straw. Different types of farming system include intensive agriculture and extensive agriculture.

The system model can be applied to all types of farming. Different types and patterns of farming result from variations in **inputs** – physical, cultural, economic, and behavioural factors. **Outputs** constitute the end product, the cost of producing which, in relation to its value, will determine the profit margin.

fast moving consumer goods FMCG, branded goods that have a quick turnover. Fast moving consumer goods include everyday products found in supermarkets such as cereals, confectionery, washing powder, and biscuits.

fault fracture in the Earth's crust, on either side of which rocks have moved past each other. Faults may occur where rocks are being pushed together (compression) or pulled apart (tension) by *plate tectonics, movements of the *plates of the Earth's crust. When large forces build up quickly in rocks, they become brittle and break; *folds result from a more gradual compression. Faults involve displacements, or offsets, ranging from the microscopic scale to hundreds of kilometres. Large offsets along a fault are the result of the accumulation of many small movements (metres or less) over long periods of time. Large movements cause detectable *earthquakes, such as those experienced along the *San Andreas Fault in California, USA.

Faults produce lines of weakness on the Earth's surface (along their strike) that are often exploited by processes of *weathering and *erosion. Coastal caves and geos (narrow inlets) often form along faults and, on a larger scale, rivers may follow the line of a fault. The Great Glen Fault in Scotland is an excellent example of a fault, and Loch Ness on this fault is an example of a fault-line feature.

Faults are planar features. The orientation of a fault is described by the inclination of the fault plane with respect to the horizontal (an angle known as *dip) and its direction in the horizontal plane (the *strike). Faults at a high angle with respect to the horizontal (in which the fault plane is steep) are classified as either **normal faults**, where the hanging wall (the body of rock above the fault) has moved down relative to the footwall (the body of rock below the fault), or **reverse faults**, where the hanging wall has moved up relative to the footwall. Normal faults occur where rocks on either side have moved apart. Reverse faults occur where rocks on either side

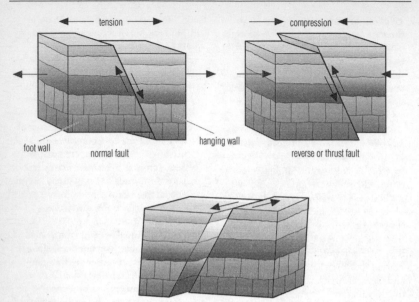

tension

compression

foot wall

hanging wall

normal fault

reverse or thrust fault

strike-slip or
transverse fault

fault Faults are caused by the movement of rock layers, producing such features as block mountains and rift valleys. A normal fault is caused by a tension or stretching force acting in the rock layers. A reverse fault is caused by compression forces. Faults can continue to move for thousands or millions of years.

have been forced together. A reverse fault that forms a low angle with the horizontal plane is called a **thrust fault**.

A **lateral fault**, or **strike-slip fault**, occurs where the relative movement along the fault plane is sideways. A **transform fault** is a major strike-slip fault along a plate boundary, that joins two other plate boundaries – two spreading centres, two subduction zones, or one spreading centre and one subduction zone. The San Andreas Fault is a transform fault.

fault gouge soft, uncemented, pulverized clay-like material found between the walls of a fault. It is created by grinding action (cataclasis) of the fault and subsequent chemical alteration of tiny mineral grains brought about by fluids that flow along the fault.

favela Brazilian makeshift housing or *shanty town.

federalism system of government in which two or more separate states unite into a federation under a common central government. A federation should be distinguished from a **confederation**, a looser union of states for mutual assistance. The USA is an example of federal government.

feldspar any of a group of *silicate minerals. Feldspars are the most abundant mineral type in the Earth's crust. They are the chief constituents of *igneous rock and are present in most metamorphic and sedimentary rocks. All feldspars contain silicon, aluminium, and oxygen, linked together to form a framework. Spaces within this framework structure are occupied by sodium, potassium, calcium, or occasionally barium, in various proportions. Feldspars form white, grey, or pink crystals and rank 6 on the *Mohs scale of hardness.

feldspathoid any of a group of silicate minerals resembling feldspars but containing less silica. Examples are nepheline ($NaAlSiO_4$ with a little potassium) and leucite ($KAlSi_2O_6$). Feldspathoids occur in igneous rocks that have relatively high proportions of sodium and potassium. Such rocks may also contain alkali feldspar, but they do not generally contain quartz because any free silica would have combined with the feldspathoid to produce more feldspar instead.

fell upland rough grazing in a *hill farming system, for example in the English Lake District. Fell land is sometimes *common land, i.e. not in the ownership of a single individual or institution. Sheep are grazed on the fells in the summer months and brought down to lower pasture during the winter.

felsic rock a *plutonic rock composed chiefly of light-coloured minerals, such as quartz, feldspar, and mica. It is derived from feldspar, lenad (meaning feldspathoid), and silica. The term felsic also applies to light-coloured minerals as a group, especially quartz, feldspar, and feldspathoids.

fen a distinctive marshland community found in the transitional area between fresh water and land. Fens are rapidly evolving communities, being dependent upon the rate at which silting and peat growth occur. Fens differ from bogs in that the former are nutrient-rich and have an alkaline soil (pH greater than 7.0). Fenland occupies upper parts of old estuaries as well as the edges of lakes. It is a particularly well developed feature of East Anglia, England, where the rivers drain from inland, chalk-rich rocks. Much fenland

has been reclaimed to be used as highly productive agriculture land.

fertilizer see *artificial fertilizer, *organic fertilizer.

fetch distance of open water over which wind can blow to create *waves. The greater the fetch the more potential power waves have when they hit the coast. In the south and west of England the fetch stretches for several thousand kilometres, all the way to South America. This combines with the southwesterly *prevailing winds to cause powerful waves and serious *coastal erosion along south- and west-facing coastlines.

field mapping mapping rocks, rock formations, and geological units while on the ground (as opposed to remote mapping, from aeroplane or satellite). Field mapping is the basis for understanding the structure of the Earth's crust and its geological history.

field studies study of ecology, geography, geology, history, archaeology, and allied subjects, in the natural environment as opposed to the laboratory.

The Council for the Promotion of Field Studies was established in Britain in 1943, in order to promote a wider knowledge and understanding of the natural environment among the public; Flatford Mill, Suffolk, was the first of its research centres to be opened.

fieldwork in anthropology, the gathering and analysis of first-hand information through direct observation of a society or social group.

It was developed by William Rivers and Bronislaw Malinowski as a more scientific method of research, but its capacity for comprehensiveness and objectivity has been questioned.

finance, sources of sources from which business organizations obtain money to start up and pay for the business, and for new capital goods. The most important source of finance in the UK is **internal finance**, the retained profit of companies. Next in importance are bank *loans and overdrafts.

Hire purchase, leasing, *factoring, and mortgages are other ways in which a company can effectively borrow money.

Sometimes, companies can obtain grants from government; for example, for investing in new technology or for setting up in a high-unemployment area of the country.

financial institution business organization that has as its core business activity the management of money. Banks, building societies, and insurance and assurance companies are all examples of financial institutions.

fiord alternative spelling of *fjord.

fire (in agriculture) the deliberate burning of natural vegetation and crop residues for agricultural management purposes. Fire was one of the first and most effective tools employed in early agriculture, and its use remains widespread throughout the world.

In tropical fallow systems fire is part of the slash-and-burn method of land clearance, and has been subject to much debate on its contribution to soil fertility. The use of fire represents a convenient labour-saving device in initial plot preparation and in the destruction of weeds and grass seeds, reducing the need for later weeding. The burning of the bulk organic matter enriches the soil with an ash that contains greater concentrations of available nutrients, especially potassium and phosphorus, and ensures a rapid release of mineralized nitrogen when the first rains fall. Heating of the soil surface also reduces the numbers of insect pests. On the other hand, burning destroys humus, and many nutrients are lost to the atmosphere. Burning offers the tropical farmer a means of regulating soil conditions so as to obtain an adequate crop yield, at least in the first year of cultivation; however, each time this expedient is used the means of building up long-term fertility is destroyed.

fire clay a *clay with refractory characteristics (resistant to high temperatures), and hence suitable for lining furnaces (firebrick). Its chemical composition consists of a high percentage of silicon and aluminium oxides, and a low percentage of the oxides of sodium, potassium, iron, and calcium.

firewood the principal fuel for some 2 billion people, mainly in the developing world. In principle a renewable energy source, firewood is being cut far faster than the trees can regenerate in many areas of Africa and Asia, leading to *deforestation.

In Mali, for example, wood provides 97% of total energy consumption, and deforestation is running at an estimated 9000 hectares a year. The heat efficiency of firewood can be increased by use of well-designed stoves, but for many people they are either unaffordable or unavailable. With wood for fuel becoming scarcer the UN Food and Agricultural Organization estimated that in the year 2000 3 billion people worldwide faced chronic problems in getting food cooked.

firn or **névé**, snow that has lain on the ground for a full calendar year. Firn is common at the tops of mountains; for example, the Alps in Europe. After many years, compaction turns firn into ice and a *glacier may form.

First World former term for the industrialized *capitalist, or free-market, countries of the West. It was used during the Cold War, along with the classifications Second World (industrialized communist countries) and Third World (non-aligned, developing countries of Africa, Asia, and Latin America). The terms originally denoted political alignment, but later took on economic connotations that are now considered derogatory.

fiscal policy that part of government policy concerning taxation and other revenues, *public spending, and government borrowing (the public sector borrowing requirement).

fish farming a form of aquaculture in which inland waters and marine environments are utilized to breed and rear fish for commercial purposes. Fish farming differs from fishing in the greater control over the raising and harvesting of fish stocks, in freeing the fisherman from the uncertainties of hunting, and by producing much higher yields per unit of area.

Ponds or temporarily flooded areas, including rice paddies, have been fish-farmed for several thousands of years in parts of eastern Asia. Complex polycultures have evolved enabling several species to be raised through exploiting different ecological niches within otherwise limited pond areas. Freshwater fish farming makes a significant contribution to the total fisheries harvest in Asian countries such as China, India and Indonesia. The Japanese pioneered marine fish farming or *mariculture* at the turn of the last century; elsewhere in the world, fish farming has developed significantly only in the last 40 years. In Africa, most progress has been made in rearing the Chinese carp and the Nile tilapia in freshwater ponds. In developed countries, the major advances have been in rearing freshwater trout in the entirely controlled environment of tanks, and salmon in cages lowered in coastal waters. Artificial fertilization has allowed fry of trout, and a species of catfish in Africa, to be raised in hatcheries before being transferred to larger ponds or enclosures. In several

tropical countries where irrigation and navigation channels have become blocked by aquatic vegetation, plant-eating species such as carp have been introduced both to clear vegetation and to provide fish stocks for human consumption.

fishing the harvesting of fish and shellfish from oceans and inland waters to provide food, industrial raw materials and as a form of recreation. Activities such as sealing, whaling and the gathering of marine plants may be separately categorized. More than three-quarters of the world catch is harvested in the northern hemisphere, and is heavily concentrated on coastal waters and the continental shelf. Only about 12% of the catch is from inland waters. The most heavily fished marine area is the North-West Pacific. The global catch has risen gradually, but individual fisheries have shown more dramatic growth rates, sometimes doubling each year as a particular species is exploited, for example, the Peruvian anchovy catch rose from less than 100 tonnes p.a. in the 1940s to 10 million tonnes in the late 1960s, before collapsing under the pressure of over-fishing. Although some species appear quite resilient to heavy exploitation, over-fishing remains a serious problem.

fission reactor the most common type of nuclear reactor in which atoms such as uranium-235 and plutonium-239 are split by neutrons thus releasing energy mainly in the form of heat. This process occurs in the reactor as a nuclear fission chain reaction. The heat generated by the process is used to raise steam which in turn drives a turbine and generates electricity.

five-year plan long-term strategic plan for the development of a country's economy. Five-year plans were from 1928 the basis of economic planning in the USSR, aimed particularly at developing heavy and light industry in a primarily agricultural country. They have since been adopted by many other countries.

fixed asset possession or valuable of a business organization that is used over a long period of time. Fixed assets include: tangible assets, such as plant and machinery, land, company cars, office equipment, and buildings; intangible assets, such as trademarks, brands, and the goodwill of the business; and investments in other companies. Over a period of time the value of fixed assets can be written off against profits by depreciating the book value of the asset in the accounts.

fixed cost or **overhead cost**. Cost which does not vary directly with output (not a variable or *direct cost), but remains constant as output increases. For example, a company may increase its output by one third; variable costs will increase in proportion with this but fixed costs will stay the same.

fjord or **fiord**, narrow sea inlet enclosed by high cliffs. Fjords are found in Norway, New Zealand, Southern Chile, and western parts of Scotland. They are formed when an overdeepened U-shaped glacial valley is drowned by a rise in sea-level. At the mouth of the fjord there is a characteristic lip causing a shallowing of the water. This is due to reduced glacial erosion and the deposition of *moraine at this point.

Fiordland is the deeply indented southwest coast of the South Island, New Zealand; one of the most beautiful inlets is Milford Sound.

flag piece of cloth used as an emblem or symbol for nationalistic, religious, or military displays, or as a means of signalling. Flags originated from the representations of animals and other objects used by ancient peoples. Many localities and public bodies, as well as shipping lines, schools, and yacht clubs, have their own distinguishing flags.

flash flood flood of water in a normally arid area brought on by a sudden downpour of rain. Flash floods are rare and usually occur in mountainous areas. They may travel many kilometres from the site of the rainfall.

Because of the suddenness of flash floods, little warning can be given of their occurrence. In 1972 a flash flood at Rapid City, South Dakota, USA, killed 238 people along Rapid Creek.

flexible manufacturing system FMS, manufacturing system that integrates robotics, automated guided vehicles, and CAM (computer-aided manufacturing). The system is totally automated and allows unattended production of parts.

flint compact, hard, brittle mineral (a variety of chert), brown, black, or grey in colour, found as nodules in limestone or shale deposits. It consists of cryptocrystalline (grains too small to be visible even under a light microscope) *silica, SiO_2, principally in the crystalline form of *quartz. Implements fashioned from flint were widely used in prehistory.

flocculation in soils, the artificially induced coupling together of particles to improve aeration and drainage. Clay soils, which have very tiny particles and are difficult to work, are often treated in this way. The method involves adding more lime to the soil.

flooding the inundation of land that is not normally covered with water. Flooding from rivers commonly takes place after heavy rainfall or in the spring after winter snows have melted. The river's *discharge (volume of water carried in a given period) becomes too great, and water spills over the banks onto the surrounding flood plain. Small floods may happen once a year – these are called **annual**

floods and are said to have a one-year return period. Much larger floods may occur on average only once every 50 years, or once every 100 years.

Flooding is least likely to occur in an efficient channel that is semicircular in shape (see *channel efficiency). Flooding can also occur at the coast in stormy conditions (see *storm surge) or when there is an exceptionally high tide. The Thames Flood Barrier was constructed in 1982 to prevent the flooding of London from the sea.

flood plain area of periodic flooding along the course of a river valley. When river discharge exceeds the capacity of the channel, water rises over the channel banks and floods the surrounding low-lying lands. As water spills out of the channel some *alluvium (silty material) will be deposited on the banks to form *levees (raised river banks). This water will slowly seep onto the flood plain, depositing a new layer of rich fertile alluvium as it does so. Many important flood plains, such as the inner Nile delta in Egypt, are in arid areas where their exceptional productivity is very important to the local economy.

A flood plain is a natural feature, flooded at regular intervals. By plotting floods that have occurred we can speak of the size of flood we would expect once every 10 years, 100 years, 500 years, and so on.

Even the most energetic flood-control plans (such as dams, dredging, and channel modification) sometimes fail, and if towns and villages are built on a flood plain there is always some risk. It is wiser to use flood plains in ways that are compatible with flooding, such as for agriculture or parks.

Flood-plain features include *meanders and *oxbow lakes.

flowering plant term generally used for angiosperms, which bear flowers with various parts, including sepals, petals, stamens, and carpels. Sometimes the term is used more broadly, to include both angiosperms and gymnosperms, in which case the cones of conifers and cycads are referred to as 'flowers'. Usually, however, the angiosperms and gymnosperms are referred to collectively as seed plants, or spermatophytes.

flow production system of organizing production so that goods are produced continuously, as on a production line, as opposed to **batch production** where goods are manufactured in batches, and **job production** where goods are made specially to order. Flow production tends to be used in capital-intensive industries where there are significant *economies of scale to be gained from production line methods of manufacturing.

flow line a diagram showing volumes of movement, e.g. of people, goods or information between places. The width of the flow line is proportional to the amount of movement, for example in portraying commuter flows into an urban centre from surrounding towns and villages:

flue-gas desulphurization process of removing harmful sulphur pollution from gases emerging from a boiler. Sulphur compounds such as sulphur dioxide are commonly produced by burning *fossil fuels, especially coal in power stations, and are the main cause of *acid rain.

fluidization making a mass of solid particles act as a fluid by agitation or by gas passing through. Much earthquake damage is attributed to fluidization of surface soils during the earthquake shock. Another example is ash flow during volcanic eruption.

fluorite or **fluorspar**, glassy, brittle halide mineral, calcium fluoride CaF_2, forming cubes and octahedra; colourless when pure, otherwise violet, blue, yellow, brown, or green.

Fluorite is used as a flux in iron and steel making; colourless fluorite is used in the manufacture of microscope lenses. It is also used for the glaze on pottery, and as a source of fluorine in the manufacture of hydrofluoric acid.

fluorocarbon compound formed by replacing the hydrogen atoms of a hydrocarbon with fluorine. Fluorocarbons are used as inert coatings, refrigerants, synthetic resins, and as propellants in aerosols.

There is concern that the release of fluorocarbons, particularly those containing chlorine (*chlorofluoro-carbons, CFCs), depletes the *ozone layer, allowing more ultraviolet light from the Sun to penetrate the Earth's atmosphere, and increasing the incidence of skin cancer in humans.

fluorspar another name for the mineral *fluorite.

fluvial of or pertaining to streams or rivers. A **fluvial deposit** is sedimentary material laid down by a stream or river, such as a sandstone or conglomerate (coarse-grained clastic sedimentary rock composed of rounded pebbles of pre-existing rock cemented in a fine-grained sand or clay matrix).

fluvioglacial of a process or landform associated with glacial meltwater. Meltwater, flowing beneath or ahead of a glacier, is capable of transporting rocky material and creating a variety of landscape features, including eskers, kames, and outwash plains.

f.o.b. or **FOB**, *abbreviation for* **free-on-board**, used in commerce to describe an export contract where the goods being shipped remain at the seller's risk until they cross the ship's rail at the docks. The term is also used to describe the value of the goods at the point of embarkation, excluding the transport and insurance costs paid for by the seller. Export values are usually expressed f.o.b. for customs and excise purposes, while imports are usually valued *c.i.f. If the goods are being sent by rail the term **f.o.r.** (free-on-rail) applies.

focus in earth science, the point within the Earth's crust at which an *earthquake originates. The point on the surface that is immediately above the focus is called the epicentre.

fodder crop a crop grown for animal feed, either for direct feeding, e.g. turnips, or for making *silage, as with grass grown for hay.

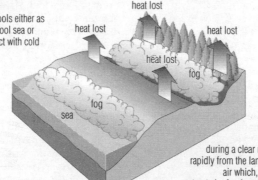

advection fog
warm moist air cools either as it passes over a cool sea or comes into contact with cold land surface

heat lost

heat lost

heat lost

heat lost

fog

fog

sea

radiation fog
during a clear night, heat is lost rapidly from the land. This cools the air which, if moist, becomes saturated – fog forms as it condenses

fog Advection fog occurs when two currents of air, one cooler than the other meet, or by warm air flowing over a cold surface. Radiation fog forms through rapid heat loss from the land, causing condensation to take place and a mist to appear.

fog   cloud that collects at the surface of the Earth, composed of water vapour that has condensed on particles of dust in the atmosphere. Cloud and fog are both caused by the air temperature falling below *dew point. The thickness of fog depends on the number of water particles it contains. Officially, fog refers to a condition when visibility is reduced to 1 km or less, and mist or haze to that giving a visibility of 1–2 km.

There are two types of fog. An **advection fog** is formed by the meeting of two currents of air, one cooler than the other, or by warm air flowing over a cold surface. Sea fogs commonly occur where warm and cold currents meet and the air above them mixes. A **radiation fog** forms on clear, calm nights when the land surface loses heat rapidly (by radiation); the air above is cooled to below its dew point and condensation takes place. A **mist** is produced by condensed water particles, and a haze by smoke or dust.

In some very dry areas, for example Baja California, Canary Islands, Cape Verde Islands, Namib Desert, and parts of Peru and Chile, coastal fogs enable plant and animal life to survive without rain and are a potential source of water for human use (by means of water collectors exploiting the effect of condensation).

Industrial areas uncontrolled by pollution laws have a continual haze of smoke over them, and if the temperature falls suddenly, a dense yellow smog forms.

föhn   or **foehn**, warm dry wind that blows down the leeward slopes of mountains.

The air heats up as it descends because of the increase in pressure, and it is dry because all the moisture was dropped on the windward side of the mountain. In the valleys of Switzerland it is regarded as a health hazard, producing migraine and high blood pressure. A similar wind, chinook, is found on the eastern slopes of the Rocky Mountains in North America.

fold   in geology, a deformation (bend) in *beds or layers of rock. Folds are caused by pressures within the Earth's crust resulting from *plate-tectonic activity. Rocks are slowly pushed and compressed together, forming folds. Such deformation usually occurs in *sedimentary layers that are softer and more flexible. If the force is more sudden, and the rock more brittle, then a *fault forms instead of a fold.

fold mountain   term no longer used to refer to mountains formed at a convergent margin. See *mountain.

Food and Agriculture Organization FAO, United Nations specialized agency that coordinates activities to

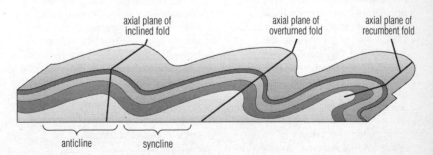

fold The folding of rock strata occurs where compression causes them to buckle. Over time, folding can assume highly complicated forms, as can sometimes be seen in the rock layers of cliff faces or deep cuttings in the rock. Folding contributed to the formation of great mountain chains such as the Himalayas.

improve food and timber production and levels of nutrition throughout the world. It is also concerned with investment in agriculture and dispersal of emergency food supplies. It has headquarters in Rome and was founded in 1945.

food chain in ecology, a sequence showing the feeding relationships between organisms in a *habitat or *ecosystem. It shows who eats whom. An organism in one food chain can belong to other food chains. This can be shown in a diagram called a **food web**.

One of the most important aspects of food is that it provides energy for an organism. So a food chain shows where each organism gets its energy. The arrow in a food chain represents the direction of energy flow. Not all of the energy in all of the organisms at one step of a food chain is available to the organisms later in the chain. In general, fewer organisms are found at each step, or *trophic level, of the chain. A *pyramid of numbers shows this clearly. Some organisms may be small but very numerous, so population size may not be a good measure of how much of an organism there is in a habitat. *Biomass – the total mass of organisms in an area – may be a more useful measure.

Some food chains start with organisms that decompose the remains of dead plants and animals. Decomposers can be fungi and bacteria and they have an important role in a habitat or ecosystem. Nutrients inside the dead organisms are released by the decomposers and are made available to other organisms. The carbon cycle shows how carbon compounds are cycled.

Energy in the form of food is shown to be transferred from autotrophs, or producers, which are principally plants and photosynthetic micro-organisms,

to a series of heterotrophs, or consumers. The heterotrophs comprise the herbivores, which feed on the producers; carnivores, which feed on the herbivores; and decomposers, which break down the dead bodies and waste products of all four groups (including their own), ready for recycling.

Consider a food chain starting with grass, then caterpillar, then blue tit, then sparrowhawk. This chain starts, as do most others, with a green plant. The reason for this is that the plant makes high-energy food (carbohydrate) using the energy of sunlight through photosynthesis. The plant is known as a producer, because it produces food. All the other members are consumers, because they consume the food made by the first organism. The second member, in this case a caterpillar, is a herbivore, because it eats plants. The third and later members are carnivores. In reality, however, organisms have varied diets, relying on different kinds of food, so the food chain is an oversimplification. The more complex food web shows a greater variety of relationships, but again emphasizes that energy passes from plants to herbivores to carnivores.

Environmentalists have used the concept of the food chain to show how poisons and other forms of pollution can pass from one animal to another, threatening rare species. For example, the pesticide DDT, which is now banned in the UK, has been found in lethal concentrations in the bodies of animals at the top of the food chain, such as the golden eagle *Aquila chrysaetos*. In the last organism it may have risen to a level that harms the animal. This is known as bioaccumulation.

food supply availability of food, usually for human consumption. Food supply can be studied at scales ranging from individual households to global patterns. Since the 1940s the industrial and agricultural aspects of food supply have become increasingly globalized. New farming, packaging, and distribution techniques mean that the seasonal aspect of food supply has been reduced in wealthier nations such as the USA and the UK. In some less developed countries there are often problems of food scarcity and distribution caused by climate-related crop failure.

food web   see *food chain.

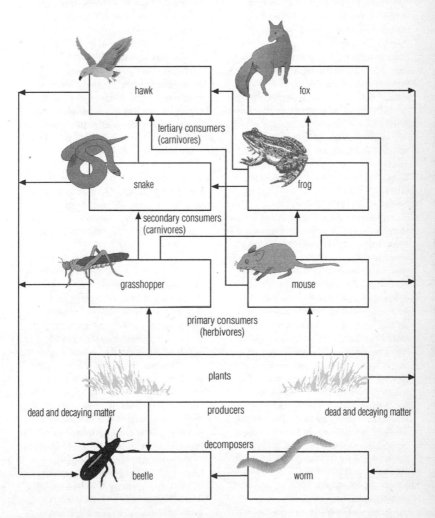

food web The complex interrelationships between animals and plants in a food web. A food web shows how different food chains are linked in an ecosystem. Note that the arrows indicate movement of energy through the web. For example, an arrow shows that energy moves from plants to the grasshopper, which eats the plants.

footloose industry   industry that can be sited in any of a number of places, often because transport costs are unimportant. Such industries may have raw materials that are commonly available, for example a bakery, or use components from a wide range of suppliers, for example the electronics industry. High-tech industries, needing a highly qualified workforce, may appear footloose, but in practice they tend to locate close to universities, research establishments, and motorways.

Fordism   mass production characterized by a high degree of job specialization, as typified by the Ford Motor Company's early use of assembly lines. Mass-production techniques were influenced by US management consultant F W Taylor's book *Principles of Scientific Management* (1911).

foredeep   in earth science, an elongated structural basin lying inland from an active mountain system and receiving sediment from the rising mountains. According to plate tectonic theory, a mountain chain forming behind a subduction zone along a continental margin develops a foredeep or gently sloping trough parallel to it on the landward side. Foredeeps form rapidly and are usually so deep initially that the sea floods them through gaps in the mountain range. As the mountain system evolves, sediments choke the foredeep, pushing out marine water. As marine sedimentation stops, only nonmarine deposits from the rapidly eroding mountains are formed. These consist of alluvial fans and also rivers, flood plains, and related environments inland.

foreign aid   see *aid.

foreign exchange   system by which the money of one country can be converted into the money of another; US dollars, French francs, and Spanish pesetas are all foreign currencies into which a holder of pounds sterling may convert their money.

forest   area where trees have grown naturally for centuries, instead of being logged at maturity (about 150–200 years). A natural, or old-growth, forest has a multistorey canopy and includes young and very old trees (this gives the canopy its range of heights). There are also fallen trees contributing to the very complex ecosystem, which may support more than 150 species of mammals and many thousands of species of insects. Globally forest is estimated to have covered around 68 million sq km during prehistoric times. By the late 1990s this is believed to have been reduced by half to 34.1 million sq km.

The Pacific forest of the west coast of North America is one of the few remaining old-growth forests in the temperate zone. It consists mainly of conifers and is threatened by logging – less than 10% of the original forest remains.

forestry   science of forest management. Recommended forestry practice aims at multipurpose crops, allowing the preservation of varied plant and animal species as well as human uses (lumbering, recreation). Forestry has often been confined to the planting of a single species, such as a rapid-growing conifer providing softwood for paper pulp and construction timber, for which world demand is greatest. In tropical countries, logging contributes to the destruction of *rainforests, causing global environmental problems. Small unplanned forests are *woodland.

The earliest planned forest dates from 1368 at Nuremberg, Germany; in Britain, planning of forests began in the 16th century. In the UK, Japan, and other countries, forestry practices have been criticized for concentration on softwood conifers to the neglect of native hardwoods.

Forestry Commission government department responsible for forestry in Britain. Established in 1919, it is responsible for over 1 million hectares of land, and is funded partly by government and partly by sales of timber. In Northern Ireland responsibility for forestry lies with the Department of Agriculture's forestry service.

formal employment any job where the employee has a contract of employment, pays taxes, and may be provided with a pension scheme. These form the majority of jobs in industrialized (developed) countries.

fossil (Latin *fossilis* 'dug up') cast, impression, or the actual remains of an animal or plant preserved in rock.

ammonite

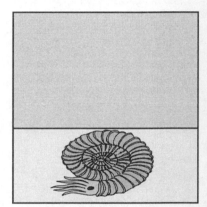

ammonite dies and sinks to sea floor where it becomes buried

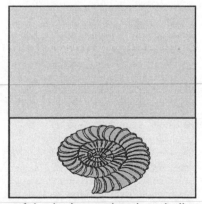

soft body decays leaving shell

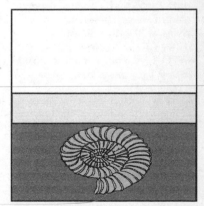

over millions of years shell is preserved as a fossil under layers of rock strata

fossil One way in which fossils were formed. When a marine animal dies it sinks to the sea floor (if it is not eaten), where it is eventually buried by sediment. The soft body parts decay and the hard parts may be preserved. Over time more and more sediment accumulates forming layers of rock.

Dead animals and plant remains that fell to the bottom of the sea bed or an inland lake were gradually buried under the accumulation of layers of sediment. Over millions of years, the sediment became *sedimentary rock and the remains preserved within the rock became fossilized. Fossils may include footprints, an internal cast, or external impression. A few fossils are preserved intact, as with mammoths fossilized in Siberian ice, or insects trapped in tree resin that is today amber. The study of fossils is called *palaeontology. Palaeontologists are able to deduce much of the geological history of a region from fossil remains. The existence of fossils is key evidence that organisms have changed with time, that is, evolved.

About 250,000 fossil species have been discovered – a figure that is believed to represent less than 1 in 20,000 of the species that ever lived. **Microfossils** are so small they can only be seen with a microscope. They include the fossils of pollen, bone fragments, bacteria, and the remains of microscopic marine animals and plants, such as foraminifera and diatoms.

fossil fuel combustible material, such as coal, lignite, oil, *peat, and natural gas, formed from the fossilized remains of plants that lived hundreds of millions of years ago. Such fuels are *non-renewable resources – once they are burnt, they cannot be replaced.

Fossil fuels are hydrocarbons (they contain atoms of carbon and hydrogen). They generate large quantities of heat when they burn in air, a process known as combustion. In this process carbon and hydrogen combine with oxygen in the air to form carbon dioxide, water vapour, and heat.

fossil magnetism see *polar reversal, *palaeomagnetic stratigraphy.

franchise in business, the right given by one company to another to manufacture, distribute, or provide its branded products. It is usual for the franchisor to impose minimum quality conditions on its franchisees to make sure that customers receive a fair deal from the franchisee and ensure that the brand image is maintained. Famous examples of franchise businesses include McDonald's and the Body Shop.

free enterprise or **free market**, economic system where private capital is used in business with profits going to private companies and individuals. The government plays a relatively small role in providing goods and services, but it is responsible for upholding laws which protect rights to own property, and for maintaining a stable currency. In practice most economies, even capitalist ones (see *capitalism), are *mixed economies – a hybrid of free and *command economies.

free market or **free enterprise**, another term for *capitalism.

free-on-board see *f.o.b.

free port port or sometimes a zone within a port, where cargo may be accepted for handling, processing, and reshipment without the imposition of tariffs or taxes. Duties and tax become payable only if the products are for consumption in the country to which the free port belongs.

free trade economic system where governments do not interfere in the movement of goods between countries; there are thus no taxes on imports. In the modern economy, free trade tends to hold within economic groups such as the European Union (EU), but not generally, despite such treaties as the *General Agreement on Tariffs and Trade (GATT) of 1948 and subsequent agreements to reduce tariffs. The opposite of free trade is *protectionism.

freeze–thaw   form of physical *weathering, common in mountains and glacial environments, caused by the expansion of water as it freezes. Water in a crack freezes and expands in volume by 9% as it turns to ice. This expansion exerts great pressure on the rock, causing the crack to enlarge. After many cycles of freeze–thaw, rock fragments may break off to form *scree slopes.

For freeze–thaw to operate effectively the temperature must fluctuate regularly above and below 0°C. It is therefore uncommon in areas of extreme and perpetual cold, such as the polar regions.

freezing point   for any given liquid, the temperature at which the liquid changes state from a liquid to a solid. The temperature remains at this point until all the liquid has solidified. It is invariable under similar conditions of pressure – for example, the freezing point of water under standard atmospheric pressure is 0°C/32°F. For a given liquid under similar conditions, the freezing point and melting point are the same temperature.

Friends of the Earth   FoE or FOE, largest international network of environmental pressure groups, established in the UK in 1971, that aims to protect the environment and to promote rational and sustainable use of the Earth's resources. It campaigns on such issues as acid rain; air, sea, river, and land pollution; recycling; disposal of toxic wastes; nuclear power and renewable energy; the destruction of rainforests; pesticides; and agriculture. FoE is represented in over 50 countries.

fringing reef   *coral reef that is attached to the coast without an intervening lagoon.

front   in meteorology, the boundary between two air masses of different temperature or humidity. A **cold front** marks the line of advance of a cold air mass from below, as it displaces a warm air mass; a **warm front** marks the advance of a warm air mass as it rises up over a cold one. Frontal systems define the weather of the mid-latitudes, where warm tropical air is continually meeting cold air from the poles.

Warm air, being lighter, tends to rise above the cold; its moisture is carried upwards and usually falls as rain or snow, hence the changeable weather conditions at fronts. Fronts are rarely stable and move with the air mass. An **occluded front** is a composite form, where a cold front catches up with a warm front and merges with it.

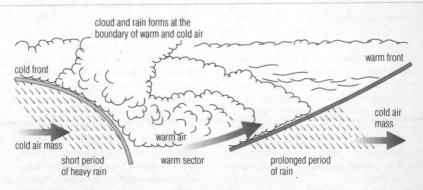

cloud and rain forms at the boundary of warm and cold air

warm front

cold front

cold air mass

cold air mass

warm air

short period of heavy rain

warm sector

prolonged period of rain

front The boundaries between two air masses of different temperature and humidity. A warm front occurs when warm air displaces cold air; if cold air replaces warm air, it is a cold front.

frost condition of the weather that occurs when the air temperature is below freezing, 0°C. Water in the atmosphere is deposited as ice crystals on the ground or on exposed objects. As cold air is heavier than warm air and sinks to the ground, ground frost is more common than hoar (air) frost, which is formed by the condensation of water particles in the air.

frost hollow depression or steep-sided valley in which cold air collects on calm, clear nights. Under clear skies, heat is lost rapidly from ground surfaces, causing the air above to cool and flow downhill (as *katabatic wind) to collect in valley bottoms. Fog may form under these conditions and, in winter, temperatures may be low enough to cause frost.

frost shattering alternative name for *freeze–thaw.

fuel any source of heat or energy, embracing the entire range of materials that burn in air (combustibles). A fuel is a substance that gives out energy when it burns. A **nuclear fuel** is any material that produces energy by nuclear fission in a nuclear reactor. *Fossil fuels are formed from the fossilized remains of plants and animals.

Crude oil (unrefined *petroleum) is purified at an oil refinery by fractional distillation into fuels such as gasoline and kerosine. The burning of fossil fuels for energy production contributes to environmental problems such as *acid rain and the *greenhouse effect.

fuel cell cell converting chemical energy directly to electrical energy. It works on the same principle as a battery but is continually fed with fuel, usually hydrogen and oxygen. Fuel cells are silent and reliable (no moving parts) but expensive to produce. They are an example of a renewable energy source.

fuelwood crisis the increasing scarcity of wood and charcoal to meet the requirements of fuel for cooking and as a source of warmth in many parts of the Third World. At least half of all the timber cut in the world is used as fuelwood. In Africa, 90% of the population use wood or charcoal for cooking. Even in urban areas where alternative energy sources are available fuelwood is favoured because of its cheapness relative to kerosene, electricity or gas, and despite the low fuel efficiency of most traditional wood burning stoves. Fuelwood is seen as reliable given the frequency of power cuts in urban areas and there is often a preference for the taste of food cooked over wood or charcoal. Long regarded as a 'free' resource, even rural areas some distance from urban centres are now finding that securing adequate supplies of fuelwood is costly. Throughout the Third World over 2 billion people are unable to obtain sufficient fuelwood to meet their minimum needs or are forced to consume wood faster than it is being replenished.

The adverse consequences of the fuelwood crisis extend beyond accelerated deforestation and the attendant problems of soil erosion and desertification, to include the increasing amounts of time, especially female labour, diverted from productive activities to the search for wood, and the use of other combustible materials such as dung and crop residues that might otherwise have enriched the soil. Despite the growing scarcity of readily accessible fuelwood supplies local perception of the problem and the response to afforestation has been poor. Fuelwood difficulties are generally seen as less pressing than food or water shortages. Solutions to the fuelwood crisis in rural areas are likely to lie less in large-scale programmes of afforestation or reforestation than in local agroforestry projects that can be shown to have tangible benefits for farming communities.

fuller's earth   soft, greenish-grey rock resembling clay, but without clay's plasticity. It is formed largely of clay minerals, rich in montmorillonite, but a great deal of silica is also present. Its absorbent properties make it suitable for removing oil and grease, and it was formerly used for cleaning fleeces ('fulling'). It is still used in the textile industry, but its chief application is in the purification of oils. Beds of fuller's earth are found in the southern USA, Germany, Japan, and the UK.

fumarole   a small vent in the Earth's surface from which steam and other volcanic gases escape. Fumarole activity is usually associated with dormant volcanoes. In Italy, some fumaroles are harnessed to produce electricity.

functions   the term used for goods and services available in a *central place; in general the number and variety of functions offered increase with the size of the *settlement. Everyday or convenience (low order) functions are available in small settlements, while these plus specialized or durable (high order) functions are available in large settlements. Small settlements may offer only two or three functions; large settlements many hundreds. See also *central place theory.

fungus   any plant of the division Fungi, lacking chlorophyll, leaves, true stems and roots. Fungi reproduce by spores and live entirely as saprophytes or parasites. The group includes moulds, yeasts, mildews, rusts and mushrooms. Some can cause disease in plants and animals, including humans.

fur   the hair of certain animals. Fur is an excellent insulating material and so has been used as clothing. This is, however, vociferously criticized by many groups on humane grounds, as the methods of breeding or trapping animals are often cruel. Mink, chinchilla, and sable are among the most valuable, the wild furs being finer than the farmed.

Fur such as mink is made up of a soft, thick, insulating layer called underfur and a top layer of longer, lustrous guard hairs.

G

gabbro mafic (consisting primarily of dark-coloured crystals) igneous rock formed deep in the Earth's crust. It contains pyroxene and calcium-rich feldspar, and may contain small amounts of olivine and amphibole. Its coarse crystals of dull minerals give it a speckled appearance.

Gabbro is the plutonic version of basalt (that is, derived from magma that has solidified below the Earth's surface), and forms in large, slow-cooling intrusions.

Gaia hypothesis theory that the Earth's living and nonliving systems form an inseparable whole that is regulated and kept adapted for life by living organisms themselves. The planet therefore functions as a single organism, or a giant cell. The hypothesis was elaborated by British scientist James Lovelock and first published in 1968.

gale strong wind, usually between force seven and ten on the *Beaufort scale (measuring 45–90 km per hour).

galena mineral consisting of lead sulphide, PbS, the chief ore of lead. It is lead-grey in colour, has a high metallic lustre and breaks into cubes because of its perfect cubic cleavage. It may contain up to 1% silver, and so the ore is sometimes mined for both metals. Galena occurs mainly among limestone deposits in Australia, Mexico, Russia, Kazakhstan, the UK, and the USA.

gangue part of an ore deposit that is not itself economically valuable; for example, calcite may occur as a gangue mineral with galena.

garnet group of *silicate minerals with the formula $X_3Y_3(SiO_4)_3$, where X is calcium, magnesium, iron, or manganese, and Y is usually aluminium or sometimes iron or chromium. Garnets are used as semi-precious gems (usually pink to deep red) and as abrasives. They occur in metamorphic rocks such as gneiss and schist.

garrigue a sclerophyllous shrub-dominated vegetation growing to a height of about 1 m and found around the Mediterranean Sea. Garrigue probably represents the most modified of the Mediterranean vegetation communities. Many generations of burning, grazing and deforestation have removed the natural mixed forest of the region and in its place is found a sparse vegetation in which geophytes and aromatic, herbaceous plants predominate. The growing period of the garrigue is unusual in that it is confined to the cooler and moister winter and spring seasons with the result that the garrigue often appears desolate in the long, hot and dry summers which characterize the Mediterranean region.

gastrolith stone that was once part of the digestive system of a dinosaur or other extinct animal. Rock fragments were swallowed to assist in the grinding process in the dinosaur digestive tract, much as some birds now swallow grit and pebbles to grind food in their crop. Once the animal has decayed, smooth round stones remain – often the only clue to their past use is the fact that they are geologically different from their surrounding strata.

GATT acronym for *General Agreement on Tariffs and Trade.

GDP *abbreviation for* *gross domestic product.

gearing ratio ratio of a company's permanent loan capital (preference shares and long-term loans) to its equity (ordinary shares plus *reserves).

gelifluction   type of *solifluction (downhill movement of water-saturated topsoil) associated with frozen ground.

gem   mineral valuable by virtue of its durability (hardness), rarity, and beauty, cut and polished for ornamental use, or engraved. Of 120 minerals known to have been used as gemstones, only about 25 are in common use in jewellery today; of these, the diamond, emerald, ruby, and sapphire are classified as precious, and all the others semi-precious; for example, the topaz, amethyst, opal, and aquamarine.

Gemeinschaft* and *Gesellschaft
German terms (roughly, 'community' and 'association') coined by Ferdinand Tönnies in 1887 to contrast social relationships in traditional rural societies with those in modern industrial societies. He saw *Gemeinschaft* (traditional) as intimate and positive, and *Gesellschaft* (modern) as impersonal and negative.

genetic engineering   all-inclusive term that describes the deliberate manipulation of genetic material by biochemical techniques. It is often achieved by the introduction of new DNA, usually by means of a virus or plasmid. This can be for pure research, gene therapy, or to breed functionally specific plants, animals, or bacteria. These organisms with a foreign gene added are said to be transgenic and the new DNA formed by this process is said to be recombinant. In most current cases the transgenic organism is a micro-organism or a plant, because ethical and safety issues are limiting its use in mammals.

General Agreement on Tariffs and Trade   GATT, agreement designed to provide an international forum to encourage regulation of international trade. The original agreement was signed in 1947, shortly after World War II. It was followed in 1948 by the creation of an international organization, within the United Nations, to support the agreement and to encourage *free trade between nations by reducing tariffs, subsidies, quotas, and regulations that discriminate against imported products. The agency GATT was effectively replaced by the *World Trade Organization (WTO) in January 1995, following the Uruguay Round. The legal agreement still exists although it was updated in 1994 to reflect a shift from trade in goods to trade in goods, services, and intellectual property. The new GATT agreements are administered by the WTO.

gentrification   the movement of higher social or economic groups into an area after it has been renovated and restored. This may result in the outmigration of the people who previously occupied the area. Often the classification of an area as a conservation area encourages gentrification. It is one strategy available to planners in urban renewal schemes within the *inner city.

geocentric   having the Earth at the centre.

geochemistry   science of chemistry as it applies to geology. It deals with the relative and absolute abundances of the chemical elements and their isotopes in the Earth, and also with the chemical changes that accompany geologic processes.

geochronology   branch of geology that deals with the dating of rocks, minerals, and fossils in order to create an accurate and precise geological history of the Earth. The *geological time scale is a result of these studies. It puts stratigraphic units in chronological order and assigns actual dates, in millions of years, to those units.

geode in geology, a subspherical cavity into which crystals have grown from the outer wall into the centre. Geodes often contain very well-formed crystals of quartz (including amethyst), calcite, or other minerals.

geodesy science of measuring and mapping Earth's surface for making maps and correlating geological, gravitational, and magnetic measurements. Geodetic surveys, formerly carried out by means of various measuring techniques on the surface, are now commonly made by using radio signals and laser beams from orbiting satellites.

geographical information system GIS, computer software that makes possible the visualization and manipulation of spatial data, and links such data with other information such as customer records.

geography study of the Earth's surface; its topography, climate, and physical conditions, and how these factors affect people and society. It is usually divided into **physical geography**, dealing with landforms and climates, and **human geography**, dealing with the distribution and activities of peoples on Earth.

history Early preclassical geographers concentrated on map-making, surveying, and exploring. In classical Greece theoretical ideas first became a characteristic of geography. Aristotle and Pythagoras believed the Earth to be a sphere, Eratosthenes was the first to calculate the circumference of the world, and Herodotus investigated the origin of the Nile floods and the relationship between climate and human behaviour.

During the medieval period the study of geography progressed little in Europe, but the Muslim world retained much of the Greek tradition, embellishing the 2nd-century maps of Ptolemy. During the early Renaissance the role of the geographer as an explorer and surveyor became important once again.

The foundation of modern geography as an academic subject stems from the writings of Friedrich Humboldt and Johann Ritter, in the late 18th and early 19th centuries, who for the first time defined geography as a major branch of scientific inquiry.

geological time time scale embracing the history of the Earth from its physical origin to the present day. Geological time is traditionally divided into eons (Archaean or Archaeozoic, Proterozoic, and Phanerozoic in ascending chronological order), which in turn are subdivided into eras, periods, epochs, ages, and finally chrons.

The terms eon, era, period, epoch, age and chron are **geochronological units** representing intervals of geological time. Rocks representing an interval of geological time comprise a **chronostratigraphic** (or **time-stratigraphic**) **unit**. Each of the hierarchical geochronological terms has a chronostratigraphic equivalent. Thus, rocks formed during an eon (a geochronological unit) are members of an eonothem (the chronostratigraphic unit equivalent of eon). Rocks of an era belong to an erathem. The chronostratigraphic equivalents of period, epoch, age, and chron are system, series, stage, and chronozone, respectively.

geology science of the Earth, its origin, composition, structure, and history. It is divided into several branches, inlcuding **mineralogy** (the minerals of Earth), **petrology** (rocks), **stratigraphy** (the deposition of successive beds of sedimentary rocks), **palaeontology** (fossils) and **tectonics** (the deformation and movement of the Earth's crust), **geophysics** (using physics to study the Earth's surface, interior, and atmosphere), and **geochemistry** (the science of chemistry as it applies to biology).

EON	ERA	PERIOD	EPOCH	TIME (my)
PHANEROZOIC	CENOZOIC *Age of mammals*	QUATERNARY *Age of man*	HOLOCENE	— 0.01 —
			PLEISTOCENE	— 1.64 —
		TERTIARY	PLIOCENE	— 5.20 —
			MIOCENE	— 23.5 —
			OLIGOCENE	— 35.5 —
			EOCENE	— 56.6 —
			PALAEOCENE	— 65.0 —
	MESOZOIC	CRETACEOUS		— 146 —
		JURASSIC *Age of Cycads*		— 208 —
		TRIASSIC		— 245 —
	PALAEOZOIC	PERMIAN *Age of Amphibians*		— 290 —
		CARBONIFEROUS *Age of Coal* *Age of Amphibians*		— 363 —
		DEVONIAN *Age of Fishes*		— 409 —
		SILURIAN *Age of Fishes*		— 439 —
		ORDOVICIAN *Age of Marine Invertebrates*		— 510 —
		CAMBRIAN *Age of Marine Invertebrates*		— 570 —
PROTEROZOIC				— 2500 —
ARCHAEOZOIC				4600

geological timescale The time column shows millions of years ago.

geomorphology branch of geology developed in the late 19th century, dealing with the morphology, or form, of the Earth's surface; nowadays it is also considered to be an integral part of physical geography. Geomorphological studies investigate the nature and origin of surface landforms, such as mountains, valleys, plains, and plateaux, and the processes that influence them. These processes include the effects of tectonic forces, *weathering, running water, waves, glacial ice, and wind, which result in the *erosion, *mass movement (landslides, rockslides, mudslides), transportation, and deposition of *rocks and *soils. In addition to the natural processes that mould landforms, human activity can produce changes, either directly or indirectly, and cause the erosion, transportation, and deposition of rocks and soils, for example by poor land management practices and techniques in farming and forestry, and in the mining and construction industries.

Geomorphology deals with changes in landforms from the present to the geologic past, and in spatial scales ranging from microscale to mountains. For example, the formation of mountain ranges takes place over millions of years, as the Earth's crust cools and solidifies and the resulting layers, or plates, are folded, uplifted or deformed by the seismic activity of the underlying magma (see *plate tectonics). The gouging out of river valleys by *glacial erosion is a gradual process that takes place over thousands of years. Conversely, volcanic eruptions, by the ejection of rocks and gases and the rapid flow of molten lava down a mountainside, create rapid changes to landforms, as with the volcanic eruptions on the island of Montserrat in the West Indies. Similarly, the eruption of undersea volcanoes can result in the sudden birth of islands, while the consequent and rapidly moving tidal waves (*tsunamis), can produce the unexpected inundation and destruction of low-lying coastal regions in their path.

geophysics branch of earth science using physics (for instance gravity, seismicity, and magnetism) to study the Earth's surface, interior, and atmosphere. Geophysics includes several sub-fields such as seismology, paleomagnetism, and remote sensing.

geopolitics study of the relationship between geographical factors and the political aspects of states. The significance of geopolitics was recognized by ancient and modern historians such as Herodotus, Thucydides, Montesquieu, Buckle, Taine, and Treitschke.

geostrophic wind a theoretical wind which flows parallel to the isobars and represents a balance between the opposing effects of the Coriolis force and pressure gradient. Upper atmospheric winds may sometimes approach perfect geostrophic flow but near the Earth's surface, friction causes winds to flow across the isobars at an oblique angle towards an area of low atmospheric pressure.

geosyncline a basin (a large *syncline) in which thick marine sediments have accumulated.

geothermal energy a method of producing power from heat contained in the lower layers of the Earth's *crust. New Zealand and Iceland both use superheated water or steam from geysers and volcanic springs to heat buildings and for hothouse cultivation and also to drive steam turbines to generate electricity. In Britain, experiments in Southampton have successfully used heated water from rocks below the city to heat shops and offices in the city centre. In the long term, scientists are hoping to be able to tap heat from the granite rocks in southwest England. Geothermal energy is an example of a renewable resource of energy (see *renewable resources, *non-renewable resources). It is relatively pollution free but hot mineral waters that are not used have to be disposed of carefully to avoid polluting surface drainage systems.

Gesellschaft (German 'society') in sociology, any group whose concerns are of a formal and practical nature. See *Gemeinschaft.

geyser natural spring that intermittently discharges an explosive column of steam and hot water into the air due to the build-up of steam in underground chambers. One of the most remarkable geysers is Old Faithful, in Yellowstone National Park, Wyoming, USA. Geysers also occur in New Zealand and Iceland.

ghetto (Old Venetian *gèto* 'foundry') any deprived area occupied by a minority group, whether voluntarily or not. Originally a ghetto was the area of a town where Jews were compelled to live, decreed by a law enforced by papal bull 1555. The term came into use 1516 when the Jews of Venice were expelled to an island within the city which contained an iron foundry. Ghettos were abolished, except in Eastern Europe, in the 19th century, but the concept and practice were revived by the Germans and Italians 1940–45.

gilgai an undulating surface of shallow depressions interspersed with ridges (up to 1 m high) which occurs on clay-rich soils. It results from alterations in the soil moisture level, which causes the clays to expand and contract.

GIS *abbreviation for* *geographical information system.

glacial advance the extension of *ice sheets and *glaciers to lower altitudes to cover large areas. This is caused by a cooling of the *climate.

glacial deposition laying-down (*deposition) of *sediment once carried by a glacier. When ice melts, it deposits the material that it has been carrying. The material deposited by a glacier is called *till, or in Britain **boulder clay**. It comprises angular particles of all sizes from boulders to clay that are unsorted and lacking in stratification (layering).

Unstratified till can be moulded by ice to form *drumlins (egg-shaped hills). At the snout of the glacier, till piles up to form a ridge called a terminal *moraine. Glacial deposits occur in many different locations – beneath the ice (subglacial), inside it (englacial), on top of it (supraglacial), at the side of it (marginal), and in front of it (proglacial).

Stratified till that has been deposited by meltwater is termed **fluvioglacial**, because it is essentially deposited by running water. Meltwater flowing away from a glacier will carry some of the till many kilometres away. This sediment will become rounded (by the water) and, when deposited, will form a gently sloping area called an **outwash plain**. Several landforms owe their existence to meltwater (**fluvioglacial landforms**) and include the long ridges called *eskers, which form parallel to the direction of the ice flow. Meltwater may fill depressions eroded by the ice to form **ribbon lakes**. Small depositional landforms may also result from glacial deposition, such as **kames** (small mounds) and **kettle holes** (small depressions, often filled with water).

In *periglacial environments on the margins of an icesheet, freeze–thaw weathering (the alternate freezing and thawing of ice in cracks in the rock) etches the outlines of rock outcrops, exploiting joints and areas of weakness, and results in aprons of *scree.

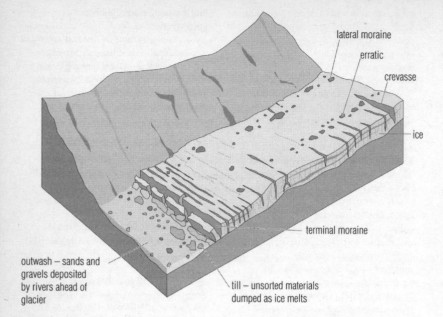

lateral moraine

erratic

crevasse

ice

terminal moraine

outwash – sands and
gravels deposited
by rivers ahead of
glacier

till – unsorted materials
dumped as ice melts

glacial deposition A glacier picks up large boulders and rock debris from the valley and deposits them at the snout of the glacier when the ice melts. Some deposited material is carried great distances by the ice to form erratics.

glacial erosion wearing-down and removal of rocks and soil by a *glacier. Glacial erosion forms impressive landscape features, including *glacial troughs (U-shaped valleys), *arêtes (steep ridges), *corries (enlarged hollows), and *pyramidal peaks (high mountain peaks with three or more arêtes).

Erosional landforms result from *abrasion and *plucking of the underlying bedrock. Abrasion is caused by the rock debris carried by a glacier, wearing away the bedrock. The action is similar to that of sandpaper attached to a block of wood. The results include the polishing and scratching of rock surfaces to form powdered rock flour, and scratches or *striations which indicate the direction of ice movement. Plucking is a form of glacial erosion restricted to the lifting

and removal of blocks of bedrock already loosened by *freeze–thaw activity.

The most extensive period of recent glacial erosion was the *Pleistocene epoch (1.6 million to 10,000 years ago) in the *Quaternary period (last 2 million years) when, over a period of 2–3 million years, the polar icecaps repeatedly advanced and retreated. More ancient glacial episodes are also preserved in the geological record, the earliest being in the middle *Precambrian era (4.6 billion–570 million years ago) and the most extensive in Permo-Carboniferous times.

glacial retreat a reduction in the area covered by *ice sheets and *glaciers, caused by a warming of the *climate. Compare *glacial advance.

glacial trough or **U-shaped valley**, steep-sided, flat-bottomed valley formed by a glacier. The erosive action

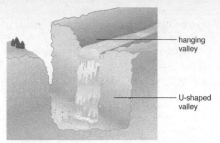

hanging valley

U-shaped valley

glacial trough Cross section of a glacial trough with a hanging valley (a smaller glacial trough). Glacial troughs are U-shaped and carved out by glaciers.

of the glacier and of the debris carried by it results in the formation not only of the trough itself but also of a number of associated features, such as *hanging valleys (smaller glacial valleys that enter the trough at a higher level than the trough floor). Features characteristic of glacial deposition, such as *drumlins, are commonly found on the floor of the trough, together with long lakes called ribbon lakes.

glaciation   a period of cold *climate during which time *ice sheets and *glaciers are the dominant forces of *denudation.

The last glaciation ended about 10,000 years ago, and much of Britain's landscape (north of a line drawn approximately between the Thames and the Severn) shows evidence of the effects of ice.

glacier   body of ice, originating in mountains in snowfields above the snowline, that moves slowly downhill and is constantly built up from its source. The geographic features produced by the erosive action of glaciers (*erosion) are characteristic and include *glacial troughs (U-shaped valleys), *corries, and *arêtes. In lowlands, the laying down of debris carried by glaciers (*glacial deposition) produces a variety of

landscape features, such as *moraines and *drumlins.

Glaciers form where annual snowfall exceeds annual melting and drainage (see *glacier budget). The area at the top of the glacier is called the zone of **accumulation**. The lower area of the glacier is called the *ablation zone. In the zone of accumulation, the snow compacts to ice under the weight of the layers above and moves downhill under the force of gravity. As the ice moves, it changes its shape and structure. Partial melting of ice at the base of the glacier also produces sliding of the glacier, as the ice travels over the bedrock. In the ablation zone, melting occurs and glacial *till is deposited.

glacier budget   in a glacier, the balance between *accumulation (the addition of snow and ice to the glacier) and *ablation (the loss of snow and ice by melting and evaporation). If accumulation exceeds ablation the glacier will advance; if ablation exceeds accumulation it will probably retreat.

The rate of advance and retreat of a glacier varies from a few centimetres a year to several metres a year.

glaucophane   in geology, a blue amphibole, $Na_2(Mg,Fe,Al)_5Si_8O_{22}(OH)_2$. It occurs in glaucophane schists (blue schists), formed from the ocean floor basalt under metamorphic conditions of high pressure and low temperature; these conditions are believed to exist in subduction systems associated with destructive plate boundaries (see *plate tectonics), and so the occurrence of glaucophane schists can indicate the location of such boundaries in geological history.

glazed frost   a type of frost which results when rain or dew comes into contact with a very cold ground surface and freezes. Glazed ice on roads, also known as *black ice*, is a serious hazard to traffic due to its clarity and hence its invisibility to road users.

gley soil   **glei soil** or **gleysol** or **meadow soil**, a compact and usually structureless soil in which one or more horizons have a grey, blue or olive colour due to intermittent waterlogging which causes anaerobic conditions. Gley soils may be caused by a high water table or by a hardpan or soil horizon which prevents the free passage of water downwards through the profile. Given adequate drainage and the addition of lime to overcome the soil acidity brought about by the anaerobic conditions, gley soils can be utilized for agriculture.

gleying   the process by which iron compounds in a soil are depleted of oxygen (*reduced*) by soil bacteria under anaerobic conditions to give a blue, grey or olive colouration. Gleying is caused by the permanent or temporary waterlogging of at least part of a soil and may take two forms:
(a) *groundwater gleying* which occurs in low-lying areas where a high water table produces anaerobic conditions;
(b) *surface water gleying* or *stagnogleying* which is due to the limited permeability of one or more soil horizons and leads to imperfect drainage.

gleysol   the term used in the FAO and Canadian soil classification systems for a *gley soil.

global brand   brand strategy where a brand is marketed in a similar way throughout the world. Although some concessions may be made to local tastes and customs, the emphasis is on standardization. Coca Cola and McDonald's are examples of global brands.

globalization   process by which different parts of the globe become interconnected by economic, social, cultural, and political means. Globalization has become increasingly rapid since the 1970s and 1980s as a result of developments in technology, communications, and trade liberalization. Critics of globalization fear the increasing power of unelected multinational corporations, financial markets, and non-government organizations (NGOs), whose decisions can have direct and rapid effects on ordinary citizens' lives. This has led to growing antiglobalization and anticapitalist protests in the 1990s and early 21st century, which have disrupted international trade talks and meetings of international finance ministers. Supporters of globalization point to the economic benefits of growing international trade and specialization.

NGOs criticized by opponents of globalization include the *World Trade Organization (WTO), *World Bank, and the *International Monetary Fund (IMF). Critics are also concerned about the potential environmental consequences, the risk of eroding distinctive local cultures, and the possible exploitation of workers. Solutions suggested to defuse opposition to globalization include making bodies such as the WTO more politically accountable, and ensuring greater protection of workers' rights.

*Global warming and the international operation of financial markets are examples of processes operating on a global scale. The spread of the Internet and satellite communication systems has helped to foster the sense of a 'global village', while the collapse of communism has led to the global spread of capitalism: China joining the WTO in 2001. Globalization has been supported by the development of several major free-trade blocs – the European Union (EU), the North American Free Trade Agreement (NAFTA), and a planned Association of South East Asian Nations (ASEAN) free-trade area by 2008.

global warming increase in average global temperature of approximately 0.5°C over the past century. Much of this is thought to be related to human activity. Global temperature has been highly variable in Earth history and many fluctuations in global temperature have occurred in historical times, but this most recent episode of warming coincides with the spread of industrialization, prompting the suggestion that it is the result of an accelerated *greenhouse effect caused by atmospheric pollutants, especially carbon dioxide gas. The melting and collapse of the Larsen Ice Shelf, Antarctica, since 1995, is a consequence of global warming. Melting of ice is expected to raise the sea level in the coming decades.

Natural, perhaps chaotic, climatic variations have not been ruled out as the cause of the current global rise in temperature, and scientists are still assessing the influence of anthropogenic (human-made) pollutants. In 1988 the World Meteorological Organization (WMO) and the United Nations (UN) set up the Intergovernmental Panel on Climate Change (IPCC), a body of more than 2000 scientists, to investigate the causes of and issue predictions regarding climate change. In June 1996 the IPCC confirmed that global warming was taking place and that human activities were probably to blame.

Assessing the impact of humankind on the global climate is complicated by the natural variability on both geological and human time scales. The present episode of global warming has thus far still left England approximately 1°C cooler than during the peak of the so-called Medieval Warm Period (1000 to 1400 AD). The latter was part of a purely natural climatic fluctuation on a global scale. The interval between this period and the recent rise in temperatures was unusually cold throughout the world, relative to historical temperatures. Scientists predict that a doubling of carbon dioxide concentrations, expected before the end of the 21st century, will increase the average global temperature by 1.4–5.8°C.

In addition to a rise in average global temperature, global warming has caused seasonal variations to be more pronounced in recent decades. Examples are the most severe winter on record in the eastern USA 1976–77, and the record heat waves in the Netherlands and Denmark the following year. Mountain glaciers have shrunk, late summer Arctic sea-ice has thinned by 40%, and sea levels have risen by 10–20 cm. Scientists have predicted a greater number of extreme weather events and sea levels are expected to rise by 9–88 cm by 2100. 1998 was the warmest year globally of the last millennium, according to US researchers who used tree rings and ice cores to determine temperatures over the past 1000 years.

A 1995 UN summit in Berlin, Germany, agreed to take action to reduce gas emissions harmful to the environment. Delegates at the summit, from more than 120 countries, approved a two-year negotiating process aimed at setting specific targets and timetables for reducing nations' emissions of carbon dioxide and other greenhouse gases after the year 2000. The *Kyoto Protocol of 1997 committed the world's industrialized countries to cutting their annual emissions of harmful gases. However, in June 2001 US president George W Bush announced that the USA would not ratify the protocol.

GMT *abbreviation for* *Greenwich Mean Time.

gneiss coarse-grained *metamorphic rock, formed under conditions of high temperature and pressure, and often occurring in association with schists and granites. It has a foliated, or layered, structure consisting of thin bands of micas and/or amphiboles dark in colour alternating with bands of granular quartz and feldspar that are light in colour. Gneisses are formed during regional *metamorphism; **paragneisses** are derived from metamorphism of sedimentary rocks and **orthogneisses** from metamorphism of granite or similar igneous rocks.

GNP *abbreviation for* *gross national product.

gold rush large influx of gold prospectors to an area where gold deposits have recently been discovered. The result is a dramatic increase in population. Cities such as Johannesburg, Melbourne, and San Francisco either originated or were considerably enlarged by gold rushes. Melbourne's population trebled from 77,000 to some 200,000 between 1851 and 1853, while San Francisco boomed from a small coastal village of a few hundred people to the largest city in the western USA during the California gold rush of 1848–56.

Gondwanaland or **Gondwana**, southern landmass formed 200 million years ago by the splitting of the single world continent *Pangaea. (The northern landmass was *Laurasia.) It later fragmented into the continents of South America, Africa, Australia, and Antarctica, which then drifted slowly to their present positions. The baobab tree found in both Africa and Australia is a relic of this ancient land mass.

A database of the entire geology of Gondwanaland has been constructed by geologists in South Africa. The database, known as Gondwana Geoscientific Indexing Database (GO-GEOID), displays information as a map of Gondwana 155 million years ago, before the continents drifted apart.

good in economics, a term often used to denote any product, including services. Equally, a good is often distinguished from a service, as in 'goods and services'. The opposite of a **normal good**, a product for which demand increases as a person's income increases, is an **inferior good**, a product for which demand decreases as income increases. A **free good** is one which an individual or organization can consume in infinite quantities at no cost, like the air we breathe. However, most goods are **economic goods**, which are scarce in supply and therefore have an *opportunity cost. In a free market, economic goods are allocated through prices.

gorge narrow steep-sided valley or canyon that may or may not have a river at the bottom. A gorge may be formed as a *waterfall retreats upstream, eroding away the rock at the base of a river valley; or it may be caused by *rejuvenation, when a river begins to cut downwards into its channel for some reason – for example, in response to a fall in sea level. Gorges are common in limestone country (see *karst), where they may be formed by the collapse of the roofs of underground caverns.

Examples of gorges in the UK are Winnats Pass in Derbyshire, Cheddar Gorge, and the Avon Gorge in Bristol.

government expenditure another name for *public spending.

graben in earth science, long block of rock that has sunk along faults so that it lies lower than the rocks on each side of it. Horsts (long raised blocks) often form the flanking highlands.

graded bedding sedimentary feature in which the sedimentary layer shows a gradual change in particle size, usually coarse at the bottom to fine at the top. It is useful for determining which way was up at the time the bed was deposited.

graded profile   the long profile of a river's course that would exist after all irregularities have been removed by *erosion.

gradient   **1.** the measure of steepness of a line or slope.
2. the measure of change in a property such as density. In *human geography gradients are found in, for example, *population density, land values and *settlement ranking.

Grand Banks   continental shelf in the North Atlantic off southeastern Newfoundland, where the shallow waters are rich fisheries, especially for cod.

granite   coarse-grained intrusive *igneous rock, typically consisting of the minerals quartz, feldspar, and biotite mica. It may be pink or grey, depending on the composition of the feldspar. Granites are chiefly used as building materials.

Granite is formed when magma (molten rock) is forced between other rocks in the Earth's crust. It cools and crystallizes deep underground. As it cools slowly large crystals are formed. Granites often form large intrusions in the core of mountain ranges, and they are usually surrounded by zones of *metamorphic rock (rock that has been altered by heat or pressure). Granite areas have characteristic moorland scenery. In exposed areas the bedrock may be weathered along joints and cracks to produce a tor, consisting of rounded blocks that appear to have been stacked upon one another.

graphite   blackish-grey, soft, flaky, crystalline form of carbon. It is used as a lubricant and as the active component of pencil lead.

Graphite, like *diamond and fullerene, is an allotrope of carbon.

The carbon atoms are strongly bonded together in sheets, but the bonds between the sheets are weak, allowing the layers to slide over one another. Graphite has a very high melting point (3500°C), and is a good conductor of heat and electricity. It absorbs neutrons and is therefore used to moderate the chain reaction in nuclear reactors.

gravel   coarse *sediment consisting of pebbles or small fragments of rock, originating in the beds of lakes and streams or on beaches. Gravel is quarried for use in road building, railway ballast, and for an aggregate in concrete. It is obtained from quarries known as gravel pits, where it is often found mixed with sand or clay.

Some gravel deposits also contain *placer deposits of metal ores (chiefly tin) or free metals (such as gold and silver).

gravimetry   measurement of the Earth's gravitational field. Small variations in the gravitational field (gravimetric anomalies) can be caused by varying densities of rocks and structure beneath the surface. Such variations are measured by a device called a gravimeter (or gravity-meter), which consists of a weighted spring that is pulled further downwards where the gravity is stronger. Gravimetry is used by geologists to map the subsurface features of the Earth's crust, such as underground masses of dense rock such as iron ore, or light rock such as salt.

Great Artesian Basin   largest area of artesian water in the world. It underlies much of Queensland, New South Wales, and South Australia, and in prehistoric times formed a sea. It has an area of 1,750,000 sq km.

great circle   circle drawn on a sphere such that the diameter of the circle is a diameter of the sphere. On the Earth, all meridians of longitude are half great circles; among the parallels of latitude, only the Equator is a great circle.

The shortest route between two points on the Earth's surface is along the arc of a great circle. These are used extensively as air routes although on maps, owing to the distortion brought about by *projection, they do not appear as straight lines.

green accounting the inclusion of economic losses caused by environmental degradation in traditional profit and loss accounting systems.

The idea arose in the 1980s when financial factors, and in particular profitability, were the main tool for judging the value of an action. By such crude measures, killing elephants for ivory, destroying tropical rainforest for hard wood, and continuing whaling, all make economic sense. However, if the future value of these resources are included, so that for example tourism, protection of biodiversity, and ecosystem stability are all given a notional economic value, it becomes plain that even in terms of pure profitability it makes no sense to destroy habitats or hunt animals to extinction.

green audit inspection of a company to assess the total environmental impact of its activities or of a particular product or process.

For example, a green audit of a manufactured product looks at the impact of production (including energy use and the extraction of raw materials used in manufacture), use (which may cause pollution and other hazards), and disposal (potential for recycling, and whether waste causes pollution).

Such 'cradle-to-grave' surveys allow a widening of the traditional scope of economics by ascribing costs to variables that are usually ignored, such as despoliation of the countryside or air pollution.

green belt an area of land, usually around the outskirts of a town or city on which building and other developments are restricted by legislation. The purpose of such planning law is to attempt to preserve open space and relatively rural *environments which would otherwise be lost with the advance of *urban sprawl.

green computing gradual movement by computer companies toward incorporating energy-saving measures in the design of systems and hardware. The increasing use of energy-saving devices, so that a computer partially shuts down during periods of inactivity, but can reactivate at the touch of a key, could play a significant role in *energy conservation.

It is estimated that worldwide electricity consumption by computers amounts to 240 billion kilowatt hours per year, equivalent to the entire annual consumption of Brazil. In the USA, carbon dioxide emissions could be reduced by 20 million tonnes per year – equivalent to the carbon dioxide output of 5 million cars – if all computers incorporated the latest 'sleep technology' (which shuts down most of the power-consuming features of a computer if it is unused for any length of time).

Although it was initially predicted that computers would mean 'paperless offices', in practice the amount of paper consumed continues to rise. Other environmentally costly features of computers include their rapid obsolescence, health problems associated with monitors and keyboards, and the unfavourable economics of component recycling.

green consumerism marketing term especially since the 1980s when consumers became increasingly concerned about the environment. Labels such as 'eco-friendly' became a common marketing tool as companies attempted to show that their goods had no negative effect on the environment.

greenfield site   area of land that has not been used for any non-agricultural development. It is therefore usually located in a rural area, or on the edge of a town or city. Greenfield sites are under threat from *urbanization – development for housing, industry, or retailing – but some greenfield sites are in protected green belt areas.

greenhouse effect   phenomenon of the Earth's atmosphere by which solar radiation, trapped by the Earth and re-emitted from the surface as long-wave infrared radiation, is prevented from escaping by various gases (the 'greenhouse gases') in the air. These gases trap heat because they readily absorb infrared radiation. As the energy cannot escape, it warms up the Earth, causing an increase in the Earth's temperature (*global warming). The main greenhouse gases are carbon dioxide, methane, and *chlorofluorocarbons (CFCs) as well as water vapour. Fossil-fuel consumption and forest fires are the principal causes of carbon dioxide build-up; methane is a by-product of agriculture (rice, cattle, sheep).

The United Nations Environment Programme estimates that by 2025, average world temperatures will have risen by 1.5°C with a consequent rise of 20 cm in sea level. Low-lying areas and entire countries would be threatened by flooding and crops would be affected by the change in climate. However, predictions about global warming and its possible climatic effects are tentative and often conflict with each other.

At the 1992 Earth Summit it was agreed that by 2000 countries would stabilize carbon dioxide emissions at

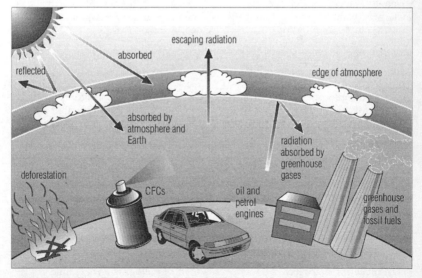

greenhouse effect   The warming effect of the Earth's atmosphere is called the greenhouse effect. Radiation from the Sun enters the atmosphere but is prevented from escaping back into space by gases such as carbon dioxide (produced for example, by the burning of fossil fuels), nitrogen oxides (from car exhausts), and CFCs (from aerosols and refrigerators). As these gases build up in the atmosphere, the Earth's average temperature is expected to rise.

1990 levels, but to halt the acceleration of global warming, emissions would probably need to be cut by 60%. Any increases in carbon dioxide emissions are expected to come from transport. The Berlin Mandate, agreed unanimously at the climate conference in Berlin in 1995, committed industrial nations to the continuing reduction of greenhouse gas emissions after 2000, when the existing pact to stabilize emissions runs out. The stabilization of carbon dioxide emissions at 1990 levels by 2000 would not be achieved by a number of developed countries, including Spain, Australia, and the USA, according to 1997 estimates. Australia is in favour of different targets for different nations, and refused to sign a communiqué at the South Pacific Forum meeting in the Cook Islands in 1997 which insisted on legally-binding reductions in greenhouse gas emissions. The United Nations Framework Convention on Climate Change (UNFCCC) adopted the *Kyoto Protocol in 1997, committing the world's industrialized countries to cutting their annual emissions of harmful gases. By July 2001 the Protocol had been signed by 84 parties and ratified by 37; the USA announced its refusal to ratify the Protocol in June 2001.

Dubbed the 'greenhouse effect' by Swedish scientist Svante Arrhenius, it was first predicted in 1827 by French mathematician Joseph Fourier.

greenmail payment made by a target company to avoid a takeover; for example, buying back a portion of its own shares from a potential predator (either a person or a company) at an inflated price.

green movement collective term for the individuals and organizations involved in efforts to protect the environment. The movement includes political parties such as the Green Party and organizations like *Friends of the Earth and *Greenpeace. See also *environmental issues.

Despite a rapid growth of public support, and membership of environmental organizations running into many millions worldwide, political green groups have failed to win significant levels of the vote in democratic societies.

Greenpeace international environmental pressure group, founded in 1971, with a policy of non-violent direct action backed by scientific research. During a protest against French atmospheric nuclear testing in the South Pacific in 1985, its ship *Rainbow Warrior* was sunk by French intelligence agents, killing a crew member. In 1995 it played a prominent role in opposing the disposal of waste from an oil rig in the North Sea, and again attempted to disrupt French nuclear tests in the Pacific. In 1997 Greenpeace had a membership in 43 'chapters' worldwide.

green pound exchange rate used by the European Union (EU) for the conversion of EU agricultural prices to sterling. The prices for all EU members are set in European Currency Units (ECUs) and are then converted into green currencies for each national currency.

green revolution the introduction of high yielding crops of rice and wheat into developing countries. These crops require fertilizers and can only be used by farmers who have appropriate funds. They have helped to increase food production.

green tax proposed tax to be levied against companies and individuals causing pollution. For example, a company emitting polluting gases would be obliged to pay a correspondingly significant tax; a company that cleans its emissions, reduces its effluent and uses energy-efficient distribution systems, would be taxed much less.

The idea is often criticized for relying on coercion rather than moral responsibility, but remains high on the political agenda, not least because of the mixed success of anti-pollution legislation (for example the *Clean Air Act).

Greenwich Mean Time GMT, local time on the zero line of longitude (the **Greenwich meridian**), which passes through the Old Royal Observatory at Greenwich, London. It was replaced in 1986 by coordinated universal time (UTC), but continued to be used to measure longitudes and the world's standard time zones.

grid reference numbering system used to specify a location on a map. The numbers representing grid lines at the bottom of the map (eastings) are given before those at the side (northings). Successive decimal digits refine the location within the grid system.

gross a particular figure or price, calculated before the deduction of specific items such as commission, discounts, interest, and taxes. The opposite is *net.

gross domestic product GDP, value of all final goods and services produced within a country within a given time period, usually one year. GDP thus includes the production of foreign-owned firms within the country, but excludes the income from domestically-owned firms located abroad. Intermediate goods, such as plastic and steel, are not included, in order to avoid double counting, because they will be turned into final goods. Household goods are included because they are intended for consumption or use rather than to be turned into other goods. GDP changes as total output and/or prices change.

A rise in total output means that an economy is growing; two consecutive quarters of decline in total output is the technical definition of recession. Optimal economic growth with full employment is considered to be in the range between 2% and 2.5%. GDP needs to be adjusted to account for inflation because it is affected by changes in prices as well as by changes in output. Inflation-adjusted GDP, known as **real GDP**, is calculated by dividing nominal GDP by the appropriate price index.

See also *gross national product (GNP).

gross national product GNP, measure of a country's total economic activity, or the wealth of the country. GNP is usually assessed quarterly or yearly, and is defined as the total value of all goods and services produced by firms owned by the country concerned. It is measured as the *gross domestic product plus income earned by domestic residents from foreign investments, minus income earned during the same period by foreign investors in the country's domestic market. GNP does not allow for inflation or for the overall value of production. It is an important indicator of an economy's strength.

groundwater water present underground in porous rock strata and soils; it emerges at the surface as springs and streams. The groundwater's upper level is called the *water table. Rock strata that are filled with groundwater that can be extracted are called **aquifers**. Aquifers must be both porous (filled with holes) and permeable (full of holes that are interconnected so that the water is able to flow).

Most groundwater near the surface moves slowly through the ground while the water table stays in the same place. The depth of the water table reflects the balance between the rate of infiltration, called recharge, and the rate of discharge at springs or rivers or pumped water wells. The force of gravity makes underground water run 'downhill' underground, just as it does above the surface. The greater the slope and the permeability, the greater the speed. Velocities vary from 100 cm per day to 0.5 cm.

Group of Eight G8; formerly Group of Seven (G7) 1975–98, the eight leading industrial nations of the world: the USA, Japan, Germany, France, the UK, Italy, Canada, and Russia, which account for more than three-fifths of global GDP. Founded as the Group of Seven (G7) in 1975, without Russia, the heads of government have met once a year to discuss economic and, increasingly, political matters. Russia attended the annual summits from 1991, and became a full member in 1998, when the name of the organization was changed. Summits are also attended by the president of the European Commission.

Group of Fifteen G15, forum of the world's industrializing states to agree cooperative policies to bridge the North–South economic divide. Leaders of 17 countries (originally 15, but the name has remained) – Algeria, Argentina, Brazil, Chile, Egypt, India, Indonesia, Jamaica, Kenya, Malaysia, Mexico, Nigeria, Peru, Senegal, Venezuela, Sri Lanka, and Zimbabwe – held their first meeting in 1990.

Group of Rio organization founded in 1987 from the Contadora Group (an alliance between Colombia, Mexico, Panama, and Venezuela) to draw up a general peace treaty for South America as a 'permanent mechanism for joint political action'. To establish the group the original Contadora members were joined by Argentina, Bolivia, Brazil, Chile, Ecuador, Paraguay, Peru, and Uruguay. Presidential-level meetings are held annually to discuss regional issues, including foreign debt and drug trafficking.

Group of Seven G7, former name 1975–98 of the *Group of Eight (G8), the eight leading industrial nations.

growth pole point within an area where economic growth is concentrated. This growth may encourage further development in the surrounding area (through the *multiplier effect), especially in areas of industrial decline or stagnation, such as *assisted areas. Growth poles are often towns (like Brasília in Brazil) or parts of towns that are gaining industry and expanding. They may be set up as part of a regional planning policy to help development in certain areas; for example, the Mezzogiorno, a region of poor economic performance in southern Italy.

groyne wooden or concrete barrier built at right angles to a beach in order to block the movement of material along the beach by *longshore drift. Groynes are usually successful in protecting individual beaches, but because they prevent beach material from passing along the coast they can mean that other beaches, starved of sand and shingle, are in danger of being eroded away by the waves. These areas are down-drift of the groyne. This happened, for example, at Barton-on-Sea in Hampshire, England, in the 1970s, following the construction of a large groyne at Bournemouth.

gryke enlarged *joint that separates blocks of limestone (clints) in a *limestone pavement.

gulf any large sea inlet.

Gulf Cooperation Council GCC, Arab organization for promoting peace in the Gulf area, established 1981. Its declared purpose is 'to bring about integration, coordination, and cooperation in economic, social, defence, and political affairs among Arab Gulf states'. Its members include Bahrain, Kuwait, Oman, Qatar, Saudi Arabia, and the United Arab Emirates; its headquarters are in Riyadh, Saudi Arabia.

Gulf Stream   warm ocean *current that flows north from the warm waters of the Gulf of Mexico along the east coast of America, from which it is separated by a channel of cold water originating in the southerly Labrador current. Off Newfoundland, part of the current is diverted east across the Atlantic, where it is known as the **North Atlantic Drift**, dividing to flow north and south, and warming what would otherwise be a colder climate in the British Isles and northwest Europe.

At its beginning the Gulf Stream is 80–150 km wide and up to 850 m deep, and moves with an average velocity of 130 km a day. Its temperature is about 26°C. As it flows northwards, the current cools and becomes broader and less rapid.

gully   long, narrow, steep-sided valley with a flat floor. Gullies are formed by water erosion and are more common in unconsolidated rock and soils that are easily eroded. They may be formed very rapidly during periods of heavy rainfall and are common in arid areas that have periods of heavy rain. Gully formation in more temperate areas is common where the vegetation cover has been destroyed or reduced, for example as a result of fire or agricultural clearance.

gust   temporary increase in *wind speed, lasting less than two minutes. Gusts are caused by rapidly moving air in higher layers of the atmosphere, mixing with slower air nearer the ground. Gusts are common in urban areas, where winds are funnelled between closely-spaced high buildings. Gusting winds do far more damage to buildings and crops than steady winds. The strongest gusts can exceed speeds of 100 m per second.

guyot   flat-topped seamount. Such undersea mountains are found throughout the abyssal plains of major ocean basins, and most of them are covered by an appreciable depth of water, sediment, and ancient coral. They are believed to have started as volcanic cones formed near mid-oceanic ridges or other hot spots, in relatively shallow water, and to have been truncated by wave action as their tops emerged above the surface. As they are transported away from the ridge or other birthplace, the ocean crust ages, cools, and sinks along with the seamounts on top.

gypsum   common sulphate *mineral, composed of hydrous calcium sulphate, $CaSO_4.2H_2O$. It ranks 2 on the *Mohs scale of hardness. Gypsum is used for making casts and moulds, and for blackboard chalk.

gyre   circular surface rotation of ocean water in each major sea (a type of *current). Gyres are large and permanent, and occupy the northern and southern halves of the three major oceans. Their movements are dictated by the prevailing winds and the *Coriolis effect. Gyres move clockwise in the northern hemisphere and anticlockwise in the southern hemisphere.

habitat   in ecology, the localized *environment in which an organism lives, and which provides for all (or almost all) of its needs. The diversity of habitats found within the Earth's ecosystem is enormous, and they are changing all the time. They may vary through the year or over many years. Many can be considered inorganic or physical; for example, the Arctic icecap, a cave, or a cliff face. Others are more complex; for instance, a woodland, or a forest floor. Some habitats are so precise that they are called **microhabitats**, such as the area under a stone where a particular type of insect lives. Most habitats provide a home for many species, which form a community.

Each species is specially adapted to life in its habitat. For example, an animal is adapted to eat other members of a *food chain or food web found in the same habitat. Some species may be found in different habitats. They may be found to have different patterns of behaviour or structure in these different habitats. For example, a plant such as the blackberry may grow in an open habitat, such as a field, or in a shaded one, such as woodland. Its leaves differ in the two habitats.

hadal zone   deepest level of the ocean, below the abyssal zone, at depths of greater than 6000 m. The ocean trenches are in the hadal zone. There is no light in this zone and pressure is over 600 times greater than atmospheric pressure.

Hadley cell   in the atmosphere, a vertical circulation of air caused by convection. The typical Hadley cell occurs in the tropics, where hot air over the Equator in the *intertropical convergence zone rises, giving the heavy rain associated with tropical rainforests. In the upper atmosphere this now dry air then spreads north and south and, cooling, descends in the latitudes of the tropics, producing the north and south tropical desert belts. After that, the air is drawn back towards the Equator, forming the northeast and southeast trade winds.

haematite   or **hematite**, principal ore of iron, consisting mainly of iron(III) oxide, Fe_2O_3. It occurs as **specular haematite** (dark, metallic lustre), **kidney ore** (reddish radiating fibres terminating in smooth, rounded surfaces), and a red earthy deposit.

hail   precipitation in the form of pellets of ice (hailstones). Water droplets freeze as they are carried upwards. As the circulation continues, layers of ice are deposited around the droplets until they become too heavy to be supported by the air current and they fall as a hailstorm. It is caused by the circulation of moisture in strong convection currents, usually within cumulonimbus *clouds.

halite   mineral form of sodium chloride, NaCl. Common salt is the mineral halite. When pure it is colourless and transparent, but it is often pink, red, or yellow. It is soft and has a low density.

halomorphic soil   any one of a range of soil types which develop under saline groundwater conditions. Solonchak and solonetz soils are typical of this group.

hamlet   small rural settlement that is more than just an isolated dwelling but not large enough to be a village. Typically it has 11–100 people.

hanging valley minor (tributary) valley that joins a larger *glacial trough at a higher level than the trough floor. During glaciation the ice in the smaller valley was unable to erode as deeply as the ice in the trough, and so the valley was left perched high on the side of the trough when the ice retreated. A river or stream flowing along the hanging valley often forms a *waterfall as it enters the trough. The Bridal Veil Falls in Yosemite National Park (USA) is an excellent example.

hardpan a hard, compact cemented layer usually found in the B horizons of soils and formed by the *illuviation of chemical compounds from the A horizon. Three main types of hardpan are recognized:
(a) *claypans* which result from the accumulation of clay-sized particles in the B horizon;
(b) *ironpans* which comprise thin crusts of mainly ferric oxide and are characteristic of podzols; and
(c) *moorpans* which may occasionally result from the deposition of humic materials.
   Hardpans restrict root formation and lead to instability amongst trees. They can impede soil drainage and often result in water-logging in the soil horizons immediately above the hardpan. If the hardpan becomes exposed at the surface by soil erosion then it can severely limit agricultural activities particularly in the low technology agricultural systems common in developing countries.

harmattan in meteorology, a dry and dusty northeast wind that blows over West Africa.

Harris and Ullman model see *multiple-nuclei theory.

harvesting strategy making a short-term profit towards the end of a product's life-cycle. Harvesting strategies usually involve cutting marketing expenditure and relying on the impetus from earlier marketing to keep consumers buying the product.

hazardous waste waste substance, usually generated by industry, that represents a hazard to the environment or to people living or working nearby. Examples include radioactive wastes, acidic resins, arsenic residues, residual hardening salts, lead from car exhausts, mercury, non-ferrous sludges, organic solvents, asbestos, chlorinated solvents, and pesticides. The cumulative effects of toxic waste can take some time to become apparent (anything from a few hours to many years), and pose a serious threat to the ecological stability of the planet; its economic disposal or recycling is the subject of research.

haze an atmospheric condition marked by a slight reduction in atmospheric visibility. Hazes may result from the formation of photochemical smog, the radiation of heat from the ground surface on hot days, or the development of a very thin mist. Visibility in a haze is usually greater than 2 km and rarely affects traffic flow or aircraft operations.

HDI see *Human Development Index.

headland area of land running out into the sea. Headlands are often high points on the coastline, usually with steep cliffs, and may be made of more resistant rock than adjacent *bays. Erosion is concentrated on the flanks of headlands due to *wave refraction. Good examples include the chalk headland between Handfast Point and Ballard Point near Swanage, England, and the area between Tennyson Down and The Needles, Isle of Wight, England.

head of water vertical distance between the top of a water stream and the point at which its energy is to be extracted.

headward erosion   the backwards erosion of material at the source (start) of a river or stream. Broken rock and soil at the source are carried away by the river, causing the source of the river to move backwards as erosion takes place in the opposite direction to the river's flow. The resulting lowering of the land behind the source may, over time, cause the river to cut backwards into a neighbouring valley to 'capture' another river (see *river capture).

heat island   large town or city that is warmer than the surrounding countryside. The difference in temperature is most pronounced during the winter, when the heat given off by the city's houses, offices, factories, and vehicles raises the temperature of the air by a few degrees. The heat island effect is also caused by the presence of surfaces such as black asphalt, that absorb rather than reflect sunlight, and the lack of vegetation, which uses sunlight to photosynthesise rather than radiating it back out as heat energy.

heave   a horizontal displacement of *strata at a fault.

heavy industry   industry that processes large amounts of bulky raw materials. Examples are the iron and steel industry, shipbuilding, and aluminium smelting. Heavy industries are often tied to locations close to their supplies of raw materials.

hedge   or **hedgerow**, row of closely planted shrubs or low trees, generally acting as a land division and windbreak. Hedges also serve as a source of food and as a refuge for wildlife, and provide a *habitat not unlike the understorey of a natural forest.

hedgerows   a feature of the landscape since Roman times. There is an estimated total of 450,600 km of hedgerows in Britain, sheltering more than 600 plant species, 1500 types of insect, 65 birds, and 20 different mammals. Among this diverse flora and fauna are 13 species which are either in very rapid decline or endangered globally. Ancient and species-rich hedgerows were among 14 key wildlife habitats on which the government and leading wildlife charities agreed rescue plans as a follow-up to the 1992 Rio Earth Summit. In 1996 the government estimated that more than 16,000 km of hedgerows were disappearing each year because of neglect, 'grubbing out', and the spray drift of pesticides. New laws were introduced in 1997 to help combat this decline.

Hedgerows are frequently mentioned in Anglo-Saxon charters, the earliest reference in England to planting a hedge being in Wiltshire in 940. During the period of enclosures, an estimated 321,800 km of hedges were planted. However, between 1984 and 1990, nearly 25% of Britain's hedgerows were destroyed.

hegemony   (Greek *hegemonia* 'authority') political dominance of one power over others in a group in which all are supposedly equal. The term was first used for the dominance of Athens over the other Greek city states, later applied to Prussia within Germany, and, in more recent times, to the USA and the USSR with regard to the rest of the world.

Henry Doubleday Research Association   British gardening group founded in 1954 by Lawrence Hills (1911–1990) to investigate organic growing techniques. It runs the **National Centre for Organic Gardening**, a 10-hectare demonstration site, at Ryton-on-Dunsmore near Coventry, England. The association is named after the person who first imported Russian comfrey, a popular green-manuring crop.

herbaceous plant   plant with very little or no wood, dying back at the end of every summer. The herbaceous perennials survive winters as underground storage organs such as bulbs and tubers.

herbivore   any animal which obtains most of its food from plants. Herbivores are the primary consumer organisms and form the second trophic level in a food chain. They are inefficient converters of plant protoplasm to animal tissue and between 80–90% of energy input is used to digest the large masses of vegetable protoplasm, to search for more food and, in mammals, to generate heat.

high   see anticyclone.

high-tech industry   any industry that makes use of advanced technology. The largest high-tech group is the fast-growing electronics industry and especially the manufacture of computers, microchips, and telecommunications equipment.

   The products of these industries have low bulk but high value, as do their components. Silicon Valley in t he USA and Silicon Glen in Scotland are two areas with high concentrations of such firms.

high-technology approach   an approach to the *development process which stresses the role of capital and sophisticated technology. It is argued that investment in large-scale *resource management schemes (e.g. dams for *hydroelectric power), and in the industrial sector in general, is the surest way to hasten national development. In Brazil, for example, development priorities have been identified in this way.

   Very often the capital for high-technology schemes in developing countries is provided by developed countries. Contrast this with the *intermediate technology (and *appropriate technology) approach.

hill farming   a system of *agriculture where sheep are grazed (and to a lesser extent cattle) on upland rough pasture.

   In Britain hill farming occurs on *marginal land in upland areas such as the Lake District, Snowdonia and the Scottish Highlands. The typical hill farm comprises three zones: the *inbye, intake* and *fell. The inbye is valley-bottom land, immediately surrounding the farm buildings; it is walled or fenced and may be cultivated for *fodder crops and sown pasture. The intake extends up the lower slopes of the surrounding fells and is an area of sheltered pasture for winter grazing and for lambing. The fell is an extensive area of upland rough grazing to which several farmers may have right of access.

hinterland   area that is served by a port or settlement (the *central place) and included in its *sphere of influence. The city of Rotterdam, the Netherlands, is the hinterland of a port.

histogram   a graph for showing values of classed data as the areas of bars.

historical inertia   another term for *industrial inertia.

hoar frost   a type of frost which results from the rapid nocturnal cooling of air above the ground. This air may freeze immediately or condense first and then gradually freeze to deposit a layer of white, needle-like ice particles on the ground and other objects such as trees and cars.

hogback   geological formation consisting of a ridge with a sharp crest and abruptly sloping sides, the outline of which resembles the back of a hog. Hogbacks are the result of differential erosion on steeply dipping rock strata composed of alternating resistant and soft beds. Exposed, almost vertical resistant beds provide the sharp crests.

Holocene epoch   period of geological time that began 10,000 years ago, and continues into the present. During this epoch the climate became warmer, the glaciers retreated, and human civilizations developed significantly.

It is the second and current epoch of the Quaternary period.

homeland or **Bantustan**, before 1980, name for the Black National States in the Republic of South Africa.

homelessness being without access to adequate *housing. The homeless include individuals and families living outdoors without shelter, persons temporarily housed in hostels, night shelters, and institutions such as psychiatric hospitals, and those temporarily accommodated by relatives or friends.

home region the area around a person's home. This might be on a small scale, e.g. the area within a short travelling distance, or on a broader scale, several counties. For example, the home region for a person living in London could be regarded as southeast England.

homeworking earning income by working from home. Homeworking was once the province of piece workers carrying out low-paid employment such as stuffing envelopes or sewing. New technology, in particular the Internet, has made homeworking a more feasible and attractive option for many companies.

honeypot site area that is of special interest or appeal to tourists. At peak times, honeypot sites may become crowded and congested, and noise and litter may eventually spoil such areas. Examples include viewpoints, museums, and even car parks.

horizon **1.** limit to which one can see across the surface of the sea or a level plain, that is, about 5 km at 1.5 m above sea level, and about 65 km at 300 m. **2.** in earth sciences, the distinct layers found in *soil. Usually three horizons are identified – A, B, C – with A being the topmost layer.

The A horizon or *topsoil contains humus and other vegetable debris. The B horizon or subsoil contains a larger proportion of inorganic material, and receives minerals washed down from the topsoil by the process of *leaching. The division between the B and C horizons is marked by a zone of decaying bedrock. In reality there will rarely be sharp divisions between zones; the A and B horizons, for example, may be mixed by the activity of worms, burrowing animals or root growth.

horizontal integration merger with, or takeover of, one company by another company operating at a similar stage in the production chain. The production and supply chain run from the sourcing of the raw material through to the sale to the consumer. In the production of chocolate bars, for example, one manufacturer might integrate horizontally with another. The benefits of this type of integration come from power in the marketplace and an *economy of scale. However, if the resulting company exerts too much control over the market it operates in, it may fall foul of legislation designed to prevent companies from obtaining a monopoly advantage.

hornblende green or black rock-forming mineral, one of the *amphiboles. It is a hydrous *silicate composed mainly of calcium, iron, magnesium, and aluminium in addition to the silicon and oxygen that are common to all silicates. Hornblende is found in both igneous and metamorphic rocks and can be recognized by its colour and prismatic shape.

hornfels *metamorphic rock formed by rocks heated by contact with a hot igneous body. It is fine-grained, brittle, and lacks foliation (a planar structure).

horse latitudes the subtropical belts of high air pressure which occur at latitudes of 30° north and south. The upper atmospheric convergence of air masses and the resulting descent and surface divergence of these masses polewards and equatorwards, causes calm weather conditions with light variable winds.

horst see *block mountain.

horticulture the growing of plants and flowers for commercial sale. It is now an international trade, for example, with orchids being grown in Southeast Asia for sale in Europe.

hot spot in earth science, area where a strong current or 'plume' of *magma rises upwards below the Earth's crust. The magma spreads horizontally in all directions, and may break through where the crust is thin. Hot spots occur within, rather than on the edges of, lithospheric *plates. However, the magma usually reaches the surface at *plate margins. Examples of hot spots include Hawaii, Iceland, and Yellowstone National Park, Wyoming, USA.

hot spot track volcanic ridge or line of volcanic centres that results when a lithospheric plate moves over a hot spot. Only the volcanoes near the hot spot are active, the others having moved away and become extinct. One major example is the Hawaii-Emperor Seamount chain. Because hot spots are relatively stable with respect to moving plates, hot spot tracks can be used to determine absolute plate motions.

household the number of people living in a single dwelling. This could be one person in a bedsit or a family of six in a house. A block of flats would be made up of several households. Information about households is required by the 10-year national *census.

household amenities utilities in a dwelling such as gas, electricity and running hot and cold water, which are important for everyday life. Older dwellings may have poorer amenities, such as a shared bathroom and an outside toilet, compared with many modern dwellings which have very good amenities, e.g. central heating. See also *neighbourhood amenities.

housing provision of residential accommodation. All countries have found some degree of state housing provision or subsidy essential, even in free-enterprise economies such as the USA. In the UK, flats and houses to rent (intended for people with low incomes) are built by local authorities or housing associations under the direction of the Secretary of State for the Environment, but houses in England and Wales would have to last 2500 years at the rate of replacement being achieved by local authorities in 1991.

housing association non-profit voluntary organization subsidizing the provision of houses for rent or sale. Housing associations receive grants and loans from the Housing Corporation and work in conjunction with local authorities to deal with the needs of special groups, especially the homeless. Expenditure on grants and loans expanded dramatically during the 1990s. There are about 3200 housing associations in England (1990); between them owning about 520,000 dwellings.

Hoyt model a model of urban structure developed by H. Hoyt in 1939 and based on an analysis of the land-use patterns of 142 American cities. See *sector theory.

human capital concept introduced by US economist Gary Becker during the 1960s to describe the skills and training present in the workforce of a company, or the population at large. Human capital can be increased by investment in training and education.

Human Development Index HDI, scheme devised by the United Nations in the 1990s as a measure of **human well-being**. It comprises a social welfare index based upon education/adult literacy, health/life expectancy, and purchasing power of income/economy. Economically more developed countries tend to have the highest human development index (over 0.9). Those that are economically less developed score lowest, (0.25 HDI).

human geography the study of people and their activities in terms of patterns and processes of population, *settlement, economic activity and *communications. There is no precise definition of such a broad subject, but the basic task of the human geographer is to try to explain distributions of people and their activities. Compare *physical geography.

human rights civil and political rights of the individual in relation to the state. Under the terms of the *United Nations Charter human rights violations by countries have become its proper concern, although the implementation of this obligation is hampered by Article 2 (7) of the charter prohibiting interference in domestic affairs. The Universal Declaration of *Human Rights, passed by the General Assembly on 10 December 1948, is based on a belief in the inherent (natural) rights, equality, and freedom of human beings, and sets out in 28 articles the fundamental freedoms – civil, political, economic – to be promoted. The declaration has considerable moral force but is not legally binding on states.

Human Rights, Universal Declaration of charter of civil and political rights drawn up by the United Nations in 1948. They include the right to life, liberty, education, and equality before the law; to freedom of movement, religion, association, and information; and to a nationality.

human species, origins of evolution of humans from ancestral primates. The African apes (gorilla and chimpanzee) are shown by anatomical and molecular comparisons to be the closest living relatives of humans. The oldest known hominids (of the human group) had been the australopithecines, found in Africa, dating from 3.5–4.4 million years ago. But in December 2000, scientists unearthed the fossilized remains of a hominid dating back 6 million years. The first hominids to use tools appeared 2 million years ago, and hominids first used fire and moved out of Africa 1.7 million years ago. Modern humans are all believed to descend from one African female of 200,000 years ago, although there is a rival theory that humans evolved in different parts of the world simultaneously.

Miocene apes Genetic studies indicate that the last common ancestor between chimpanzees and humans lived 5–10 million years ago. There are only fragmentary remains of ape and hominid fossils from this period. Dispute continues over the hominid status of *Ramapithecus*, the jaws and teeth of which have been found in India and Kenya in late Miocene deposits, dating from between 14 and 10 million years ago. The lower jaw of a fossil ape found in the Otavi Mountains, Namibia, comes from deposits dated between 10 and 15 million years ago, and is similar to finds from East Africa and Turkey. It is thought to be close to the initial divergence of the great apes and humans.

australopithecines *Australopithecus afarensis*, found in Ethiopia and Kenya, date from 3.9–4.4 million years ago. These hominids walked upright and they were either direct ancestors or an offshoot of the line that led to modern humans. They may have been the ancestors of *Homo habilis* (considered by some to be a species of *Australopithecus*), who appeared about 2 million years later, had slightly larger bodies and brains, and were probably the first to use stone tools. Also living in Africa at the same time was *A. africanus*, a gracile hominid thought to be a meat-eater, and *A. robustus*, a hominid with robust bones, large teeth, heavy jaws, and thought to be a vegetarian. They are not generally considered to be our ancestors.

A new species of *Australopithecus* was discovered in Ethiopia in 1999. Named *A. garhi*, the fossils date from 2.5 million years ago and also share anatomical features with *Homo* species. The most complete australopithecine skeleton to date was found in South Africa in April 2000. It is about 1.8 million years old and from a female *A. robustus*.

The skull of an unknown hominid species, *Kenyanthropus platyops*, was discovered in Kenya in March 2001. Approximately 3.5 million years old, it is contemporary with the australopithecines, previously the oldest known hominids, leading to the suggestion that humans are descended from *K. platyops*, rather than the australopithecines as has been thought.

Homo erectus Over 1.7 million years ago, *Homo erectus*, believed by some to be descended from *H. habilis*, appeared in Africa. *H. erectus* had prominent brow ridges, a flattened cranium, with the widest part of the skull low down, and jaws with a rounded tooth row, but the chin, characteristic of modern humans, is lacking. They also had much larger brains (900–1200 cu cm), and were probably the first to use fire and the first to move out of Africa. Their remains are found as far afield as China, West Asia, Spain, and southern Britain. Modern human *H. sapiens sapiens* and the Neanderthals *H. sapiens*

neanderthulensis are probably descended from *H. erectus*.

Australian palaeontologists announced the discovery of stone tools dated at about 800,000–900,000 years old and belonging to *H. erectus* on Flores, an island near Bali, in 1998. The discovery provided strong evidence that *H. erectus* were seafarers and had the language abilities and social structure to organize the movements of large groups to colonize new islands. In 2000 Japanese archaeologists discovered that *H. erectus* were probably building hut-like shelters around 500,000 years ago, the oldest known artificial structures.

Neanderthals Neanderthals were large-brained and heavily built, probably adapted to the cold conditions of the ice ages. They lived in Europe and the Middle East, and disappeared about 40,000 years ago, leaving *H. sapiens sapiens* as the only remaining species of the hominid group. Possible intermediate forms between Neanderthals and *H. sapiens sapiens* have been found at Mount Carmel in Israel and at Broken Hill in Zambia, but it seems that *H. sapiens sapiens* appeared in Europe quite rapidly and either wiped out the Neanderthals or interbred with them.

modern humans There are currently two major views of human evolution: the 'out of Africa' model, according to which *H. sapiens* emerged from *H. erectus*, or a descendant species, in Africa and then spread throughout the world; and the multiregional model, according to which selection pressures led to the emergence of similar advanced types of *H. sapiens* from *H. erectus* in different parts of the world at around the same time. Analysis of DNA in recent human populations suggests that *H. sapiens* originated about 200,000 years ago in Africa from a single female ancestor,

'Eve'. The oldest known fossils of *H. sapiens* also come from Africa, dating from 100,000–150,000 years ago. Separation of human populations occurred later, with separation of Asian, European, and Australian populations taking place between 100,000 and 50,000 years ago.

The human genome consists of between 27,000–40,000 genes. Of these only about 1.5% differ between humans and the great apes.

Humans are distinguished from apes by the complexity of their brain and its size relative to body size; by their small jaw, which is situated under the face and is correlated with reduction in the size of the anterior teeth, especially the canines, which no longer project beyond the tooth row; by their bipedalism, which affects the position of the head on the vertebral column, the lumbar and cervical curvature of the vertebral column, and the structure of the pelvis, knee joint, and foot; by their complex language; and by their elaborate culture.

The broad characteristics of human behaviour are a continuation of primate behaviour rather than a departure from it. For example, tool use, once a criterion for human status, has been found regularly in gorillas, orang-utans, and chimpanzees, and sporadically in baboons and macaques. Chimpanzees even make tools. In hominid evolution, manual dexterity has increased so that more precise tools can be made. Cooperation in hunting, also once thought to be a unique human characteristic, has been found in chimpanzees, and some gorillas and chimpanzees have been taught to use sign language to communicate.

Humboldt Current   former name of the *Peru Current.

hum, environmental   disturbing sound of frequency about 40 Hz, heard by individuals sensitive to this range, but inaudible to the rest of the population. It may be caused by industrial noise pollution or have a more exotic origin, such as the jet stream, a fast-flowing high-altitude (about 15,000 m) mass of air.

humidity   the amount of water vapour in the atmosphere. At a specific temperature there is a maximum limit to the quantity of moisture that can be held by a body of air. When this state has been reached the air is said to be *saturated*. The proportion of water vapour present relative to the maximum quantity possible is the relative humidity value, and is expressed as a percentage. Relative humidity can change due to a gradual diffusion of water into an air mass, or from a change in the air temperature. The warmer the air mass then the greater its moisture-holding capacity.

The actual quantity of moisture held in the air is the absolute humidity level and is the weight of water vapour contained in a given volume of air measured in grams per cubic metre. Due to the constant changes in air temperature then the absolute humidity of an air mass is liable to rapid fluctuation.

humification   the decomposition of organic material at the top of a *soil profile and the subsequent mixing of humus with mineral soil. This results from physical processes such as eluviation, chemical processes such as leaching and biological processes such as the activity of burrowing animals.

humus   component of *soil consisting of decomposed or partly decomposed organic matter, dark in colour and usually richer towards the surface. It has a higher carbon content than the original material and a lower nitrogen content, and is an important source of minerals in soil fertility.

hunting and gathering   living by hunting animals and gathering seeds, nuts, roots, and berries for consumption rather than trade. Hunting and gathering was the primary means of subsistence for 99% of human history. With the development of agriculture and animal domestication in the Neolithic period (from 9000 BC), hunting and gathering gradually declined in importance. The Australian Aborigines, Inuit, Kung, and Pygmies are among the few remaining peoples who live chiefly by hunting and gathering.

Hunter-gatherers obtain ample food for the expenditure of much less effort than is required to obtain the same result in an agricultural economy. Hunting, which is done chiefly by men, supplies only 30–40% of the necessary calories and protein.

hurricane or **tropical cyclone** or **typhoon**, a severe *depression (region of very low atmospheric pressure) in tropical regions, called **typhoon** in the North Pacific. It is a revolving storm originating at latitudes between 5° and 20° north or south of the Equator,

when the surface temperature of the ocean is above 27°C. A central calm area, called the eye, is surrounded by inwardly spiralling winds (anticlockwise in the northern hemisphere and clockwise in the southern hemisphere) of up to 320 kph. A hurricane is accompanied by lightning and torrential rain, and can cause extensive damage. In meteorology, a hurricane is a wind of force 12 or more on the *Beaufort scale.

During 1995 the Atlantic Ocean region suffered 19 tropical storms, 11 of them hurricanes. This was the third-worst season since 1871, causing 137 deaths. The most intense hurricane recorded in the Caribbean/Atlantic sector was Hurricane Gilbert in 1988, with sustained winds of 280 kph and gusts of over 320 kph.

In October 1987 and January 1990, winds of near-hurricane strength were experienced in southern England. Although not technically hurricanes, they were the strongest winds there for three centuries.

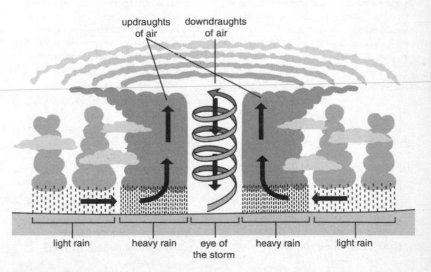

hurricane Hurricanes, also called typhoons, are violent tropical cyclones. The calm centre of the hurricane, known as the 'eye of the storm', is encircled by inwardly spiralling high winds.

The naming of hurricanes began in the 1940s with female names. Owing to public opinion that using female names was sexist, the practice was changed in 1978 to using both male and female names alternately.

HWM *abbreviation for* **high water mark**.

hydration in earth science, a form of *chemical weathering caused by the expansion of certain minerals as they absorb water. The expansion weakens the parent rock and may cause it to break up.

hydraulic action in earth science, the erosive force exerted by water (as distinct from erosion by the rock particles that are carried by water). It can wear away the banks of a river, particularly at the outer curve of a *meander (bend in the river), where the current flows most strongly.

Hydraulic action occurs as a river tumbles over a *waterfall to crash onto the rocks below. It will lead to the formation of a plunge pool below the waterfall. The hydraulic action of ocean waves and turbulent currents forces air into rock cracks, and therefore brings about erosion by *cavitation. In coastal areas hydraulic action is often the most important form of *erosion.

hydraulic radius measure of a river's *channel efficiency (its ability to move water and sediment), used by water engineers to assess the likelihood of flooding. The hydraulic radius of a channel is defined as the ratio of its cross-sectional area to its wetted perimeter (the part of the cross-section – bed and bank – that is in contact with the water).

hydroelectric power the generation of electricity by turbines driven by flowing water. Hydroelectricity is most efficiently generated in rugged *topography where a head of water can most easily be created, or on a large river where a dam can create similar conditions. Whatever the location, the principle remains the same – that water descending via conduits from an upper storage area passes through turbines and thus creates electricity.

hydrograph graph showing how the discharge of a river varies with time (generally over a matter of days). By studying hydrographs, water engineers can predict when flooding is likely and take action to prevent it.

A hydrograph shows the lag time, or delay, between peak rainfall and the resultant peak in discharge, and the

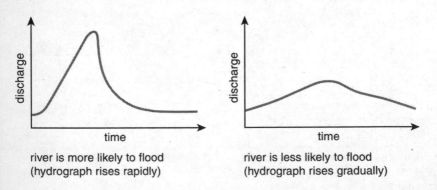

river is more likely to flood
(hydrograph rises rapidly)

river is less likely to flood
(hydrograph rises gradually)

hydrograph A hydrograph shows how the amount of water flowing in a river (the discharge) changes over time. Hydrographs can help river engineers to predict when flooding will take place.

length of time taken for that discharge to peak. The shorter the lag time and the higher the peak, the more likely it is that flooding will occur. The peak flow is equal to the **groundwater flow** plus the **storm flow**. Factors likely to give short lag times and high peaks include heavy rainstorms, steep slopes, deforestation, poor soil quality, and the covering of surfaces with impermeable substances such as tarmac and concrete. Actions taken by water engineers to increase lag times and lower peaks includes planting trees in the drainage basin of a river.

hydrography   study and charting of Earth's surface waters in seas, lakes, and rivers.

hydrological cycle   also known as the *water cycle, by which water is circulated between the Earth's surface and its atmosphere.

hydrology   study of the location and movement of inland water, both frozen and liquid, above and below ground. It is applied to major civil engineering projects such as irrigation schemes, dams, and hydroelectric power, and in planning water supply. Hydrologic studies are also undertaken to assess drinking water supplies, to track water underground, and to understand the role of water in geological processes such as fault movement and mineral deposition.

hydrolysis   in earth science, a form of *chemical weathering caused by the chemical alteration of certain minerals as they react with water. For example, the mineral feldspar in granite reacts with water to form a white clay called *china clay.

hydromorphic soil   any one of a range of soil types whose pedogenic processes are dominated by the presence of abundant water. Hydromorphic soils may be seasonally or permanently waterlogged. If adequate drainage can be provided, hydromorphic soils can become productive agricultural soils; this is due to their high fertility which has been achieved by the flushing in of nutrients. See *gleying.

hydroponics   cultivation of plants without soil, using specially prepared solutions of mineral salts. Beginning in the 1930s, large crops were grown by hydroponic methods, at first in California but since then in many other parts of the world.

hydrosere   see *sere.

hydrosphere   portion of the Earth made of water, ice, and water vapour, including the oceans, seas, rivers, streams, swamps, lakes, groundwater, and atmospheric water vapour. In some cases its definition is extended to include the water locked up in the Earth's crust and mantle.

hydrothermal   in geology, pertaining to a fluid whose principal component is hot water, or to a mineral deposit believed to be precipitated from such a fluid.

hydrothermal vein   crack in rock filled with minerals precipitated through the action of circulating high-temperature fluids. Igneous activity often gives rise to the circulation of heated fluids that migrate outwards and move through the surrounding rock. When such solutions carry metallic ions, ore-mineral deposition occurs in the new surroundings on cooling.

hydrothermal vent   or **smoker**, crack in the ocean floor, commonly associated with an *ocean ridge, through which hot, mineral-rich water flows into the cold ocean water, forming thick clouds of suspended material. The clouds may be dark or light, depending on the mineral content, thus producing 'white smokers' or 'black smokers'. In some cases the water is clear.

Sea water percolating through the sediments and crust is heated by the hot rocks and magma below and then dissolves minerals from the hot rocks. The water gets so hot that its increased buoyancy drives it back out into the ocean via a hydrothermal ('hot water') vent. When the water, anywhere from 60°C to over 400°C (kept liquid by the pressure of the ocean above) comes into contact with the frigid sea water, the sudden cooling causes these minerals to precipitate from solution, so forming the suspension. These minerals settle out and crystallize, forming stalagmite-like 'chimneys'. The chemical-rich water around a smoker gives rise to colonies of primitive bacteria that use the chemicals in the water, rather than the sunlight, for energy. Strange animals that live in such regions include huge tube worms 2 m long, giant clams, and species of crab, anemone, and shrimp found nowhere else.

hygrometer   an instrument for measuring the relative humidity of the *atmosphere. It comprises two thermometers, one of which is kept moist by a wick inserted in a water reservoir. Evaporation from the wick reduces the temperature of the 'wet bulb' thermometer, and the difference between the dry and the wet bulb temperatures is used to calculate relative humidity from standard tables.

hygroscopic nuclei   minute particles in the atmosphere, such as salt particles, which actively attract condensing water vapour. Without hygroscopic nuclei for water droplets to form around, the scale of atmospheric condensation and resulting meteorological phenomena such as precipitation, clouds and fog, would be greatly reduced.

hygroscopic water   see *wilting point.

hyperinflation   rapid and uncontrolled inflation, or increases in prices, usually associated with political and / or social instability, as in Germany in the 1920s.

hypermarket   very large *supermarket.

Iapetus Ocean or **Proto-Atlantic**, sea that existed in early *Palaeozoic times between the continent that was to become Europe and that which was to become North America. The continents moved together in the late Palaeozoic, obliterating the ocean. When they moved apart once more, they formed the Atlantic.

ice age any period of extensive glaciation (in which icesheets and icecaps expand over the Earth) occurring in the Earth's history, but particularly that in the *Pleistocene epoch (last 2 million years), immediately preceding historic times. On the North American continent, *glaciers reached as far south as the Great Lakes, and an icesheet spread over northern Europe, leaving its remains as far south as Switzerland. In Britain ice reached as far south as a line from Bristol to Banbury to Exeter. There were several glacial advances separated by interglacial (warm) stages, during which the ice melted and temperatures were higher than today. We are currently in an interglacial phase of an ice age.

Other ice ages have occurred throughout geological time: there were four in the Precambrian era, one in the Ordovician, and one at the end of the Carboniferous and beginning of the Permian. The occurrence of an ice age is governed by a combination of factors (the **Milankovitch hypothesis**): (1) the Earth's change of attitude in relation to the Sun – that is, the way it tilts in a 41,000-year cycle and at the same time wobbles on its axis in a 22,000-year cycle, making the time of its closest approach to the Sun come at different seasons; and

(2) the 92,000-year cycle of eccentricity in its orbit around the Sun, changing it from an elliptical to a near circular orbit, the severest period of an ice age coinciding with the approach to circularity. There is a possibility that the Pleistocene ice age is not yet over. It may reach another maximum in another 60,000 years.

The theory of ice ages was first proposed in the 19th century by, among others, Swiss civil engineer Ignace Venetz 1821 and Swiss naturalist Louis Agassiz 1837. (Before, most geologists had believed that the rocks and sediment they left behind were caused by the biblical flood.) The term 'ice age' was first used by botanist Karl Schimper in 1837.

Ice Age, Little period of particularly severe winters that gripped northern Europe between the 13th and 17th centuries. Contemporary writings and paintings show that Alpine glaciers were much more extensive than at present, and rivers such as the Thames, which do not ice over today, were so frozen that festivals could be held on them.

iceberg floating mass of ice, about 80% of which is submerged, rising sometimes to 100 m above sea level. Glaciers that reach the coast become extended into a broad foot; as this enters the sea, masses break off and drift towards temperate latitudes, becoming a danger to shipping.

icecap body of ice that is larger than a glacier but smaller than an ice sheet. Such ice masses cover mountain ranges, such as the Alps, or small islands. Glaciers often originate from icecaps.

ice fall an area of fractured ice in a *glacier where a change of *gradient occurs.

Iceland spar form of *calcite, $CaCO_3$, originally found in Iceland. In its pure form Iceland spar is transparent and exhibits the peculiar phenomenon of producing two images of anything seen through it (birefringence or double refraction). It is used in optical instruments. The crystals cleave into perfect rhombohedra.

ice sheet body of ice that covers a large land mass or continent; it is larger than an ice cap. During the last *ice age, ice sheets spread over large parts of Europe and North America. Today there are two ice sheets, covering much of Antarctica and Greenland. About 96% of all present-day ice is in the form of ice sheets. The ice sheet covering western Greenland increased in thickness by 2 m in 1981–93; this increase is the equivalent of a 10% rise in global sea levels.

igneous rock a *rock which originated as *magma (molten rock) at depth in or below the Earth's *crust. Igneous rocks are generally classified according to crystal size, colour and mineral composition; intrusive and extrusive types are also recognized:
(a) *batholith: a large body of magma intruded into the Earth's crust; this cools slowly at depth to form igneous rocks with large crystals such as *granite.
(b) *dyke: vertical or semi-vertical sheet of igneous rock; a minor intrusion compared with a batholith. Dolerite is a common dyke rock.
(c) *sill: horizontal or semi-horizontal minor intrusion; sills and dykes exploit lines of weakness (e.g. *joints, bedding planes, *faults) in the crustal rocks. See *bed.
(d) *lava flow: extrusive igneous rocks are those which reach the Earth's surface via some form of volcanic

eruption. Such rocks have small crystals due to rapid cooling on the Earth's surface. *Basalt is a common example.

The terms volcanic, *hypabyssal* and plutonic are also used to describe igneous rocks: *volcanic rocks are those which are extruded onto the surface; hypabyssal rocks are those of intrusions such as sills and dykes; and *plutonic rocks are those of deep intrusions such as batholiths.

ignis fatuus another name for *will-o'-the-wisp.

illuviation the precipitation and deposition of organic materials and soluble salts, in particular iron and aluminium compounds, within the B horizon of a soil. These materials have been washed down from the A horizon by the process of eluviation. Illuviation is one of the processes of translocation and may result in the formation of a hardpan. The occurrence and rate of illuviation is usually determined by the character of the soil (especially the texture) and climatic conditions.

ilmenite oxide of iron and titanium, iron titanate ($FeTiO_3$); an ore of titanium. The mineral is black, with a metallic lustre. It is found as an accessory mineral in mafic igneous rocks and in sands.

IMF *abbreviation for* *International Monetary Fund.

immigration in ecology, movement of individuals into a population. Immigration contributes to the increase in numbers of a population.

immigration and emigration movement of people from one country to another. Immigration is movement to a country; emigration is movement from a country. Immigration or emigration on a large scale is often for economic reasons or because of religious, political, or social persecution (which may create *refugees), and often results in restrictive legislation by individual countries. The USA has received immigrants on a larger scale than any other country, more than 50 million during its history.

imperfect competition competition between firms that supply branded products. Firms therefore compete not just on price, as in *perfect competition, but on the type of good they supply. In an *oligopoly, the market is dominated by a few firms offering strongly branded products and new firms find it difficult to establish themselves in the industry, whereas in monopolistic competition there are many small firms, branding is weaker, and entry to the industry is easier.

imperialism policy of extending the power and rule of a government beyond its own boundaries. A country may attempt to dominate others by direct rule and settlement – the establishment of a *colony – or by less obvious means such as control of markets for goods or raw materials. These less obvious means are often called *neocolonialism.

impermeable rock rock that does not allow water to pass through it – for example, clay, shale, and slate. Unlike *permeable rocks, which absorb water, impermeable rocks can support rivers. They therefore experience considerable erosion (unless, like slate, they are very hard) and commonly form lowland areas.

impervious rock a non-porous rock with no cracks or fissures through which water might pass. An *impermeable rock such as granite may be pervious due to the presence of *joints.

import product or service that one country purchases from another for domestic consumption, or for processing and re-exporting (Hong Kong, for example, is heavily dependent on imports for its export business). Imports may be visible (goods) or invisible (services). If an importing country does not have a counterbalancing value of exports, it may experience balance-of-payments difficulties and accordingly consider restricting imports by some form of protectionism (such as an import tariff or import quotas).

import control control that limits the number of imports entering the country. One type of import control is an import *quota.

inbye see *hill farming.

inch (Gaelic **innis**) Scottish and Irish geographical term denoting a small island or land by a river, or occasionally rising ground on a plain. It is found in place names such as the Inches of Perth on the River Tay, local water meadows, and the islands of Inchcolm, Inch Garvie, and Inchkeith in the Firth of Forth, Scotland.

incised meander in a river, a deep steep-sided *meander (bend) formed by the severe downwards erosion of an existing meander. Such erosion is usually brought about by the *rejuvenation of a river (for example, in response to a fall in sea level).

There are several incised meanders along the middle course of the River Wye, near Chepstow, Gwent.

income earnings of an individual or business organization over a period of time. Gross earnings are earnings before tax and other deductions, while net earnings are earnings after tax. Earned income is income received from working, while unearned income is income such as interest and dividends from financial and other wealth. Part of the income for a company is the value of its turnover. The income of a whole economy is often measured by *gross national product (GNP) or *gross domestic product (GDP). The study of the allocation of a country's national income is known as *distribution theory.

incomes policy government-initiated exercise to curb inflation by restraining rises in incomes, on either a voluntary or a compulsory basis; often linked with action to control prices, in which case it becomes a prices and incomes policy.

index in economics, an indicator of a general movement in wages and prices over a specified period.

indicator species plant or animal whose presence or absence in an area indicates certain environmental conditions, such as soil type, high levels of pollution, or, in rivers, low levels of dissolved oxygen. Many plants show a preference for either alkaline or acid soil conditions, while certain trees require aluminium, and are found only in soils where it is present. Some lichens are sensitive to sulphur dioxide in the air, and absence of these species indicates atmospheric pollution.

indigenous the people, animals, or plants that are native to a country, but especially a people whose territory has been colonized by others (particularly Europeans). In 1995 it was estimated that there were approximately 220 million indigenous people in the world. Examples of indigenous peoples include Australian Aborigines and American Indians. A World Council of Indigenous Peoples is based in Canada. The United Nations declared 1993 the International Year of Indigenous Peoples.

indirect cost or **semi-variable cost** in business studies, cost which changes as output changes, but not in direct proportion. For example, the costs of running an office are indirect costs for a manufacturing company. Day-to-day changes in output in the factory will not affect office costs. However, a permanent 20% increase in output might lead to a need for more office staff and hence higher indirect costs.

individualism in politics, a view in which the individual takes precedence over the collective: the opposite of *collectivism. The term **possessive individualism** has been applied to the writings of John Locke and Jeremy Bentham, describing society as comprising individuals interacting through market relations.

industrial estate area planned for industry, where space is available for large buildings and further expansion. Industrial estates often have good internal road layouts and occupy accessible sites near main road junctions but away from the *central business district (CBD).

industrial inertia tendency of an industry or group of industries to remain in one place even though its original locational needs have changed. For example, in Sheffield, England, iron and steel are still produced, even though the local supplies of iron ore ran out some time ago. The cost of moving is one reason why industries may stay in a particular place.

industrialization process by which an increasing proportion of a country's economic activity is involved in industry. It is essential for economic development and largely responsible for the growth of cities (*urbanization).

industrial location the optimum location for an industry is one where the costs of transport are minimized, with respect to both *raw materials and finished products. The standard theory of industrial location is that of Alfred Weber's *locational triangle* in which one corner represents the source of raw materials (factor 1), another represents the location of the labour supply (factor 2) and the other represents location of the market (factor 3). An industry may locate anywhere within the triangle, depending on the balance of transport costs relating to the three factors.

Various examples can be envisaged, e.g. an industry with bulky raw materials which are expensive to transport will locate close to factor 1.

According to the theory, an industry such as bread manufacture will locate close to the market, since distribution costs are the most important consideration.

An industry for which all transport costs are similar will locate anywhere in the triangle. In reality of course the locational decision is considerably more complex: factors such as the availability of power and the suitability of the land will also be taken into account, as will less tangible issues such as managers' personal preferences and outlook. Government intervention, for example through *development area policy, may also affect the locational decision. See *iron and steel industry for an example of a shifting location over time.

Industrial Revolution   the period in Britain's history from approximately 1780–1900, when the invention and application of industrial processes led to the establishment of the world's first urban/industrial society. The invention of the steam engine provided the key to a number of crucial industrial innovations in mining, transport and manufacture. Prior to the industrial revolution, successes within agriculture had been due to increased mechanization and efficiency, plus greater profits. This released labour to work in the new industrial cities. Throughout the 19th century large-scale *migration to urban areas continued, and it was during this period that the major industrial concentrations were established, for example in West Yorkshire,

South Lancashire, the West Midlands and Central Scotland.

industrial sector   any of the different groups into which industries may be divided: primary, secondary, tertiary, and quaternary. ***Primary** industries extract or use raw materials; for example, mining and agriculture. **Secondary** industries are manufacturing industries, where raw materials are processed or components are assembled. **Tertiary** industries supply services such as retailing. The **quaternary** sector of industry is concerned with the professions and those services that require a high level of skill, expertise, and specialization. It includes education, research and development, administration, and financial services such as accountancy.

inelastic demand   demand where a proportionate change in price (say 10%) leads to a lesser proportionate change in quantity demanded (say 5%). Formally, it is when the elasticity of demand is between 0 and 1.

infiltration   passage of water into the soil. The rate of absorption of surface water by soil (the infiltration capacity) depends on the intensity of rainfall, the permeability and compactness of the soil, and the extent to which it is already saturated with water. Once in the soil, water may pass into the bedrock to form *groundwater.

informal employment   paid work on a casual basis. Jobs are irregular and people are often self-employed without pensions and without paying taxes. This sort of employment is common in the urban areas of developing countries; for example, in Mexico City. It may involve service jobs of the tertiary *industrial sector – shoe cleaning or selling bottled water – as well as craft industries. Informal employment also includes illegal activities such as theft, prostitution, and selling drugs. In some Third

World cities it is estimated that the informal sector employs between 40% and 60% of the working population.

information superhighway popular collective name for the Internet and other related large-scale computer networks. The term was first used in 1993 by US vice-president Al Gore in a speech outlining plans to build a high-speed national data communications network.

information technology IT, collective term for the various technologies involved in processing and transmitting information. They include computing, telecommunications, and microelectronics. The term became popular in the UK after the Government's 'Information Technology Year' in 1978.

infrared radiation i.r., electromagnetic radiation of wavelength between about 700 nanometres and 1 millimetre – that is, between the limit of the red end of the visible spectrum and the shortest microwaves. All bodies above the *absolute zero of temperature absorb and radiate infrared radiation. Infrared radiation is used in medical photography and treatment, and in industry, astronomy, and criminology. The human eye cannot detect infrared, but its effects can be demonstrated. For example, an electric hob operates at high temperatures and only the visible light it gives out can be seen. As it cools down, the visible light is no longer seen. However, the heat (infrared radiation) that continues to be given out can be felt. Infrared absorption spectra are used in chemical analysis, particularly for organic compounds. Objects that radiate infrared radiation can be photographed or made visible in the dark on specially sensitized emulsions. This is important for military purposes and in detecting people buried under rubble. The strong absorption by many

substances of infrared radiation is a useful method of applying heat.

infrastructure relatively permanent facilities that serve an industrial economy. Infrastructure usually includes roads, railways, other communication networks, energy and water supply, and education and training facilities. Some definitions also include sociocultural installations such as health-care and leisure facilities.

ingrown meander an *incised meander formed by both lateral and vertical fluvial erosion. River valleys containing ingrown meanders usually have one steep valley side and another gentler slope, resulting in an asymetrical cross-section. Examples of ingrown meanders can often be found in rejuvenated river systems such as the River Rheidol in Wales and the lower reaches of the Seine in France.

inner city the ring of buildings around the *CBD of a town or city. These buildings are usually *terraced houses of the early industrial period (many of which have been modernized – see *gentrification), blocks of high and low-rise flats (to replace terraced houses which have become dilapidated), as well as offices and small *light industries. The buildings in the inner city tend to be high density, i.e. many buildings in a relatively small area.

innovation creating something new. Process innovation is a new way of doing something, and product innovation is the introduction of a new product or the modification of an existing one. Product innovation often brings competitive advantage in product markets that are relatively static, such as the 'air' training shoe developed by Nike that transformed the training shoe from sports accessory to mainstream street fashion. Process innovation, such as the development of production line manufacturing, can lead to significant cost reductions and licensing opportunities.

inorganic fraction that proportion of the *soil which is composed of *rock and mineral fragments and particles deriving from the *weathering of bedrock. Compare *organic fraction. See also *horizon.

inselberg or **kopje**, (German 'island mountain') prominent steep-sided hill of resistant solid rock, such as granite, rising out of a plain, usually in a tropical area. Its rounded appearance is caused by so-called onion-skin *weathering (exfoliation), in which the surface is eroded in successive layers.

The Sugar Loaf in Rio de Janeiro harbour in Brazil, and Ayers Rock in Northern Territory, Australia, are well-known examples.

insolation amount of solar radiation (heat energy from the Sun) that reaches the Earth's surface. Insolation varies with season and latitude, being greatest at the Equator and least at the poles. At the Equator the Sun is consistently high in the sky: its rays strike the equatorial region directly and are therefore more intense. At the poles the tilt of the Earth means that the Sun is low in the sky, and so its rays are slanted and spread out. Winds and ocean currents help to balance out the uneven spread of radiation.

insolation weathering type of physical or mechanical weathering that involves the alternate heating and cooling of rocks and minerals. This causes expansion and contraction, particularly of dark minerals (because they absorb more heat), setting up stresses in rocks. As rocks are poor conductors of heat only the surface layer is affected. This may contribute to exfoliation (breaking away of the outer layer of rock).

intangible asset *asset that does not exist physically and so cannot be touched or seen. Examples of intangibles include patents, copyright, brand names, and other intellectual property, as well as goodwill.

integrated pest management IPM, the use of a coordinated array of methods to control pests, including biological control, chemical *pesticides, crop rotation, and avoiding monoculture. By cutting back on the level of chemicals used the system can be both economical and beneficial to health and the environment.

integrated steelworks modern industrial complex where all the steelmaking processes – such as iron smelting and steel shaping – take place on the same site.

integration merger of two firms. **Vertical integration** occurs if the two firms are at different stages of the production process. **Forward integration** occurs when a firm takes over a company which buys its products, such as a car manufacturer taking over a car dealership. **Backward integration** occurs when one firm takes over another which supplies it, such as a car manufacturer taking over a car components company. **Horizontal integration** occurs if the two firms are at the same stage of the production process, such as one car manufacturer merging with another car manufacturer.

intensive farming a system of *agriculture where relatively large amounts of capital and/or labour are invested on relatively small areas of land. An example of intensive farming is *market gardening, where large investment, in the form of greenhouses, fertilization, *irrigation and heating systems, occurs on small holdings of land close to urban centres. In such a case yields per unit area are high. The small land-holdings reflect the cost of land close to market. See *Von Thünen theory and *agribusiness. Compare *extensive farming.

Inter-American Development Bank
IADB, bank founded in 1959, at the instigation of the Organization of American States (OAS), to finance economic and social development, particularly in the less wealthy regions of the Americas. Its membership includes the states of Central and Southern America, the Caribbean, and the USA, as well as Austria, Belgium, Canada, Denmark, Finland, France, Germany, Israel, Italy, Japan, the Netherlands, Norway, Portugal, Spain, Sweden, Switzerland, and the UK. Its headquarters are in Washington DC.

interception see *rainfall interception.

interdependence in economics, situation where an individual, business organization, or economy is economically reliant on others. Interdependence always occurs when *specialization occurs. For example, the UK is dependent on India for tea. The USA is dependent on the UK for consumption of Rolls Royce cars. A teacher is dependent on a farmer for food.

interglacial a warm period between two periods of *glaciation and cold *climate. The present interglacial began about 10,000 years ago.

interlocking spur one of a series of spurs (ridges of land) jutting out from alternate sides of a river valley. During glaciation its tip may be sheared off by erosion, creating a *truncated spur.

intermediate technology equipment and facilities of simple, cheap, practical design directly relevant to the immediate needs of a community in the *developing world. For example, rivers may be regulated by a series of small earthen dams to provide local flood control and water for *irrigation. Such dams can be built with local labour and equipment, can easily be repaired and are cheap. In contrast, the *high technology approach of sophisticated dams for *hydroelectric

power and industrial development could involve foreign capital, equipment and expertise, and might not be appropriate for a developing country's needs. Intermediate technology, being labour intensive rather than capital intensive, does not leave the country with a big burden of debt in conditions of scarce capital. See also *appropriate technology.

Intermediate Technology Development Group UK-based international aid organization established in 1965 by E F Schumacher to give advice and assistance on the appropriate choice of technologies for the rural poor of the developing world. It is an independent charity financed through donations.

internal economies of scale *economies of scale that arise because of the change in the scale of production by a firm, which can be caused by automation, specialization, or bulk purchasing power.

internal processes landscape-forming processes which originate below the Earth's surface in the movements connected with *plate tectonics: folding, faulting and **vulcanicity**.

International Atomic Energy Agency **IAEA**, agency of the United Nations established in 1957 to advise and assist member countries in the development and peaceful application of nuclear power, and to guard against its misuse. It has its headquarters in Vienna, and is responsible for research centres in Austria and Monaco, and the International Centre for Theoretical Physics, Trieste, Italy, established in 1964. It conducts inspections of nuclear installations in countries suspected of developing nuclear weapons, for example Iraq and North Korea.

International Bank for Reconstruction and Development specialized agency of the United Nations. Its popular name is the *World Bank.

International Date Line IDL, imaginary line that approximately follows the 180° line of longitude. The date is put forward a day when crossing the line going west, and back a day when going east. The IDL was chosen at the International Meridian Conference in 1884.

International Development Association IDA, agency of the United Nations, established 1960 and affiliated to the *World Bank.

International Development, Department for DFID, UK government department with responsibility for promoting international development and the reduction of poverty and for managing the UK's programme of assistance to poorer countries. It superseded the Overseas Development Administration in 1997.

International Fund for Agricultural Development agency of the *United Nations, established in 1977, to provide funds for benefiting the poor in developing countries.

International Labour Organization ILO, specialized agency of the United Nations, originally established in 1919, which formulates standards for labour and social conditions. Its headquarters are in Geneva, Switzerland. It was awarded the Nobel Peace Prize in 1969. By 1997, the agency was responsible for over 70 international labour conventions.

International Monetary Fund IMF, specialized agency of the United Nations, headquarters Washington, DC, established under the 1944 Bretton Woods agreement and operational since 1947. It seeks to promote international monetary cooperation and the growth of world trade, and to smooth payment arrangements among member states. IMF standby loans are available to members in balance-of-payments difficulties (the amount being governed by the member's quota), usually on the basis that the country must agree to take certain corrective measures.

international trade the exchange of goods and services between countries. Ideally, a country should aim to export more than it imports, thus establishing a favourable *balance of payments*. Countries in the *developing world tend to import more than they export, thus incurring large debts to developed countries.

International Union for the Conservation of Nature IUCN, alternative name for the *World Conservation Union.

interplanetary matter gas, dust, and charged particles from a variety of sources that occupies the space between the planets. Dust left over from the formation of the Solar System, or from disintegrating comets, orbits the Sun in or near the ecliptic plane and is responsible for scattering sunlight to cause the zodiacal light (mainly the smaller particles) and for meteor showers in the Earth's atmosphere (larger particles). The charged particles mostly originate in the solar wind, but some are cosmic rays from deep space.

interquartile range in statistics, a measure of the range of data, equalling the difference in value between the upper and lower *quartiles. It provides information on the spread of data in the middle 50% of a distribution and is not affected by freak extreme values.

intertropical convergence zone ITCZ, area of heavy rainfall found in the tropics and formed as the trade winds converge and rise to form cloud and rain. It moves a few degrees northwards during the northern summer and a few degrees southwards during the southern summer, following the apparent movement of the Sun. The ITCZ is responsible for most of the rain that falls in Africa. The *doldrums are also associated with this zone.

interview face-to-face meeting where one party, the interviewer, wishes to collect information from another party, the interviewee. Interviews are commonly used in selecting people for jobs. Banks and building societies also use interviews for processing applications for loans, and market researchers conduct interviews to gather information about the market they are investigating.

intrapreneur individual within an organization who is allowed to operate as an *entrepreneur. Intrapreneurs can be senior managers in companies that have decided to introduce internal competition where areas of the business are run as profit centres. This approach allows the manager to contract out activities that they would normally have to source and carry out within the company. With this approach there is a trade-off between cost and control. Intrapreneurs might also be individuals within an organization selected for their particular aptitude and encouraged to develop business ideas within an organization as if they were in an external new venture. Intrapreneurs may become entrepreneurs if their profit centre is spun out of its parent company.

intrazonal soil a type of soil for which local features such as drainage, slope and parent material are the major features of the soil-forming process, rather than climate and vegetation. The concept of the intrazonal soil is too simplistic to explain the multitude of soil types to be found in a region and is considered outdated.

intrusion mass of *igneous rock that has formed by 'injection' of molten rock, or magma, into existing cracks beneath the surface of the Earth, as distinct from a volcanic rock mass which has erupted from the surface. Intrusion features include vertical

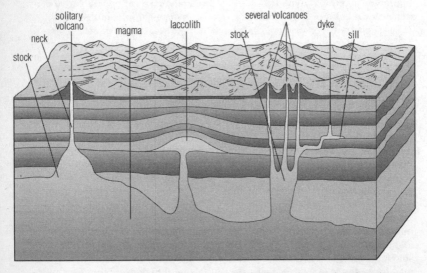

intrusion Igneous intrusions can be a variety of shapes and sizes. Laccoliths are domed circular shapes, and can be many miles across. Sills are intrusions that flow between rock layers. Pipes or necks connect the underlying magma chamber to surface volcanoes.

cylindrical structures such as stocks, pipes, and necks; sheet structures such as dykes that cut across the strata and sills that push between them; laccoliths, which are blisters that push up the overlying rock; and batholiths, which represent chambers of solidified magma and contain vast volumes of rock.

intrusive rock or **plutonic rock**, *igneous rock formed beneath the Earth's surface. Magma, or molten rock, cools slowly at these depths to form coarse-grained rocks, such as granite, with large crystals. (*Extrusive rocks, which are formed on the surface, are generally fine-grained.) A mass of intrusive rock is called an intrusion.

investment in economics, the purchase of any asset with the potential to yield future financial benefit to the purchaser (such as a house, a work of art, stocks and shares, or even a private education).

invisible in economics, term describing a service on the *balance of payments account. Invisible exports are exports of services and invisible imports are imports of services.

inward investment financial investment in a country by a *transnational corporation. The investment may be used to develop industrial or agricultural projects in developing countries and may provide much-needed employment opportunities. The investment may also be made in developed countries, with the host government providing incentives (e.g. tax benefits or a funding contribution) to the transnational corporation. Such investments are usually made in regions requiring new economic activity. However, the outputs of such projects are usually intended for the transnational corporation's country of origin.

ionosphere ionized layer of Earth's outer *atmosphere (60–1000 km) that contains sufficient free electrons to

modify the way in which radio waves are propagated, for instance by reflecting them back to Earth. The ionosphere is thought to be produced by absorption of the Sun's ultraviolet radiation. The British Antarctic Survey estimates that the ionosphere is decreasing at a rate of 1 km every five years, based on an analysis of data from 1960–1998. Global warming is the probable cause.

iridium anomaly unusually high concentrations of the element iridium found worldwide in sediments that were deposited at the Cretaceous-Tertiary boundary (*K-T boundary) 65 million years ago. Since iridium is more abundant in extraterrestrial material, its presence is thought to be evidence for a large meteorite impact that may have caused the extinction of the dinosaurs and other life at the end of the Cretaceous.

iron and steel industry the extraction of iron from ore and the manufacture of steel, which together are a key element in the heavy industrial structure of a nation. In Britain, the location of the iron and steel industry is an example of changing optimum location over time. The total pattern of location has three elements. Firstly, there are the coalfield locations such as Sheffield, dating from the early establishment of iron and steel manufacture using *coal and blackband iron ore, found in the coal measures, as *raw materials. Secondly, there are the orefield locations such as Scunthorpe (and formerly Corby), developed in the mid-20th century to exploit the iron ore deposits of Jurassic rocks in eastern England. By this stage the steel-making process depended less on coal, owing to the development of the electric arc furnace and, in blast furnaces, better designs which required less coal to produce iron. Thirdly, there are the most recently established coastal locations of the iron and steel industry, for example at Port Talbot in South Wales and on Teesside in northeast England. These locations reflect the dominant role today of imported ores from places such as Scandinavia and Canada. Thus the three locational elements of the overall distribution of the iron and steel industry reflect the three stages of its evolution. Steel-making in general is in decline, and several inland steelworks in the UK have closed (e.g. Consett in Co. Durham, Ravenscraig in Lanarkshire and Corby in Northamptonshire).

iron ore   any mineral from which iron is extracted. The chief iron ores are *magnetite, a black oxide; *haematite, or kidney ore, a reddish oxide; *limonite, brown, impure oxyhydroxides of iron; and siderite, a brownish carbonate.

Iron ores are found in a number of different forms, including distinct layers in igneous intrusions, as components of contact metamorphic rocks, and as sedimentary beds. Much of the world's iron is extracted in Russia, Kazakhstan, and the Ukraine. Other important producers are the USA, Australia, France, Brazil, and Canada; over 40 countries produce significant quantities of ore.

ironpan   see *hardpan.

irrigation   artificial water supply for dry agricultural areas by means of dams and channels. Drawbacks are that it tends to concentrate salts at the surface, ultimately causing soil infertility, and that rich river silt is retained at dams, to the impoverishment of the land and fisheries below them.

Irrigation has been practised for thousands of years, in Eurasia as well as the Americas. An example is the channelling of the annual Nile flood in Egypt, which has been done from earliest times to its present control by the Aswan High Dam.

island   area of land surrounded entirely by water. Australia is classed as a continent rather than an island, because of its size.

Islands can be formed in many ways. **Continental islands** were once part of the mainland, but became isolated (by tectonic movement, erosion, or a rise in sea level, for example). **Volcanic islands**, such as Japan, were formed by the explosion of underwater volcanoes. **Coral islands** consist mainly of *coral, built up over many years. An **atoll** is a circular coral reef surrounding a lagoon; atolls were formed when a coral reef grew up around a volcanic island that subsequently sank or was submerged by a rise in sea level. **Barrier islands** are found by the shore in shallow water, and are formed by the deposition of sediment eroded from the shoreline.

island arc   curved chain of volcanic islands. Island arcs are common in the Pacific where they ring the ocean on both sides; the Aleutian Islands off Alaska are an example. The volcanism that forms island arcs is a result of subduction of an oceanic plate beneath another plate, as evidenced by the presence of ocean trenches on the convex side of the arc, and the Benioff zone of high seismic activity beneath.

Such island arcs are often later incorporated into continental margins during mountain-building episodes.

isobar   line drawn on maps and weather charts linking all places with the same atmospheric pressure (usually measured in millibars). When used in weather forecasting, the distance between the isobars is an indication of the barometric gradient (the rate of change in pressure).

Where the isobars are close together, cyclonic weather is indicated, bringing strong winds and a depression, and where far apart anticyclonic, bringing calmer, settled conditions.

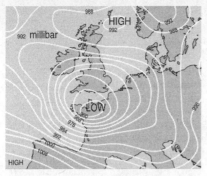

isobar The isobars around a low-pressure area or depression. In the northern hemisphere, winds blow anticlockwise around lows, approximately parallel to the isobars, and clockwise around highs. In the southern hemisphere, the winds blow in the opposite directions.

isobleth term meaning a line on a *map which connects points of equal value. Specific types of isobleth include: a contour line, connecting points of equal altitude; an *isohyet, connecting points of equal rainfall; an *isotherm, connecting points of equal temperature at a given time; an *isobath, connecting points of equal distance below a water body; and an *isobar, connecting points of equal atmospheric pressure.

isochrone on a map, a line that joins places that are equal in terms of the time it takes to reach them.

isohyet on a map, a line joining points of equal rainfall.

isolationism in politics, concentration on internal rather than foreign affairs; a foreign policy having no interest in international affairs that do not affect the country's own interests.

isoline on a map, a line that joins places of equal value. Examples are contour lines (joining places of equal height), *isobars (for equal pressure), *isotherms (for equal temperature), and *isohyets (for equal rainfall). Isolines are most effective when values change gradually and when there is plenty of data.

isostasy condition of gravitational equilibrium of all parts of the Earth's *crust. The crust is in isostatic equilibrium if, below a certain depth, the weight and thus pressure of rocks above is the same everywhere. The idea is that the lithosphere floats on the asthenosphere as a piece of wood floats on water. A thick piece of wood floats lower than a thin piece, and a denser piece of wood floats lower than a less dense piece. There are two theories of the mechanism of isostasy, the Airy hypothesis and the Pratt hypothesis, both of which have validity. In the **Airy hypothesis** crustal blocks have the same density but different thicknesses: like ice cubes floating in water, higher mountains have deeper roots. In the **Pratt hypothesis**, crustal blocks have different densities allowing the depth of crustal material to be the same. In practice, both mechanisms are at work.

isotherm line on a map, linking all places having the same temperature at a given time.

isthmus narrow strip of land joining two larger land masses. The Isthmus of Panama joins North and South America.

IUCN *abbreviation for* International Union for the Conservation of Nature, an alternative name for the *World Conservation Union.

Ivermectin pesticide used by fish farmers in Scotland to control salmon lice. It is controversial because environmentalists claim it contaminates the water surrounding fish farms, killing other marine organisms, particularly polychaete worms such as lugworms.

ivory hard white substance of which the teeth and tusks of certain mammals are made. Among the most valuable are elephants' tusks, which are of unusual hardness and density. Ivory is used in carving and other decorative work, and is so valuable that poachers continue to illegally destroy the remaining wild elephant herds in Africa to obtain it.

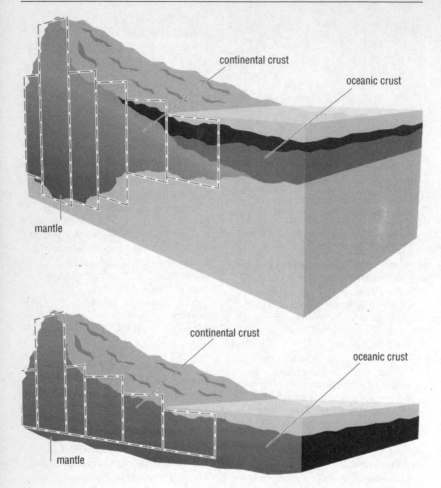

isostasy Isostasy explains the vertical distribution of Earth's crust. George Bedell Airy proposed that the density of the crust is everywhere the same and the thickness of crustal material varies. Higher mountains are compensated by deeper roots. This explains the high elevations of most major mountain chains, such as the Himalayas. G H Pratt hypothesized that the density of the crust varies, allowing the base of the crust to be the same everywhere. Sections of crust with high mountains, therefore, would be less dense than sections of crust where there are lowlands. This applies to instances where density varies, such as the difference between continental and oceanic crust.

J

jacinth or **hyacinth**, red or yellowish-red gem, a variety of zircon, $ZrSiO_4$.

jade semi-precious stone consisting of either jadeite, $NaAlSi_2O_6$ (a pyroxene), or nephrite, $Ca_2(Mg,Fe)_5Si_8O_{22}(OH,F)_2$ (an amphibole), ranging from colourless through shades of green to black according to the iron content. Jade ranks 5.5–6.5 on the Mohs scale of hardness.

Japan Current or **Kuroshio**, warm ocean *current flowing from Japan to North America.

jasper hard, compact variety of *chalcedony SiO_2, usually coloured red, brown, or yellow. Jasper can be used as a gem.

jet in earth science, hard, black variety of lignite, a type of coal. It is cut and polished for use in jewellery and ornaments. Articles made of jet have been found in Bronze Age tombs.

jet stream narrow band of very fast wind (velocities of over 150 kph) found at altitudes of 10–16 km in the upper troposphere or lower stratosphere. Jet streams usually occur about the latitudes of the Westerlies (35–60°).

The jet stream may be used by high flying aircraft to speed their journeys. Their discovery of the existence of the jet stream allowed the Japanese to send gas-filled balloons carrying bombs to the northwestern USA during World War II.

joint in earth science, a vertical crack in a rock, often formed by compression; it is usually several metres in length. A joint differs from a *fault in that no displacement of the rocks on either side has taken place. The weathering of joints in rocks such as limestone and granite is responsible for the formation of features such as *limestone pavements and *tors. Joints in coastal rocks are often exploited by the sea to form erosion features such as caves and geos.

joint venture in business, an undertaking in which an individual or legal entity of one country forms a company with those of another country, with risks being shared.

jungle popular name for *rainforest.

Jurassic period period of geological time 208–146 million years ago; the middle period of the Mesozoic era. Climates worldwide were equable, creating forests of conifers and ferns; dinosaurs were abundant, birds evolved, and limestones and iron ores were deposited.

The name comes from the Jura Mountains in France and Switzerland, where the rocks formed during this period were first studied.

just-in-time JIT, production management practice requiring that incoming supplies arrive at the time when they are needed by the customer, most typically in a manufacturer's assembly operations. JIT requires considerable cooperation between supplier and customer, but can reduce expenses and improve efficiency, for example by reducing stock levels and by increasing the quality of goods supplied.

Kainozoic era see *Cenozoic Era.

kame geological feature, usually in the form of a mound or ridge, formed by the deposition of rocky material carried by a stream of glacial meltwater. Kames are commonly laid down in front of or at the edge of a glacier (kame terrace), and are associated with the disintegration of glaciers at the end of an ice age.

Kames are made of well-sorted rocky material, usually sands and gravels. The rock particles tend to be rounded (by attrition) because they have been transported by water.

kaolin or **china clay**, group of clay minerals, such as *kaolinite, $Al_2Si_2O_5(OH)_4$, derived from the alteration of aluminium silicate minerals, such as *feldspars and *mica. It is used in medicine to treat digestive upsets, and in poultices.

Kaolinite is economically important in the ceramic and paper industries. It is mined in the UK, the USA, France, and the Czech Republic.

kaolinite white or greyish *clay mineral, hydrated aluminium silicate, $Al_2Si_2O_5(OH)_4$, formed mainly by the decomposition of feldspar in granite. It is made up of platelike crystals, the atoms of which are bonded together in two-dimensional sheets, between which the bonds are weak, so that they are able to slip over one another, a process made more easy by a layer of water. China clay (kaolin) is derived from it. It is mined in France, the UK, Germany, China, and the USA.

karst landscape characterized by remarkable surface and underground forms, created as a result of the action of water on permeable limestone. The feature takes its name from the Karst (meaning **dry**) region on the Adriatic coast of Slovenia and Croatia, but the name is applied to landscapes throughout the world, the most dramatic of which is found near the city of Guilin in the Guangxi province of China. Karst landscapes are characterized by underground features such as caves, caverns, stalactites, and stalagmites. On the surface, clints, grikes, gorges, and swallow holes are common features.

Limestone is soluble in the weak acid of rainwater. Erosion takes place most swiftly along cracks and joints in the limestone and these open up into gullies called grikes. The rounded blocks left upstanding between them are called clints.

katabatic wind cool wind that blows down a valley on calm clear nights. (By contrast, an *anabatic wind is warm and moves up a valley in the early morning.) When the sky is clear, heat escapes rapidly from ground surfaces, and the air above the ground becomes chilled. The cold dense air moves downhill, forming a wind that tends to blow most strongly just before dawn.

Cold air blown by a katabatic wind may collect in a depression or valley bottom to create a *frost hollow. Katabatic winds are most likely to occur in the late spring and autumn because of the greater daily temperature differences.

Kennewick Man 9300-year-old skeleton found along the banks of the Columbia River in Kennewick, Washington, USA, in July 1996. The skeleton, the oldest ever found in the

Northwest USA, is 90% complete – one of the most complete skeletons ever unearthed in the Americas. It belongs to a 45-year-old male about 175 cm tall, who died with a projectile point embedded in his pelvis. It also displays Caucasoid features and has raised questions about the origin of humans in the Americas.

kettle hole or **kettle**, pit or depression formed when a block of ice from a receding glacier becomes isolated and buried in glacial debris (till). As the block melts the till collapses to form a hollow, which may become filled with water to form a kettle lake or pond. Kettle holes range from 5–13 km in diameter, and may exceed 33 m in depth.

As time passes, water sometimes fills the kettle holes to form lakes or swamps, features found throughout much of northern North America. Lake Ronkonkoma, the largest lake on Long Island, New York, is an example.

key small, flat, low-lying island, normally composed of sand, formed on a coral-reef platform. Keys usually reach just a couple of metres above high-tide level. Initially formed from reef debris, they become stabilized by vegetation growth, but always remain susceptible to erosion. Keys are most commonly found in the Gulf of Mexico, particularly off the south coast of Florida, and in the Caribbean (for example, the Abacos islands).

Keynesian economics the economic theory of English economist John Maynard Keynes, which argues that a fall in national income, lack of demand for goods, and rising unemployment should be countered by increased government expenditure to stimulate the economy. It is opposed by monetarists (see *monetarism).

khamsin hot southeasterly wind that blows from the Sahara desert over Egypt and parts of the Middle East from late March to May or June. It is called *sharav* in Israel.

kimberlite igneous rock that is ultramafic (containing very little silica); a type of alkaline *peridotite with a porphyritic texture (larger crystals in a fine-grained matrix), containing mica in addition to olivine and other minerals. Kimberlite represents the world's principal source of diamonds.

Kimberlite is found in carrot-shaped pipelike *intrusions called **diatremes**, where mobile material from very deep in the Earth's crust has forced itself upwards, expanding in its ascent. The material, brought upwards from near the boundary between crust and mantle, often altered and fragmented, includes diamonds. Diatremes are found principally near Kimberley, South Africa, from which the name of the rock is derived, and in the Yakut area of Siberia, Russia.

kinship in anthropology, human relationship based on blood or marriage, and sanctified by law and custom. Kinship forms the basis for most human societies and for such social groupings as the family, clan, or tribe.

knickpoint a marked change of slope in the long profile of a river caused by rejuvenation or the outcropping of a resistant rock formation such as a dyke.

Koh-i-noor (Persian 'mountain of light') diamond, originally part of the Aurangzeb treasure, seized in 1739 by the shah of Iran from the Moguls in India, taken back by Sikhs, and acquired by Britain in 1849 when the Punjab was annexed.

komatiite oldest volcanic rock with three times as much magnesium as other volcanic rocks. Unlike basaltic lavas, which comprise oceanic crust, komatiites have the chemical composition of *peridotite, the primary constituent rock of the upper *mantle. Komatiites were extruded as a liquid at high temperatures, perhaps more than 1600°C. They have low titanium and high magnesium, nickel, and chromium content.

Kondratieff cycle 50-year cycle of economic upturn and downturn. At the beginning of the 20th century, Russian economist Nikolai Kondratieff proposed that, in addition to the commonly accepted five- to ten-year cycle, there existed a longer cycle of about 50 years. Dismissed at the time, the theory has been re-evaluated in the light of the depressions of the 1880s and 1930s and the recessions between 1970 and the mid-1990s. It is thought that these longer cycles, if they do exist, are related to major technological change and the turmoil that follows.

Koonalda Cave cave in southwestern South Australia below the Nullarbor Plain. Anthropologists in the 1950s and 1960s discovered evidence of flint-quarrying and human markings that have been dated as 20,000 years old.

Kow Swamp area in northern Victoria, Australia, west of the town of Echuca, where the remains of about 40 humans have been found, buried in shallow graves. These bones have mostly been dated as 9000–14,000 years old and were accompanied by human artefacts and objects such as shells. They are of a larger and more robust group of humans than those found from an earlier period at Lake Mungo and are similar to other finds from widely scattered sites in Australia such as Talgai in Queensland.

k-species any species of plant or animal characterized by a large body size, a slow rate of development, considerable competitive ability, a reproductive phase occurring mid-way or late in the life history and which can be undertaken several or many times over. Such species often live for many years and are characterized by terrestrial vertebrates.

K-T boundary geologists' shorthand for the boundary between the rocks of the *Cretaceous and the *Tertiary periods 65 million years ago. It coincides with the end of the extinction of the dinosaurs and in many places is marked by a layer of clay or rock enriched in the element iridium. Extinction of the dinosaurs at the K-T boundary and deposition of the iridium layer are thought to be the result of either impact of an asteroid or comet that crashed into the Yucatán Peninsula (forming the **Chicxulub crater**), perhaps combined with a period of intense volcanism on the continent of India.

Kuroshio or **Japan Current**, warm ocean *current flowing from Japan to North America.

kyanite aluminium silicate, Al_2SiO_5, a pale-blue mineral occurring as blade-shaped crystals. It is an indicator of high-pressure conditions in metamorphic rocks formed from clay sediments. Andalusite, kyanite, and sillimanite are all polymorphs (see *polymorphism).

Kyoto Protocol international protocol to the United Nations Framework Convention on Climate Change (UNFCCC) that was agreed at Kyoto, Japan, in December 1997. It commits the 186 signatory countries to binding limits on carbon dioxide and other heat-trapping 'greenhouse gases', which many scientists believe contribute to *global warming. For industrialized nations, Kyoto requires cuts in greenhouse gas emissions to an average of 5.2% below 1990 levels by 2012. Developing countries are also committed to emissions targets. The text of the UNFCCC was adopted in 1992 and promoted at the climate summit held in Rio de Janeiro, Brazil, in June 1992. The convention entered into force in 1994, with 166 countries as signatories. The protocol was adopted at the December 1997 Kyoto conference on the UNFCCC. It will come into force on the 90th day after it is ratified by at least 55 parties to the convention which accounted in total for at least 55% of global carbon dioxide emissions in 1990.

A controversial feature of the protocol is that it allows countries to count, as reducing emissions, 'carbon sinks' (forests and grasslands) added to soak up emissions of greenhouse gases (which include carbon dioxide, methane, and nitrous oxide).The accord was dealt a blow in March 2001 when US president George W Bush announced that the USA, which emits 25% of world greenhouse gases with a 15% increase in emission levels over the last ten years, would not ratify Kyoto as mandatory pollution reductions would harm US economic interests. A European Union (EU) summit held in Göteborg, Sweden, in June 2001, centred on the EU's attempt to persuade the USA to accept the Kyoto Protocol on climate change. However, five hours of talks between President George W Bush and EU leaders yielded no movement from the USA, and the EU member states declared that they would ratify the protocol without US involvement. However, the international effort to save the protocol was dealt two severe blows in July 2001, when first Japan and then Australia both said they would not sign up to any agreement that did not include the USA. The decisions had come despite the efforts of a European Union (EU) delegation that had sought to secure Japanese and Australian involvement. An environmental summit held in Bonn, Germany, in July 2001, ended with a compromise deal based on a more flexible version of the protocol. All sides gave ground in order to salvage the treaty, but the biggest compromises were made by the EU, who eventually conceded substantial 'carbon sinks' to Canada, Japan, and Russia. The EU's earlier objection to the widespread use of 'sink' forests had led to the withdrawal of the USA from the protocol: however, despite the concessions, the USA maintained its opposition. In March 2002, the EU agreed to be legally bound by the terms of the protocol.

labour one of the factors of production, used to produce goods and provide services. Wages are the reward for labour. The quantity of labour in a modern economy is determined by the size of the population and the extent to which young and old people and women are prepared to take paid work.

labour-intensive production production where a large amount of labour is used relative to capital. For example, window-cleaning or brick-laying are labour-intensive jobs. Service industries tend to be far more labour-intensive than either primary or secondary industries.

labour market market that determines the cost and conditions of the work force, taking into consideration the demand of employers, the levels and availability of skills, and social conditions.

labour mobility degree to which labour is willing to move in order to obtain employment. Moving from one part of the country to another, or from one country to another, is known as geographical mobility. Movement from one type of employment to another is known as occupational mobility. Factors that affect geographical mobility include regional property prices, local ties, and communications infrastructure.

labour theory of value in classical economics, the theory that the price (value) of a product directly reflects the amount of labour it involves. According to the theory, if the price of a product falls, either the share of labour in that product has declined or that expended in the production of other goods has risen.

laccolith intruded mass of igneous rock that forces apart two strata and forms a round lens-shaped mass many times wider than thick. The overlying layers are often pushed upward to form a dome. A classic development of laccoliths is illustrated in the Henry, La Sal, and Abajo mountains of southeastern Utah, found on the Colorado Plateau.

lagoon coastal body of shallow salt water, usually with limited access to the sea. The term is normally used to describe the shallow sea area cut off by a *coral reef or barrier islands.

lahar mudflow formed of a fluid mixture of water and volcanic ash. During a volcanic eruption, melting ice may combine with ash to form a powerful flow capable of causing great destruction. The lahars created by the eruption of Nevado del Ruiz in Colombia, South America, in 1985 buried 22,000 people in 8 m of mud.

lake body of still water lying in depressed ground without direct communication with the sea. Lakes are common in formerly glaciated regions, along the courses of slow rivers, and in low land near the sea. The main classifications are by origin: **glacial lakes**, formed by glacial scouring; **barrier lakes**, formed by *landslides and glacial *moraines; **crater lakes**, found in *volcanoes; and **tectonic lakes**, occurring in natural fissures.

Crater lakes form in the *calderas of extinct volcanoes, for example Crater Lake, Oregon, USA. Subsidence of the roofs of limestone caves in *karst landscape exposes the subterranean stream network and provides a cavity in which a lake can develop. Tectonic lakes form during tectonic movement, as when a *rift valley is formed. Lake Tanganyika was created in conjunction with the East African Great Rift Valley. Glaciers produce several distinct types of lake, such as the lochs of Scotland and the Great Lakes of North America.

Lakes are mainly freshwater, but salt and bitter lakes are found in areas of low annual rainfall and little surface run-off, so that the rate of evaporation exceeds the rate of inflow, allowing mineral salts to accumulate. The Dead Sea has a salinity of about 250 parts per 1000 and the Great Salt Lake, Utah, about 220 parts per 1000. Salinity can also be caused by volcanic gases or fluids, for example Lake Natron, Tanzania.

From the 20th century large artificial lakes have been created in connection with hydroelectric and other works. Some lakes have become polluted as a result of human activity. Sometimes *eutrophication (a state of over-nourishment) occurs, when agricultural fertilizers leaching into lakes cause an explosion of aquatic life, which then depletes the lake's oxygen supply until it is no longer able to support life.

Land plural **Länder**, federal state of Germany or Austria.

land in economics, the factor of production which comprises not just land itself but all natural resources. Shoals of fish, natural forests, the atmosphere, and rivers are examples of land. The reward paid to owners of land is rent.

land breeze a cool, gentle breeze blowing from the land towards the sea during summer and autumn, affecting coastal areas. It occurs at night when the sea is relatively warmer than the land and when, as a result, pressure is relatively lower over the sea. During the day opposite conditions prevail: the land is relatively warmer and the pressure gradient is from land to sea; a sea breeze then occurs. Such conditions

arise because the land heats up and cools down more quickly than the sea.

landfill site large holes in the ground used for dumping household and commercial waste. Landfill disposal has been the preferred option in the UK and the USA for many years, with up to 85% of household waste being dumped in this fashion. However, the sites can be dangerous, releasing toxins and other leachates (see *leaching) into the soil and the policy is itself wasteful both in terms of the materials dumped and land usage.

landform the shape, size, form, nature, and characteristics of a specific feature of the land's surface, whether above ground or underwater. The term *geomorphology is used to describe the study of landforms.

land reclamation conversion of derelict or otherwise unusable areas into productive land. For example, where industrial or agricultural activities, such as sand and gravel extraction or open-cast mining, have created large areas of derelict or waste ground, the companies involved are usually required to improve the land so that it can be used.

land reform the process of redistributing land, especially in developing nations where current circumstances may be against fair access to land and against agricultural improvement. For example, in parts of Latin America the traditional *land tenure system is made up of large commercial estates (known as *estancias*) and small peasant holdings. Often the estate occupies the best farming land whilst the peasants farm in difficult conditions such as on valley sides. This inequality has led to pressure for land reform, sometimes by violent revolution. The aim of recent land-reform schemes has been to provide typical farming families with security, reasonable farming land, and an incentive to improve productivity. Concern for land reform occurs not only in situations of unequal distribution as described above, but also in traditional rural economies where land is allocated through the chieftainship, often in small and fragmented parcels, as occurs in parts of Africa. Here the necessary reforms would be consolidation of holdings, the provision of secure tenure, and some facility for the progressive farmer to improve productivity. However, clumsy land reform may do more harm than good: land in many rural societies is much more than a commodity, it is an integral part of the culture and heritage.

landslide sudden downward movement (see *mass movement) of a mass of soil or rocks from a cliff or steep slope. Landslides happen when a slope becomes unstable, usually because the base has been undercut or because materials within the mass have become wet and slippery.

A **mudflow** happens when soil or loose material is soaked so that it no longer adheres to the slope; it forms a tongue of mud that reaches downhill from a semicircular hollow. A **slump** occurs when the material stays together as a large mass, or several smaller masses, and these may form a tilted steplike structure as they slide. A **landslip** is formed when *beds of rock dipping towards a cliff slide along a lower bed. Earthquakes may precipitate landslides.

land tenure the relation of a farmer to the land farmed. Farmers may be owner-occupiers, tenants, landless labourers, or state employees.

The *Latifundia system is common in Latin America. Land is organized into large, centrally managed estates worked by landless labourers for low wages. Crops are produced for local use. In the 1980s 70% of Brazil's land

mudflow landslide

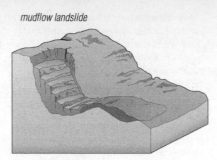

slump landslide

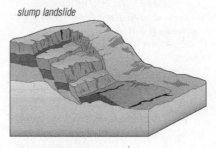

landslip landslide

landslide Types of landslide. A mudflow is a tongue of mud that slides downhill. A slump is a fall of a large mass that stays together after the fall. A landslip occurs when beds of rock move along a lower bed.

was owned by 3% of the population.

Peasant farmers may have limited access to land, which they may or own or rent from the local landowner. This type of tenancy takes the form of **cash crops**, where up to 80% of the farmer's income is given to the landowner as rent, and **share crops**, where part of the farmer's crop is given directly to the landowner.

The **plantation** is a variant form of the large estate system and is usually operated commercially, producing crops for the world market rather than for local use. Labourers may be landless and receive a fixed wage.

land use   the way in which a given area of land is used. Land is often classified according to its use, for example, for agriculture, industry, residential buildings, and recreation. The first land use surveys in the UK were conducted during the 1930s.

land-value gradient   the decline of average land values per unit area with increasing distance from the *CBD. Whilst this is usually true, some suburban areas with good *accessibility may prove to be good locations for business and retail centres. Therefore the land-value gradient may show peaks within suburban areas.

lapis lazuli   rock containing the blue mineral lazurite in a matrix of white calcite with small amounts of other minerals. It occurs in silica-poor igneous rocks and metamorphic limestones found in Afghanistan, Siberia, Iran, and Chile. Lapis lazuli was a valuable pigment of the Middle Ages, also used as a gemstone and in inlaying and ornamental work.

lapse rate   the rate at which temperature changes with altitude. In the *troposphere, it is usual for temperature to decrease with altitude at an average rate of 0.6°C per 100 metres. In certain conditions of very still air a temperature inversion happens, with temperature increasing with altitude. Such an inversion can trap pollutants close to the surface of the Earth. Beyond the **tropopause**, in the upper part of the *stratosphere, temperature increases with height.

lat.   *abbreviation for* *latitude.

lateral moraine   linear ridge of rocky debris deposited near the edge of a *glacier. Much of the debris is material that has fallen from the valley side onto the glacier's edge, having been weathered by *freeze–thaw (the alternate freezing and thawing of ice in cracks); it will, therefore, tend to be angular in nature. Where two glaciers merge, two lateral moraines may join together to form a **medial moraine** running along the centre of the merged glacier.

laterite   red residual soil characteristic of tropical rainforests. It is formed by the weathering of basalts, granites, and shales and contains a high percentage of aluminium and iron hydroxides. It may form an impermeable and infertile layer that hinders plant growth.

latifundia   large estates most commonly found in Central and South America on which much or all of the land is worked by means of sharecropping and day labourers. Latifundia are symptomatic of the gross inequalities in land distribution found in many Latin American countries. In many cases the land owners' influence is all-pervasive and extends beyond the farming activities of their employees to include effective social control over the lives of the mass of the rural population.   .

Latin American Economic System LAES; Spanish **Sistema Económico Latino-Americana (SELA)**, international coordinating body for economic, technological, and scientific cooperation in Latin America and the Caribbean, aiming to create and promote multinational enterprises in the region and provide markets. Founded in 1975 as the successor to the Latin American Economic Coordination Commission, its members include Argentina, Barbados, Bolivia, Brazil, Chile, Colombia, Costa Rica, Cuba, Dominican Republic, Ecuador, El Salvador, Grenada, Guatemala, Guyana, Haiti, Honduras, Mexico, Jamaica,

Nicaragua, Panama, Paraguay, Peru, Spain (from 1979), Suriname, Trinidad and Tobago, Uruguay, and Venezuela. Its headquarters are in Caracas, Venezuela.

Latin American Integration Association Spanish **Asociación Latino-Americana de Integración (ALADI)**, organization aiming to create a common market in Latin America; to promote trade it applies tariff reductions preferentially on the basis of the different stages of economic development that individual member countries have reached. Formed in 1980 to replace the Latin American Free Trade Association (formed in 1961), it has 11 members: Argentina, Bolivia, Brazil, Chile, Colombia, Ecuador, Mexico, Paraguay, Peru, Uruguay, and Venezuela. Its headquarters are in Bogotá, Colombia.

latitude and longitude imaginary lines used to locate position on the globe. Lines of latitude are drawn parallel to the Equator, with 0° at the Equator and 90° at the north and south poles. Lines of longitude are drawn at right-angles to these, with 0° (the Prime Meridian) passing through Greenwich, England.

The 0-degree line of latitude is defined by Earth's Equator, a characteristic definable by astronomical observation. It was determined as early as AD 150 by Egyptian astronomer Ptolemy in his world atlas. The prime meridian, or 0-degree line of longitude, is a matter of convention rather than physics. Prior to the latter half of the 18th century, sailors navigated by referring to their position east or west of any arbitrary meridian. When Nevil Maskelyne (1732–1811), English astronomer and fifth Astronomer Royal, published the *Nautical Almanac* he referred all of his lunar–stellar distance tables to the Greenwich meridian. These tables were relied

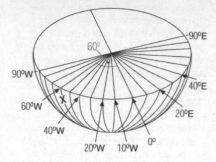

Point X lies on longitude 60°W

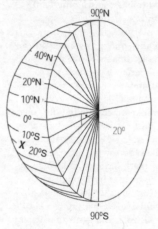

Point X lies on latitude 20°S

latitude and longitude Locating a point on a globe using latitude and longitude. Longitude is the angle between the terrestrial meridian through a place and the standard meridian 0° passing through Greenwich, England. Latitude is the angular distance of a place from the equator.

upon for computing longitudinal position and so the Greenwich meridian became widely accepted.

Chronometers, time-keeping devices with sufficient accuracy for longitude determination, invented by English instrument-maker John Harrison (1693–1776) and perfected in 1759, would gradually replace the lunar distance method for navigation, but reliance on the Greenwich meridian persisted because the *Nautical Almanac* was used by sailors to verify their position. The Greenwich meridian was officially adopted as the Prime Meridian by the International Meridian Conference held in Washington, DC, in 1884.

Laurasia northern landmass formed 200 million years ago by the splitting of the single world continent *Pangaea. (The southern landmass was *Gondwanaland.) It consisted of what was to become North America, Greenland, Europe, and Asia, and is believed to have broken up about 100 million years ago with the separation of North America from Europe.

lava molten *magma that erupts from a *volcano and cools to form extrusive *igneous rock. Lava types differ in composition, temperature, gas content, and viscosity (resistance to flow).

The three major lava types are basalt (dark, fluid, and relatively low silica content), rhyolite (light, viscous, high silica content), and andesite (an intermediate lava).

The viscosity of lava was once ascribed to whether it was **acidic** or **basic**. However, the terms *acid rock and *basic rock are misleading, and are no longer used as classifications.

lava flow a stream of *lava issuing from some form of volcanic eruption. See also *viscous lava.

lava plateau a relatively flat upland composed of layer upon layer of approximately horizontally bedded lavas. An example of this is the Deccan Plateau of India.

leaching process by which substances are washed through or out of the soil. Fertilizers leached out of the soil drain into rivers, lakes, and ponds and cause *water pollution. In tropical areas, leaching of the soil after the destruction of forests removes scarce nutrients and can lead to a dramatic

loss of soil fertility. The leaching of soluble minerals in soils can lead to the formation of distinct soil horizons as different minerals are deposited at successively lower levels.

lead ore any of several minerals from which lead is extracted. The primary ore is galena or lead sulphite PbS. This is unstable, and on prolonged exposure to the atmosphere it oxidizes into the minerals cerussite $PbCO_3$ and anglesite $PbSO_4$. Lead ores are usually associated with other metals, particularly silver – which can be mined at the same time – and zinc, which can cause problems during smelting.

Most commercial deposits of lead ore are in the form of veins, where hot fluids have leached the ore from cooling *igneous masses and deposited it in cracks in the surrounding country rock, and in thermal *metamorphic zones, where the heat of igneous intrusions has altered the minerals of surrounding rocks. Lead is mined in over 40 countries, but half of the world's output comes from the USA, Canada, Russia, Kazakhstan, Uzbekistan, Canada, and Australia.

leaf litter in ecology, accumulation of leaves and other organic detritus (debris) on a woodland floor following leaf fall. It is a rich source of food for detritovores (animals that feed on detritus) and decomposers. Their activities eventually lead to the formation of *humus, which improves the texture and structure of soil.

learning curve curve reflecting the reduction in time taken to perform a task as experience in performing that task increases. Successful companies and individuals are those that progress up the learning curve the quickest. The expression 'steep learning curve' is commonly used to describe situations where an individual has to become familiar with a new task in a very short time.

least-cost location   the optimum location for an industry. The least-cost location may change over time with transport developments, changes in the sources of raw materials and changing markets. The least-cost location for iron and steel manufacture, for example, has shifted from inland coalfields to coastal sites as imported *raw materials have increased in importance. See also *industrial location.

less developed country   or **least developed country; LDC,** any country late in developing an industrial base, and dependent on cash crops and unprocessed minerals; part of the *developing world. The terms 'less developed', 'least developed', and 'developing' imply that industrial development is desirable or inevitable.

The Group of 77 was established in 1964 to pressure industrialized countries into giving greater aid to less developed countries.

leucite   silicate mineral, $KAlSi_2O_6$, occurring frequently in some potassium-rich volcanic rocks. It is dull white to grey, and usually opaque. It is used as a source of potassium for fertilizer.

levee   naturally formed raised bank along the side of a river channel. When a river overflows its banks, the rate of flow is less than that in the channel, and silt is deposited on the banks. With each successive flood the levee increases in height so that eventually the river may be above the surface of the surrounding flood plain. Notable levees are found on the lower reaches of the Mississippi in the USA, and along the Po in Italy. The Huang He in China also has well-developed levees.

liberalism   political and social theory that supports representative government, freedom of the press, speech, and worship, the abolition of class privileges, the use of state resources to protect the welfare of the individual, and international

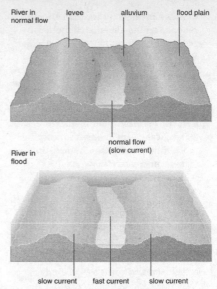

levee Levees are formed on the banks of meandering rivers when the river floods over its flood plain. The flow of the flood water is slower over the levees than it is in the river bed and so silt is deposited there.

*free trade. It is historically associated with the Liberal Party in the UK and the Democratic Party in the USA.

libertarianism   political theory that upholds the rights of the individual above all other considerations and seeks to minimize the power of the state to the safeguarding of those rights.

lichen desert   any geographical area from which lichens are absent as a result of atmospheric pollution.

life-cycle analysis   assessment of the environmental impact of a product, taking into account all aspects of production (including resources used), packaging, distribution and ultimate end.

This 'cradle-to-grave' approach can expose inconsistencies in many so-called 'eco-friendly' labels, applied to products such as soap powders, which may be biodegradable but which are perhaps contained in nonrecyclable containers.

life expectancy   average lifespan that can be presumed of a person at birth. It depends on nutrition, disease control, environmental contaminants, war, stress, and living standards in general.

light industry   industry on a relatively small scale, as contrasted with *heavy industry. Light industry includes industries such as the manufacture of electrical goods, printing and publishing, distribution trades and clothing manufacture.

Light industry is generally to be found on purpose-built estates, often on the edge of an urban area where congestion is less and *communications are good. Light industry is generally cleaner and less polluting than heavy industry, and is organized mainly for the manufacture of consumer durables.

lightning   high-voltage electrical discharge between two rainclouds or between a cloud and the Earth, caused by the build-up of electrical charges. Air in the path of lightning ionizes (becomes a conductor), and expands; the accompanying noise is heard as thunder. Currents of 20,000 amperes and temperatures of 30,000°C are common. Lightning causes nitrogen oxides to form in the atmosphere and approximately 25% of atmospheric nitrogen oxides are formed in this way.

lightning conductor   device that protects a tall building from lightning strike, by providing an easier path for current to flow to earth than through the building. It consists of a thick copper strip of very low resistance connected to the ground below. A good connection to the ground is essential and is made by burying a large metal plate deep in the damp earth. In the event of a direct lightning strike, the current in the conductor may be so great as to melt or even vaporize the metal, but the damage to the building will nevertheless be limited.

lignite   type of *coal that is brown and fibrous, with a relatively low carbon content. As a fuel it is less efficient because more of it must be burned to produce the same amount of energy generated by bituminous coal. Lignite also has a high sulphur content and is more polluting. It is burned to generate power in Scandinavia and some European countries because it is the only fuel resource available without importing.

lime   or **quicklime**; technical name **calcium oxide**; CaO, white powdery substance used in making mortar and cement. It is made commercially by heating calcium carbonate ($CaCO_3$), obtained from limestone or chalk, in a lime kiln. Quicklime readily absorbs water to become calcium hydroxide $Ca(OH)_2$, known as slaked lime, which is used to reduce soil acidity.

limestone   sedimentary rock composed chiefly of calcium carbonate ($CaCO_3$), either derived from the shells of marine organisms or precipitated from solution, mostly in the ocean. Various types of limestone are used as building stone.

*Karst is a type of limestone landscape. Caves commonly occur in limestone. *Marble is metamorphosed limestone. Certain so-called marbles are not in fact marbles but fine-grained fossiliferous limestones that have been polished.

limestone pavement   bare rock surface resembling a block of chocolate, found on limestone plateaus. It is formed by the weathering of limestone into individual upstanding blocks, called clints, separated from each other by joints, called grykes. The weathering process is thought to entail a combination of freeze–thaw (the alternate freezing and thawing of ice in cracks) and carbonation (the dissolving of minerals in the limestone by weakly acidic rainwater). Malham Tarn in North Yorkshire is an example of a limestone pavement.

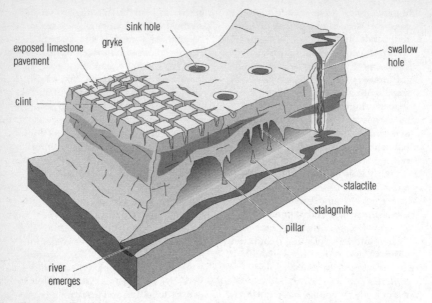

limestone The physical weathering and erosion of a limestone landscape. The freezing and thawing of rain and its mild acidic properties cause cracks and joints to enlarge, forming limestone pavements, potholes, caves, and caverns.

limits to growth the belief, based on computer calculations, that the steadily rising population growth combined with the rapid depletion of the Earth's natural resources will lead to environmental catastrophe in the 21st century. A book with this title was published in the 1970s outlining this theory.

limnology study of lakes and other bodies of open fresh water, in terms of their plant and animal biology, chemistry, and physical properties.

limonite iron ore, mostly poorly crystalline iron oxyhydroxide, but usually mixed with *haematite and other iron oxides. Also known as brown iron ore, it is often found in bog deposits.

linear development or **ribbon development**, housing that has grown up along a route such as a road. Many settlements show this pattern, since roads offer improved access to the central business district and other areas. Linear development may result in *urban sprawl.

line of best fit on a *scatter diagram, line drawn as near as possible to the various points so as to best represent the trend being graphed. The sums of the displacements of the points on either side of the line should be equal.

line transect in ecology, method used to map plants (or sedentary animals) along a line tied between two trees or other suitable upright supports. The line is marked at 1-m intervals. At each metre mark the distance to the ground is measured and the species recorded.

link the connection between two *nodes in a communication *network.

liquefaction in earth science, the conversion of a soft deposit, such as clay, to a jellylike state by severe shaking. During an earthquake buildings and lines of communication built on materials prone to liquefaction will sink and topple. In the Alaskan earthquake of 1964 liquefaction led to the destruction of much of the city of Anchorage.

liquid asset   possession or valuable that is in the form of cash or can easily be turned into cash. Liquidity is important for a company because if it does not have enough liquid assets, it can go bankrupt as a result of a *cash flow crisis.

lithification   conversion of an unconsolidated (loose) sediment into solid sedimentary rock by **compaction** of mineral grains that make up the sediment, **cementation** by crystallization of new minerals from percolating water solutions, and new growth of the original mineral grains. The term is less commonly used to refer to solidification of *magma to form igneous rock.

lithosere   see *sere.

lithosphere   upper rocky layer of the Earth that forms the jigsaw of plates that take part in the movements of *plate tectonics. The lithosphere comprises the *crust and a portion of the upper *mantle. It is regarded as being rigid and brittle and moves about on the more plastic and less brittle *asthenosphere. The lithosphere ranges in thickness from 2–3 km at mid-ocean ridges to 150 km beneath old ocean crust, to 250 km under *cratons.

littoral region   in ecology, region of seashore that lies between the high-water mark and low-water mark. Its zones – the upper, middle, and lower shores – each support a community of organisms characterized by their ability to withstand greater or lesser periods of exposure at low tide.

load   the *sediment transported by the agents of *erosion – rivers, moving ice, and the sea. The size and volume of load transported depends upon the

power of the transporting medium. In a river system, for example, more load is carried in times of high *discharge. A river's load comprises material rolled or bounced along the bed (bedload), material carried in suspension (suspended load), and material carried in solution. The finest sediment is carried the greatest distances and may contribute to the formation of, for example, a *delta.

loam   type of fertile soil, a mixture of sand, silt, clay, and organic material. It is porous, which allows for good air circulation and retention of moisture.

loan   form of borrowing by individuals, businesses, and governments. Individuals and companies usually obtain loans from banks. The loan with interest is typically paid back in fixed monthly instalments over a period of between one and five years in the UK, although longer-term loans and different repayment conditions may be arranged. Specific forms of loan include mortgages (transfers of property as security for loans), and debentures (loans raised by a company, paying a fixed rate of interest, and secured on the assets of the company). In business, loans are the second most important way (after retained profit) in which firms finance their expansion.

local authority   government at a local level in the UK. Examples include county councils, with district councils below them, in shire counties and metropolitan borough councils in city areas. Since 1997 most are unitary authorities, each providing a full range of local services. Such services already include education, social services, local roads, refuse collection, and fire protection. Around 85% of their funds come from grants from central government but they also receive monies from local business rates and the *council tax.

local government that part of government dealing mainly with matters concerning the inhabitants of a particular area or town, usually financed at least in part by local taxes. In the USA and UK, local government has comparatively large powers and responsibilities.

local option right granted by a government to the electors of each particular area to decide whether the sale of intoxicants shall be permitted. Such a system has been tried in certain states of the USA, in certain Canadian provinces, and in Norway and Sweden.

location the position of population, settlement and economic activity in an area or areas. Location is a basic theme in *human geography.

location triangle see *industrial location.

lode geological deposit rich in certain minerals, generally consisting of a large vein or set of veins containing ore minerals. A system of veins that can be mined directly forms a lode, for example the mother lode of the California gold rush.

Lodes form because hot hydrothermal liquids and gases from magmas penetrate surrounding rocks, especially when these are limestones; on cooling, veins of ores formed from the magma then extend from the igneous mass into the local rock.

loess yellow loam, derived from glacial meltwater deposits and accumulated by wind in periglacial regions during the *ice ages. Loess usually attains considerable depths, and the soil derived from it is very fertile. There are large deposits in central Europe (Hungary), China, and North America. It was first described in 1821 in the Rhine area, and takes its name from a village in Alsace.

Lomé Convention convention in 1975 that established economic cooperation between the *European Economic Community and developing countries of Africa, the Caribbean, and the Pacific (ACP). It was renewed in 1979, 1985, 1989, and 2000.

long. *abbreviation for* **longitude**; see *latitude and longitude.

longitude see *latitude and longitude.

direction of longshore drift and therefore of movement along the beach

swash

backwash

waves strike the shore at an angle

longshore drift Waves sometimes hit the beach at an angle. The incoming waves (swash) carry sand and shingle up onto the shore and the outgoing wave takes some material away with it. Gradually material is carried down the shoreline in the same direction as the longshore current.

longshore drift   movement of material along a *beach. When a wave breaks at an angle to the beach, pebbles are carried up the beach in the direction of the wave (**swash**). The wave returns to the sea at right angles to the beach (**backwash**) because that is the steepest gradient, carrying some pebbles with it. In this way, material moves in a zigzag fashion along a beach. Longshore drift is responsible for the erosion of beaches and the formation of *spits (ridges of sand or shingle projecting into the water). Attempts are often made to halt longshore drift by erecting barriers, or *groynes, at right angles to the shore.

lopolith   a basin-shaped igneous *intrusion. Many lopoliths are small, though the Bushveldt lopolith of South Africa is exposed over an area of 65,000 km^2.

low   see *depression.

lower course   see *river.

low order   any settlement, service, or good that is at the bottom of its hierarchy. Low-order features are frequently provided and require only low *threshold populations to sustain them. Their *range is limited and their *sphere of influence small.

LWM   *abbreviation for* **low water mark**.

Maastricht Treaty treaty establishing the *European Union (EU). Agreed in 1991 and signed in 1992, the treaty took effect on 1 November 1993 following ratification by member states. It advanced the commitment of member states to *economic and monetary union (but included an opt-out clause for the United Kingdom); provided for intergovernmental arrangements for a common foreign and security policy; increased cooperation on justice and home affairs policy issues (though the Social Chapter was rejected by the UK until a change of government in 1997); introduced the concept of EU citizenship (as a supplement to national citizenship); established new regional development bodies; increased the powers of the European Parliament; and accepted the principle of *subsidiarity (a controversial term defining the limits of European Community involvement in national affairs).

macchia see *maquis.

machair low-lying, fertile, sandy coastal plain integral to the crofting agricultural system (a form of subsistence farming) of the Hebrides and parts of the northern Highlands of Scotland.

macroeconomics division of economics concerned with the study of whole (aggregate) economies or systems, including such aspects as government income and expenditure, the balance of payments, fiscal policy, investment, inflation, and unemployment. It seeks to understand the influence of all relevant economic factors on each other and thus to quantify and predict aggregate national income.

maelstrom whirlpool off the Lofoten Islands, Norway, also known as the Moskenesstraumen, which gave its name to whirlpools in general.

mafic rock plutonic rock composed chiefly of dark-coloured minerals such as olivine and pyroxene that contain abundant magnesium and iron. It is derived from **magnesium** and **ferric** (iron). The term **mafic** also applies to dark-coloured minerals rich in iron and magnesium as a group. 'Mafic rocks' usually refers to dark-coloured igneous rocks such as basalt, but can also refer to their metamorphic counterparts.

magma molten rock material that originates in the lower part of the Earth's crust, or *mantle, where it reaches temperatures as high as 1000°C. *Igneous rocks are formed from magma. *Lava is magma that has extruded onto the surface.

magnetic dip see *dip, magnetic and *angle of dip.

magnetic pole region of a magnet in which its magnetic properties are strongest. Every magnet has two poles, called north and south. The north (or north-seeking) pole is so named because a freely-suspended magnet will turn so that this pole points towards the Earth's magnetic north pole.

magnetic storm in meteorology, a sudden disturbance affecting the Earth's magnetic field, causing anomalies in radio transmissions and magnetic compasses. It is probably caused by sunspot activity.

magnetite black, strongly magnetic opaque mineral, Fe_3O_4, of the spinel group, an important ore of iron.

Widely distributed, magnetite is found in nearly all igneous and metamorphic rocks. Some deposits, called lodestone, are permanently magnetized. Lodestone has been used as a compass since the first millennium BC. Today the orientations of magnetite grains in rocks are used in the study of the Earth's magnetic field (see *palaeomagnetism).

magnetometer   device for measuring the intensity and orientation of the magnetic field of a particular rock or of a certain area. In geology, magnetometers are used to determine the original orientation of a rock formation (or the orientation when the magnetic signature was locked in), which allows for past plate reconstruction. They are also used to delineate 'magnetic striping' on the sea floor in order to make plate reconstruction and to prospect for ore bodies such as iron ore, which can disrupt the local magnetic field.

magnetopause   in earth science, narrow region that lies between the magnetosphere and outer space (the interplanetary medium).

magnetosphere   volume of space, surrounding a planet, in which the planet's magnetic field has a significant influence. The Earth's magnetosphere extends 64,000 km towards the Sun, but many times this distance on the side away from the Sun. That of Jupiter is much larger, and, if it were visible, would appear from the Earth to have roughly the same extent as the full Moon. The Russian-led space missions Coronas-I (launched in 1994) and Coronas-F (launched in 2001) were designed to investigate the magnetosphere of the Sun.

Magnox reactor   a type of gas-cooled nuclear reactor in which the moderator is graphite and the core coolant gas is carbon dioxide; they operate at the relatively low temperature of 300–330°C. The enclosure of the uranium fuel in a magnesium alloy called magnox gives the reactor type its name. Magnox reactors were first built in the UK (1956) and a total of nine commercial electricity stations were constructed around this type of reactor; a further seven were built in France. They have now reached the end of their commercial life. Magnox reactors were inefficient in terms of their useful generating capacity per tonne of fuel used but had an excellent safety record.

malachite   common *copper ore, basic copper carbonate, $Cu_2CO_3(OH)_2$. It is a source of green pigment and is used as an antifungal agent in fish farming, as well as being polished for use in jewellery, ornaments, and art objects.

malnutrition   condition resulting from a defective diet where certain important food nutrients (such as proteins, vitamins, or carbohydrates) are absent. It can lead to deficiency diseases. A related problem is *undernourishment. A high global death rate linked to malnutrition has arisen from famine situations caused by global warming, droughts, and the greenhouse effect, as well as by socio-political factors, such as alcohol and drug abuse, poverty, and war.

Malthus theory   projection of population growth made by Thomas Malthus. He based his theory on the *population explosion that was already becoming evident in the 18th century, and argued that the number of people would increase faster than the food supply. Population would eventually reach a resource limit (*overpopulation). Any further increase would result in a population crash, caused by famine, disease, or war.

mandate   in history, a territory whose administration was entrusted to

Allied states by the League of Nations under the Treaty of Versailles after World War I. Mandated territories were former German and Turkish possessions (including Iraq, Syria, Lebanon, and Palestine). When the United Nations replaced the League of Nations in 1945, mandates that had not achieved independence became known as *trust territories.

manganese ore   any mineral from which manganese is produced. The main ores are the oxides, such as **pyrolusite**, MnO_2; **hausmannite**, Mn_3O_4; and **manganite**, $MnO(OH)$.

Manganese ores may accumulate in metamorphic rocks or as sedimentary deposits, frequently forming nodules on the sea floor (since the 1970s many schemes have been put forward to harvest deep-sea manganese nodules). The world's main producers are Georgia, Ukraine, South Africa, Brazil, Gabon, and India.

mangrove swamp   muddy swamp found on tropical coasts and estuaries, characterized by dense thickets of mangrove trees. These low trees are adapted to live in creeks of salt water and send down special breathing roots from their branches to take in oxygen from the air. The roots trap silt and mud, creating a firmer drier environment over time. Mangrove swamps are common in the Amazon delta and along the coasts of West Africa, northern Australia, and Florida, USA.

mantle   intermediate zone of the Earth between the *crust and the *core, accounting for 82% of the Earth's volume. The crust, made up of separate tectonic *plates, floats on the mantle which is made of dark semi-liquid rock that is rich in magnesium and silicon. The temperature of the mantle can be as high as 3700°C. Heat generated in the core causes convection currents in the semi-liquid mantle; rock rises and then slowly

sinks again as it cools, causing the movements of the tectonic plates. The boundary (junction) between the mantle and the crust is called the *Mohorovicic discontinuity, which lies at an average depth of 32 km. The boundary between the mantle and the core is called the Gutenburg discontinuity, and lies at an average depth of 2900 km.

The mantle is subdivided into **upper mantle**, **transition zone**, and **lower mantle**, based upon the different velocities with which seismic waves travel through these regions. The upper mantle includes a zone characterized by low velocities of seismic waves, called the **low-velocity zone**, at 72–250 km depth. This zone corresponds to the *asthenosphere upon which the Earth's lithospheric plates glide. Seismic velocities in the upper mantle are overall less than those in the transition zone, and those of the transition zone are in turn less than those of the lower mantle. Faster propagation of seismic waves in the lower mantle implies that the lower mantle is more dense than the upper mantle.

The mantle is composed primarily of magnesium, silicon, and oxygen in the form of *silicate minerals. In the upper mantle, the silicon in silicate minerals, such as olivine, is surrounded by four oxygen atoms. Deeper in the transition zone greater pressures promote denser packing of oxygen such that some silicon is surrounded by six oxygen atoms, resulting in magnesium silicates with garnet and pyroxene structures. Deeper still, all silicon is surrounded by six oxygen atoms so that the mineral perovskite $MgSiO_3$ predominates.

mantle keel   in earth science, relatively cold slab of mantle material attached to the underside of a continental craton (core of a continent composed of old, highly deformed metamorphic rock), and protruding down into the mantle like the keel of a boat. Their presence suggests that tectonic processes may have been different at the time the cratons were formed.

manufacturing base share of the total output in a country's economy contributed by the manufacturing sector. This sector has greater potential for productivity growth than the service sector, which is labour-intensive; in manufacturing, productivity can be increased by replacing workers with technically advanced capital equipment. It is also significant because of its contribution to exports.

manufacturing industry or **secondary industry**, industry that involves the processing of raw materials or the assembly of components. Examples are aluminium smelting, car assembly, and computer assembly. In the UK many traditional manufacturing industries, built up in the Industrial Revolution, are now declining in importance; for example, shipbuilding. Developing countries may lack the capital (equipment and money) and expert knowledge necessary for these industries.

Maoism form of communism based on the ideas and teachings of the Chinese communist leader Mao Zedong. It involves an adaptation of *Marxism to suit conditions in China and apportions a much greater role to agriculture and the peasantry in the building of socialism, thus effectively bypassing the capitalist (industrial) stage envisaged by Marx. In addition, Maoism stresses ideological, as well as economic, transformation, based on regular contact between party members and the general population.

map diagrammatic representation of an area, for example part of the Earth's surface or the distribution of the stars. Modern maps of the Earth are made using satellites in low orbit to take a series of overlapping stereoscopic photographs from which a three-dimensional image can be prepared. The earliest accurate large-scale maps appeared about 1580 (see *atlas).

Conventional aerial photography, laser beams, microwaves, and infrared equipment are also used for land surveying. Many different kinds of *map projection (the means by which a three-dimensional body is shown in two dimensions) are used in map-making. Detailed maps requiring constant updating are kept in digital form on computer so that minor revisions can be made without redrafting.

The *Ordnance Survey is the official body responsible for the mapping of Britain; it produces maps in a variety of scales, such as the Landranger series (scale 1:50,000). Large-scale maps (for example, 1:25,000) show greater detail at a local level than small-scale maps (for example, 1:100,000).

map projection way of showing the Earth's spherical surface on a flat piece of paper. The most common approach has been to redraw the Earth's surface within a rectangular boundary. The main weakness of this is that countries in high latitudes are shown disproportionately large. The most famous cylindrical projection is the Mercator projection, which dates from 1569. Although it gives an exaggerated view of the size of northern continents, it is the best map for navigation because a constant bearing appears as a straight line.

In 1973 German historian Arno Peters devised the **Peters projection** in which the countries of the world retain their relative areas. In other projections, lines of longitude and latitude appear distorted, or the Earth's surface is shown as a series of segments joined along the Equator.

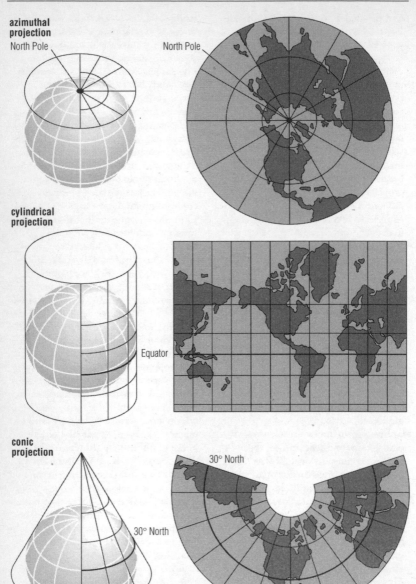

azimuthal
projection
North Pole

North Pole

cylindrical
projection

Equator

conic
projection

30° North

30° North

map projection Three widely used map projections. If a light were placed at the centre of a transparent Earth, the shapes of the countries would be thrown as shadows on a sheet of paper. If the paper is flat, the azimuthal projection results; if it is wrapped around a cylinder or in the form of a cone, the cylindrical or conic projections result.

In 1992 the US physicist Mitchell Feigenbaum devised the **optimal conformal** projection, using a computer program designed to take data about the boundary of a given area and calculate the projection that produces the minimum of inaccuracies.

maps   see chronology for some highlights of mapping.

maquis   **mattoral** or **macchia**, a mixed, sclerophyllous vegetation of the Mediterranean basin in which pines, oak, wild olive, carob and lentisk tree grow to a height of about 5 m. Beneath the tree layer is found a wide variety of aromatic herbaceous plants, the luxuriance and diversity of which is dependent upon the annual rainfall and the degree of human interference brought about by grazing pressure and the frequency of burning. Vegetation is interspersed with extensive areas of bare ground. The maquis is drab and desolate in the long dry summer season although in the cooler, moister spring time it is transformed into a short lasting blaze of colour when broom, heathers, gorse and asphodel along with members of the *Labiatae* and *Thymus* families complete their life cycles.

Maralinga   site in the southern Australian desert about 1000 km north of Adelaide, used by the British government for nuclear testing between 1952 and 1963. Nine atomic explosions were conducted, resulting in the contamination of 3000 sq km of land. In 1993 Britain agreed to contribute 45% of the cost of cleaning the area. In 1994 the Australian government agreed to compensate the Aborigines, who were removed from their traditional homelands before the tests started, and who can never return because of the contamination of the land.

marble   rock formed by metamorphosis of sedimentary *limestone. It takes and retains a good polish, and is used in building and sculpture. In its pure form it is white and consists almost entirely of calcite ($CaCO_3$). Mineral impurities give it various colours and patterns. Carrara, Italy, is known for white marble.

marginal analysis   a major theoretical building block of neoclassical *economics, marginal analysis suggests that economic decisions are made at the margin, for example a consumer might decide to buy one more apple if the price was reduced by 5p, or a business might decide to buy one more van if the cost was reduced by £1000. The extra cost of an extra unit is known as the marginal cost.

marginal cost   or **contribution cost** for a business, the cost of producing an extra unit of output. For example, if a bakery increased its production from 10,000 loaves a day to 11,000 loaves a day, and its costs increased from £2000 to £2100, then the marginal cost of the extra 1000 loaves would be £100. So the marginal cost of a loaf would be 10p.

marginal cost pricing   in economics, the setting of a price based on the additional cost to a firm of producing one more unit of output (the marginal cost), rather than the actual average cost per unit (total production costs divided by the total number of units produced). In this way, the price of an item is kept to a minimum, reflecting only the extra cost of labour and materials.

marginal efficiency of capital   in economics, effectively the rate of return on investment in a given business project compared with the rate of return if the capital were invested at prevailing interest rates.

marginal land   in farming, poor-quality land that is likely to yield a poor return. It is the last land to be brought into production and the first land to be abandoned. Examples are desert fringes in Africa and mountain areas in the UK.

marginal theory in economics, the study of the effect of increasing a factor by one more unit (known as the marginal unit). For example, if a firm's production is increased by one unit, its costs will increase also; the increase in costs is called the marginal cost of production. Marginal theory is a central tool of microeconomics.

marginal utility in economics, the measure of additional satisfaction (utility) gained by a consumer who receives one additional unit of a product or service. The concept is used to explain why consumers buy more of a product when the price falls.

Mariana Trench lowest region on the Earth's surface; the deepest part of the sea floor. The trench is 2400 km long and is situated 300 km east of the Mariana Islands, in the northwestern Pacific Ocean. Its deepest part is the gorge known as the Challenger Deep, which extends 11,034 m below sea level.

maritime climate a *temperate climate that is affected by the closeness of the sea, giving a small annual range of temperatures – a coolish summer and a mild winter – and rainfall throughout the year. Britain has a maritime climate. Compare *extreme climate.

market any situation where buyers and sellers are in contact with each other. This could be a street market or it could be a world market where buyers and sellers communicate via letters, faxes, telephones, and representatives.

In a perfect or **free market**, there are many buyers and sellers, so that no single buyer or seller is able to influence the price of the product; there is therefore *perfect competition in the market. In an **imperfect market** either a few buyers or sellers (or even just one) dominates the market.

market area the area from which consumers will travel into a *central place in order to obtain goods and services. The size of the market area is determined by the position of the central place in the *settlement hierarchy: the market area (or *sphere of influence) for low-order central places is small, since individual travel tolerance for everyday goods and services is low. Conversely, the market area for high-order central places is large, since individual travel tolerance for specialized goods and services is high.

market economy economy in which most resources are allocated through markets rather than through state planning. See *free enterprise.

market forces in economics, the forces of demand (a want backed by the ability to pay) and supply (the willingness and ability to supply).

market gardening an intensive type of *agriculture traditionally located on the margins of urban areas to supply fresh produce on a daily basis to the city population. Typical market-garden produce includes salad crops such as tomatoes, lettuce, cucumber, etc., cut flowers, fruit and some green vegetables. The classic market garden is characterized by a small landholding, high capital investment in the form of equipment such as greenhouses and fertilizers, plus high yields: these conditions are in response to the high value of land near urban areas and the perishability of the produce (see *Von Thünen theory). However, modern developments in *communications and transport technology, and indeed in food processing and consumer tastes, have led to a relaxation of the traditional locational factors behind market gardening. The production of, e.g. peas, beans and cauliflowers in the English fens, primarily for canning and freezing, is an example of the 'new' market gardening; some produce is marketed direct to London and to supermarkets via the excellent communications of eastern lowland England.

Although market gardening is still important in Great Britain, improved air communications have meant that much market-garden produce can be flown in from abroad, where it is grown more economically. For example, tomatoes are flown in from Spain, new potatoes from Egypt, and vegetables from Kenya (providing fresh food out of its normal season).

market mechanism way in which the forces of *demand and *supply allocate resources through a system of markets. Customers send signals to producers through their spending. Producers who fail to supply goods which customers want to buy will be forced to produce more successful lines or go out of business. In a *free enterprise economy, most resources are allocated via the market mechanism.

market orientation production of goods carefully researched and designed to appeal to customers in the market. This contrasts with **product orientation** where the product is designed and produced with little or no market research background in the hope that customers will find it attractive.

market penetration strategy for entering a new market by adopting aggressive pricing. Goods are priced at a very low, gross profit margin, relying on high turnover to cover overheads. The aim is to penetrate the market with the new product and quickly to capture a large market share. After market penetration is achieved, prices can be raised if there is sufficient brand loyalty. However, there is a risk that low prices may be permanently associated with the product by consumers.

market research process of gaining information about customers in a market through field research or desk research. Field research involves collecting primary data by interviewing customers or completing questionnaires. Desk research involves collecting secondary data by looking at information and statistics collected by others and published, for example, by the government.

market saturation position in a market when all consumers who want a product already have one. Once a market is saturated, the majority of sales, in the case of consumer durables, will come from replacement of items.

market segmentation dividing a market into consumer categories. The idea is that by segmenting the market according to consumer needs it may be possible to identify an unfulfilled market niche. Markets can be segmented in several ways: psychographic, according to attitudes, tastes, and personality traits such as aspirational, homely, or image-conscious; geographic, according to location; and demographic, according to age, sex, or class, such as baby boomers or Generation X.

market share the proportion of a market taken by one producer. Market share can be measured in terms of volume of sales, for example company X sold 40% of the cars sold last month in the UK market. Or it can be measured as the value of sales, for example company X sold 30% of the total value of all cars sold in the UK market last month. Or it could be measured in terms of output by volume or value. For example, company X could have produced 50% of all the cars made in the UK even though its UK market share was 30% because it exported cars to other markets while some cars sold in the UK were produced in other countries.

market size   total sales of all companies operating in a particular market, or the total number of units sold in the market. The information is important to companies thinking of entering a market in order for them to assess the feasibility of entering that market, and also to companies already operating in the market in order for them to assess their market share.

marl   crumbling sedimentary rock, sometimes called **clayey limestone**, including various types of calcareous *clays and fine-grained *limestones. Marls are often laid down in freshwater lakes and are usually soft, earthy, and of a white, grey, or brownish colour. They are used in cement-making and as fertilizer.

marsh   low-lying wetland. Freshwater marshes are common wherever groundwater, surface springs, streams, or run-off cause frequent flooding, or more or less permanent shallow water. A marsh is alkaline whereas a *peat bog is acid. Marshes develop on inorganic silt or clay soils. Rushes are typical marsh plants. Large marshes dominated by papyrus, cattail, and reeds, with standing water throughout the year, are commonly called *swamps. Near the sea, *salt marshes may form.

Marxism   philosophical system, developed by the 19th-century German social theorists Marx and Engels, also known as **dialectical materialism**, under which matter gives rise to mind (materialism) and all is subject to change (from dialectic). As applied to history, it supposes that the succession of feudalism, capitalism, socialism, and finally the classless society is inevitable. The stubborn resistance of any existing system to change necessitates its complete overthrow in the **class struggle** – in the case of capitalism, by the proletariat – rather than gradual modification.

mass extinction   event that produces the extinction of many species at about the same time. One notable example is the boundary between the Cretaceous and Tertiary periods (known as the *K-T boundary) that saw the extinction of the dinosaurs and other large reptiles, and many of the marine invertebrates as well. Mass extinctions have taken place frequently during Earth's history.

There have been five major mass extinctions, in which 75% or more of the world's species have been wiped out: End Ordovician period (440 million years ago) in which about 85% of species were destroyed (second most severe); Late Devonian period (365 million years ago) which took place in two waves a million years apart, and was the third most severe, with marine species particularly badly hit; Late Permian period (251 million years ago), the gravest mass Late Triassic (205 million years ago), in which about 76% of species were destroyed, mainly marine; Late Cretaceous period (65 million years ago), in which 75–80% of species became extinct, including dinosaurs.

mass movement   processes that move weathered material on slopes without the action of running water. Movement is due to gravity, and may be on a large or small scale. Different types of mass movement include rockfalls, *landslides, *soil creep, and *solifluction.

mass spectrometer   apparatus for analysing chemical composition by separating ions by their mass; the ions may be elements, isotopes, or molecular compounds. Positive ions (charged particles) of a substance are separated by an electromagnetic system, designed to focus particles of equal mass to a point where they can be detected. This permits accurate measurement of the relative concentrations of the various ionic masses present. A mass spectrometer can be used to identify compounds or to measure the relative abundance of compounds in a sample.

matriarchy form of society where domestic and political life is dominated by women, where kinship is traced exclusively through the female line, and where religion is centred around the cult of a mother goddess. A society dominated by men is known as a **patriarchy**.

matrix model organizational structure adopted by many multinational companies in an attempt to deal with the difficulties of managing large organizations across different regional markets. The concept was developed by the electronics company Phillips after World War II. Matrix management represents a middle way between centralization and decentralization, and provides for multiple reporting along geographical and functional lines. A marketing manager in one country will report to a boss in his own country, as well as the marketing head in the company's home country. Instead of a linear chain of command the structure takes the form of a multi-dimensional matrix. An organization using the matrix model may include regional managers, functional managers, country managers, and business sector managers.

mean in statistics, a measure of the average of a number of terms or quantities. The simple **arithmetic mean** is the average value of the quantities, that is, the sum of the quantities divided by their number. The **weighted mean** takes into account the frequency of the terms that are summed; it is calculated by multiplying each term by the number of times it occurs, summing the results and dividing this total by the total number of

occurrences. The **geometric mean** of n quantities is the nth root of their product. In statistics, it is a measure of central tendency of a set of data, that is one measure used to express the frequency distribution of a number of recorded events.

meander loop-shaped curve in a *river flowing sinuously across flat country. As a river flows, any curve in its course is accentuated (intensified) by the current. On the outside of the curve the velocity, and therefore the erosion, of the current is greatest. Here the river cuts into the outside bank, producing a **river cliff**. On the inside of the curve the current is slow and so it deposits any transported material, building up a gentle slip-off slope. As each meander migrates in the direction of its outer bank, the river gradually changes its course across the flood plain.

A loop in a river's flow may become so exaggerated that eventually it is cut off from the normal course and forms an *oxbow lake. Meanders are common where the gradient is gentle, the discharge fairly steady (not subject to extremes), and the material that is carried is fine sediment. The word *meander comes from the name of the River Menderes in Turkey.

mechanical weathering in earth science, an alternative name for *physical weathering.

medial moraine linear ridge of rocky debris running along the centre of a glacier. Medial moraines are commonly formed by the joining of two *lateral moraines when two glaciers merge.

Mediterranean climate climate characterized by hot, dry summers and warm, wet winters. Mediterranean zones are situated in either hemisphere on the western side of continents, between latitudes of 30° and 60°.

During the winter, rain is brought by the *westerlies, which blow from the sea. In summer, Mediterranean climate areas are under the influence of the *trade winds which bring arid conditions with a longer dry season towards the desert margins. Most areas are backed by coastal mountains and so the combined effects of *orographic rainfall and precipitation on the weather *front give high seasonal rainfall totals. The Mediterranean Sea region is noted for its local winds. The *sirocco and *khamsin are two of the hot, dry winds that blow from the Sahara, and can raise temperatures to over 40°C.

meerschaum aggregate of minerals, usually the soft white clay mineral **sepiolite**, hydrous magnesium silicate. It floats on water and is used for making pipe bowls.

melting point temperature at which a substance melts, or changes from solid to liquid form. A pure substance under standard conditions of pressure (usually one atmosphere) has a definite melting point. If heat is supplied to a solid at its melting point, the temperature does not change until the melting process is complete. The melting point of ice is 0°C.

meltwater water produced by the melting of snow and ice, particularly in glaciated areas. Streams of meltwater flowing from glaciers transport rocky materials away from the ice to form *outwash. Features formed by the deposition of debris carried by meltwater or by its erosive action are called **fluvioglacial features**; they include eskers, kames, and outwash plains.

menagerie small collection of wild animals kept in captivity for display. Rulers of early times used to bring back wild animals from abroad for public exhibition. Private collections later became common, and until the early 19th century one was maintained at the Tower of London, England.

From these collections developed present-day zoos.

MEP *abbreviation for* member of the European Parliament.

Mercalli scale qualitative scale describing the **intensity** of an *earthquake. It differs from the *Richter scale, which indicates earthquake **magnitude** and is quantitative. It is named after the Italian seismologist Giuseppe Mercalli (1850–1914).

Intensity is a subjective value, based on observed phenomena, and varies from place to place even when describing the same earthquake.

Mercator's projection map projection devised by Flemish mapmaker Gerardus Mercator (1512–1594) in which the parallels and meridians on maps are drawn uniformly at 90°. The projection continues to be used, in particular for navigational charts, because compass courses can be drawn as straight lines, but the true area of countries is increasingly distorted the further north or south they are from the Equator.

Mercosur or **South American Common Market**; Portuguese **Mercosul**, (Spanish *Mercado del Sur* 'Market of the South') free-trade organization, founded in March 1991 on signature of the Asunción Treaty by Argentina, Brazil, Paraguay, and Uruguay, and formally inaugurated on 1 January 1995. With a GNP of $800,000 million and a population of more than 190 million, Mercosur constitutes the world's fourth-largest free-trade bloc after the *European Economic Area, the *North American Free Trade Agreement, and the *Asia-Pacific Economic Cooperation Conference.

meridian half a *great circle drawn on the Earth's surface passing through both poles and thus through all places with the same longitude. Terrestrial longitudes are usually measured from the Greenwich Meridian.

An astronomical meridian is a great circle passing through the celestial pole and the zenith (the point immediately overhead).

meritocracy system (of, for example, education or government) in which selection is by performance or competitive examinations. Such a system favours intelligence and ability rather than social position or wealth.

mesa (Spanish 'table') flat-topped, steep-sided plateau, consisting of horizontal weak layers of rock topped by a resistant formation; in particular, those found in the desert areas of the USA and Mexico. A small mesa is called a butte.

mesopause layer in the Earth's atmosphere at an altitude of 80–85 km, at the top of the mesosphere.

mesophyte in ecology, plant characterized by having only moderate requirements for water.

mesosphere layer in the Earth's *atmosphere above the stratosphere and below the thermosphere. It lies between about 50 km and 80 km above the ground.

Mesozoic era of geological time 245–65 million years ago, consisting of the Triassic, Jurassic, and Cretaceous periods. At the beginning of the era, the continents were joined together as Pangaea; dinosaurs and other giant reptiles dominated the sea and air; and ferns, horsetails, and cycads thrived in a warm climate worldwide. By the end of the Mesozoic era, the continents had begun to assume their present positions, flowering plants were dominant, and many of the large reptiles and marine fauna were becoming extinct.

metamorphic rock rock altered in structure, composition, and texture by pressure or heat after its original formation. (Rock that actually melts

under heat is called *igneous rock upon cooling.) For example, limestone can be metamorphosed by heat into marble, and shale by pressure into slate. The term was coined in 1833 by Scottish geologist Charles Lyell. Metamorphism is part of the *rock cycle, the gradual formation, change, and re-formation of rocks over millions of years.

The mineral assemblage present in a metamorphic rock depends on the composition of the starting material (which may be sedimentary or igneous) and the temperature and pressure conditions to which it is subjected. There are two main types of metamorphism. **Thermal metamorphism**, or contact metamorphism, is brought about by the baking of solid rocks in the vicinity of an igneous intrusion (molten rock, or *magma, in a crack in the Earth's crust). It is responsible, for example, for the conversion of limestone to marble. **Regional metamorphism** results from the heat and intense pressures associated with burial and the movements and collision of tectonic plates (see *plate tectonics). It brings about the conversion of shale to slate, for example. A third type, **shock metamorphism**, occurs when a rock is very quickly subjected to high pressures such as those brought about by a meteorite impact.

Metamorphic rocks have essentially the same chemical composition as their original components, but because different minerals are stable at different temperatures and pressures, they are commonly composed of different, metamorphic minerals. Metamorphism also produces changes in the texture of the rock, for example rocks become foliated (or layered), and crystals grow larger.

metamorphism geological term referring to the changes in rocks of the

Earth's crust caused by increasing pressure and temperature. The resulting rocks are metamorphic rocks. All metamorphic changes take place in solid rocks. If the rocks melt and then harden, they are considered *igneous rocks.

meteoritics branch of geology dealing with the study of meteorites, their composition, texture, formation, and origin, and applying these observations to a more thorough understanding of the evolution of the Solar System and planets, including Earth.

meteoroid small natural object in interplanetary space. Meteoroids are smaller than asteroids, ranging from the size of a pebble up, and move through space at high speeds. There is no official distinction between meteoroids and asteroids, although the term 'asteroid' is generally reserved for objects larger than 1.6 km in diameter.

Meteorological Office or **Met Office**, organization based in Bracknell, England, producing weather reports for the media and conducting specialist work for industry, agriculture, and transport. Established in London, England, in 1855, it is a partner in the British National Space Centre (BNSC).

The first weather map in England, showing the trade winds and monsoons, was made in 1688, and the first telegraphic weather report appeared in 1848. The first daily telegraphic weather map was prepared at the Great Exhibition in 1851. The first regular daily collections of weather observations by telegraph and the first British daily weather reports were made in 1860, and the first daily printed maps appeared 1868.

meteorology scientific observation and study of the *atmosphere, so that *weather can be accurately forecast.

Data from meteorological stations and weather satellites are collated by computers at central agencies, and forecast and weather maps based on current readings are issued at regular intervals. Modern analysis, employing some of the most powerful computers, can give useful forecasts for up to six days ahead.

At meteorological stations readings are taken of the factors determining weather conditions: atmospheric pressure, temperature, humidity, wind (using the *Beaufort scale), cloud cover (measuring both type of cloud and coverage), and precipitation such as rain, snow, and hail (measured at 12-hour intervals). Satellites are used either to relay information transmitted from the Earth-based stations, or to send pictures of cloud development, indicating wind patterns, and snow and ice cover.

history Apart from some observations included by Aristotle in his book *Meteorologia*, meteorology did not become a precise science until the end of the 16th century, when Galileo and the Florentine academicians constructed the first thermometer of any importance, and when Evangelista Torricelli in 1643 discovered the principle of the barometer. Robert Boyle's work on gases, and that of his assistant, Robert Hooke, on barometers, advanced the physics necessary for the understanding of the weather. Gabriel Fahrenheit's invention of a superior mercury thermometer provided further means for temperature recording.

weather maps In the early 19th century a chain of meteorological stations was established in France, and weather maps were constructed from the data collected. The first weather map in England, showing the trade winds and monsoons, was made in 1688, and the first telegraphic weather report appeared on 31 August 1848.

The first daily telegraphic weather map was prepared at the Great Exhibition in 1851, but the Meteorological Office was not established in London until 1855. The first regular daily collections of weather observations by telegraph and the first British daily weather reports were made in 1860, and the first daily printed maps appeared in 1868.

collecting data Observations can be collected not only from land stations, but also from weather ships, aircraft, and self-recording and automatic transmitting stations, such as the radiosonde. Radar may be used to map clouds and storms. Satellites have played an important role in televising pictures of global cloud distribution.

As well as supplying reports for the media, the Meteorological Office in Bracknell, near London, does specialist work for industry, agriculture, and transport. Kew is the main meteorological observatory in the British Isles, but there are other observatories, for example at Eskdalemuir in the southern uplands of Scotland, Lerwick in the Shetlands, and Valentia in southwestern Ireland. Climatic information from British climatological reporting stations is published in the *Monthly Weather Report*, and periodically in tables of averages and frequencies. The British Meteorological Office's *Daily Weather Report* contains a detailed map of the weather over the British Isles and a less detailed map of the weather over the northern hemisphere, and the *Daily Aerological Record* contains full reports of radiosonde ascents made over the British Isles and from some of the ocean weather ships, together with maps of the heights of the 700 mb, 500 mb, and 300 mb pressure surfaces, giving a picture of the winds at 3048 m, 5182 m, and 9144 m; there is also a map of the height of the tropopause.

Ships' reports are plotted on the same charts using the same symbolic form. Data from radiosondes and aircraft are plotted on upper-air charts and on temperature–height diagrams, the diagram in use in Britain being the tephigram. With the help of this diagram it is possible to predict the formation or otherwise of clouds, showers, or thunderstorms, and sometimes to identify the source region of the air mass.

Observation stations may be classified as follows:

Observatories where reliable standard and absolute measurements are made as far as possible with autographic instruments, which are often duplicated for checking and research purposes. Elements such as atmospheric electricity and solar radiation are measured only at observatories and other research establishments.

Climatological reporting stations report the general daily weather conditions and make observations at standard hours during the day to provide cumulative data, such as average temperature, maximum and minimum temperatures, rainfall, sunshine, mean pressure, days of fog, frost, snowfall, and the extent and persistence of snow cover. After statistical analysis, climatic charts and tables are constructed showing the frequency of the different weather elements, such as gales and frost.

Crop weather stations make observations for use in agricultural meteorology, or micrometeorology, where the elements of the weather need to be studied in detail. It is necessary to have detailed information on temperature, humidity, and wind at heights below the average crop height in order to study and control plant diseases spread by aphids or wind-borne viruses. Details of frost hollows, wind breaks, and the degree of frost that will damage plants must all be studied.

Rainfall stations measure the amount of rain that falls. Most stations measure the daily amount, while those in remote areas measure the monthly rainfall.

Synoptic reporting stations, where observations are made simultaneously throughout the world, report in a mutually agreed form so that data can be directly compared between them. Observations are restricted to the elements required for forecasting. Reports are received at a national centre and a selection is broadcast for use by other countries.

analysis The huge mass of synoptic data collected and disseminated for forecasting purposes is plotted on synoptic weather charts. Modified copies of these charts using standard symbols are published daily by most meteorological services.

measuring and describing conditions Meteorological observations, for whatever purpose, must be clear, precise, and strictly comparable between stations. It is easy to decide whether it is fine or cloudy, or if there is a thunderstorm; the distinctions between rain, snow, and hail are obvious; sleet is wet snow, melting snow, or a mixture of rain and snow; soft hail is halfway between snow and hail; and drizzle, which consists of very small drops, is halfway between rain and cloud, the water drops being just large enough to fall to the ground. It is useful also to describe rain as showery, intermittent, or continuous, as light, moderate, or heavy. The rate of rainfall and the total rainfall during a given period can be measured with a rain gauge. Wind strength and direction can be measured accurately by anemometers and wind vanes. Clouds are observed

carefully, since they are closely related to other weather conditions. Fog and other hindrances to visibility indicate approaching weather conditions as well as being of great practical importance.

air pressure Although the condition of the air overhead can be partly deduced from cloud observations, measurements of the physical state of the atmosphere are made at as great a height as possible. The weight of the air above any point presses downwards (see *atmospheric pressure) and the force this produces in all directions is called the air pressure; it is measured by barometers and barographs, the unit of measurement being the millibar (mb). With increase in height there is less air above and therefore pressure decreases with height by about a factor of 10 for every increase of height of 16 km. If the temperature of the air is known the decrease in pressure can be calculated: near sea level it amounts to about 1 mb in every 10 m. In order to compare pressures between many stations at a constant level, pressures are reduced to sea level – that is, barometer readings are adjusted to show what the pressure would be at sea level.

temperature Measuring the temperature of the air can be difficult, because a thermometer measures its own temperature, not necessarily that of its surroundings. In addition, temperature varies irregularly with height, particularly in the first few metres. Thus the temperature at 1 m above the ground may easily be 5°C lower than the temperature closer to the ground, whereas on a following clear night the reverse usually occurs. On the other hand the temperature of air which rises 100 m only cools about 1°C by reason of its change of height and consequent expansion because of decrease in pressure. Temperatures are therefore read at a standard height (usually 1.2 m) above the ground, and not reduced to sea level. A thermometer is kept at the same temperature as the surrounding air by sheltering it in a Stevenson screen.

humidity The humidity of the air is found by comparing readings taken by an ordinary thermometer with readings from a thermometer whose bulb is covered with moist muslin (a hygrometer). Temperature and humidity in the upper air are measured by attaching instruments to an aircraft, to a small balloon, or even to a rocket. With aircraft, measurements have been made up to more than 15 km above the Earth, with balloons to 30 km, and with rockets to more than 150 km. Initially, instruments carried by balloon had to be recovered before readings could be obtained; now, radiosonde balloons transmit observations back to Earth. Balloons can also be used to measure winds since they are carried by the wind, and direction-finding radio or radar enable their drift to be calculated, thus determining the speed and strength of winds at the different heights through which the balloon passes.

weather ships Reports from ships follow the same patterns as those from land stations, but also include the state and temperature of the sea. Weather ships are equipped to cruise at fixed points making observations, including radiosonde ascents, at standard intervals. These ships are also used as navigational beacons for aircraft.

methane or **marsh gas**, an ordourless, colourless hydrocarbon gas produced either by natural or artificial anaerobic decomposition of organic material. Formula: CH_4. It is nearly insoluble in water, burns with a pale flame to produce water and carbon dioxide and releases no hazardous air pollutants. It is the principle constituent of natural

gas and can be used as a fuel. Methane gas can be produced from a biogas digester. Some intensive agricultural systems, such as stall-fed cattle rearing, have such high densities of animals per unit area and generate such considerable amounts of animal urea (a major source of methane) that they have been cited as a major cause of the increase in carbon in the atmosphere.

methyl bromide or **bromoethane**; CH_3Br, pesticide gas used to fumigate soil. It is a major *ozone depleter. Industry produces 50,000 tonnes of methyl bromide annually (1995). The European Union (EU) promised a 2 5% reduction in manufacture by 1998, and the USA intends to ban use by 2001. The EU proposed a total ban on usage by 2001 at a meeting in July 1998. Some countries, however, have failed to significantly reduce their ozone-depleting emissions, such as China, which at the end of 1999 was producing 10% of the world's methyl bromide.

metropolitan area another term for *conurbation.

metropolitan county in England, a group of six counties established under the Local Government Act of 1972 in the largest urban areas outside London: Tyne and Wear, South Yorkshire, Merseyside, West Midlands, Greater Manchester, and West Yorkshire. Their elected assemblies (county councils) were abolished in 1986 when most of their responsibilities reverted to metropolitan borough councils.

MFA *abbreviation for* *Multi-Fibre Arrangement.

mica any of a group of silicate minerals that split easily into thin flakes along lines of weakness in their crystal structure (perfect basal cleavage). They are glossy, have a pearly lustre, and are found in many igneous and metamorphic rocks. Their good thermal and electrical

insulation qualities make them
valuable in industry.

Their chemical composition is
complicated, but they are silicates
with silicon–oxygen tetrahedra
arranged in continuous sheets, with
weak bonding between the layers,
resulting in perfect cleavage.

A common example of mica is
muscovite (white mica),
$KAl_2Si_3AlO_{10}(OH,F)_2$.

microclimate climate of a small area,
such as a woodland, lake, or even a
hedgerow. Significant differences can
exist between the climates of two
neighbouring areas – for example,
a town is usually warmer than the
surrounding countryside (forming a
heat island), and a woodland cooler,
darker, and less windy than an area
of open land.

Microclimates play a significant
role in agriculture and horticulture,
as different crops require different
growing conditions.

microeconomics the division of
economics concerned with the study
of individual decision-making units
within an economy: a consumer, firm,
or industry. Unlike macroeconomics,
it looks at how individual markets
work and how individual producers
and consumers make their choices and
with what consequences. This is done
by analysing how relevant prices of
goods are determined and the
quantities that will be bought and sold.

microhabitat in ecology, localized
environmental conditions of a
population or even an individual.
The surface of a pond or the underside
of a leaf are examples.

Mid-Atlantic Ridge *mid-ocean ridge
that runs along the centre of the
Atlantic Ocean, parallel to its edges,
for some 14,000 km – almost from the
Arctic to the Antarctic. Like other
ocean ridges, the Mid-Atlantic Ridge is
essentially a linear, segmented volcano.

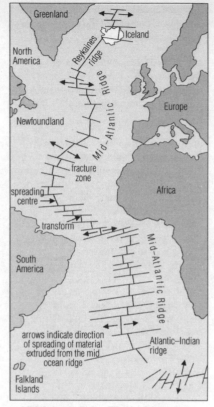

Mid-Atlantic Ridge The Mid-Atlantic Ridge
is the boundary between the crustal plates
that form America, and Europe and Africa.
An oceanic ridge cannot be curved since the
material welling up to form the ridge flows
at a right angle to the ridge. The ridge takes
the shape of small straight sections offset
by fractures transverse to the main ridge.

The Mid-Atlantic Ridge runs down
the centre of the ocean because the
ocean crust has continually grown
outwards from the ridge at a steady
rate during the past 200 million years.
Iceland straddles the ridge and was
formed by volcanic outpourings.

middle course see *river.

midnight sun constant appearance
of the Sun (within the Arctic and
Antarctic circles) above the *horizon
during the summer.

mid-ocean ridge   long submarine mountain range that winds along the middle of the ocean floor. The mid-ocean ridge system is essentially a segmented, linear *shield volcano. There are a number of major ridges, including the *Mid-Atlantic Ridge, which runs down the centre of the Atlantic; the East Pacific Rise in the southeast Pacific; and the Southeast Indian Ridge. These ridges are now known to be spreading centres, or divergent margins, where two plates of oceanic lithosphere are moving away from one another (see *plate tectonics). Ocean ridges can rise thousands of metres above the surrounding seabed, and extend for up to 60,000 km in length.

Ocean ridges usually have a *rift valley along their crests, indicating where the flanks are being pulled apart by the growth of the plates of the *lithosphere beneath. The crests are generally free of sediment; increasing depths of sediment are found with increasing distance down the flanks.

migrant labour   people who move from place to place to work. Economic or political pressures often cause people to leave their homelands to earn wages in this way, but some families live this way for several generations. As economic development has taken place at different rates in different countries, the supplies of and need for labour have been uneven.

migration   movement of population away from the home region, either from one country to another (**international** migration) or from one part of a country to another (**internal** migration). Migrations may be temporary (for example, holiday-makers), seasonal (transhumance), or permanent (people moving to cities to find employment). For people, migration is often a permanent (or long-term) movement, which involves the break-up of a person's residential and social environment. People leave areas due to *push factors (negative factors such as overcrowding and lack of employment) and are drawn to areas by *pull factors (such as better housing, better jobs, and improved facilities). Barriers such as cost, language, politics, and knowledge also influence migration.

Milankovitch hypothesis   combination of factors governing the occurrence of *ice ages proposed in 1930 by the Yugoslav geophysicist M Milankovitch

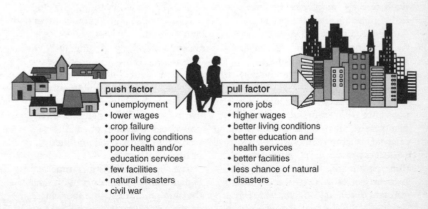

migration Some causes of migration. Factors that tend to repel people from an area, such as high unemployment and poor housing, are called push factors. Factors that tend to attract people to an area, such as a wide range of facilities or well-paid jobs, are called pull factors.

(1879–1958). These include the variation in the angle of the Earth's axis, and the geometry of the Earth's orbit around the Sun.

millibar unit of pressure, equal to one-thousandth of a *bar.

millimetre of mercury symbol mmHg, unit of pressure, defined as the pressure exerted by a column of mercury one millimetre high, under the action of gravity.

millionaire city or **million city**, city with more than 1 million inhabitants. In 1985 there were 273 millionaire cities in the world, compared with just two in 1850. Most of these are now found in developing countries, whereas before 1970 most were in industrialized countries.

Minamata Japanese town, site of an ecological disaster 1953–56, where 43 people died after eating fish poisoned with dimethyl mercury. The poison had been released as *effluent from a local plastics factory, and became concentrated in the flesh of sea organisms. Many townspeople suffered long-terms effects, including paralysis, tremors, paralysis and brain damage. In 1996 the remaining 1500 uncompensated victims dropped their case against the chemicals company in return for individual payments of £16,000.

The mercury was released as a stable mercury compound but was converted to methyl mercury by anaerobic bacteria in the sediments at the bottom of the bay.

mineral naturally formed inorganic substance with a particular chemical composition and a regularly repeating internal structure. Either in their perfect crystalline form or otherwise, minerals are the constituents of *rocks. In more general usage, a mineral is any substance economically valuable for mining (including coal and oil, despite their organic origins).

Mineral-forming processes include:

melting of pre-existing rock and subsequent crystallization of a mineral to form igneous or volcanic rocks; weathering of rocks exposed at the land surface, with subsequent transport and grading by surface waters, ice, or wind to form sediments; and recrystallization through increasing temperature and pressure with depth to form *metamorphic rocks. The transformation and recycling of the minerals of the Earth's outer layers is known as the rock cycle.

Minerals are usually classified as magmatic, *sedimentary, or *metamorphic. The magmatic minerals, in *igneous rock, include the feldspars, quartz, pyroxenes, amphiboles, micas, and olivines that crystallize from silica-rich rock melts within the crust or from extruded lavas.

The most commonly occurring sedimentary minerals are either pure concentrates or mixtures of sand, clay minerals, and carbonates (chiefly calcite, aragonite, and dolomite).

Minerals typical of metamorphism include andalusite, cordierite, garnet, tremolite, lawsonite, pumpellyite, glaucophane, wollastonite, chlorite, micas, hornblende, staurolite, kyanite, and diopside.

mineralogy study of minerals. The classification of minerals is based chiefly on their chemical composition and the kind of chemical bonding that holds their atoms together. The mineralogist also studies their crystallographic and physical characters, occurrence, and mode of formation.

The systematic study of minerals began in the 18th century, with the division of minerals into four classes: earths, metals, salts, and bituminous substances, distinguished by their reactions to heat and water.

mineral oil   oil obtained from mineral sources, for example coal or petroleum, as distinct from oil obtained from vegetable or animal sources.

minifundia   small farms commonly found in Central and South America. Some are independently owned smallholdings but most are tenanted farms or squatter holdings too small to support the family unit. Compare *latifundia.

Miocene   ('middle recent') fourth epoch of the Tertiary period of geological time, 23.5–5.2 million years ago. At this time grasslands spread over the interior of continents, and hoofed mammals rapidly evolved.

mirage   illusion seen in hot weather of water on the horizon, or of distant objects being enlarged. The effect is caused by the refraction, or bending, of light.

misfit valley   large valley with a small stream. In most cases the valley was formed when a much larger river flowed through it. The valleys of the Cotswolds, England, were mostly formed by large rivers of meltwater at the end of the last glacial period. They now contain small streams such as the Evenlode and the Windrush, which do not have the volume of water or power to erode them.

mission   office of a government representative accredited to another government. Britain's missions include high commissions headed by British high commissioners in Commonwealth countries and embassies headed by ambassadors in other countries. Reciprocally Commonwealth governments are represented in London by high commissioners and other governments by ambassadors.

Mississippian   US term for the Lower or Early *Carboniferous period of geological time, 363–323 million years ago. It is named after the state of Mississippi.

mist   low cloud caused by the condensation of water vapour in the lower part of the *atmosphere. Mist is less thick than *fog, visibility being 1–2 km.

mistral   cold, dry, northerly wind that occasionally blows during the winter on the Mediterranean coast of France, particularly concentrated along the Rhône valley. It has been known to reach a velocity of 145 kph.

mixed economy   type of economic structure that combines the private enterprise of capitalism with a degree of state monopoly. In mixed economies, governments seek to control the public services, the basic industries, and those industries that cannot raise sufficient capital investment from private sources. Thus a measure of economic planning can be combined with a measure of free enterprise. A notable example was US president Franklin D Roosevelt's New Deal in the 1930s.

mixed farming   a system of *agriculture which comprises a variety of arable and pastoral elements.

mmHg   symbol for *millimetre of mercury.

mobility   in economics, the degree of movement of the factors of production from one occupation to another (occupational mobility) or from one region to another (regional mobility). The labour mobility of unskilled workers is not very great in the UK, for example, because of the lack of knowledge about jobs between one part of the country and another and because it is so difficult to find affordable housing in some areas.

model   simplified version of some aspect of the real world. Models are produced to show the relationships between two or more factors, such as land use and the distance from the centre of a town (for example, *concentric-ring theory). Because models are idealized, they give only a general guide to what may happen.

Mohorovicic discontinuity or **Moho** or **M-discontinuity**, seismic discontinuity, marked by a rapid increase in the speed of earthquake waves, that is taken to represent the boundary between the Earth's crust and mantle. It follows the variations in the thickness of the crust and is found approximately 35–40 km below the continents and about 10 km below the oceans. It is named after the Croatian geophysicist Andrija Mohorovicic, who suspected its presence after analysing seismic waves from the Kulpa Valley earthquake in 1909. The 'Moho' is as deep as 70 km beneath high mountain ranges.

Mohs scale scale of hardness for minerals (in ascending order): 1 talc; 2 gypsum; 3 calcite; 4 fluorite; 5 apatite; 6 orthoclase; 7 quartz; 8 topaz; 9 corundum; 10 diamond.

The scale is useful in mineral identification because any mineral will scratch any other mineral lower on the scale than itself, and similarly it will be scratched by any other mineral higher on the scale.

molybdenite molybdenum sulphide, MoS_2, the chief ore mineral of molybdenum. It possesses a hexagonal crystal structure similar to graphite, has a blue metallic lustre, and is very soft (1–1.5 on the *Mohs scale).

monadock in earth science, isolated hill or mountain of generally hard rock that rises above a lowland (*peneplain) levelled by water erosion.

monazite mineral, $(Ce,La,Th)PO_4$, yellow to red, valued as a source of lanthanides or rare earths, including cerium and europium; generally found in placer deposit (alluvial) sands.

monetarism economic policy that proposes control of a country's money supply to keep it in step with the country's ability to produce goods, with the aim of controlling inflation.

Cutting government spending is advised, and the long-term aim is to return as much of the economy as possible to the private sector, which is said to be in the interests of efficiency. Monetarism was first put forward by the economist Milton Friedman and the Chicago school of economists.

monoculture the growing of a single crop. Traditional grape cultivation for wine in the south of France is an example. Monoculture can be unsound in two ways: firstly the *soil is progressively drained of specific nutrients, and secondly dependence on a single crop may be dangerous in the market if the crop fails or if consumer tastes change. Monoculture may also cause an increase in plant-specific pests and diseases.

monopoly in economics, the domination of a market for a particular product or service by a single company, which can therefore restrict competition and keep prices high. In practice, a company can be said to have a monopoly when it controls a significant proportion of the market (technically an *oligopoly). In a communist country the state itself has the overall monopoly; in capitalist countries some services, such as transport or electricity supply, may be state monopolies.

monsoon the term strictly means 'seasonal wind' and is used generally to describe a situation where there is a reversal of wind direction from one season to another. This is especially the case in South and Southeast Asia, where two monsoon winds occur, both related to the extreme pressure gradients created by the large land mass of the Asian continent. In summer (April to September), the intense heating of the land leads to the development of low pressure over northwestern India, and southwesterly winds are drawn in over the Indian ocean. This southwest monsoon brings heavy rain to India and Southwest Asia in general, and especially to those areas where the orographic effect is operating (see *orographic rainfall). In winter (October to March), the chilling of the Asian interior leads to high pressure and the establishment of the northeast monsoon: cold dry winds which bring little rain. In northern China they may also carry dust from the deserts of the continental interior.

The monsoon cycle is believed to have started about 12 million years ago with the uplift of the Himalayas.

Montréal Protocol   international agreement, signed in 1987, to stop the production of chemicals that are *ozone depleters by the year 2000.

Originally the agreement was to reduce the production of ozone depleters by 35% by 1999. The green movement criticized the agreement as inadequate, arguing that an 85% reduction in ozone depleters would be necessary just to stabilize the ozone layer at 1987 levels. The protocol (under the Vienna Convention for the Protection of the Ozone Layer) was reviewed in 1992. Amendments added another 11 chemicals to the original list of eight chemicals suspected of harming the ozone layer. A controversial amendment concerns a fund established to pay for the transfer of ozone-safe technology to poor countries.

Moon   natural satellite of Earth, 3,476 km in diameter, with a mass 0.012 (approximately one-eightieth) that of Earth.

Its surface gravity is only 0.16 (one-sixth) that of Earth. Its average distance from Earth is 384,400 km, and it orbits in a west-to-east direction every 27.32 days (the **sidereal month**). It spins on its axis with one side permanently turned towards Earth. The Moon has no atmosphere and was

thought to have no water until ice was discovered on its surface in 1998.

phases The Moon is illuminated by sunlight, and goes through a cycle of phases of shadow, waxing from **new** (dark) via **first quarter** (half Moon) to **full**, and waning back again to new every 29.53 days (the **synodic month**, also known as a **lunation**). On its sunlit side, temperatures reach 110°C, but during the two-week lunar night the surface temperature drops to –170°C.

origins The origin of the Moon is still open to debate. Scientists suggest the following theories: that it split from the Earth; that it was a separate body captured by Earth's gravity; that it formed in orbit around Earth; or that it was formed from debris thrown off when a body the size of Mars struck Earth.

research 70% of the far side of the Moon was photographed from the Soviet Lunik 3 in October 1959. Much of our information about the Moon has been derived from this and other photographs and measurements taken by US and Soviet Moon probes, from geological samples brought back by US *Apollo* astronauts and by Soviet Luna probes, and from experiments set up by US astronauts 1969–72. US astronaut Neil Armstrong (1930–) became the first person to set foot on the Moon on 20 July 1969. The Moon landing was part of the *Apollo* project. The US probe Lunar Prospector, launched in January 1998, examined the composition of the lunar crust, recorded gamma rays, and mapped the lunar magnetic field. It also discovered the ice on the moon in March 1998.

composition The Moon is rocky, with a surface heavily scarred by meteorite impacts that have formed craters up to 240 km/150 mi across. Seismic observations indicate that the Moon's surface extends downwards for tens of kilometres; below this crust is a solid mantle about 1,100 km thick, and below that a silicate core, part of which may be molten. Rocks brought back by astronauts show that the Moon is 4.6 billion years old, the same age as Earth. It is made up of the same chemical elements as Earth, but in different proportions, and differs from Earth in that most of the Moon's surface features were formed within the first billion years of its history when it was hit repeatedly by meteorites.

The youngest craters are surrounded by bright rays of ejected rock. The largest scars have been filled by dark lava to produce the lowland plains called seas, or **maria** (plural of mare). These dark patches form the so-called 'man-in-the-Moon' pattern. Inside some craters that are permanently in shadow is up to 300 million tonnes of ice existing as a thin layer of crystals.

moonstone translucent, pearly variety of potassium sodium *feldspar, found in Sri Lanka and Myanmar, and distinguished by a blue, silvery, or red opalescent tint. It is valued as a gem.

moor in earth science, a stretch of land, usually at a height, which is characterized by a vegetation of heather, coarse grass, and bracken. A moor may be poorly drained and contain boggy hollows.

More than 50% of Scotland is regarded as moorland.

moorpan see *hardpan.

mor an acidic humus layer (pH generally less than 4.5) common on poorly aerated soils. Mor is especially common in climatic regions with long, cold and wet winters and mild, damp summers and on older Palaeozoic rocks. The adverse soil climate precludes an inactive soil fauna, hence decomposition of plant litter is slow and incomplete. A thick organic layer (peat) can develop on the surface of the mineral soil which reduces the supply of oxygen and further

diminishes the activity rate of the soil fauna. Typical soil types on which mor can be found include peaty brown earth, peaty gley soils, and peaty podzols. Compare *mull.

moraine a collective term for rocky debris or *till deposited on or by *glaciers and ice bodies in general. Moraine differs from fluvial *sediment in being unsorted and composed of angular fragments. Several types of moraine are recognized: **lateral moraine** forms along the edges of a valley glacier where debris eroded from the valley sides, or weathered from the slopes above the glacier, collects; **medial moraine** forms where two lateral moraines meet at a glacier junction; **englacial moraine** is material which is trapped within the body of the glacier – when this is exposed at the surface due to partial melting it becomes an **ablation moraine**; and **ground moraine** is material eroded from the floor of the valley and used by the glacier as an abrasive tool. A *terminal moraine is material bulldozed by the glacier during its advance and deposited at its maximum down-valley extent. **Recessional moraines** may be deposited at standstills during a period of general *glacial retreat.

mortlake see *oxbow lake.

mountain natural upward projection of the Earth's surface, higher and steeper than a hill. Mountains are at least 330 m above the surrounding topography. The existing rock below a high mountain may be subjected to high temperatures and pressures, causing metamorphism. Plutonic activity also can accompany mountain building.

mud in earth science, mixture of water and particles of clay or silt that may be semi-liquid or soft and plastic. The fine-grained underwater sediment on a continental shelf is also known as mud.

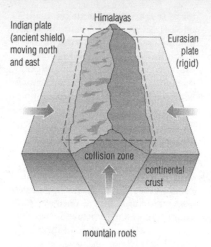

Indian plate (ancient shield) moving north and east

Himalayas

Eurasian plate (rigid)

collision zone

continental crust

mountain roots

mountain Mountains are created when two continental plates collide and no subduction takes place, resulting in the land at the collision zone being squeezed together and thrust upwards.

mudflow downhill movement (mass movement) of muddy sediment containing a large proportion of water. Mudflows can be fast and destructive: in 1966 coal waste saturated with water engulfed a school in Aberfan, South Wales, killing 116 children. A *lahar is a form of mudflow associated with volcanic activity.

mudstone fine-grained sedimentary rock made up of clay- to silt-sized particles (up to 0.0625 mm).

mull a nutrient-rich humus usually found in well-aerated soils such as those in European lowland deciduous forests. The pH of mull is generally greater than 4.5 and the mineral soil which occurs beneath the humus layer can show a neutral or even slightly alkaline reaction (pH 7.0). Long hot summers and short cold winters encourages microbial activity in the soil and a large earthworm population ensures the mixing of organic material with the mineral layers. Mull-rich soils are usually very fertile and have long been used as

agricultural sites. Typical soils associated with mull include deep brown earth, brown forest soil, and brown earth with surface gley soils. Compare *mor.

Multi-Fibre Arrangement MFA, worldwide system of managed trade in textiles and clothing which came into force in 1974. It has been revised four times to take into account changing trends in production, consumption, and world trading conditions. MFA IV (1986–91) included silk, linen, ramie, and jute in an attempt to control trade in all products.

multilateralism trade among more than two countries without discrimination over origin or destination and regardless of whether a large trade gap is involved.

multinational corporation company or enterprise operating in several countries, usually defined as one that has 25% or more of its output capacity located outside its country of origin.

multiple-nuclei theory a model of urban structure devised by Harris and Ullman in 1945, stating that most large cities develop around a number of separate centres or nuclei, rather than round a single centre (compare *Burgess model and *sector model). Different land uses are therefore situated around the city, creating a cellular structure. The pattern of these cells or nuclei reflects the unique factors of the site and/or history of any particular city. See also *shanty town.

multiple-purpose resource management a strategy for resource management which attempts to provide maximum availability of a *resource to as wide as possible a variety of users, without endangering the quantity or quality of the resource for any particular consumer. The management of water resources is a case in point: multiple-purpose dams can cater for recreation, *hydroelectric power generation, flood control and fishing. In practice, multiple-purpose resource management may be an elusive goal due to the problems of accommodating widely different demands.

multiplier effect process whereby one change sets in motion a sequence of events that results in decline or growth. For example, in South Wales, pit closures in the coal industry resulted in unemployment, depopulation, closure of services, and disinvestment. This in turn led to further unemployment. Multiplier effects are also important in *new towns, where industry is needed to attract people and create further wealth.

muscovite white *mica, $KAl_2Si_3AlO_{10}(OH,F)_2$, a common silicate mineral. It is colourless to silvery white with shiny surfaces, and like all micas it splits into thin flakes along its one perfect cleavage. Muscovite is a metamorphic mineral occurring mainly in schists; it is also found in some granites, and appears as shiny flakes on bedding planes of some sandstones.

mylonite metamorphic rock composed mainly of feldspar, quartz, and micas and comprising fine-grained highly-deformed layers alternating with layers of less-deformed relict grains often ovoidal in shape. The varied degree of deformation among bands produces a typically streaky appearance. They are found in areas in which rock layers have been deformed by sliding past one another parallel to their boundaries (shear strain).

N *abbreviation for* **north**; newton, a unit of force; the chemical symbol for **nitrogen**.

NAFTA acronym for *North American Free Trade Agreement.

Narmada Valley Project controversial US$5 billion project to build more than 3000 dams on the Narmada River, which runs for 1245 km through western India, to supply water to irrigate farmland for 30 million people and provide hydroelectric power. The project has been fiercely opposed by environmentalists in India and abroad. They believe it will lead to the displacement of 300,000 people, chiefly from small tribal communities; disrupt downstream fisheries; increase the risk of earthquakes; submerge forest land; increase the spread of insect-borne diseases; and threaten the fragile regional ecosystem through reducing, by two-thirds, the flow of water from the Narmada into the Arabian Sea.

national accounts statistical report on the value of income, expenditure, and production in the economy of a country.

national debt debt incurred by the central government of a country to its own people and institutions and also to overseas creditors. A government can borrow from the public by means of selling interest-bearing bonds, for example, or from abroad. Traditionally, a major cause of national debt was the cost of war but in recent decades governments have borrowed heavily in order to finance development or nationalization, to support an ailing currency, or to avoid raising taxes.

national income the total income of a state in one year, including both the wages of individuals and the profits of companies. It is equal to the value of the output of all goods and services during the same period. National income is equal to gross national product (the value of a country's total output) minus an allowance for replacement of ageing capital stock.

nationalization policy of bringing a country's essential services and industries under public ownership. It was pursued, for example, by the UK Labour government 1945–51. Assets in the hands of foreign governments or companies may also be nationalized; for example, Iran's oil industry, the Suez Canal, and US-owned fruit plantations in Guatemala, all in the 1950s.

national park land available for public enjoyment. National parks include not only the most scenic places, but also places distinguished for their historic, prehistoric, or scientific interest, or for their superior recreational assets. They range from areas the size of small countries to pockets of just a few hectares. The first was Yellowstone National Park, USA, established in 1872. In the USA, Republic of Ireland, South Africa – and most countries – national parks are owned by the government. In the UK, by contrast, most of the land in national parks is owned by a mix of farmers, the Forestry Commission, Ministry of Defence, county councils, and so on. In the UK national parks are not wholly wilderness or conservation areas, but merely places where planning controls on development are stricter than elsewhere.

In some countries national parks are *wilderness areas, with no motorized traffic, no overflying aircraft, no hotels, hostels, shops or cafés, no industry, and the minimum of management.

In England and Wales under the National Park Act 1949 ten national parks were established including the Peak District, the Lake District, and Snowdonia. The first national park in Scotland was established around Loch Lomond in 2002. National parks in Britain are protected from large-scale development, but from time to time pressure to develop land for agriculture, quarrying, or tourism, or to improve amenities for the local community means that conflicts of interest arise between land users.

Other protected areas include Areas of Outstanding Natural Beauty (AONBs) and *Sites of Special Scientific Interest (SSSIs).

National Rivers Authority NRA, UK government agency launched in 1989. It had responsibility for managing water resources, investigating and regulating pollution, and taking over flood controls and land drainage from the former ten regional water authorities of England and Wales. In April 1996 the NRA was replaced by the Environment Agency, having begun to establish a reputation for being supportive to wildlife projects and tough on polluters.

national grid the network of cables and wires, carried overhead on pylons or buried under the ground, that connects consumers of electrical energy to power stations, and interconnects the power stations. It ensures that power can be made available to all customers at any time, allowing demand to be shared by several power stations, and particular power stations to be shut down for maintenance work from time to time.

native metal or **free metal**, any of the metallic elements that occur in nature in the chemically uncombined or elemental form (in addition to any combined form). They include bismuth, cobalt, copper, gold, iridium, iron, lead, mercury, nickel, osmium, palladium, platinum, ruthenium, rhodium, tin, and silver. Some are commonly found in the free state, such as gold; others occur almost exclusively in the combined state, but under unusual conditions do occur as native metals, such as mercury. Examples of native non-metals are carbon and sulphur.

Natural Environment Research Council NERC, UK organization established by royal charter in 1965 to undertake and support research in the earth sciences, to give advice both on exploiting natural resources and on protecting the environment, and to support education and training of scientists in these fields of study. Research areas include geothermal energy, industrial pollution, waste disposal, satellite surveying, acid rain, biotechnology, atmospheric circulation, and climate. Research is carried out principally within the UK but also in Antarctica and in many developing countries. It comprises 13 research bodies.

natural gas mixture of flammable gases found in the Earth's crust (often in association with petroleum). It is one of the world's three main fossil fuels (with coal and oil).

natural hazard naturally occurring phenomenon capable of causing destruction, injury, disease, or death. Examples include earthquakes, floods, hurricanes, or famine. Natural hazards occur globally and can play an important role in shaping the landscape. The events only become hazards where people are affected. Because of this, natural hazards are usually measured in terms of the damage they cause to persons or property. Human activities can trigger natural hazards, for example skiers crossing the top of a snowpack may cause an avalanche.

natural increase in demography, the rise in population caused by *birth rate exceeding death rate. Rates of natural increase vary considerably throughout the world. The highest rates are found in poor countries, but with industrialization they undergo *demographic transition. Natural increase excludes any population change due to migration.

natural radioactivity radioactivity generated by those radioactive elements that exist in the Earth's crust. All the elements from polonium (atomic number 84) to uranium (atomic number 92) are radioactive. Radioisotopes of some lighter elements are also found in nature (for example potassium-40).

natural resource see *resource.

natural selection process by which gene frequencies in a population change through certain individuals producing more descendants than others because they are better able to survive and reproduce in their environment. The accumulated effect of natural selection is to produce adaptations such as the insulating coat of a polar bear or the spadelike forelimbs of a mole. The process is slow, relying firstly on random variation in the genes of an organism being produced by mutation and secondly on the genetic recombination of sexual reproduction. It was recognized by English naturalist Charles Darwin and Welsh naturalist Alfred Russel Wallace as the main process driving *evolution.

Natural Step environmental organization founded in Sweden in 1989, with branches in the USA, Australia, and the Netherlands. NS focuses on industrialists and policymakers, encouraging them to cooperate towards more ecological practices. The ten largest Swedish companies have formed links with NS.

Nature Conservancy Council NCC, former name of UK government agency divided in 1991 into English Nature, Scottish Natural Heritage, and the Countryside Council for Wales.

nature reserve area set aside to protect a habitat and the wildlife that lives within it, with only restricted admission for the public. A nature reserve often provides a sanctuary for rare species and rare habitats, such as marshland. The world's largest is Etosha Reserve, Namibia; area 99,520 sq km.

In Britain, under the National Parks Act (1949) the (now defunct) *Nature Conservancy Council (NCC) was given the power to designate nature reserves; this is now under the control of the Joint Nature Conservation Committee (JNCC). There are both officially designated nature reserves – managed by *English Nature, the *Countryside Council for Wales, and *Scottish Natural Heritage – and those run by a variety of voluntary conservation organizations. In 1997 there were 343 National Nature Reserves (covering more than 490,000 acres); 3 Marine Nature Reserves; over 500 Local Nature Reserves; and nearly 62,000 *Sites of Special Scientific Interest (SSSIs).

neap tides see *tide.

nearest neighbour analysis see *spatial distribution.

neighbourhood amenities useful facilities in the local area. Many neighbourhood amenities are provided by the local council, e.g. swimming pools, park benches, bus shelters and street lights. See also *household amenities.

neighbourhood unit in the UK, an area within a *new town planned to serve the local needs of families. These units typically have a primary school, a *low order shopping centre, a church, and a public house. Main roads form the boundaries. Neighbourhood units were designed to give a sense of community to people migrating to the new towns.

neoclassical economics school of economic thought based on the work of 19th-century economists such as Alfred Marshall, using *marginal theory to modify classical economic theories, and placing greater emphasis on mathematical techniques and theories of the firm. Neoclassicists believed competition to be the regulator of economic activity that would establish equilibrium between demand and supply through the operation of market forces.

neocolonialism disguised form of *imperialism, by which a country may grant independence to another country but continue to dominate it by control of markets for goods or raw materials. Examples of countries that have used economic pressure to secure and protect their interests internationally are the USA and Japan.

neoconservatism version of conservatism that emerged in the USA in opposition to the liberal social and political attitudes of the 1960s. It advocates a narrow, patriarchal approach to morality and family life, extols the virtues of Western capitalism as a system that encourages individual initiative and freedom, and attacks the notion of the state as the promoter of equality and as a provider of welfare.

neo-Darwinism modern theory of evolution, built up since the 1930s by integrating the 19th-century English scientist Charles Darwin's theory of evolution through natural selection with the theory of genetic inheritance founded on the work of the Austrian biologist Gregor Mendel.

Neolithic literally 'New Stone', the last period of the Stone Age. It was

characterized by settled agricultural communities who kept domesticated animals, and made pottery and sophisticated, finely finished stone tools.

The Neolithic period began and ended at different times in different parts of the world. For example, the earliest Neolithic communities appeared about 9000 BC in the Middle East, and were followed by those in Egypt, India, and China. In Europe farming began in about 6500 BC in the Balkans and Aegean Sea areas, spreading north and east by 1000 BC. The Neolithic period ended with the start of the Bronze Age, when people began using metals. Some Stone Age cultures persisted into the 20th century, notably in remote parts of New Guinea.

NERC *abbreviation for* *Natural Environment Research Council.

ness (Old English *næse* 'nose') geographical term signifying promontory. Names with this suffix are common in the Orkney and Shetland Islands, on the coast of Caithness, Scotland, and along the east coast of England as far as Dungeness in Kent. It has been suggested that its use may indicate Scandinavian colonization, *naes* being its Nordic counterpart.

net of a particular figure or price, calculated after the deduction of specific items such as commission, discounts, interest, and taxes. The opposite is *gross.

network system of nodes (junctions) and links (transport routes) through which goods, services, people, money, or information flow. Networks are often shown on topological maps.

neutrality the legal status of a country that decides not to choose sides in a war. Certain states, notably Switzerland and Austria, have opted for permanent neutrality. Neutrality always has a legal connotation. In peacetime, neutrality towards the big power alliances is called **non-alignment** (see *non-aligned movement).

nevé compact snow. In a *corrie icefield, for example, four layers are recognized: blue and white ice at the bottom of the ice mass; nevé overlying the ice, and powder snow on the surface.

newly industrialized country NIC, country formerly classified as less developed, but which is becoming rapidly industrialized. Usually the capital for such developments comes from outside the country. The first wave of countries to be identified as newly industrializing included Hong Kong, South Korea, Singapore, and Taiwan. These countries underwent rapid industrial growth in the 1970s and 1980s, attracting significant financial investment, and are now associated with high-technology industries. More recently, Thailand, China, and Malaysia have been classified as newly industrializing countries.

New Madrid seismic fault zone the largest system of geological faults in the eastern USA, centred on New Madrid, Missouri. There are several hundred earthquakes along the fault every year, more than anywhere else in the USA east of the Rocky mountains; but most of them are too small to be felt. A series of very severe earthquakes in 1811–12 created a lake 22 km long and was felt as far away as Washington, DC, and Canada. Strong earthquakes would cause much damage because the solid continental rocks would transmit the vibrations over a wide area, and buildings in the region have not been designed with earthquakes in mind.

new town in the UK, centrally planned urban area. New towns such as Milton Keynes and Stevenage were built after World War II to accommodate the overspill from cities and large towns, notably London, at a time when the population was rapidly expanding and inner-city centres had either decayed or been destroyed. In 1976 the policy, which had been criticized for disrupting family groupings and local communities, destroying small shops and specialist industries, and furthering the decay of city centres, was abandoned.

New towns are characterized by a regular street pattern and the presence of a number of self-contained neighbourhood units, consisting of houses, shops, and other local services. Modern industrial estates are located on the outskirts of towns where they are well served by main roads and motorways.

In order to stimulate employment in depressed areas, 14 original new towns were planned between 1946 and 1950, with populations of 25,000–60,000, among them Cwmbran (Wales) and Peterlee (in County Durham), and eight near London to relieve congestion there. Another 15, with populations up to 250,000, were established 1951–1975, but by then a static population and cuts in government spending halted their creation.

niche in ecology, the 'place' occupied by a species in its habitat, including all chemical, physical, and biological components, such as what it eats, the time of day at which the species feeds, temperature, moisture, the parts of the habitat that it uses (for example, trees or open grassland), the way it reproduces, and how it behaves.

nickel ore any mineral ore from which nickel is obtained. The main minerals are arsenides such as chloanthite ($NiAs_2$), and the sulphides millerite (NiS) and pentlandite ((Ni,Fe)$_9S_8$), the commonest ore. The chief nickel-producing countries are Canada, Russia, Kazakhstan, Cuba, and Australia.

nitrate pollution contamination of water by nitrates. Increased use of artificial fertilizers and land cultivation means that higher levels of nitrates are being washed from the soil into rivers, lakes, and aquifers. There they cause an excessive enrichment of the water (*eutrophication), leading to a rapid growth of algae, which in turn darkens the water and reduces its oxygen content. The water is expensive to purify and many plants and animals die. High levels are now found in drinking water in arable areas. These may be harmful to newborn babies, and it is possible that they contribute to stomach cancer, although the evidence for this is unproven.

nitre or **saltpetre**, potassium nitrate, KNO_3, a mineral found on and just under the ground in desert regions; used in explosives. Nitre occurs in Bihar, India, Iran, and Cape Province,

South Africa. The salt was formerly used for the manufacture of gunpowder, but the supply of nitre for explosives is today largely met by making the salt from nitratine (also called Chile saltpetre, $NaNO_3$). Saltpetre is a preservative and is widely used for curing meats.

nitrogen cycle process of nitrogen passing through the ecosystem. Nitrogen, in the form of inorganic compounds (such as nitrates) in the soil, is absorbed by plants and turned into organic compounds (such as proteins) in plant tissue. A proportion of this nitrogen is eaten by herbivores, with some of this in turn being passed on to the carnivores, which feed on the herbivores. The nitrogen is ultimately returned to the soil as excrement and when organisms die, and is converted back to inorganic forms by decomposers.

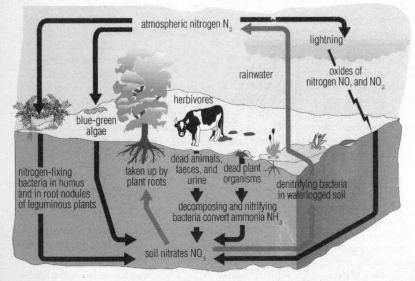

nitrogen cycle The nitrogen cycle is one of a number of cycles during which the chemicals necessary for life are recycled. The carbon, sulphur, and phosphorus cycles are others. Since there is only a limited amount of these chemicals in the Earth and its atmosphere, the chemicals must be continuously recycled if life is to go on.

Although about 78% of the atmosphere is nitrogen, this cannot be used directly by most organisms. However, certain bacteria and cyanobacteria (blue-green algae) are capable of nitrogen fixation. Some nitrogen-fixing bacteria live mutually with leguminous plants (peas and beans) or other plants (for example, alder), where they form characteristic nodules on the roots. The presence of such plants increases the nitrate content, and hence the fertility, of the soil.

nivation   set of processes, operating beneath or next to snow, believed to be responsible for the development of the hollows in which snow collects. It is also thought to play a role in the early formation of *corries. The processes involved include *freeze–thaw (weathering by the alternate freezing and melting of ice), *mass movement (the downhill movement of substances under gravity), and *erosion by meltwater.

noctilucent cloud   clouds of ice forming in the upper atmosphere at around 83 km. They are visible on summer nights, particularly when sunspot activity is low.

nodal region   an area defined by communication links radiating from a *node. For example, in the nodal region of the Paris Basin all communication links spread out from the French capital city, Paris, which thus dominates the region.

node   1. point where routes meet. It may therefore be the same as a route centre. In a topological network, a node may be the start or crossing point of routes, also called a **vertex**.
2. in oceanography, the point on a stationary wave (an oscillating wave with no progressive motion) at which vertical motion is least and horizontal motion is greatest. In earth science, it is the point on a *fault where the

apparent motion has changed direction.

nodule   in geology, a lump of mineral or other matter found within rocks or formed on the seabed surface; mining technology is being developed to exploit them.

noise   unwanted sound. Permanent, incurable loss of hearing can be caused by prolonged exposure to high noise levels (above 85 decibels). Over 55 decibels on a daily outdoor basis is regarded as an unacceptable level. In scientific and engineering terms, a noise is any random, unpredictable signal.

nomad   (Greek *nomas* 'roaming') person whose way of life involves freely moving from place to place according to the state of pasturage or food availability. Nomads believe that land is not an object of property. Nomads fall into two main groups: herders and hunter-gatherers. Those who move from place to place selling their skills or trading are also nomads; for example, the Romany people.

Both hunter-gatherers and pastoralists are threatened by enclosure of land and by habitat degradation and destruction, as well as by the social and economic pressures of a money economy. Remaining examples of hunter-gatherers are the Australian Aborigines, many Amazon Indian peoples, and the Kung and San of the Kalahari Desert in South Africa.

nomadic pastoralism   a system of *agriculture in dry grassland regions; examples of societies which are based on nomadic pastoralism include the Masai of East Africa and the Fulani of northern Nigeria. People and stock (cattle, sheep, goats) are continually moving in search of pasture and water: the low rainfall of much of the African *savannah and the scanty nature of the grazing vegetation require a nomadic cycle of activity. The pastoralists subsist on meat, milk and other animal products.

Nomadic pastoralism is under pressure from a variety of causes, for example severe drought in the Sahel region (the southern margins of the Sahara desert) in the late 1970s led to destitution among thousands of wandering herdspeople. Overgrazing and the trampling of the soil surface around boreholes has hastened the process of *desertification.

Various schemes sponsored, for example, by the United Nations, have been introduced in an attempt to improve conditions for the pastoralists, for example the introduction of high-quality stock, parasite control and pasture improvement. The chief problem is that many of these schemes require a fundamental change in the nomad's traditional lifestyle: permanent *settlement is an integral part of most development schemes.

non-aligned movement countries with a strategic and political position of neutrality ('non-alignment') towards major powers, specifically the USA and former USSR. The movement emerged in the 1960s during the Cold War between East and West 1949–89. Although originally used by poorer states, the non-aligned position was later adopted by oil-producing nations. Its 113 members hold more than half the world's population and 85% of oil resources, but only 7% of global GDP (1995).

non-governmental organization NGO, independent, service-providing, not-for-profit organization involved in a range of activities, including the provision of aid to less developed countries. They are often driven by issues, and the services and support they provide include funding volunteers, and organizing action to promote peace, environmental protection, human rights, social and

economic justice, education, sustainable and equitable development, health, and aid. Examples of such NGOs include Oxfam, the Red Cross, and Greenpeace. NGOs do not belong to and are usually not linked with a government. Governments are involved in the formation of some NGOs, appointing panel or board members. These are known as quasi-autonomous non-governmental organizations (QUANGOs) and, while not directly part of government, they may report to a government minister.

non-renewable resource natural resource, such as coal, oil, or natural gas, that takes millions of years to form naturally and therefore cannot be replaced once it is consumed; it will eventually be used up. The main energy sources used by humans are non-renewable; *renewable resources, such as solar, tidal, wind, and geothermal power, have so far been less exploited.

Fossil fuels like coal, oil, and gas generate a considerable amount of energy when they are burnt (the process of combustion). Non-renewable resources have a high carbon content because their origin lies in the photosynthetic activity of plants millions of years ago. The fuels release this carbon back into the atmosphere as carbon dioxide. The rate at which such fuels are being burnt is thus resulting in a rise in the concentration of carbon dioxide in the atmosphere, a cause of the *greenhouse effect.

normal curve bell-shaped curve of a normal distribution. This shows that extreme values are infrequent. Most of the values lie towards the centre of the symmetrical normal curve with the **arithmetic mean**, median, and mode lying at its very centre.

normal good good for which demand rises as income rises. For example, a rise in incomes leads to a rise in demand for cars and CDs. Hence cars and CDs are normal goods. Normal goods have a positive income *elasticity of demand (demand increases as incomes rise).

North American Free Trade Agreement NAFTA, trade agreement between the USA, Canada, and Mexico, intended to promote trade and investment between the signatories, agreed in August 1992, and effective from January 1994. The first trade pact of its kind to link two highly-industrialized countries to a developing one, it created a free market of 375 million people, with a total GDP of $6.8 trillion (equivalent to 30% of the world's GDP). Tariffs were to be progressively eliminated over a 10–15 year period (tariffs on trade in originating goods from Mexico and Canada are to be eliminated by 2008) and investment into low-wage Mexico by Canada and the USA progressively increased. Another aim of the agreement was to make provisions on transacting business in the free trade area. The NAFTA Centre is located in Dallas, Texas.

North American geology North America is made up of three major structural provinces: the **North American craton**, an ancient Precambrian shield area which occupies the major part of the north of the continent, tapering towards the south in an inverted triangle; the **Appalachian fold belt**, which abuts the southeast margin of the craton; and the **Cordilleran fold belt**, which lies along the southwest margin of the craton. The Precambrian rocks of the shield are well exposed in the north of the continent, where they form the Canadian Shield, but to the south, in the central plains of the USA, they are largely covered by later Phanerozoic sediments.

Canadian Shield The Canadian Shield is a vast area of Precambrian rocks, which can be divided into a number of structural provinces, each of which represents a mobile area active during a different part of Precambrian time. The provinces are in fact fragments of complexes which were originally of far greater extent.

The oldest Precambrian rocks (formed more than 3000 million years ago) occur as small isolated masses within larger areas of younger rocks.

Superior province The Superior province is a huge area of rocks of middle Precambrian age. It is made up of a basement of granites, gneisses, and migmatites, within which are elongate greenstone belts – belts where the rocks have been less highly metamorphosed and contain many primary features. These belts have an east–west trend and contain metamorphosed volcanics and sediments; they represent thick accumulations of sediment in unstable basins of deposition. They carry deposits of gold and other metal ores, and form some of Canada's major mining resources. This cycle of Precambrian activity reached a climax in the Kenoran *orogeny, 2600–2400 million years ago, during which time there was widespread regional metamorphism and deformation and granites were intruded. After this orogenic phase there was a relatively stable phase, during which an extensive dyke swarm was intruded.

Churchill province The Churchill province, which adjoins the north margin of the Superior province, is made up of rocks formed during a later Precambrian orogenic cycle, the Hudsonian cycle. The Hudsonian zones of mobile belt activity occur to the north (in the Churchill province), northeast (in the Nain province), and south (in the Southern province) of the stable Superior province. The rocks in these belts are volcanics and sediments, and contain thick banded iron formations, one of the main sources of iron ore in North America. Uranium, copper, lead, zinc, and silver are other important ores found in these rocks. Within the mobile belts the rocks have suffered intense deformation and regional metamorphism during the Hudsonian orogeny, 1800–1650 million years ago, but the sediments deposited on the foreland outside the mobile belts are often only gently folded. The end of the Hudsonian cycle was followed by the welding of the new and stabilized belts onto the old Superior craton to form a much larger shield area. A thick succession of plateau basalts (the Keweenawan succession) was poured out onto the southern part of the craton 1300–1050 million years ago; this succession is up to 15 km thick, and is characterized by copper mineralization.

Grenville province The Grenville province is the youngest province of the Canadian Shield. It is a clearly defined belt lying in the east of the Canadian Shield and extending southwest through the USA to the Mexican border. The Grenville mobile belt is characterized by a high grade of regional metamorphism, and it is therefore difficult to distinguish the original rock types occurring in the belt. This belt suffered its major metamorphism and folding in the late Precambrian, 1100–900 million years ago.

Appalachian belt This mobile belt came into existence at the end of the Precambrian, and forms a belt 3000 km long adjoining the Precambrian craton of North America, running from Newfoundland to Alabama, and partially superimposed on the earlier Grenville belt. The Appalachian cycle lasted from late Precambrian to Permian times, and sediments were laid down in a typically geosynclinal environment. The earliest basins of deposition were initiated in the eugeosynclinal zone, an area which extends the whole length of the belt and which includes detrital sediments, volcanics, and ultrabasic intrusions. The rocks in this zone are usually metamorphosed. From early Cambrian to mid-Ordovician times deposition took place both in the eugeosyncline and in the miogeosyncline – the latter being an area in the northwest part of the belt, adjacent to the foreland. This zone contains abundant carbonate sediments and is unmetamorphosed. Orogenic movements during the Ordovician raised several massifs in the central and southeast part of the belt, and restricted the later basins of deposition. By the late Devonian, much of the northern Appalachian belt was a mountain system, and deposition continued only in intermontane basins. Further south, stabilization occurred at the end of the Palaeozoic. Coal is of immense economic importance in the Appalachian belt, and is preserved at the margin of the mobile belt and on the adjoining foreland. Large oil reserves occur in mid-Devonian to Carboniferous rocks in the Allegheny Plateau.

During the Palaeozoic, shallow-water sandstones and carbonate rocks were laid down in thin layers on the interior craton; later in the Palaeozoic, as the new Appalachian mountain ranges were rising to the east and south, detrital continental sediments were deposited on the craton.

Cordilleran belt This is an extremely long-lived mobile belt initiated in the late Precambrian and still active at the present day. It extends along the whole west coast of North America from Alaska to California and Guatemala.

The earliest sediments were laid down in basins in late Precambrian times, and from the early Palaeozoic onwards geosynclinal sediments accumulated at an active continental margin. As in the Appalachian belt, a eugeosynclinal and miogeosynclinal zone can be distinguished, this time with the miogeosyncline on the northeast side adjoining the craton. The first orogenic disturbances occurred in late Mesozoic and Tertiary times, when there was widespread uplift and intrusion of granites. This uplifted zone continued to form a highland region throughout the Tertiary and Quaternary eras. To the west of the Cordilleran belt is the narrow Pacific belt, where the main orogeny occurred in mid-Tertiary or later times. The Cordillera contains rich mineral deposits, many of which were associated with Tertiary volcanic activity. These include gold, lead, zinc, silver, tin, and copper.

   Colorado Plateau The Colorado Plateau and the Basin and Range region, Nevada, represent part of the old craton adjoining the Cordilleran mobile belt which became reactivated in Mesozoic and Tertiary times and was eventually incorporated in the new orogenic belt. The central and southern Cordillera contain huge thicknesses of lavas and pyroclastic rocks laid down in Tertiary to recent times.

North and South   a way of dividing the industrialized nations, found predominantly in the north – Europe, North America and Japan, plus Australia – from those less developed nations in the south – South America, Africa and parts of Asia. The term 'North and South' is taken from the Brandt Report on the inequalities of wealth between nations, published in 1980, which, by using a number of indicators, distinguished the richer countries of the world from the poorer countries. The gap which exists

between the rich 'North' and the poor 'South' is called the *development gap*.

North Atlantic Drift   warm *ocean current in the North Atlantic Ocean; an extension of the *Gulf Stream. It flows east across the Atlantic and has a mellowing effect on the climate of northwestern Europe, particularly the British Isles and Scandinavia.

North Atlantic Treaty   agreement signed on 4 April 1949 by Belgium, Canada, Denmark, France, Iceland, Italy, Luxembourg, the Netherlands, Norway, Portugal, the UK, and the USA, in response to the Soviet blockade of Berlin June 1948–May 1949. They agreed that 'an armed attack against one or more of them in Europe or North America shall be considered an attack against them all'. The *North Atlantic Treaty Organization (NATO), which other countries have joined since, is based on this agreement.

North Atlantic Treaty Organization NATO, military association of major Western European and North American states set up under the *North Atlantic Treaty of 4 April 1949. The original signatories were Belgium, Canada, Denmark, France, Iceland, Italy, Luxembourg, Netherlands, Norway, Portugal, the UK, and the USA. Greece and Turkey were admitted to NATO in 1952, West Germany in 1955, Spain in 1982, and Poland, Hungary, and the Czech Republic in 1999. NATO has been the basis of the defence of the Western world since 1949. During the Cold War (1945–89), NATO stood in opposition to the perceived threat of communist Eastern Europe, led by the USSR and later allied under the military Warsaw Pact (1955–91). Having outlasted the Warsaw Pact, NATO has increasingly redefined itself as an agent of international peace-keeping and enforcement.

northing an element of a *grid reference. See *easting.

nuclear energy or **atomic energy**, energy released from the inner core, or nucleus, of the atom. Energy produced by nuclear fission (the splitting of certain atomic nuclei) has been harnessed since the 1950s to generate electricity, and research continues into the possible controlled use of nuclear fusion (the fusing, or combining, of atomic nuclei).

In nuclear power stations, fission of radioactive substances takes place, releasing large amounts of heat energy. The heat is used to produce the steam that drives turbines and generators, producing electrical power.

The Sun is an example of a natural nuclear reactor. Millions of atoms of hydrogen fuse together to form millions of atoms of helium, generating a continuous supply of heat and light energy. This is called a fusion reaction.

nuclear family the basic family unit of mother, father, and children. This is the familial norm of industrial societies, in contrast to countries with traditional economies, where the extended family (nuclear family plus assorted kin) is more common.

nuclear fission a process in which an atom of an element is struck by a neutron causing the atomic nucleus to split apart and release other neutrons. A chain reaction results as these neutrons then strike other atomic nuclei. In so doing large amounts of energy are released. In *nuclear reactors the chain reaction caused by the fission of uranium-235 is regulated by the use of control rods which absorb quantities of free neutrons. Compare *nuclear fusion.

nuclear fusion a process in which two nuclei of light elements such as hydrogen are combined to form a heavier nucleus such as helium with a substantial release of energy. This process occurs only at exceedingly high temperatures (in excess of 100 million degrees Celsius). Nuclear fusion power is seen as the ultimate source of abundant, cheap and pollution-free electricity with none of the present disadvantages of nuclear fission or problems associated with the burning of fossil fuels. However, all efforts to harness fusion power for commercial energy production have so far been unsuccessful.

Nuclear Non-Proliferation Treaty treaty signed 1968 to limit the spread of nuclear weapons. Under the terms of the treaty, those signatories declared to be nuclear powers (China, France, Russia, the UK, and the USA) pledged to work towards nuclear disarmament and not to supply military nuclear technology to non-nuclear countries, while other signatories pledged not to develop or acquire their own nuclear weapons. The treaty was renewed and extended indefinitely May 1995.

nuclear power or **atomic power**, energy generated by a *nuclear reactor primarily by *nuclear fission. There is almost unanimous agreement that fission power is one of the most hazardous methods of producing power (usually in the form of electrical power) for use by industry, transport and domestic users. Nuclear power presents major technical, moral and ethical problems for humanity. The problems of nuclear waste disposal, nuclear accidents, and possible terrorist attacks on reactors have caused many nations to rethink their nuclear power programme.

nuclear power station an electricity-generating plant using nuclear fuel as an alternative to the conventional *fossil fuels of *coal, oil and gas. Nuclear power stations, while expensive to construct, are relatively cheap to run but have to be built in remote (often coastal) locations well away from population concentrations. This is partly in response to public anxiety over the safety of such stations and partly due to the problems of radioactive waste disposal. The fuel used in most nuclear power stations is the element uranium, world *reserves (2) of which are extensive.

nuclear reactor   device for producing *nuclear energy in a controlled manner. There are various types of reactor in use, all using nuclear fission. In a **gas-cooled reactor**, a circulating gas under pressure (such as carbon dioxide) removes heat from the core of the reactor, which usually contains natural uranium. The efficiency of the fission process is increased by slowing neutrons in the core by using a moderator such as carbon. The reaction is controlled with neutron-absorbing rods made of boron. An **advanced gas-cooled reactor** (AGR) generally has enriched uranium as its fuel. A **water-cooled reactor**, such as the steam-generating heavy water (deuterium oxide) reactor, has water circulating through the hot core. The water is converted to steam, which drives turbo-alternators for generating electricity. The most widely used reactor is the **pressurized-water reactor** (PWR), which contains a sealed system of pressurized water that is heated to form steam in heat exchangers in an external circuit. The **fast reactor** has no moderator and uses fast neutrons to bring about fission. It uses a mixture of plutonium and uranium oxide as fuel. When operating, uranium is converted to plutonium, which can be extracted and used later as fuel. It is also called the fast breeder or breeder reactor because it produces more plutonium than it consumes. Heat is removed from the reactor by a coolant of liquid sodium.

nuclear safety   measures to avoid accidents in the operation of nuclear reactors and in the production and disposal of nuclear weapons and of nuclear waste. There are no guarantees of the safety of any of the various methods of disposal. Nuclear safety is a controversial subject – some governments do not acknowledge the hazards of atomic radiation and radiation sickness.

nuclear accidents   Windscale (now Sellafield), Cumbria, England. In 1957 fire destroyed the core of a reactor, releasing large quantities of radioactive fumes into the atmosphere. In 1990 a scientific study revealed an increased risk of leukaemia in children whose fathers had worked at Sellafield between 1950 and 1985.

Ticonderoga, 130 km off the coast of Japan. In 1965 a US Navy Skyhawk jet bomber fell off the deck of this ship, sinking in 4900 m of water. It carried a one-megaton hydrogen bomb. The accident was only revealed in 1989.

Three Mile Island, Harrisburg, Pennsylvania, USA. In 1979 a combination of mechanical and electrical failure, as well as operator error, caused a pressurized water reactor to leak radioactive matter.

Church Rock, New Mexico, USA. In July 1979, 380 million litres of radioactive water containing uranium leaked from a pond into the Rio Purco, causing the water to become over 6500 times as radioactive as safety standards allow for drinking water.

Chernobyl, Ukraine. In April 1986 there was an explosive leak, caused by overheating, from a nonpressurized boiling-water reactor, one of the largest in Europe. The resulting clouds of radioactive material spread as far as the UK. Thirty-one people were killed in the explosion, and thousands of square kilometres of land were contaminated by fallout. By June 1992, seven times as many children in the Ukraine and Belarus were contracting thyroid cancer as before the accident, and the incidence of leukaemia was rising; it was estimated that more than 6000 people had died as a result of the accident, and that the death toll in the Ukraine alone would eventually reach 40,000.

Tomsk, Siberia, Russia. In April 1993 a tank exploded at a uranium reprocessing plant, sending a cloud of radioactive particles into the air.

nuclear testing   detonation of nuclear devices to verify their reliability, power, and destructive abilities. Although carried out secretly in remote regions of the world, such tests are easily detected by the seismic shock waves produced. The first tests were carried out in the atmosphere during the 1950s by the USA, the USSR, and the UK. The Comprehensive Test Ban Treaty was signed by 149 countries in 1996.

nuclear waste   the radioactive and toxic by-products of the nuclear energy and nuclear weapons industries. Nuclear waste may have an active life of several thousand years. Reactor waste is of three types: **high-level** spent fuel, or the residue when nuclear fuel has been removed from a reactor and reprocessed; **intermediate**, which may be long-or short-lived; and **low-level**, but bulky, waste from reactors, which has only short-lived radioactivity. Disposal, by burial on land or at sea, has raised problems of safety, environmental pollution, and security.

nuée ardente   rapidly flowing, glowing white-hot cloud of ash and gas emitted by a volcano during a violent eruption. The ash and other pyroclastics in the lower part of the cloud behave like an ash flow. In 1902 a *nuée ardente* produced by the eruption of Mount Pelee in Martinique swept down the volcano in a matter of seconds and killed 28,000 people in the nearby town of St Pierre.

nugget   piece of gold found as a lump of the *native metal. Nuggets occur in *alluvial deposits where river-borne particles of the metal have adhered to one another.

nunatak   mountain peak protruding through an ice sheet. Such peaks are common in Antarctica.

nutrient cycle   transfer of nutrients from one part of an *ecosystem to another. Trees, for example, take up nutrients such as calcium and potassium from the soil through their root systems and store them in leaves. When the leaves fall they are decomposed by bacteria and the nutrients are released back into the soil where they become available for root uptake again.

OAPEC *abbreviation for* *Organization of Arab Petroleum Exporting Countries.

oasis area of land made fertile by the presence of water near the surface in an otherwise arid region. The occurrence of oases affects the distribution of plants, animals, and people in the desert regions of the world.

obsidian black or dark-coloured glassy volcanic rock, chemically similar to *granite, but formed by cooling rapidly on the Earth's surface at low pressure.

The glassy texture is the result of rapid cooling, which inhibits the growth of crystals. Obsidian was valued by the early civilizations of Mexico for making sharp-edged tools and ceremonial sculptures.

obsidian hydration-rim dating in archaeology, a method of dating artefacts made from the volcanic glass obsidian. Water molecules absorbed by inward diffusion through cut surfaces cause the outer areas of an obsidian article to convert to the mineral perlite. An object may be dated by measuring the thickness of this perlite – the **hydration** (combined with water) rim.

Only a molecule-thick water film is required at the surface to maintain the process, an amount available even in the near-arid zones of Egypt. Temperature, sunlight, and different chemical compositions cause variation in the hydration rate; therefore the method needs to be calibrated against an established chronological sequence for absolute dating. The method has been applied to many periods, including the Aztec age in Mexico, the pre-ceramic era of Japan (about 23,000 BC), and the tribal-war periods of Easter Island before the arrival of traders and missionaries in 1722. Obsidian hydration-rim dating was stimulated by the early research of Irving Friedman of the US Geological Survey from 1955.

OCAM acronym for *Organization Commune Africaine et Mauricienne, body for economic cooperation in Africa.

occluded front weather *front formed when a cold front catches up with a warm front. It brings clouds and rain as air is forced to rise upwards along the front, cooling and condensing as it does so.

ocean great mass of salt water. Geographically speaking three oceans exist – the Atlantic, Indian, and Pacific – to which the Arctic is often added. They cover approximately 70% or 363,000,000 sq km of the total surface area of the Earth. According to figures released in August 2001, the total volume of the world's oceans is 1370 million cubic km. Water levels recorded in the world's oceans have shown an increase of 10–15 cm over the past 100 years.

depth (average) 3660 m, but shallow ledges (continental shelves) 180 m run out from the continents, beyond which the continental slope reaches down to the *abyssal zone, the largest area, ranging from 2000–6000 m. Only the *deep-sea trenches go deeper, the deepest recorded being 11,034 m (by the *Vityaz*, USSR) in the Mariana Trench of the western Pacific in 1957

features deep trenches (off eastern and southeastern Asia, and western South America), volcanic belts (in the western Pacific and eastern Indian Ocean), and ocean ridges (in the mid-

Atlantic, eastern Pacific, and Indian Ocean).

temperature varies on the surface with latitude (–2°C to +29°C); decreases rapidly to 370 m, then more slowly to 2200 m; and hardly at all beyond that

seawater contains about 3% dissolved salts, the most abundant being sodium chloride; salts come from the weathering of rocks on land; rainwater flowing over rocks, soils, and organic matter on land dissolves small amounts of substances, which pass into rivers to be carried to the sea. Salt concentration in the oceans remains remarkably constant as water is evaporated by the Sun and fresh water added by rivers. Positive ions present in sea water include sodium, magnesium, potassium, and calcium; negative ions include chloride, sulphate, hydrogencarbonate, and bromide

commercial extraction of minerals includes bromine, magnesium, potassium, salt (sodium chloride); those potentially recoverable include aluminium, calcium, copper, gold, manganese, silver.

pollution Oceans have always been used as a dumping area for human waste, but as the quantity of waste increases, and land areas for dumping diminish, the problem is exacerbated. Today ocean pollutants include airborne emissions from land (33% by weight of total marine pollution); oil from both shipping and land-based sources; toxins from industrial, agricultural, and domestic uses; sewage; sediments from mining, forestry, and farming; plastic litter; and radioactive isotopes. Thermal pollution by cooling water from power plants or other industry is also a problem, killing coral and other temperature-sensitive sedentary species.

ocean current fast-flowing body of seawater forced by the wind or by variations in water density (as a result of temperature or salinity variations) between two areas. Ocean currents are partly responsible for transferring heat from the Equator to the poles and thereby evening out the global heat imbalance.

Ocean Drilling Program ODP; formerly **Deep-Sea Drilling Project** (1968–85), research project initiated in

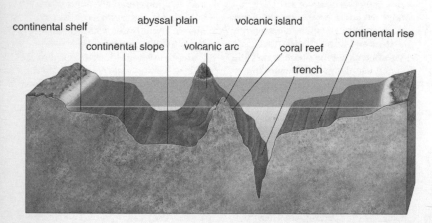

ocean Profile of an ocean floor. The ocean trench is the deepest part of the ocean and the abyssal plains constitute most of the ocean bed.

the USA to sample the rocks of the ocean *crust. Initially under the direction of Scripps Institution of Oceanography, the project was planned and administered by the Joint Oceanographic Institutions for Deep Earth Sampling (JOIDES). The operation became international in 1975, when Britain, France, West Germany, Japan, and the USSR also became involved.

Boreholes were drilled in all the oceans using the JOIDES ships *Glomar Challenger* and *Resolution*. Knowledge of the nature and history of the ocean basins was increased dramatically. The technical difficulty of drilling the seabed to a depth of 2000 m was overcome by keeping the ship in position with side-thrusting propellers and satellite navigation, and by guiding the drill using a radiolocation system. The project is intended to continue until 2005.

oceanic crust   see *crust.

oceanic ridge   mountain range that extends along the centre of the bed of an ocean; see *mid-ocean ridge.

oceanography   study of the oceans. Its subdivisions deal with each ocean's extent and depth, the water's evolution and composition, its physics and chemistry, the bottom topography, currents and wind, tidal ranges, biology, and the various aspects of human use. Computer simulations are widely used in oceanography to plot the possible movements of the waters, and many studies are carried out by remote sensing.

Oceanography involves the study of water movements – currents, waves, and tides – and the chemical and physical properties of the seawater. It deals with the origin and topography of the ocean floor – ocean trenches and ridges formed by *plate tectonics, and continental shelves from the submerged portions of the continents.

ocean trench   submarine valley. Ocean trenches are characterized by the presence of a volcanic arc on the concave side of the trench. Trenches are now known to be related to subduction zones, places where a plate of oceanic *lithosphere dives beneath another plate of either oceanic or continental lithosphere. Ocean trenches are found around the edge of the Pacific Ocean and the northeastern Indian Ocean; minor ones occur in the Caribbean and near the Falkland Islands.

Ocean trenches represent the deepest parts of the ocean floor, the deepest being the *Mariana Trench which has a depth of 11,034 m. At depths of below 6 km there is no light and very high pressure; ocean trenches are inhabited by crustaceans, coelenterates (for example, sea anemones), polychaetes (a type of worm), molluscs, and echinoderms.

ODA   *abbreviation for* *Overseas Development Administration.

odour   irritating smell, which causes a nuisance but is not actually dangerous. Odours are a frequent cause of complaints made by householders living near some types of industry or agriculture.

OECD   *abbreviation for* *Organization for Economic Cooperation and Development.

OEEC   *abbreviation for* **Organization for European Economic Cooperation.**

offshore bar   a low bank of sand and shingle lying some distance offshore and exposed at high tide. The offshore bar is created when a very gently shelving seabed causes waves to break well away from the actual shoreline. The Cape Hatteras coastline of the Atlantic coast of the USA is an area where such conditions prevail.

oil spill   oil discharged from an ocean-going tanker, pipeline, or oil installation, often as a result of damage. An oil spill kills shore life, clogging the feathers of birds and suffocating other creatures. At sea, toxic chemicals spread into the water below, poisoning sea life. Mixed with dust, the oil forms globules that sink to the seabed, poisoning sea life there as well. Oil spills are broken up by the use of detergents but such chemicals can themselves damage wildlife. The annual spillage of oil is 8 million barrels (280 million gallons) a year. At any given time tankers are carrying 500 million barrels (17.5 billion gallons).

The amount of oil entering oceans from shipping operations decreased by 60% 1981–91.

In March 1989 the *Exxon Valdez* (belonging to the Exxon Corporation) ran aground and spilled oil in Alaska's Prince William Sound, covering 12,400 sq km and killing at least 34,400 sea birds, 10,000 sea otters, and up to 16 whales. The incident led to the US Oil Pollution Act of 1990, which requires tankers operating in US waters to have double hulls.

The world's largest oil spill was in the Gulf in 1991 as a direct result of hostilities during the Gulf War. Around 6–8 million barrels (210–280 million gallons) of oil were spilled, polluting 675 km of Saudi coastline. In some places, the oil was 30 cm deep in the sand.

oligarchy (Greek *oligarchia* 'government of the few') rule of the few, in their own interests. It was first identified as a form of government by the Greek philosopher Aristotle. In modern times there have been a number of oligarchies, sometimes posing as democracies; the paramilitary rule of the Duvalier family in Haiti, 1957–1986, is an example.

Oligocene epoch third epoch of the Tertiary period of geological time, 35.5–3.25 million years ago. The name,

from Greek, means 'a little recent', referring to the presence of the remains of some modern types of animals existing at that time.

oligopoly in economics, a situation in which a few companies control the major part of a particular market. For example, in the UK the two largest soap-powder companies, Procter & Gamble and Unilever, control over 85% of the market. In an oligopolistic market, firms may well join together in a *cartel, colluding to fix high prices. This collusion, an example of a restrictive trade practice, is illegal in the UK and the European Union (EU).

oligotrophic in ecology, describing ponds, lakes, and other bodies of still water that lack in nutrients.

olivenite basic copper arsenate, $Cu_2AsO_4(OH)$, occurring as a mineral in olive-green prisms.

olivine greenish mineral, magnesium iron silicate, $(Mg,Fe)_2SiO_4$. It is a rock-forming mineral, present in, for example, peridotite, gabbro, and basalt. Olivine is called **peridot** when pale green and transparent, and used in jewellery.

omnivore animal that feeds on both plant and animal material. Omnivores have digestive adaptations intermediate between those of *herbivores and *carnivores, with relatively unspecialized digestive systems and gut micro-organisms that can digest a variety of foodstuffs. Omnivores include humans, the chimpanzee, the cockroach, and the ant.

one-party state state in which one political party dominates, constitutionally or unofficially, to the point where there is no effective opposition. There may be no legal alternative parties, as, for example, in Cuba. In other instances, a few token members of an opposition party may be tolerated, or one party may be permanently in power, with no elections. The one-party state differs from the 'dominant-party' state, where one party controls government for an extended period, as the Liberal Democrats did in Japan 1955–93, but where there are openly-democratic competitive elections.

onyx   semi-precious variety of chalcedonic *silica (SiO_2) in which the crystals are too fine to be detected under a microscope, a state known as cryptocrystalline. It has straight parallel bands of different colours: milk-white, black, and red.

Sardonyx, an onyx variety, has layers of brown or red carnelian alternating with lighter layers of onyx. It can be carved into cameos.

oolite   limestone made up of tiny spherical carbonate particles, called **ooliths**, cemented together. Ooliths have a concentric structure with a diameter up to 2 mm. They were formed by chemical precipitation and accumulation on ancient sea floors.

The surface texture of oolites is rather like that of fish roe. The late Jurassic limestones of the British Isles are mostly oolitic in nature.

ooze   sediment of fine texture consisting mainly of organic matter found on the ocean floor at depths greater than 2000 m. Several kinds of ooze exist, each named after its constituents.

Siliceous ooze is composed of the *silica shells of tiny marine plants (diatoms) and animals (radiolarians). Calcareous ooze is formed from the *calcite shells of microscopic animals (foraminifera) and floating algae (coccoliths).

opal   form of hydrous *silica ($SiO_2.nH_2O$), often occurring as stalactites and found in many types of rock. The common opal is translucent, milk-white, yellow, red, blue, or green, and lustrous. Precious opal is opalescent, the characteristic play of colours being caused by close-packed silica spheres diffracting light rays within the stone.

Opal is cryptocrystalline, that is, the crystals are too fine to be detected under an optical microscope. Opals are found in Hungary; New South Wales, Australia (black opals were first discovered there in 1905); and Mexico (red fire opals).

OPEC   acronym for *Organization of Petroleum-Exporting Countries.

opencast mining   a type of mining where the mineral is extracted by direct excavation rather than by shaft or drift methods. For example, in parts of the Yorkshire coalfield the coal measures occur very close to the surface and the superficial overburden is relatively easily removed. The *coal is then excavated by mechanical grabs and removed by trucks.

Such mining creates extensive scars in the landscape which, if left unmanaged, represent serious environmental deterioration. Mining companies are required to undertake *landscaping* after the cessation of mining, usually involving the infilling of the opencast site and the planting of vegetation on the reclaimed surface. The vegetation also helps to stabilize the infill.

open shop   factory or other business employing men and women not belonging to trade unions, as opposed to the closed shop, which employs trade unionists only.

opportunity cost   in economics, that which has been forgone in order to achieve an objective. A family may choose to buy a new television set and forgo their annual holiday; the holiday represents the opportunity cost.

optimum population   the number of people that will produce the highest per capita economic return given the resources available, and their full utilization. Should the population rise or fall from the optimum the output per capita, and standard of living, will fall.

Ordnance Survey OS, official body responsible for the mapping of Britain. It was established in 1791 as the **Trigonometrical Survey** to continue work initiated in 1784 by Scottish military surveyor General William Roy (1726–1790). Its first accurate maps appeared in 1830, drawn to a scale of 1 in to the mile (1:63,000). In 1858 the OS settled on a scale of 1:2500 for the mapping of Great Britain and Ireland (higher for urban areas, lower for uncultivated areas).

Subsequent revisions and editions include the 1:50,000 Landranger series of 1971–86. In 1989, the OS began using a computerized system for the creation and continuous revision of maps. Customers can now have maps drafted to their own specifications, choosing from over 50 features (such as houses, roads, and vegetation).

Ordovician period period of geological time 510–439 million years ago; the second period of the *Palaeozoic era. Animal life was confined to the sea: reef-building algae and the first jawless fish are characteristic.

The period is named after the Ordovices, an ancient Welsh people, because the system of rocks formed in the Ordovician period was first studied in Wales.

ore body of rock, a vein within it, or a deposit of sediment, worth mining for the economically valuable mineral it contains. The term is usually applied to sources of metals. Occasionally metals are found uncombined (native metals), but more often they occur as compounds such as carbonates, sulphides, or oxides. The ores often contain unwanted impurities that must be removed when the metal is extracted.

Commercially valuable ores include bauxite (aluminium oxide, Al_2O_3) haematite (iron(III) oxide, Fe_2O_3), zinc blende (zinc sulphide, ZnS), and rutile (titanium dioxide, TiO_2).

Hydrothermal ore deposits are formed from fluids such as saline water passing through fissures in the host rock at an elevated temperature. Examples are the 'porphyry copper' deposits of Chile and Bolivia, the submarine copper–zinc–iron sulphide deposits recently discovered on the East Pacific Rise, and the limestone lead–zinc deposits that occur in the southern USA and in the Pennines of Britain.

Other ores are concentrated by igneous processes, causing the ore metals to become segregated from a magma – for example, the chromite- and platinum-rich bands within the bushveld, South Africa. Erosion and transportation in rivers of material from an existing rock source can lead to further concentration of heavy minerals in a deposit – for example, Malaysian tin deposits.

Weathering of rocks in situ can result in residual metal-rich soils, such as the nickel-bearing laterites of New Caledonia.

organic farming a system of farming that avoids the use of any *artificial fertilizers or chemical pesticides, using only *organic fertilizers and pesticides derived directly from animal or vegetable matter. Yields from organic farming are lower, but the products are sold at a premium price.

organic fertilizer a fertilizer composed of organic material, e.g. horse manure, farmyard manure, seaweed derivatives and bonemeal. Compare *artificial fertilizer.

organic fraction that proportion of the *soil which is composed of material derived from the breakdown of vegetation or other organic matter. Compare *inorganic fraction.

organophosphate insecticide insecticidal compounds whose mechanism of action is very toxic to humans and which must be used with great care. Malathion and permethrin may be used to control lice in humans and have many applications in veterinary medicine and agriculture. In 1998 organophosphates were the most widely used insecticides, with 40% of the global market.

Organization Commune Africaine et Mauricienne OCAM; French 'Joint African and Mauritian Organization', organization founded in 1965 to strengthen the solidarity and close ties between member states, raise living standards, and coordinate economic policies. The membership includes Benin, Burkina Faso, Central African Republic, Côte d'Ivoire, Niger, Rwanda, Senegal, and Togo. Through the organization, members share an airline, a merchant fleet, and a common postal and communications system. The headquarters of OCAM are in Bangui in the Central African Republic.

Organization for Economic Cooperation and Development OECD, international organization of 29 industrialized countries that provides a forum for discussion and coordination of member states' economic and social policies. Founded in 1961, with its headquarters in Paris, the OECD replaced the Organization for European Economic Cooperation (OEEC), which had been established in 1948 to implement the Marshall Plan. The Commission of the European Union also takes part in the OECD's work.

Organization for Security and Cooperation in Europe OSCE; formerly the **Conference on Security and Cooperation in Europe (CSCE)** (until 1995), regional security organization whose 55 participating states are from Europe, Central Asia, and North America. Its main areas of action are early warning and conflict prevention, crisis management, post-conflict rehabilitation, and supervising elections. It was founded in 1975 as the Conference on Security and Cooperation in Europe (CSCE) at the Helsinki Conference in Finland, under the Helsinki Final Act on East-West Relations, committing members to increasing consultation and cooperation. By mid-1995, having admitted the former republics of the USSR, as well as other new nations from the former communist bloc, its membership had risen to 55 states (including the Federal Republic of Yugoslavia, whose membership was suspended July 1992–November 2000).

Organization of American States OAS, association founded in 1948 at Bogotá, Colombia by a charter signed by representatives of North, Central, and South American states. It aims to maintain peace and solidarity within the hemisphere, and is also concerned with the social and economic development of Latin America.

Organization of Arab Petroleum Exporting Countries OAPEC, body established in 1968 to safeguard the interests of its members and encourage economic cooperation within the petroleum industry. Its members are Algeria, Bahrain, Egypt, Iraq, Kuwait, Libya, Qatar, Saudi Arabia, Syria, and the United Arab Emirates; together they account for more than 25% of the world's oil output. The organization's headquarters are in Kuwait.

Organization of Petroleum-Exporting Countries OPEC, body established in 1960 to coordinate price and supply policies of oil-producing states, protecting its members' interests by manipulating oil production and the price of crude oil. Its concerted action in raising prices in the 1970s triggered worldwide recession but also lessened demand so that its influence was reduced by the mid-1980s. However, continued reliance on oil re-strengthened its influence in the late 1990s. OPEC members are: Algeria, Gabon, Indonesia, Iran, Iraq, Kuwait, Libya, Nigeria, Qatar, Saudi Arabia, the United Arab Emirates, and Venezuela. Ecuador, formerly a member, withdrew in 1993. OPEC's secretary general is Rilwanu Lukman of Nigeria.

Organization of the Islamic Conference OIC, association of 44 states in the Middle East, Africa, and Asia, established in 1971 in Rabat, Morocco, to promote Islamic solidarity between member countries, and to consolidate economic, social, cultural, and scientific cooperation. Its head-quarters are in Jeddah, Saudi Arabia.

organophosphate insecticide insecticidal compounds that cause the irreversible inhibition of the cholinesterase enzymes that break down acetylcholine. As this mechanism of action is very toxic to humans, the compounds should be used with great care. Malathion and permethrin may be used to control lice in humans and have many applications in veterinary medicine and agriculture. In 1998 organophosphates were the most widely used insecticides, with 40% of the global market.

According to the results of a survey by a UK team of psychiatrists, announced in 1998, one in ten farmers regularly exposed to organophosphates will suffer irreversible physical and mental damage. Sheep dipping is the most common source of exposure. Reported symptoms include tiredness and speech problems. **Organophosphate-induced delayed neuropathy** is a condition of nerve and muscle damage through prolonged contact with organophosphates, which can lead to paralysis.

orogenesis in its original, literal sense, orogenesis means 'mountain building', but today it more specifically refers to the tectonics of mountain building (as opposed to mountain building by erosion).

Orogenesis is brought about by the movements of the rigid plates making up the Earth's crust and uppermost mantle (described by *plate tectonics). Where two plates collide at a destructive margin rocks become folded and lifted to form chains of mountains (such as the Himalayas). Processes associated with orogeny are faulting and thrusting (see *fault), folding, metamorphism, and plutonism (see *plutonic rock). However, many topographical features of mountains – cirques, U-shaped valleys – are the result of *non-orogenic* processes, such as weathering, erosion, and glaciation. *Isostasy (uplift due to the buoyancy of the Earth's crust) can also influence mountain physiography. There are three generally recognized orogenies: the Alpine or Tertiary orogeny (approx. 50 million years ago), examples being the Rockies, the Andes, the Alps and the Himalayas; the Hercynean orogeny (approx. 300 million years ago), an example being the Appalachians; and the Caledonian orogeny (approx 400 million years ago), an example being the Scottish Highlands.

orographic rainfall rainfall that occurs when an airmass is forced to rise over a mountain range. As the air rises, it cools. The amount of moisture that air can hold decreases with decreasing temperature. So the water vapour in the rising airstream condenses, and rain falls on the windward side of the mountain. The air descending on the leeward side contains less moisture, resulting in a **rainshadow** where there is little or no rain.

Ottawa agreements trade agreements concluded at the Imperial Economic Conference, held in Ottawa, Canada 1932, between Britain and its dependent territories, lowering tariffs on British manufactured goods and increasing duties on non-Dominion produce.

outback the inland region of Australia. Its main inhabitants are Aborigines, miners (including opal miners), and cattle ranchers. Its harsh beauty has been recorded by such artists as Sidney Nolan.

out-of-town shopping purchasing activity provided for by the construction of a complex of shops situated out of a town or city centre.

output quantity of goods and services produced or provided by a business organization or economy.

outsourcing situation where an organization passes the provision of a service, or other task previously undertaken in-house, to an external third party for them to perform on behalf of the original organization.

outwash sands and gravels deposited by streams of meltwater (water produced by the melting of a glacier). Such material may be laid down ahead of the glacier's snout to form a large flat expanse called an *outwash plain.

Outwash is usually well sorted, the particles being deposited by the meltwater according to their size – the largest are deposited next to the snout while finer particles are deposited further downstream.

outwash plain or **sandur**, an area of stratified glacial till deposited by meltwater streams sloping away from the margins of a glacier or icesheet. In an enclosed valley outwash plains are known as *valley trains*. In parts of northern Europe and the American Midwest outwash plains form productive agricultural land. Sand and gravel extraction is a common feature of many outwash plains.

overcapacity situation where total production capability outstrips demand. This may be a temporary situation caused by normal market cycles, or a brief economic downturn. The reasons may also be structural, such as the decline of a market. In this case companies in the market will have to restructure, diversify, or go in to liquidation.

overfishing fishing at rates that exceed the *sustained-yield cropping of fish species, resulting in a net population decline. For example, in the North Atlantic, herring has been fished to the verge of extinction, and the cod and haddock populations are severely depleted. In the developing world, use of huge factory ships, often by fisheries from industrialized countries, has depleted stocks for local people who cannot obtain protein in any other way.

Ecologists have long been concerned at the wider implications of overfishing, in particular the devastation wrought on oceanic *food chains. The United Nations Food and Agriculture Organization estimates that worldwide overfishing has damaged oceanic ecosystems to such an extent that potential catches are on average reduced by 20%. With better management of fishing programmes the fishing catch could in principle be increased; it is estimated that, annually, 20 million tonnes of fish are discarded from fishing vessels at sea, because they are not the species sought.

According to an estimate by the Food and Agriculture Organization in 1993, nine of the world's 17 main fishing grounds were suffering a potentially catastrophic decline in some species. In 1994 approximately 17% of fishing waters off the coast of New England, USA, was closed in an attempt to restore dwindling stocks. The area affected covered 17,000 sq km and lay within the Georges Bank region of the Atlantic Ocean.

overland flow another term for *surface run-off of water after rain.

overpopulation too many people for the resources available in an area (such as food, land, and water). The consequences were first set out by English economist Thomas Malthus at the start of the population explosion.

Overseas Development Administration ODA, see *International Development Department (IDD).

overspill that part of an urban population or employment which has moved to areas outside the main settlement. This may be either in response to *push factors, such as *congestion and overcrowding, or as a result of planning policies perhaps connected with inner-city redevelopment schemes. *New towns may be built to accommodate this outflow, as Stevenage, Hertfordshire, and Harlow, Essex.

oxbow lake curved lake found on the flood plain of a river. Oxbows are caused by the loops of *meanders that are cut off at times of flood and the river subsequently adopts a shorter course. In the USA, the term bayou is often used.

oxidation in earth science, a form of *chemical weathering caused by the chemical reaction that takes place between certain iron-rich minerals in rock and the oxygen in water. It tends to result in the formation of a red-coloured soil or deposit. The inside walls of canal tunnels and bridges often have deposits formed in this way.

oxyfuel fuel enriched with oxygen to decrease carbon monoxide (CO) emissions. Oxygen is added in the form of chemicals such as methyl tertiary butyl ether (MTBE) and ethanol.

Cars produce CO when there is insufficient oxygen present to convert

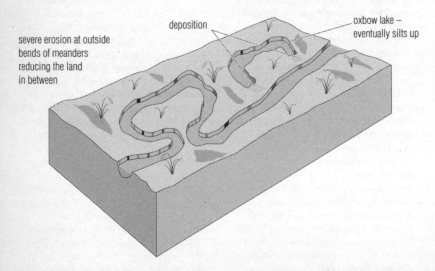

severe erosion at outside bends of meanders reducing the land in between

deposition

oxbow lake – eventually silts up

oxbow lake The formation of an oxbow lake. As a river meanders across a flood plain, the outer bends are gradually eroded and the water channel deepens; as the loops widen, the neck of the loop narrows and finally gives way, allowing the water to flow in a more direct route, isolating the old water channel and forming an oxbow lake.

all the carbon in the petrol to CO_2. This occurs mostly at low temperatures, such as during the first five minutes of starting the engine and in cold weather. CO emissions are reduced by the addition of oxygen-rich chemicals. The use of oxyfuels in winter is compulsory in 35 US cities. There are fears, however, that MTBE can cause health problems, including nausea, headaches, and skin rashes.

ozone   a form of oxygen found in a layer in the *stratosphere, where it protects the Earth's surface from ultraviolet rays. There has been recent concern about holes appearing in the ozone layer over the polar regions. There are many possible reasons why these holes might occur, but it is known that the ozone layer is damaged by the chlorine released by *chlorofluorocarbons. An increase in the ultraviolet rays reaching the Earth could result in an increase in skin cancers, and damage to crops, vegetation, livestock and wildlife in countries near the polar regions.

Ozone is a highly reactive pale-blue gas with a penetrating odour.

ozone depleter   any chemical that destroys the ozone in the stratosphere. Most ozone depleters are chemically stable compounds containing chlorine or bromine, which remain unchanged for long enough to drift up to the upper atmosphere. The best known are *chlorofluorocarbons (CFCs).

Other ozone depleters include halons, used in some fire extinguishers; methyl chloroform and carbon tetrachloride, both solvents; some CFC substitutes; and the pesticide methyl bromide.

CFCs accounted for approximately 75% of ozone depletion in 1995, whereas methyl chloroform (atmospheric concentrations of which had markedly decreased during 1990–94) accounted for an estimated

12.5%. The ozone depletion rate overall is now decreasing as international agreements to curb the use of ozone-depleting chemicals begin to take effect. In 1996 there was a decrease in ozone depleters in the lower atmosphere. This trend is expected to continue into the stratosphere over the next few years.

ozone layer   thin layer of the gas *ozone in the upper atmosphere that shields the Earth from harmful ultraviolet rays. A continent-sized hole has formed over Antarctica as a result of damage to the ozone layer. This has been caused in part by *chlorofluorocarbons (CFCs), but many reactions destroy ozone in the stratosphere: nitric oxide, chlorine, and bromine atoms are implicated.

It is believed that the ozone layer is depleting at a rate of about 5% every 10 years over northern Europe, with depletion extending south to the Mediterranean and southern USA. However, ozone depletion over the polar regions is the most dramatic manifestation of a general global effect. Ozone levels over the Arctic in spring 1997 fell over 10% since 1987, despite the reduction in the concentration of CFCs and other industrial compounds which destroy the ozone when exposed to sunlight. It is thought that this may be because of an expanding vortex of cold air forming in the lower stratosphere above the Arctic, leading to increased ozone loss. It is expected that an Arctic hole as large as that over Antarctica could remain a threat to the northern hemisphere for several decades.

The size of the hole in the ozone layer in October 1998 was three times the size of the USA, larger than it had ever been before. In autumn 2000 the hole in the ozone layer was at its largest ever. Observers had hoped that its 1998 level was due to *El Niño and would not be exceeded.

In April 1991, satellite data from the US space agency NASA revealed that the ozone layer had depleted by 4–8% in the northern hemisphere and by 6–10% in the southern hemisphere between 1978 and 1990. Between 1989 and 1996 the ozone layer over the Arctic diminished by around 40%. The ozone layer above the USA was reduced by 12.6% in 1993; that above the UK was reduced by almost 50% in 1996. By September 1996, levels in the upper atmosphere had reached a record low.

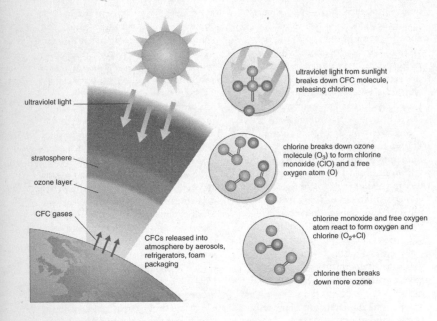

ultraviolet light

stratosphere

ozone layer

CFC gases

ultraviolet light from sunlight breaks down CFC molecule, releasing chlorine

chlorine breaks down ozone molecule (O_3) to form chlorine monoxide (ClO) and a free oxygen atom (O)

chlorine monoxide and free oxygen atom react to form oxygen and chlorine (O_2+Cl)

CFCs released into atmosphere by aerosols, refrigerators, foam packaging

chlorine then breaks down more ozone

ozone layer The destruction of the ozone layer by chlorofluorocarbons (CFCs). CFCs discharged into the atmosphere break down in sunlight releasing chlorine, which breaks down the ozone to form chlorine monoxide and a free oxygen atom. These products react together to form oxygen and chlorine, leaving the chlorine to break down another ozone molecule, and so on.

P

Pacific Community PC; formerly **South Pacific Commission** until 1998, organization to promote economic and social cooperation in the region, including dialogue between Pacific countries and those, such as France and the UK, that have dependencies in the region. It was established in February 1947. Its members include American Samoa, Australia, Cook Islands, Federated States of Micronesia, Fiji Islands, France, French Polynesia, Guam, Kiribati, Marshall Islands, Nauru, New Caledonia, New Zealand, Niue, Northern Marianas, Palau, Papua New Guinea, Pitcairn Islands, Samoa, Solomon Islands, Tokelau, Tonga, Tuvalu, United Kingdom, United States of America, Vanuatu, and Wallis and Futuna; headquarters are in Nouméa, New Caledonia.

Pacific Islands Forum PIF; formerly called **South Pacific Forum** until 2000, association of states in the region to discuss common interests and develop common policies, created in 1971 as an offshoot of the South Pacific Commission, now the *Pacific Community. Its 26 member countries include Australia, Cook Islands, Fiji Islands, Kiribati, Marshall Islands, the Federated States of Micronesia, Nauru, New Zealand, Niue, Papua New Guinea, Samoa, Solomon Islands, Tonga, Tuvalu, and Vanuatu, with New Caledonia an observer since 1999.

Pacific Ocean world's largest ocean, extending from Antarctica to the Bering Strait; area 166,242,500 sq km; greatest breadth 16,000 km; length 11,000 km; average depth 4188 m; greatest depth of any ocean is the found in the Mariana Trench, in the northwest Pacific, with a depth of 11,034 m.

depth The Pacific is the deepest ocean; the western and northern areas are deeper than the east and south. Some of the greatest depths lie alongside islands, such as the Mariana Trench (11,034 m), alongside the Mariana Islands; the Tuscarora Deep (8500 m), alongside Japan and the Kurils and extending for 640 km; and the Aldrich Deep, east of New Zealand (9400 m).

islands There are over 2500 islands in the central and western regions, of volcanic or coral origin, many being atolls. The Pacific is ringed by an area of volcanic activity, with accompanying earthquakes.

currents Winds in the northern Pacific produce generally clockwise currents; a northward current from the Equator flows past the Philippines and is joined by currents from the East Indies and China Sea at Taiwan to form the Kuroshio. This branches opposite Vancouver to flow south as the California current, and north around the Alaskan coast. A cold current from the Bering Sea enters the Okhotsk and Japan seas, causing freezing in winter. In the South Pacific the trade winds cause anticlockwise equatorial currents which branch opposite southern Chile, to flow north as the cooling *Peru Current (Humboldt Current), and south round Cape Horn.

European exploration Vasco Núñez de Balboa was the first European to see the Pacific Ocean from Panama in 1513. Ferdinand Magellan sailed through the Strait of Magellan in 1520, and gave the name to the ocean (because of its calmness during his voyage). In 1577 Francis Drake, the first Englishman to enter the Pacific, sailed north to California and across to the Moluccas. The Australasian region was explored by Europeans in the 17th century.

Pacific Security Treaty   military alliance agreement between Australia, New Zealand, and the USA, signed in 1951. Military cooperation between the USA and New Zealand has been restricted by the latter's policy of banning ships that might be carrying nuclear weapons or nuclear power sources.

packaging   material, usually of metal, paper, plastic, or glass, used to protect products, make them easier to display, and as a form of advertising. Environmentalists have criticized packaging materials as being wasteful of energy and resources. Recycling bins are being placed in residential areas to facilitate the collection of surplus packaging.

paddy field   a small flooded field where *wet-rice is grown.

pahoehoe   or **ropy lava**, type of lava that has a smooth, wavy, or ropy surface.

Palaeocene epoch   (Greek 'old' + 'recent') first epoch of the Tertiary period of geological time, 65–56.5 million years ago. Many types of mammals spread rapidly after the disappearance of the great reptiles of the Mesozoic. Flying mammals replaced the flying reptiles, swimming mammals replaced the swimming reptiles, and all the ecological niches vacated by the reptiles were adopted by mammals.

At the end of the Palaeocene there was a mass extinction that caused more than half of all bottom-dwelling organisms to disappear worldwide, over a period of around 1000 years. Surface-dwelling organisms remained unaffected, as did those on land.

The cause of this extinction remains unknown, though US palaeontologists have found evidence (released in 1998) that it may have been caused by the Earth releasing tonnes of methane into the oceans causing increased water temperatures.

palaeomagnetic stratigraphy   use of distinctive sequences of magnetic polarity reversals to date rocks. Magnetism retained in rocks at the time of their formation are matched with known dated sequences of *polar reversals or with known patterns of *secular variation.

palaeomagnetism   study of the magnetic properties of rocks in order to reconstruct the Earth's ancient magnetic field and the former positions of the continents, using traces left by the Earth's magenetic field in igneous rocks before they cool. Palaeomagnetism shows that the Earth's magnetic field has reversed itself – the magnetic north pole becoming the magnetic south pole, and vice versa – at approximate half-million-year intervals, with shorter reversal periods in between the major spans.

Starting in the 1960s, this known pattern of magnetic reversals was used to demonstrate seafloor spreading or the formation of new ocean crust on either side of mid-oceanic ridges. As new material hardened on either side of a ridge, it would retain the imprint of the magnetic field, furnishing datable proof that material was spreading steadily outward. Palaeomagnetism is also used to demonstrate *continental drift by determining the direction of the magnetic field of dated rocks from different continents.

palaeontology   the study of ancient life, encompassing the structure of ancient organisms and their environment, evolution, and ecology, as revealed by their *fossils and the rocks in which those fossils are found. The practical aspects of palaeontology are based on using the presence of different fossils to date particular rock strata and to identify rocks that were laid down under particular conditions; for instance, giving rise to the formation of oil.

The use of fossils to trace the age of rocks was pioneered in Germany by Johann Friedrich Blumenbach (1752–1830) at Göttingen, followed by Georges Cuvier and Alexandre Brongniart in France in 1811.

Palaeozoic era era of geological time 570–245 million years ago. It comprises the Cambrian, Ordovician, Silurian, Devonian, Carboniferous, and Permian periods. The Cambrian, Ordovician, and Silurian constitute the Lower or Early Palaeozoic; the Devonian, Carboniferous, and Permian make up the Upper or Late Palaeozoic. The era includes the evolution of hard-shelled multicellular life forms in the sea; the invasion of land by plants and animals; and the evolution of fish, amphibians, and early reptiles.

Pan-American Union former name (1910–48) of the *Organization of American States.

Pangaea or **Pangea**, (Greek 'all-land') single land mass, made up of all the present continents, believed to have existed between 300 and 200 million years ago; the rest of the Earth was covered by the Panthalassa ocean. Pangaea split into two land masses – *Laurasia in the north and *Gondwanaland in the south – which subsequently broke up into several continents. These then moved slowly to their present positions (see *plate tectonics).

The former existence of a single 'supercontinent' was proposed by German meteorologist Alfred Wegener in 1912.

Panthalassa ocean that covered the surface of the Earth not occupied by the world continent *Pangaea between 300 and 200 million years ago.

Pareto's rule or **80/20 rule**, originally a rule applied to income distribution by Italian economist Vilfredo Pareto. Pareto stated that 80% of a nation's income would only benefit 20% of the population. This principle was subsequently extended to a range of other examples such as 80% of a company's profits coming from 20% of its products or services, or 80% of a company's computer problems stemming from 20% of its machines.

Paris Club international forum dating from the 1950s for the rescheduling of debts granted or guaranteed by official bilateral creditors; it has no fixed membership nor an institutional structure. In the 1980s it was closely involved in seeking solutions to the serious debt crises affecting many developing countries.

parish council lowest neighbourhood unit of local government in England and Wales, based on church parishes. They developed as units for local government with the introduction of the Poor Law in the 17th century. In Wales and Scotland they are commonly called **community councils**. In England approximately 8200 out of the 10,000 parishes have elected councils. There are 730 community councils in Wales and about 1000 in Scotland, which, unlike their English and Welsh counterparts, do not have statutory powers.

park and ride town-planning scheme in which parking space is provided (often free) some distance away from the central business district. Shoppers are taken by bus to the central area, which may be traffic-free (*pedestrianization). Park and ride is one of the planning strategies that can be used to combat *congestion.

Park and ride became widespread in the 1980s.

parliament (French 'speaking') legislative (law-making) body of a country. The world's oldest parliament is the Icelandic Althing, which dates from about 930. The UK Parliament is usually dated from 1265. The legislature of the USA is called Congress and comprises the House of Representatives and the Senate.

parliamentary government or **cabinet government**, form of government in which the executive (administration) is drawn from and is constitutionally responsible to the legislature (law-making body). This is known as the 'fusion of powers' as distinct from the 'separation of powers', in which the three branches of government the executive, legislature, and judiciary (courts system) are separated in terms of personnel and constitutional powers.

partition division of a country into two or more nations. Ireland was divided into Northern Ireland and the Irish Republic under the Government of Ireland Act 1920. The division of the Indian subcontinent into India and Pakistan took place in 1947. Other examples of partition include Korea 1953 and Vietnam 1954.

passive margin in *plate tectonics, a boundary between oceanic and continental crust that is within a single tectonic plate rather than at the boundary between two plates. Very little tectonic activity occurs at passive margins. Examples include the east coast of North America and the entire coast of Australia.

passport document issued by a national government authorizing the bearer to go abroad and guaranteeing the bearer the state's protection. Some countries require an intending visitor to obtain a special endorsement or visa.

pastoral farming a system of farming in which the raising of livestock is the dominant element. In the commercial context this may be, for example, dairy farming in Britain or sheep rearing in Australia, while subsistence pastoralism occurs in many parts of the *Third World. See also *nomadic pastoralism.

patterned ground a series of distinct, geometric patterns formed on the soil surface of tundra regions as a result of the natural sorting of stones and fine material. The regular freezing and thawing of the soil surface results in the finer materials being moved towards the centre of a roughly circular, or polygonal shape which may vary in size from 1–15 m in diameter. Patterned ground is best developed on undisturbed flat, or very gently sloping ground. On slopes with an angle greater than 5° the polygons become elongated and eventually form into *stone stripes*.

peasant agriculture the growing of crops or raising of animals, partly for subsistence needs and partly for market sale. Peasant agriculture is thus an intermediate stage between subsistence and commercial farming.

peat organic matter found in bogs and formed by the incomplete decomposition of plants such as sphagnum moss. Northern Asia, Canada, Finland, Ireland, and other places have large deposits, which have been dried and used as fuel from ancient times. Peat can also be used as a soil additive.

Peat bogs began to be formed when glaciers retreated, about 9000 years ago. They grow at the rate of only a millimetre a year, and large-scale digging can result in destruction both of the bog and of specialized plants growing there. The destruction of peat bogs is responsible for diminishing fish stocks in coastal waters; the run-off from the peatlands carries high concentrations of iron, which affects the growth of the plankton on which the fish feed.

Approximately 60% of the world's wetlands are peat. In May 1999 the Ramsar Convention on the Conservation of Wetlands approved a peatlands action plan that should have a major impact on the conservation of peat bogs.

peat bog type of wetland where slow-growing mosses have accumulated over centuries to form thick blankets of a material called **peat** or turf. Peat can be dried and used for fuel, or ground and used to enrich poor soils. Peat bogs once covered 17–20% of Ireland.

Bogs are waterlogged habitats that are strongly acid with a pH of 3.2–4.2. Sphagnum mosses are the predominant plant in bogs and there are more than 30 species growing in Ireland, but peat is also built up of the roots, leaves, flowers, and seeds of heathers, grasses, and sedges. The bogs started growing at the end of the last glaciation about 10,000 years ago, with living plants growing on layer upon layer of dead plant material, forming peat depths of up to 12 m.

Generally referred to as 'turf' in Ireland, peat is brownish-black in colour and in its natural state is 90% water and only 10% solid material. When cut into slabs and dried it burns like wood with a bright flame. Harvesting of peat accelerated as Ireland's population grew, and by the 1940s large-scale mechanized peat extraction developed under the state-owned company Bord na Móna. Today, the Bog of Allen, occupying 958 sq km of the counties of Offaly, Laois, and Kildare is the country's main source of peat.

Intensification of farming has also served to deplete Ireland's bogs and today only about 20% of the peatland resource remains relatively intact. Only 8% of the original peatland area can be considered suitable for scientific study and conservation purposes.

ped a naturally formed aggregate of soil particles. The ped is the smallest identifiable structural unit in a soil.

pedalfer a group of soils which occur in humid regions where leaching, *eluviation and *illuviation are the major pedogenic processes. Most pedalfers are characterized by the predominance of iron and aluminium minerals that remain after other soluble salts have been washed from the soil. Pedalfers, which incorporate soils such as podzols and laterites, form one of the major soil groups, the other being the pedocals.

pedestal rock see *Zeugen.

pedestrianization the closing of an area to traffic, making it more suitable for people on foot. It is now common in many town shopping centres, since cars and people often obstruct one another. This restricts *accessibility and causes *congestion. Sometimes service vehicles (such as buses and taxis) are allowed access.

pediment broad, gently inclined erosion surface formed at the base of a mountain as it erodes and retreats. Pediments consist of bedrock and are often covered with a thin layer of sediments, called alluvium, which have been eroded off the mountain.

pediplain in earth science, plain at the base of a mountain in arid climates, either bare or covered with a thin layer of alluvial deposits.

pedocal a group of soils that occur in dry regions (especially where evaporation exceeds precipitation) and where *calcification, particularly in the B horizon, is the dominant pedogenic process, leading to the characteristic predominance of calcium carbonate. *leaching, *illuviation and *eluviation occur only slightly in pedocals.

Pedocals form one of the major soil groups, the other being the pedalfers.

pedogenesis the combined effect of a number of interconnected processes which result in the formation and development of a *soil. Climate, topography, parent material, vegetation and the activities of animals (including humans) are all important factors in the creation of a soil.

pedology the scientific study of the formation, characteristics, distribution and use of soils.

pegmatite extremely coarse-grained *igneous rock of any composition found in veins; pegmatites are usually associated with large granite masses.

pelagic of or pertaining to the open ocean, as opposed to bottom or shore areas. **Pelagic sediment** is fine-grained fragmental material that has settled from the surface waters, usually the siliceous and calcareous skeletal remains (see *ooze) of marine organisms, such as radiolarians and foraminifera.

peneplain in earth science, plain to which the features of a landscape are hypothetically reduced by prolonged erosion.

peninsula land surrounded on three sides by water but still attached to a larger landmass. Florida, USA, is an example.

Pennsylvanian period US term for the Upper or Late *Carboniferous period of geological time, 323–290 million years ago; it is named after the US state, which contains vast coal deposits.

per capita income the GNP (gross national product or national income) of a country divided by the size of its population. It gives the average income per head of the population if the national income were shared out equally. Per capita income comparisons are used as one indicator of levels of economic development.

percolation gradual movement or transfer of water thorough porous substances (such as porous rocks or soil).

perfect competition in economics, a market in which there are many potential and actual buyers and sellers, each being too small to be an individual influence on the price; there are no barriers to entry or exit; and the products being traded are identical. At the same time, the producers are seeking the maximum profit and consumers the best value for money. Consumers have perfect knowledge of this type of market.

Pergau Dam hydroelectric dam on the Pergau River in Malaysia, near the Thai border. Building work began in 1991 with money from the UK foreign aid budget. Concurrently, the Malaysian government bought around £1 billion worth of arms from the UK. The suggested linkage of arms deals to aid became the subject of a UK government enquiry from March 1994. In November 1994 a High Court ruled as illegal British foreign secretary Douglas Hurd's allocation of £234 million towards the funding of the dam, on the grounds that it was not of economic or humanitarian benefit to the Malaysian people.

peridot pale-green, transparent gem variety of the mineral *olivine.

peridotite rock consisting largely of the mineral olivine; pyroxene and other minerals may also be present. Peridotite is an ultramafic rock containing less than 45% silica by weight. It is believed to be one of the rock types making up the Earth's upper mantle, and is sometimes brought from the depths to the surface by major movements, or as inclusions in lavas.

periglacial environment bordering a glacial area but not actually covered by ice all year round, or having similar climatic and environmental characteristics, such as in mountainous areas. Periglacial areas today include parts of Siberia, Greenland, and North America. The rock and soil in these areas is frozen to a depth of several metres (*permafrost) with only the top few centimetres thawing during the brief summer (the active layer). The vegetation is characteristic of *tundra.

During the last ice age all of southern England was periglacial. Weathering by *freeze–thaw (the alternate freezing and thawing of ice in rock cracks) would have been severe, and *solifluction (movement of soil that is saturated by water) would have taken place on a large scale, causing wet topsoil to slip from valley sides.

periphery a remote and/or underprivileged region as in the core/periphery model (see *core (2)). Such regions are generally lacking in resources and offer little development opportunity, and as such are the last to be integrated into the national development process.

permafrost condition in which a deep layer of soil does not thaw out during the summer. Permafrost occurs under *periglacial conditions. It is claimed that 26% of the world's land surface is permafrost.

Permafrost gives rise to a poorly drained form of grassland typical of northern Canada, Siberia, and Alaska known as *tundra.

permeable rock rock which through which water can pass either via a network of spaces between particles or along bedding planes, cracks, and fissures. Permeable rocks can become saturated. Examples of permeable rocks include limestone (which is heavily jointed) and chalk (porous).

Unlike *impermeable rocks, which do not allow water to pass through, permeable rocks rarely support rivers and are therefore subject to less

erosion. As a result they commonly form upland areas (such as the chalk downs of southeastern England, and the limestone Pennines of northern England).

Permian period of geological time 290–245 million years ago, the last period of the Palaeozoic era. Its end was marked by a dramatic change in marine life – the greatest mass extinction in geological history – including the extinction of many corals and trilobites. Deserts were widespread, terrestrial amphibians and mammal-like reptiles flourished, and cone-bearing plants (gymnosperms) came to prominence. In the oceans, 49% of families and 72% of genera vanished in the late Permian. On land, 78% of reptile families and 67% of amphibian families disappeared.

perovskite yellow, brown, or greyish-black orthorhombic mineral, $CaTiO_3$, which sometimes contains cerium. Other minerals that have a similar structure are said to have the **perovskite structure**. The term also refers to $MgSiO_3$ with the perovskite structure, the principal mineral that makes up the Earth's lower *mantle.

Peru Current formerly **Humboldt Current**, cold ocean *current flowing north from the Antarctic along the west coast of South America to southern Ecuador, then west. It reduces the coastal temperature, making the western slopes of the Andes arid because winds are already chilled and dry when they meet the coast.

pervious rock rock that allows water to pass through via bedding planes, cracks, and fissures, but not through pores within the rock (that is, they are non-porous). See *permeable rock.

pesticide any chemical used in farming, gardening, or in the home to combat pests. Pesticides are of three main types: **insecticides** (to kill insects), **fungicides** (to kill fungal diseases), and **herbicides** (to kill plants, mainly those considered weeds). Pesticides cause a number of pollution problems through spray drift onto surrounding areas, direct contamination of users or the public, and as residues on food. The World Health Organization (WHO) estimated in 1999 that 20,000 people die annually worldwide from pesticide poisoning incidents.

The safest pesticides include those made from plants, such as the insecticides pyrethrum and derris. Pyrethrins are safe and insects do not develop resistance to them. Their impact on the environment is very small as the ingredients break down harmlessly.

More potent are synthetic products, such as chlorinated hydrocarbons. These products, including DDT and dieldrin, are highly toxic to wildlife and often to humans, so their use is now restricted by law in some areas and is declining. Safer pesticides such as malathion are based on organic phosphorus compounds, but they still present hazards to health. An international treaty to ban persistent organic pollutants (POPs), including pesticides such as DDT, was signed in Stockholm, Sweden, in May 2001. The United Nations Food and Agriculture Organization reported in the same month that more than 500,000 tonnes of pesticide waste resides in dumps worldwide.

petroleum or **crude oil**, natural mineral oil, a thick greenish-brown flammable liquid found underground in permeable rocks. Petroleum consists of hydrocarbons mixed with oxygen, sulphur, nitrogen, and other elements in varying proportions. It is thought to be derived from ancient organic material that has been converted by, first, bacterial action, then heat, and pressure (but its origin may be chemical also).

From crude petroleum, various products are made by fractional distillation and other processes; for example, fuel oil, petrol, kerosene, diesel, and lubricating oil. Petroleum products and chemicals are used in large quantities in the manufacture of detergents, artificial fibres, plastics, insecticides, fertilizers, pharmaceuticals, toiletries, and synthetic rubber.

Petroleum was formed from the remains of marine plant and animal life which existed many millions of years ago (hence it is known as a *fossil fuel). Some of these remains were deposited along with rock-forming sediments under the sea where they were decomposed anaerobically (without oxygen) by bacteria which changed the fats in the sediments into fatty acids which were then changed into an asphaltic material called kerogen. This was then converted over millions of years into petroleum by the combined action of heat and pressure. At an early stage the organic material was squeezed out of its original sedimentary mud into adjacent sandstones. Small globules of oil collected together in the pores of the rock and eventually migrated upwards through layers of porous rock by the action of the oil's own surface tension (capillary action), by the force of water movement within the rock, and by gas pressure. This migration ended either when the petroleum emerged through a fissure as a seepage of gas or oil onto the Earth's surface, or when it was trapped in porous reservoir rocks, such as sandstone or limestone, in anticlines and other traps below impervious rock layers.

The modern oil industry originates in the discovery of oil in western Ontario in 1857 followed by Edwin Drake's discovery in Pennsylvania in 1859. Drake used a steam engine to drive a punching tool to 21 m below the surface where he struck oil and started an oil boom. Rapid development followed in other parts of the USA, Canada, Mexico, and then Venezuela where commercial production began in 1878. Oil was found in Romania in 1860, Iran in 1908, Iraq in 1923, Bahrain in 1932, and Saudi Arabia and Kuwait in 1938.

The USA led in production until the 1960s, when the Middle East outproduced other areas, their immense reserves leading to a worldwide dependence on cheap oil for transport and industry. In 1961 the Organization of the Petroleum Exporting Countries (OPEC) was established to avoid exploitation of member countries; after OPEC's price rises in 1973, the International Energy Agency (IEA) was established in 1974 to protect the interests of oil-consuming countries. New technologies were introduced to pump oil from offshore and from the Arctic (the Alaska pipeline) in an effort to avoid a monopoly by OPEC. Global consumption of petroleum in 1993 was 23 billion barrels.

As shallow-water oil reserves dwindle, multinational companies have been developing deep-water oilfields at the edge of the continental shelf in the Gulf of Mexico. Shell has developed Mars, a 500-million-barrel

marine plants and animals die and are trapped beneath layers of sediment where they are broken down by anaerobic bacteria

increasing heat and pressure changes the fats into fatty acids, which are then changed into an asphaltic material, keragen

further increases in temperature and pressure cause petroleum to form

natural gas collects above oil

petroleum The formation of oil and natural gas. Oil forms when marine plants and animals die and accumulate in stagnant water lacking in oxygen. They are quickly buried by clay and so do not completely decay but form hydrocarbon-rich muds, broken down by anaerobic bacteria. Increasing heat and pressure transform the hydrocarbons into fatty acids, which are then changed into an asphaltic material, keragen. Further increases in temperature and pressure cause oil to form and natural gas collects above the oil.

distillation tower for separating
components of crude oil

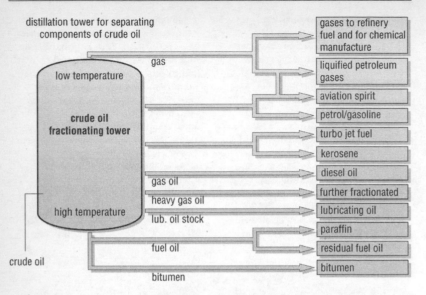

petroleum Refining petroleum using a distillation column. The crude petroleum is fed in at the
bottom of the column where the temperature is high. The gases produced rise up the column, cooling
as they travel. At different heights up the column, different gases condense to liquids called fractions,
and are drawn off.

oilfield, in 900 m of water, and the oil
companies now have the technology
to drill wells of up to 3075 m under the
sea. It is estimated that the deep waters
of Mexico could yield 8–15 million
barrels in total; it could overtake the
North Sea in importance as an oil source.

In Asia, the oil pipeline from
Azerbaijan through Russia to the West,
which is the only major pipeline from
the Caspian Sea, closed during
Russia's conflict with Chechnya but
reopened in 1997.

pollution The burning of petroleum
fuel is one cause of *air pollution.
The transport of oil can lead to
catastrophes – for example, the *Torrey
Canyon* tanker lost off southwestern
England in 1967, which led to an
agreement by the international oil
companies in 1968 to pay
compensation for massive shore
pollution. The 1989 oil spill in Alaska
from the *Exxon Valdez* damaged the

area's fragile environment, despite
clean-up efforts. Drilling for oil
involves the risks of accidental spillage
and drilling-rig accidents. The
problems associated with oil have
led to the various alternative *energy
technologies.

A new kind of bacterium was
developed during the 1970s in the
USA, capable of 'eating' oil as a means
of countering oil spills.

petrology branch of geology that deals
with the study of rocks, their mineral
compositions, their textures, and their
origins.

pH a measure of acidity/alkalinity.
A pH value of 7.0 is regarded as
neutral, while pH values of less than
7.0 indicate acidic conditions. pH values
greater than 7.0 indicate increasingly
alkaline conditions. pH tests are used
to assess the acidity of soil and of
water. See *acid rain. The optimum soil
pH for cereal growth is about 6.5.

Phanerozoic eon (Greek *phanero* 'visible') eon in Earth history, consisting of the most recent 570 million years. It comprises the Palaeozoic, Mesozoic, and Cenozoic eras. The vast majority of fossils come from this eon, owing to the evolution of hard shells and internal skeletons. The name means 'interval of well-displayed life'.

Phillips curve graph showing the relationship between percentage changes in wages and unemployment, and indicating that wages rise faster during periods of low unemployment as employers compete for labour. The implication is that there is a stable trade-off between inflation and unemployment, and that the dual objectives of low unemployment and low inflation are inconsistent. The concept has been widely questioned since the early 1960s because of the apparent instability of the relationship between wages and unemployment, and since then the Phillips curve has been widely regarded as misleading in its explanation of inflation. The curve was developed by the New Zealand-born British economist Alban William Phillips, who graphically plotted wage and unemployment changes between 1861 and 1957.

photosynthesis the process by which green plants make carbohydrates from carbon dioxide and water, and give off oxygen. Photosynthesis balances *respiration. However, the burning of *fossil fuels has greatly increased the amount of carbon dioxide in the atmosphere, and *deforestation has seriously reduced the number of trees available to release oxygen, so that the natural balance has been lost.

See *global warming.

phyllite *metamorphic rock produced under increasing temperature and pressure, in which minute mica crystals are aligned so that the rock splits along their plane of orientation, the resulting break being shiny and smooth. Intermediate between slate and schist, its silky sheen is an identifying characteristic.

physical geography the study of our *environment, comprising such elements as geomorphology, hydrology, pedology, meteorology, climatology and biogeography.

physical weathering or **mechanical weathering,** form of *weathering responsible for the mechanical breakdown of rocks but involving no chemical change.

Forces acting on rock exposed on the Earth's surface open up any weak points in the rock and cause pieces to be broken off. Piles of jagged rock fragments called *scree are formed. Processes involved include *freeze–thaw (the alternate freezing and melting of ice in rock cracks or pores) and *exfoliation (the alternate expansion and contraction of rocks in response to extreme changes in temperature). Physical weathering is also brought about by the growth of plants. A seed falling into a crack in a rock may germinate and the growing plant forces the crack to widen. Similarly, a tree root may grow down into a crack and split the rock. Physical weathering is one of the processes involved in the *rock cycle, the formation, change, and re-formation of the Earth's outer layers.

piece rate form of payment for work done where the amount received is calculated per item produced. For example, a worker may be paid 1p per skirt hemmed by machine. If a worker hems 2000 skirts on Tuesday, then his or her pay would be £20. If the same worker hems 2200 skirts on Wednesday, the pay would be £22. Piece rates are paid where it is easy to identify the contribution of a worker to a task, usually to manual workers. They are used to provide an incentive to workers to work hard.

pie chart a circular graph for displaying values as proportions.

piggy-back export scheme in business, a firm already established in the export field that makes its services available without charge to a small firm just entering the market. The small firm thus obtains the assistance of the large firm's good will, experience, and know-how, and is saved the trouble and expense of setting up its own export department.

pillow lava type of usually basaltic lava composed of ellipsoidal or spherical structures up to a metre across. It forms when hot molten lava cools rapidly on coming into contact with water, as when a stream of lava flows into the sea.

pingo landscape feature of *tundra terrain consisting of a hemispherical mound about 30 m high, covered with soil that is cracked at the top. The core consists of ice, probably formed from the water of a former lake. The lake that forms when such a feature melts after an ice age is also called a pingo.

pioneer species in ecology, those species that are the first to colonize and thrive in new areas. Coal tips, recently cleared woodland, and new roadsides are areas where pioneer species will quickly appear. As the habitat matures other species take over, a process known as **succession**.

pipeflow movement of water through natural pipes in the soil; it is a form of throughflow and may be very rapid. The pipes used may be of animal or plant origin – for example, worm burrows or gaps created by tree roots.

pipeline see *petroleum.

pitfall trap simple device for trapping small invertebrates. In its simplest form a beaker or jam jar is buried in the ground so that the rim of the jar is flush with the soil. Beetles, millipedes, spiders, and other arthropods tumble into the jar and are unable to escape.

placer deposit detrital concentration of an economically important mineral, such as gold, but also other minerals such as cassiterite, chromite, and platinum metals. The mineral grains become concentrated during transport by water or wind because they are more dense than other detrital minerals such as quartz, and (like quartz) they are relatively resistant to chemical breakdown. Examples are the Witwatersrand gold deposits of South Africa, which are gold- and uranium-bearing conglomerates laid down by ancient rivers, and the placer tin deposits of the Malay Peninsula.

plagiosere see *sere.

plain or **grassland**, land, usually flat, upon which grass predominates. The plains cover large areas of the Earth's surface, especially between the *deserts of the tropics and the *rainforests of the Equator, and have rain in one season only. In such regions the *climate belts move north and south during the year, bringing rainforest conditions at one time and desert conditions at another. Temperate plains include the North European Plain, the High Plains of the USA and Canada, and the Russian Plain (also known as the *steppe).

planetary geology study of the structure, composition, and geological history of rocky planets, satellites, and asteroids (including, for instance, both the Earth and Moon) and the processes that have shaped them.

planned economy another term for *command economy.

plantation large farm or estate where commercial production of one crop – such as rubber (in Malaysia), palm oil (in Nigeria), or tea (in Sri Lanka) – is carried out. Many plantations were established in countries under colonial rule, using slave labour.

plantation agriculture a system of *agriculture located in a tropical or semitropical *environment, producing

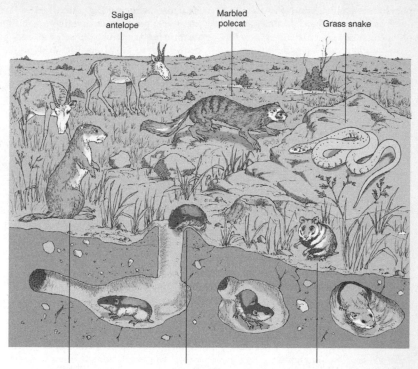

Saiga antelope

Marbled polecat

Grass snake

Suslik

Lemming

Black Bellied hamster

plain The Eurasian steppe or plain is largely treeless and characterized by a climate ranging from extreme winter cold to great summer heat. The larger animals found here overcome this problem by being nomadic and the smaller ones by burrowing. At one time the steppe carried immense herds of herbivores including wild ass, wild horse, wild camel, and saiga, but owing to human intervention these species have declined, some almost to the point of extinction.

commodities for export to Europe, North America and other industrialized regions. Coffee, tea, bananas, rubber and sisal are examples of plantation crops.

Plantation agriculture is distinctive in that it is a form of *commercial agriculture located in a generally subsistence or peasant environment: it is an extension of the commercial agriculture of the developed world into a mainly *Third World environment. Some plantations are run and financed by *transnational corporations and the profits from such operations are generally channelled back to Europe or North America. As a result many plantations are institutions of *neocolonialism. There is a worry that plantations often take up valuable farmland, growing commodities required by the richer countries. Thus more valuable local crops are forced onto poorer land. In areas where unemployment is high, plantations have traditionally paid low wages.

PLATE 304

On a more positive side, many plantation operators provide such facilities as housing, education and health care for their workers, as well as a plot of land. But it is difficult to avoid the conclusion that plantation agriculture in its traditional form is unacceptable in view of contemporary development priorities in Third World nations.

plate or **tectonic plate** or **lithospheric plate**, one of several relatively distinct sections of the *lithosphere, approximately 100 km thick, which together comprise the outermost layer of the Earth (like the pieces of the cracked shell of a hard-boiled egg).

The plates are made up of two types of crustal material: oceanic crust (sima) and continental crust (sial), both of which are underlain by a solid layer of *mantle. Dense **oceanic crust** lies beneath Earth's oceans and consists largely of *basalt. **Continental crust**, which underlies the continents and the continental shelves, is thicker, less dense, and consists of rocks that are rich in silica and aluminium.

Due to convection in the Earth's mantle (see *plate tectonics) these pieces of lithosphere are in motion, riding on a more plastic layer of the mantle, called the asthenosphere. Mountains, volcanoes, earthquakes, and other geological features and events all come about as a result of interaction between these plates.

plateau elevated area of fairly flat land, or a mountainous region in which the peaks are at the same height. An **intermontane plateau** is one surrounded by mountains. A **piedmont plateau** is one that lies between the mountains and low-lying land. A **continental plateau** rises abruptly from low-lying lands or the sea. Examples are the Tibetan Plateau and the Massif Central in France.

plate margin or **plate boundary**, the meeting place of one *plate (plates make up the top layer of the Earth's structure) with another plate. There are four types of plate margin – destructive, constructive, collision, and conservative. A *volcano may be found along two of the types of plate margin, and an *earthquake may occur at all four plate margins.

destructive or convergent plate margins At a *destructive margin an oceanic plate moves towards (and disappears into the *mantle of) a continental plate or another oceanic plate. This is the **subduction zone**. As it is forced downwards, pressure at the margins increases, and this can result in violent earthquakes. The heat produced by friction turns the crust into *magma (liquid rock). The magma tries to rise to the surface and, if it succeeds, violent volcanic eruptions occur.

constructive or divergent plate margins At a *constructive margin the Earth's crust is forced apart. Magma rises and solidifies to create a new oceanic crust and forms a mid-ocean ridge. This ridge is made from *igneous rock; such ridges usually form below sea level on the sea bed (an exception to this is in Iceland).

collision plate margins A collision margin occurs when two plates moving together are both made from continental crust. Continental crust cannot sink or be destroyed, and as a result the land between them is pushed upwards to form high 'fold' mountains like the Himalayas. Earthquakes are common along collision margins but there are no volcanic eruptions.

conservative plate margins At a conservative margin two plates try to slide past each other slowly. Quite often, the two plates stick and pressure builds up; the release of this pressure creates a severe earthquake. There are no volcanic eruptions along conservative plate margins because the crust is neither being created nor destroyed. The *San Andreas Fault in California lies above the North American and Pacific plates, and is an example of a conservative plate margin.

plate reconstruction in geology, an illustration of positions of continents and oceans sometime in the geological past.

plate tectonics theory formulated in the 1960s to explain the phenomena of *continental drift and sea-floor spreading, and the formation of the major physical features of the Earth's surface. The Earth's outermost layer, the *lithosphere, is seen as a jigsaw puzzle of rigid major and minor plates that move relative to each other, probably under the influence of convection currents in the *mantle beneath. At the margins of the plates, where they collide or move apart or slide past one another, major landforms such as *mountains, *rift valleys, *volcanoes, *ocean trenches, and **mid-ocean ridges** are created. The rate of plate movement is on average 2–3 cm per year and at most 15 cm per year.

The concept of plate tectonics brings together under one unifying theory many phenomena observed in the Earth's crust that were previously thought to be unrelated. The size of the crust plates is variable, as they are constantly changing, but six or seven large plates now cover much of the Earth's surface, the remainder being occupied by a number of smaller plates. Each large plate may include both continental and ocean lithosphere. As a result of seismic studies it is known that the lithosphere is a rigid layer extending to depths of about 50–100 km, overlying the upper part of the mantle (the *asthenosphere), which is composed of rocks very close to melting point. This zone of mechanical weakness allows the movement of the overlying plates. The margins of the plates are defined

by major earthquake zones and belts of volcanic and tectonic activity. Almost all earthquake, volcanic, and tectonic activity is confined to the margins of plates, and shows that the plates are in constant motion (see *plate margin).

sea-floor spreading New plate material is generated along the mid-ocean ridges, where basaltic lava is poured out by submarine volcanoes. The theory of sea-floor spreading has demonstrated the way in which the basaltic lava spreads outwards away from the ridge crest at 1–6 cm per year. Plate material is consumed at a rate of 5–15 cm per year at the site of the deep ocean trenches, for example along the Pacific coast of South America. The trenches are sites where two plates of lithosphere meet; the one bearing ocean-floor basalts plunges beneath the adjacent continental mass at an angle of 45°, giving rise to shallow earthquakes near the coast and progressively deeper earthquakes inland. In places the sinking plate may descend beneath an island arc of offshore islands, as in the Aleutian Islands (Alaska) and Japan, and in this case the shallow earthquakes occur beneath the island arc. The destruction of ocean crust in this way accounts for another well-known geological fact – that there are no old rocks found in the ocean basins. The oldest sediments found are 200 million years old, but the vast majority are less than 80 million years old. This suggests that plate tectonics has been operating for at least the last 200 million years. In other areas plates slide past each other along fault zones, giving rise to shallow earthquakes. Sites where three plates meet are known as triple junctions.

causes of plate movement The causes of plate movement are all very hypothetical. It has been known for

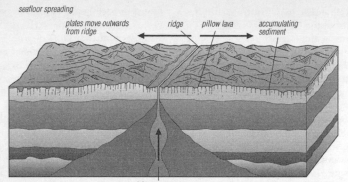

seafloor spreading

plates move outwards
from ridge ridge pillow lava accumulating
sediment

rising magma

subduction zone

one plate slides
under another magma

collision zone

continental crust collides
and is partly subducted younger folded mountains older folded mountains

plate tectonics Constructive and destructive action in plate tectonics. (top) Seafloor spreading. The upwelling of magma forces apart the crust plates, producing new crust at the joint. Rapid extrusion of magma produces a domed ridge; more gentle spreading produces a central valley. (middle) The drawing downwards of an oceanic plate beneath a continent produces a range of volcanic fold mountains parallel to the plate edge. (bottom) Collision of continental plates produces immense fold mountains, such as the Himalayas.

Younger mountains are found near the coast with older ranges inland. The plates of the Earth's lithosphere are always changing in size and shape of as material is added at constructive margins and removed at destructive margins. The process is extremely slow, but it means that the tectonic history of the Earth cannot be traced back further than about 200 million years.

some time that heat flow from the interior of the Earth is high over the mid-ocean ridges, and so various models of thermal convection in the mantle have been proposed; the geometry of the flow in any convective system must be complex, as there is no symmetry to the arrangement of ridges and trench systems over the Earth's surface. It seems likely that a plume of hot, molten material rises below the ridges and is extruded at the surface as basaltic lava. In zones of descending flow, at deep ocean trenches, the surface sediment is scraped off the descending plate onto the margin of the static plate, causing it to grow outwards towards the ocean, while the basaltic rocks of the descending plate, together with any remaining sediment, undergo partial fusion as they descend. This gives rise to large volumes of molten rock material, or *magma, which ascends to form andesitic lavas and intrusions of diorite or granodiorite at the margin of the overlying continent. These theories of plate tectonics may provide explanations for the formation of the Earth's crust; oceanic crust is generated at the mid-ocean ridges by partial fusion of the mantle, while below active continental margins partial fusion generates the more silica-rich continental rocks. Oceanic crust is continually produced by, and returned to, the mantle, but the continental crust rocks, once formed, remain on the surface and are not returned to the mantle.

development of plate tectonics theory The concept of continental drift was set out in 1915 in a book entitled *The Origin of Continents and Oceans* by the German meteorologist Alfred Wegener, who recognized that continental plates rupture, drift apart, and eventually collide with one another. Wegener's theory explained why the shape of the east coast of the

Americas and that of the west coast of Africa seem to fit together like pieces of a jigsaw puzzle; evidence for the drift came from the presence of certain rock deposits which indicated that continents have changed position over time. In the early 1960s scientists discovered that most earthquakes occur along lines that are parallel to ocean trenches and ridges, and in 1965 the theory of plate tectonics was formulated by Canadian geophysicist John Tuzo Wilson; it has now gained widespread acceptance among earth scientists who have traced the movements of tectonic plates millions of years into the past. The widely accepted belief is that all the continents originally formed part of an enormous single land mass, known as Pangaea. This land was surrounded by a giant ocean known as Panthalassa. About 200 million years ago, Pangaea began to break up into two large masses called Gondwanaland and Laurasia, which in turn separated into the continents as they are today, and which have drifted to their present locations. In 1995 US and French geophysicists produced the first direct evidence that the Indo-Australian plate has split in two in the middle of the Indian Ocean, just south of the Equator. They believe the split began about 8 million years ago.

playa temporary lake in a region of interior drainage (it has no outlet to the sea). Such lakes are common features in arid desert basins that are fed by intermittent streams. The streams bring dissolved salts to the playa, and when it shrinks during dry spells, the salts precipitate as evaporite deposits.

plebiscite (Latin *plebiscitium* 'ordinance, decree') referendum or direct vote by all the electors of a country or district on a specific question. Since the 18th century plebiscites have been employed on many occasions to decide to what country a particular area should belong; for example, in Upper Silesia and elsewhere after World War I, and in the Saar in 1935.

The term fell into disuse during the 1930s, after the widespread abuse by the Nazis in Germany to legitimize their regime.

Pleistocene epoch first part of the Quaternary period of geological time, beginning 1.64 million years ago and ending 10,000 years ago. The polar ice caps were extensive and glaciers were abundant during the ice age of this period, and humans evolved into modern *Homo sapiens sapiens* about 100,000 years ago.

plenipotentiary or **envoy extraordinary**, person accredited to a sovereign power and invested with unlimited power to negotiate a treaty or transact other diplomatic business. It is usual for the sovereign powers who are parties to a treaty to ratify the treaty, even though it is signed by a plenipotentiary.

Pliocene Epoch ('almost recent') fifth and last epoch of the Tertiary period of geological time, 5.2–1.64 million years ago. The earliest hominid, the humanlike ape *Australopithecus*, evolved in Africa.

plucking in earth science, a process of glacial erosion. Water beneath a glacier will freeze fragments of loose rock to the base of the ice. When the ice moves, the rock fragment is 'plucked' away (literally ripped) from the underlying bedrock. Plucking is thought to be responsible for the formation of steep, jagged slopes such as the backwall of the corrie and the downslope-side of the roche moutonnée.

plug the solidified material which seals the vent of a *volcano after an eruption. A volcanic plug is thus

responsible for the build-up of pressure which may result in an explosive eruption at some later stage. *Viscous lavas produce the most effective plugs and, because of their resistance to *erosion, volcanic plugs tend to stand out in the landscape when softer surrounding material has been worn away. A well known example is Edinburgh Castle Rock.

plumbago alternative name for the mineral *graphite.

plunge pool deep pool at the bottom of a *waterfall. It is formed by the hydraulic action of the water as it crashes down onto the river bed from a height.

plutonic rock igneous rock derived from magma that has cooled and solidified deep in the crust of the Earth; granites and gabbros are examples of plutonic rocks.

PM10 *abbreviation for* particulate matter less than 10 micrometres (10 mm) across, clusters of small particles, such as carbon particles, in the air that come mostly from vehicle exhausts. There is a link between increase in PM10 levels and a rise in death rate, increased hospital admissions, and asthma incidence. The elderly and those with chronic heart or lung disease are most at risk.

podzol or **podsol**, type of light-coloured soil found predominantly under coniferous forests and on moorlands in cool regions where rainfall exceeds evaporation. The constant downward movement of water leaches nutrients from the upper layers, making podzols poor agricultural soils. Podzols are very acidic soils.

The leaching of minerals such as iron and alumina leads to the formation of a bleached zone, which is often also depleted of clay.

These minerals can accumulate lower down the soil profile to form a hard, impermeable layer which restricts the drainage of water through the soil.

polar front a large-scale *front in the North Atlantic and North Pacific oceans along which northward-moving tropical maritime air masses meet southward-moving polar maritime air masses. The polar front moves to the north during the winter and to the south in the summer. The formation of small wave-like irregularities in the polar front is responsible for development of temperate latitude depressions.

polar reversal or **magnetic reversal**, change in polarity of Earth's magnetic field. Like all magnets, Earth's magnetic field has two opposing regions, or poles, positioned approximately near geographical North and South Poles. During a period of normal polarity the region of attraction corresponds with the North Pole. Today, a compass needle, like other magnetic materials, aligns itself parallel to the magnetizing force and points to the North Pole. During a period of reversed polarity, the region of attraction would change to the South Pole and the needle of a compass would point south.

Studies of the magnetism retained in rocks at the time of their formation (like small compasses frozen in time) have shown that the polarity of the magnetic field has reversed repeatedly throughout geological time.

The reason for polar reversals is not known. Although the average time between reversals over the last 10 million years has been 250,000 years, the rate of reversal has changed continuously over geological time. The most recent reversal was 780,000 years ago; scientists have no way of predicting when the next reversal will occur. The reversal process probably takes a few thousand years. Dating rocks using distinctive sequences of magnetic reversals is called *magnetic stratigraphy.

polder area of flat reclaimed land that used to be covered by a river, lake, or the sea. Polders have been artificially drained and protected from flooding by building dykes. They are common in the Netherlands, where the total land area has been increased by nearly one-fifth since AD 1200. Such schemes as the Zuider Zee project have provided some of the best agricultural land in the country.

pole either of the geographic north and south points of the axis about which the Earth rotates. The geographic poles differ from the magnetic poles, which are the points towards which a freely suspended magnetic needle will point.

In 1985 the magnetic north pole was some 350 km northwest of Resolute Bay, Northwest Territories, Canada. It moves northwards about 10 km each year, although it can vary in a day about 80 km from its average position. It is relocated every decade in order to update navigational charts.

It is thought that periodic changes in the Earth's core cause a reversal of the magnetic poles (see *polar reversal). Many animals, including migrating birds and fish, are believed to orient themselves partly using the Earth's magnetic field. A permanent scientific base collects data at the South Pole.

pole, magnetic see *magnetic pole.

polje a steep-sided, flat-floored linear depression in *karst landscapes, up to 65 km in length and 10 km wide. The origin of polje is unclear but may be due to solution weathering (see *chemical weathering) along lines of weakness, the collapse of caverns formed by underground drainage, or, in some instances, may be the result of late glacial meltwater streams which carved out a channel. The floors of polje are often damp, clay-covered areas that can be used for agriculture in otherwise barren limestone regions. They can also provide valuable lines of communication in otherwise impenetrable terrain, for example through the northern limestone massif of Mallorca in the Balearic Islands.

pollarding   type of pruning whereby the young branches of a tree are severely cut back, about 2–4 m above the ground, to produce a stumplike trunk with a rounded, bushy head of thin new branches. It is similar to *coppicing.

poll tax   tax levied on every individual, without reference to income or property. Being simple to administer, it was among the earliest sorts of tax (introduced in England in 1379), but because of its indiscriminate nature (it is a regressive tax, in that it falls proportionately more heavily on poorer people) it has often proved unpopular.

polluter-pays principle   the idea that whoever causes pollution is responsible for the cost of repairing any damage. The principle is accepted in British law but has in practice often been ignored; for example, farmers causing the death of fish through slurry pollution have not been fined the full costs of restocking the river.

pollution   harmful effect on the environment of by-products of human activity, principally industrial and agricultural processes – for example, noise, smoke, car emissions, pesticides, radiation, *sewage disposal, household waste, and chemical and radioactive effluents in air, seas, and rivers. *Air pollution contributes to the *greenhouse effect.

Pollution control involves higher production costs for the industries concerned, but failure to implement adequate controls may result in irreversible environmental damage and an increase in the incidence of diseases such as cancer. For example, in agriculture the mismanagement of fertilizers may result in *eutrophication, an excessive enrichment of lakes and rivers caused by *nitrate pollution; the subsequent rapid growth of algae darkens the water and eventually depletes its oxygen, causing plants and animals to die. Nitrate pollution has also been linked to stomach cancer, although this is unproven. Radioactive pollution results from inadequate *nuclear safety.

Transboundary pollution is when the pollution generated in one country affects another, for example as occurs with *acid rain. Natural disasters may also cause pollution; volcanic eruptions, for example, cause ash to be ejected into the atmosphere and deposited on land surfaces.

The greatest single cause of **air pollution** in the UK is the car, which is responsible for 85% of the carbon monoxide and 45% of the oxides of nitrogen present in the atmosphere. In 1987 carbon monoxide emission from road transport was measured at 5.26 million tonnes. According to a UK government report in 1998, air pollution causes up to 24,000 deaths in Britain per year.

Water pollution is controlled by the Environment Agency (from 1996). In 1989 the regional water authorities of England and Wales were privatized to form ten water and sewerage companies. Following concern that some of these companies were failing to meet EU drinking-water standards on nitrate and pesticide levels, the companies were served with enforcement notices by the government Drinking Water Inspectorate. The existence of 1300 toxic waste tips in the UK poses a considerable threat for increased water pollution.

polymorphism   in mineralogy, the ability of a substance to adopt different internal structures and external forms, in response to different conditions of temperature and/or pressure. For example, diamond and graphite are both forms of the element carbon, but they have very different properties and appearance.

Silica (SiO_2) also has several polymorphs, including quartz, tridymite, cristobalite, and stishovite (the latter a very high-pressure form found in meteoritic impact craters).

pools and riffles   alternating deeps (pools) and shallows (riffles) along the course of a river. There is evidence to suggest a link between pools and riffles and the occurrence of *meanders (bends in a river), although it is not certain whether they are responsible for meander formation.

pooter   small device for collecting invertebrates, consisting of a jar to which two tubes are attached. A sharp suck on one of the tubes, while the other is held just above an insect, will propel the animal into the jar. A filter wrapped around the mouth tube, prevents debris or organisms from being swallowed.

population   in biology and ecology, a group of organisms of one species, living in a certain area. The organisms are able to interbreed. It also refers to the members of a given species in a *community of living things. The area can be small. For example, one can refer to the population of duckweed (a small floating plant found on the surface of ponds) on a pond. Since the pond is a *habitat, one can consider the population of duckweed in a habitat and forming part of the community of plants and animals there. However, it is also possible to use the term population for all the organisms of one species in a large geographical area, for example the elephant population in Africa. It could also be used to describe all the organisms of that species on Earth, for example the world population of humans. Population sizes in habitats change over a period of time. The timescale may be daily, seasonal, or there may be changes over the years.

The success of an organism can be determined by measuring the size of a population or by measuring *biomass. Measuring the size of a population is quite difficult and requires careful sampling of the habitat and careful calculation to estimate population size. It is rarely possible to count directly all the individuals in a population. Typically, sampling techniques require random sampling to get a fair estimate of the population and may use equipment such as quadrats (a way of defining a square sampling area of a certain size). However, these techniques have their limitations – for example, animals move around and may also be difficult to find. Species may also be difficult to identify. There may not be time to take a large enough sample to make the estimate accurate.

population, human   the number of people living in a specific area or region, such as a town or country, at any one time. The study of populations, their distribution and structure, resources, and patterns of migration, is called *demography. Information on population is obtained in a number of ways, such as through the registration of births and deaths. These figures are known as 'vital statistics'. However, more detailed information on *population distribution, *population density, and change is necessary to enable governments to plan for education, health, housing, and transport on local and national levels. This information is usually obtained from *censuses (population counts), which provide data on sex, age, occupation, and nationality.

The word *census* comes from a Latin word meaning to count or assess, and the censuses conducted by the Romans were mainly for purposes of tax and the recruitment of armies. Nowadays, censuses are carried out by most countries on a regular basis; in the USA a census has been taken every ten years since 1790. In Europe the first national censuses were taken in 1800 and 1801 and provided population statistics for Ireland, Italy, Spain, and the UK, and for the cities of Berlin, London, Paris, and Vienna. A census of the population of New York was also taken at that time. Most countries in the world have taken at least one census within the last decade.

population control measures taken by some governments to limit the growth of their countries' populations by trying to reduce *birth rates. Propaganda, freely available contraception, and tax disincentives for large families are some of the measures that have been tried.

The population-control policies introduced by the Chinese government are the best known. In 1979 the government introduced a 'one-child policy' that encouraged family planning and penalized couples who have more than one child. It has been only partially successful since it has been difficult to administer, especially in rural areas, and has in some cases led to the killing of girls in favour of sons as heirs.

population density the number of people living in a given area, usually expressed as people per square kilometre. It is calculated by dividing the population of a region by its area.

Population density provides a useful means for comparing *population distribution. Densities vary considerably over the globe.

High population densities may amount to *overpopulation where resources are scarce.

The population density in the UK is 239 per square kilometre. By comparison, Norway has a population density of 14 per square kilometre.

population distribution the location of people within an area. Areas may be compared by looking at variations in *population density. Population is unevenly distributed for a number of reasons. *Pull factors (which attract people) include mineral resources, temperate climate, the availability of water, and fertile, flat land. *Push factors (which repel people) include dense vegetation, limited accessibility, and political or religious oppression.

Over 90% of the world's population is found in the northern hemisphere, where there is most land. Some 60% of the world's land surface is unpopulated. At the other extreme, urban areas such as Mong Kok in Hong Kong may have as many as 160,000 people living in one square kilometre and Calcutta slums up to 80,000 per square kilometre.

population explosion the rapid and dramatic rise in world population that has occurred over the last few hundred years. Between 1959 and 2000, the world's population increased from 2.5 billion–6.1 billion people. According to United Nations projections, the world population will be between 7.9 billion and 10.9 billion by 2050.

Most of the growth is currently taking place in the *developing world, where rates of natural increase are much higher than in industrialized countries. Concern that this might lead to *overpopulation has led some countries to adopt *population control policies. However, since people in developing countries consume far less, especially of non-renewable resources, per head of population than people in industrialized countries, it has been argued that the West should set an example in population control instead of giving, for example, universal child benefit.

The rate of population growth tends to be inversely related to the level of women's education; except in industrialized countries, where there is a somewhat higher birth rate in such countries as Sweden, with good childcare facilities and equality for women in the workplace, than in, for example, the UK, where the paucity of such provision may force women to choose between motherhood and a career.

population growth   an increase in the population of a given region. This may be the result of natural increase (more births than deaths) or of in-migration, or both.

population migration   see *migration.

population pyramid   a type of bar graph used to show population structure, i.e. the age and sex composition of the population for a given region or nation. A population pyramid has data for males plotted on one side and data for females on the other.

The shape of a population pyramid is a useful indicator of the stage of development reached by a nation (see *demographic transition) and can also indicate government policy on population planning and the impact of major events such as wars.

pore space   a minute space between soil particles and rock grains. Pore spaces are usually interconnected thus allowing the movement of soil water and soil atmosphere.

porosity   the amount of space between soil particles or rock grains. Porosity is usually expressed as the ratio of the volume of space between the soil particles to the total volume of a soil or rock sample, given as a percentage.

porphyry   any *igneous rock containing large crystals in a finer matrix.

port   point where goods are loaded or unloaded from a water-based to a land-based form of transport. Most ports are coastal, though inland ports on rivers also exist. Ports often have specialized equipment to handle cargo in large quantities (for example, container or roll-on/roll-off facilities).

post-industrial   characteristic of an economy that is no longer based on *heavy industry but is increasingly dominated by microelectronics, automation, and the service and information sectors. See *industrial sector.

potassium-argon dating   or **K-Ar dating,** isotopic dating method based on the radioactive decay of potassium-40 (^{40}K) to the stable isotope argon-40 (^{40}Ar). Ages are based on the known half-life of ^{40}K, and the ratio of ^{40}K to ^{40}Ar. The method is routinely applied to rock samples about 100,000–30 million years old.

The method is used primarily to date volcanic layers in stratigraphic sequences with archaeological deposits, and the palaeomagnetic-reversal timescale. Complicating factors, such as sample contamination by argon absorbed from the atmosphere, and argon gas loss by diffusion out of the mineral, limit the application of this technique.

pothole   small hollow in the rock bed of a river. Potholes are formed by the erosive action of rocky material carried by the river (corrasion), and are commonly found along the river's upper course, where it tends to flow directly over solid bedrock.

poverty   condition in which the basic needs of human beings (shelter, food, and clothing) are not being met. Over one-fifth of the world's population was living in extreme poverty in 1995, of which around 70% were women. Nearly 13.5 million children under five die each year from poverty-related illness (measles, diarrhoea, malaria,

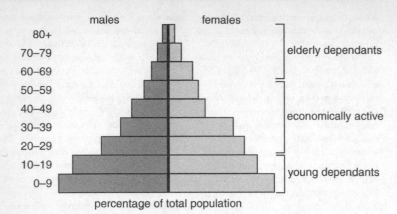

percentage of total population

wide base and narrow top indicate high birth rate and low life expectancy

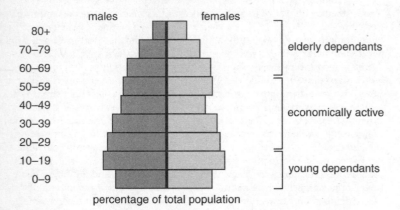

percentage of total population

pillar-shaped structure indicates a low death rate and a high life expectancy

population pyramid Population pyramids are a useful way of summarizing population development and can be used to predict future changes. The 'pyramid' shape with a wide base narrowing upwards is typical of less economically developed countries (LEDCs). The more rectangular, pillar-shaped 'pyramid' is typical of more economically developed countries (MEDCs).

pneumonia, and *malnutrition). In its annual report, the UN Children's Fund (UNICEF) said that 600 million children continue to live in poverty. There are different definitions of the standard of living considered to be the minimum adequate level (known as the **poverty level**). The European Union (EU) definition of poverty is an income of less than half the EU average (£150 a week in 1993). By this definition, there were 50 million poor in the EU in 1993.

poverty cycle set of factors or events by which poverty, once started, is likely to continue unless there is outside intervention. Once an area or a person has become poor, this tends to lead to other disadvantages, which may in turn result in further poverty. The situation is often found in *inner city areas and *shanty towns. Applied to countries, the poverty cycle is often called the **development trap**.

poverty trap situation where a person reduces his or her net income by taking a job, or gaining a higher wage, which disqualifies him/her from claiming social security benefits or raises his/her tax liability.

power station building where electrical energy is generated from a fuel or from another form of energy. Fuels used include *fossil fuels such as coal, gas, and oil, and the nuclear fuel uranium. Renewable sources of energy include gravitational potential energy, used to produce *hydroelectric power, and *wind power.

pozzolan or **pozzolana**; or **puzzolan**, silica-rich material, such as volcanic tuff or chert, which can be ground up and mixed to form a cement highly resistant to corrosion by salt water. Currently used in Portland cement, it was named after the town of Pozzuoli, Italy, where a nearby volcanic tuff was used as a basis for cement in Roman times.

prairie central North American plain, formerly grass-covered, extending over most of the region between the Rocky Mountains to the west, and the Great Lakes and Ohio River to the east.

The term was first applied by French explorers to vast, largely level grasslands in central North America, centred on the Mississippi River valley, which extend from the Gulf of Mexico to central Alberta, Canada, and from west of the Appalachian system into the Great Plains. When first seen by explorers, the prairies were characterized by unbroken, waist-high, coarse grasses.

Trees were common only along rivers and streams, or in occasional depressions in the land. This prairie is now almost gone, altered by farming to become what is known as the 'Corn Belt', much of the 'Wheat Belt', and other ploughed lands. Its humus-rich black *loess soils, adequate rainfall, and warm summers foster heavily productive agriculture.

In the west – west Kansas, Nebraska, and the Dakotas – is the **short-grass prairie**, occupying large parts of the Great Plains. Higher, drier land here has been used primarily for wheat production (aided by deep-well irrigation) and stock raising.

The prairies were formerly the primary habitat of the American bison; other prominent species include prairie dogs, deer and antelope, grasshoppers, and a variety of prairie birds.

Precambrian era in geology, the time from the formation of the Earth (4.6 billion years ago) up to 570 million years ago. Its boundary with the succeeding Cambrian period marks the time when animals first developed hard outer parts (exoskeletons) and so left abundant fossil remains. It comprises about 85% of geological time and is divided into two eons: the Archaean and the Proterozoic.

precipitation in meteorology, water that falls to the Earth from the atmosphere. It is part of the *water (hydrological) cycle. Forms of precipitation include *rain, *snow, *sleet, *hail, *dew, and *frost.

The amount of precipitation in any one area depends on *climate, *weather, and phenomena like trade winds and ocean currents. The cyclical change in the Peruvian Current off the coasts of Ecuador and Peru, known as *El Niño, causes dramatic shifts in the amount of precipitation in South and Central America and throughout the Pacific region.

Precipitation can also be influenced by people. In urban areas dust, smoke, and other particulate pollution that comprise **condensation nuclei**, cause water in the air to condense more readily. *Fog is one example. Precipitation also can react chemically with airborne pollutants to produce *acid rain.

predation in ecology, taking of prey by a predator.

Preferential Trade Area for Eastern and Southern Africa PTA, organization established in 1981 with the object of increasing economic and commercial cooperation between member states, harmonizing tariffs, and reducing trade barriers, with the eventual aim of creating a common market. Its headquarters were in Lusaka, Zambia. It was replaced by *Common Market for Eastern and Southern Africa (COMESA) in December 1994.

preindustrial a term used to describe the early stages of the *development process. Largely agricultural economies in which the foundations for development are being established, such as agricultural extension projects and improved *communications, would be described as preindustrial. The implication of such terminology is that *industrialization can be equated with development; while this has been true in recent history, it may be that alternative models will emerge as *resource shortage and *pollution become recognized globally.

pressure gradient or **barometric gradient**, the rate of horizontal change in air pressure from a particular point in any direction, as indicated by the location of *isobars on a weather chart. The maximum gradient occurs at right angles to the isobars. Strong winds are usually associated with steep gradients, which are portrayed by closely spaced isobars.

prevailing wind direction from which the wind most commonly blows in a locality. In northwestern Europe, for example, the prevailing wind is southwesterly, blowing from the Atlantic Ocean in the southwest and bringing moist and warm conditions.

price value put on a commodity at the point of exchange. In a free market it is determined by the market forces of demand and supply. In an imperfect market, firms face a trade-off between charging a higher price and losing sales, or charging a lower price and gaining sales.

price elasticity of demand responsiveness of changes in quantity demanded to a change in price of the product. It is measured by the formula: percentage change in quantity demanded/percentage change in price. For example, if the price of butter is reduced by 10% and the demand increases by 20%, the price elasticity of demand is 2.

pricing strategy the decision a business organization has to make about the price at which it will sell its products. Pricing strategies include creaming, penetration pricing, profit maximization, price capturing, price discrimination, range pricing, and loss leading.

primary consumer in ecology, those animals (herbivores) in the *food chain that eat plants.

primary data information that has been collected at first hand. It involves measurement of some sort, whether by taking readings off instruments, sketching, counting, or conducting interviews (using questionnaires).

primary industry any extractive industry, including mining and quarrying. Agriculture, fishing, and forestry are also included in this category since they involve the extraction of natural resources (see *industrial sector).

Developing countries often have a higher proportion of their workforce involved in primary industries than developed countries, where secondary and tertiary industries (see *industrial sector) are of more importance.

primate city   city that is by far the largest within a country or area. Such a city holds a larger proportion of the population, economic activity, and social functions than other settlements within that area. It is also likely to dominate politically together with the surrounding core area.

prior informed consent   informal policy whereby companies who sell *pesticides to developing countries agree to suspend exporting the product if there is an objection from the government of the receiving country and to inform the government of the nature of the pesticide. The situation arises frequently because some pesticides banned in the developed world may be bought by agricultural operations or companies in the developing world, perhaps unaware of any health implications. The policy was adopted by the FAO in 1989, and has since been made binding by the EU on its member states.

private enterprise   sector of the economy or business unit where economic activities are in private hands, set up through private capital and carried out for private profit, as opposed to national, municipal, or cooperative ownership.

private sector   the part of the economy that is owned and controlled by private individuals and business organizations such as private and public limited companies. In a *free enterprise economy, the private sector is responsible for allocating most of the resources within the economy. This contrasts with the *public sector, where economic resources are owned and controlled by the state.

privatization   policy or process of selling or transferring state-owned or public assets and services (notably nationalized industries) to private investors. Privatization of services involves the government giving contracts to private firms to supply services previously supplied by public authorities.

probability tree   diagrammatic representation of possible outcomes of series of events. A probability tree to calculate the chances of flipping a coin and coming up heads three times in a row would have three levels. The first reflects the chances of throwing either heads or tails; the second level reflects the chances of throwing heads or tails after throwing heads the first time, and the chances of throwing heads or tails after throwing tails the first time. The third level shows the chances of throwing heads or tails after all the possible outcomes of the first two throws. The series of probabilities can be multiplied to give the overall probability of a possible event occurring.

producer goods   goods that are purchased by business organizations rather than consumers. They include *capital goods such as buildings and machinery, stocks, and nondurables such as stationery for offices.

production   the process of making a good or service. Job production, batch production, and *flow production are three different ways in which production is organized. Production can be classified by industrial sector: primary production (mining and agriculture), secondary production (manufacturing), and tertiary production (services).

productivity   in economics, the output produced by a given quantity of labour, usually measured as output per person employed in the firm, industry, sector, or economy concerned. Productivity is determined by the quality and quantity of the fixed *capital used by labour, and the effort of the workers concerned.

product life cycle   the stages through which a product passes from development to being withdrawn from the market. The first stage is the **development** stage, when a product is designed. Then comes the **launch** of the product, which is likely to be associated with informative advertising and promotion because consumers need to be made aware that the product is now available on the market and what its purpose is. The next stage is the **growth** stage, as sales and revenue increase. Then comes a period of **maturity** for the product when sales and revenue level off. Competitors may have entered the market, taking away growth of the product. Or the market may have become saturated. Finally, the product goes into **decline** as sales fall.

product mix   range of products sold by a firm. For example, a supermarket sells food, but it may also sell clothes, electrical equipment, beauty products, and stationery; an electric appliance manufacturer may sell washing machines, dishwashers, refrigerators, televisions, and videos. A firm may expand its product mix by offering different products for sale; for example, a clothes shop may add underwear to its range. It may extend existing ranges; for example, a car manufacturer may bring out a special edition of an existing car model. It may change existing products; for example, a washing powder manufacturer may add more blue flakes to an existing brand and declare it to be 'new' and 'improved', or a food company may repackage a product.

profit   difference between the selling price and the production cost. This means production cost in its wide sense, that is not only the cost of manufacturing a product, but all the fixed and variable costs incurred in the process of producing and delivering the product or service. A more refined definition of profit is that of net profit. This is the income remaining after all costs have been subtracted. The net profit figure may be stated as being before or after tax. Operating profit is a term used to define profit (or loss) arising from the principal trading activity of a company. Operating profit is calculated by deducting operating expenses – expenses vital to core activity – from trading profit – profit before deduction of items such as auditors fees, interest etc.

projection   representation of the surface of the Earth on paper, see *map projection.

proportional representation   PR, electoral system in which share of party seats corresponds to their proportion of the total votes cast, and minority votes are not wasted (as opposed to a simple majority, or 'first past the post', system).

protectionism   in economics, the imposition of heavy *duties or import *quotas by a government as a means of discouraging the import of foreign goods likely to compete with domestic products. Price controls, quota systems, and the reduction of surpluses are among the measures taken for agricultural products in the European Union. The opposite practice is *free trade.

protectorate   formerly in international law, a small state under the direct or indirect control of a larger one. The 20th-century equivalent was a *trust territory.

Proterozoic Eon   eon of geological time, 3.5 billion–570 million years ago, the second division of the Precambrian. It is defined as the time of simple life, since many rocks dating from this eon show traces of biological activity, and some contain the fossils of bacteria and algae.

provenance   in earth science, area from which sedimentary materials are derived.

psammosere   a vegetation succession which begins on sand dunes or dry, weathered rock particles. Psammosere species show extreme specialization to withstand drought and extremes of nutrient availability.

pseudomorph   mineral that has replaced another in situ and has retained the external crystal shape of the original mineral.

public corporation   company structure that is similar in organization to a public limited company but with no shareholder rights. Such corporations are established to carry out state-owned activities, but are financially independent of the state and are run by a board. The first public corporation to be formed in the UK was the Central Electricity Board in the 1920s.

public good   in economics, a service, resource, or facility, such as street lighting, that is equally accessible to everyone for unlimited use.

public sector   the part of the economy that is owned and controlled by the state, namely central government, local government, and government enterprises. In a *command economy, the public sector provides most of the resources in the economy. The opposite of the public sector is the *private sector, where resources are provided by private individuals and business organizations.

public spending   expenditure by government, covering the military, health, education, infrastructure, development projects, and the cost of servicing (paying off the interest on) overseas borrowing.

pueblo   (Spanish 'village') settlement of flat-roofed stone or adobe houses that are the communal dwelling houses of the Hopi, Zuni, and other American Indians of Arizona and New Mexico. The word has also come to refer to the pueblo-dwelling American Indians of the southwest themselves.

pull factor   any factor that tends to attract people to an area. Examples include higher wages, better housing, and better educational opportunities.

pumice   light volcanic rock produced by the frothing action of expanding gases during the solidification of lava. It has the texture of a hard sponge and is used as an abrasive.

pumped storage   water pumped back up to the storage lake of a *hydroelectric power station, using surplus 'off-peak' electricity. The water can then be used again for power generation.

purchasing-power parity   PPP, system for comparing standards of living between different countries. Comparing the gross domestic product of different countries involves first converting them to a common currency (usually US dollars or pounds sterling), a conversion which is subject to large fluctuations with variations in exchange rates. Purchasing-power parity aims to overcome this by measuring how much money in the currency of those countries is required to buy a comparable range of goods and services.

punctuated equilibrium model   evolutionary theory developed by Niles Eldredge and US palaeontologist and writer Stephen Jay Gould (1941–2002) in 1972 to explain discontinuities in the fossil record. It claims that evolution continues through periods of rapid change alternating with periods of relative stability (stasis), and that the appearance of new lineages is a separate process from the gradual evolution of adaptive changes within a species.

push factor   any factor that tends to repel people from an area. Examples are crop failure, pollution, natural disasters, and poor living standards.

P-wave   *abbreviation of* primary wave, in seismology, a class of *seismic wave that passes through the Earth in the form of longitudinal pressure waves at speeds of 6–7 kps in the crust and up to 13 kps in deeper layers, the speed depending on the density of the rock. P-waves from an earthquake travel faster than S-waves and are the first to arrive at monitoring stations (hence primary waves). They can travel through both solid rock and the liquid outer core of the Earth.

pyramidal peak   angular mountain peak with three or more arêtes found in glaciated areas; for example, the Matterhorn in Switzerland. It is formed when three or four *corries (steep-sided hollows) are eroded, back-to-back, around the sides of a mountain, leaving an isolated peak in the middle.

pyramid of numbers   in ecology, a diagram that shows quantities of plants and animals at different levels (steps) of a *food chain. This may be measured in terms of numbers (how many animals) or biomass (total mass of living matter), though in terms of showing transfer of food, biomass is a more useful measure. Where biomass

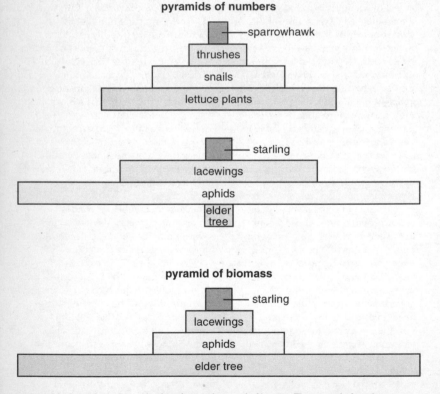

pyramids of numbers

pyramid of biomass

pyramid of numbers Pyramids of numbers and pyramid of biomass. The pyramid of numbers is a useful way of representing a food chain as it shows how the number of consumers at each level decreases, with plants being the most numerous at the base of the pyramid and top carnivores the smallest group. Where the plant being eaten is a tree, however, the pyramid no longer works as a useful model. This is rectified by the use of the pyramid of biomass, where it is the mass of the levels rather than numbers that are represented.

is measured, the diagram is often termed a pyramid of biomass. There is always far less biomass, or fewer organisms, at the top of the chain than at the bottom, because only about 10% of the food (energy) an animal eats is turned into flesh – the rest is lost through metabolism and excretion. The amount of food flowing through the chain therefore drops with each step up the chain, supporting fewer organisms, hence giving the characteristic 'pyramid' shape.

In a pyramid of biomass, the primary producers (usually plants) are represented at the bottom by a broad band, the plant-eaters are shown above by a narrower band, and the animals that prey on them by a narrower band still. At the top of the pyramid are the few 'top carnivores' such as lions and sharks.

pyrite   or **fool's gold**, iron sulphide FeS_2. It has a yellow metallic lustre and a hardness of 6–6.5 on the Mohs scale. It is used in the production of sulphuric acid.

pyroclastic   describing fragments of solidified volcanic magma, ranging in size from fine ash to large boulders, that are extruded during an explosive volcanic eruption; also the rocks that are formed by consolidation of such material. Pyroclastic rocks include tuff (ash deposit) and agglomerate (volcanic breccia).

pyroclastic deposit   deposit made up of fragments of rock, ranging in size from fine ash to large boulders, ejected during an explosive volcanic eruption.

pyroxene   any one of a group of minerals, silicates of calcium, iron, and magnesium with a general formula X,YSi_2O_6, found in igneous and metamorphic rocks. The internal structure is based on single chains of silicon and oxygen. Diopside ($X = Ca, Y = Mg$) and augite ($X = Ca, Y = Mg,Fe,Al$) are common pyroxenes.

quadrat in environmental studies, a square structure used to study the distribution of plants in a particular place, for instance a field, rocky shore, or mountainside. The size varies, but is usually 0.5 or 1 metre square, small enough to be carried easily. The quadrat is placed on the ground and the abundance of species estimated. By making such measurements a reliable understanding of species distribution is obtained.

qualitative research research that relies on opinions and beliefs rather than statistical data. Qualitative research usually involves interviews with small sample groups from target markets. These are often conducted by a sociologist or psychologist who endeavours to ascertain the motivation for decisions made by individuals within the group. Why, for example, does one person buy one brand of coffee instead of another? Although expensive to conduct, qualitative research enables companies to keep in touch with their customers' needs.

quality circle in business, a small group of production workers concerned with problems relating to the quality, safety, and efficiency of their product. Key characteristics of quality circles are size (8–12 members); voluntary membership; natural work groups, rather than artificially created ones; autonomy in setting their own agenda; access to senior managers; and a relatively permanent existence. Quality circles were popularized in Japan.

quality of life the level of wellbeing of a community and of the area in which the community lives.

quantitative research research that gives rise to statistically valid data. Quantitative market research uses sufficiently large samples to give statistically meaningful results. Typically it will involve the use of a questionnaire that asks closed questions. Quantitative research questionnaires are often based on information derived from previously conducted qualitative research.

quartile in statistics, any one of the three values that divide data into four equal parts. They comprise the **lower quartile**, below which lies the lowest 25% of the data; the **median**, which is the middle 50%, half way through the data; and the **upper quartile**, above which lies the top 25%. The difference of value between the upper and lower quartiles is known as the interquartile range, which is a useful measure of the dispersion of a statistical distribution because it is not affected by freak extreme values. These values are usually found using a cumulative frequency diagram.

quartz crystalline form of *silica SiO_2, one of the most abundant minerals of the Earth's crust (12% by volume). Quartz occurs in many different kinds of rock, including sandstone and granite. It ranks 7 on the Mohs scale of hardness and is resistant to chemical or mechanical breakdown. Quartzes vary according to the size and purity of their crystals. Crystals of pure quartz are coarse, colourless, transparent, show no cleavage, and fracture unevenly; this form is usually called rock crystal. Impure coloured varieties, often used as gemstones, include *agate, citrine quartz, and *amethyst. Quartz is also used as a general name for the cryptocrystalline and noncrystalline varieties of silica, such as chalcedony, chert, and opal.

Quartz is used in ornamental work and industry, where its reaction to electricity makes it valuable in electronic instruments. Quartz can also be made synthetically.

quartzite *metamorphic rock consisting of pure quartz sandstone that has recrystallized under increasing heat and pressure.

quartzite in sedimentology, an unmetamorphosed sandstone composed chiefly of quartz grains held together by silica that was precipitated after the original sand was deposited.

Quaternary Period period of geological time from 1.64 million years ago through to the present. It is divided into the *Pleistocene (1.64 million to 10,000 years ago) and *Holocene (last 10,000 years) epochs.

quaternary sector that sector of the economy providing information and expertise. This includes the microchip and microelectronics industries. Highly developed economies are seeing an increasing number of their workforce employed in this sector. See *industrial sector.

quota in international trade, a limitation on the amount of a commodity that may be exported, imported, or produced. Restrictions may be imposed forcibly or voluntarily.

race term sometimes applied to a physically distinctive group of people, on the basis of their difference from other groups in skin colour, head shape, hair type, and physique. Formerly, anthropologists divided the human race into three hypothetical racial groups: Caucasoid, Mongoloid, and Negroid. Others postulated from 6–30 races. Scientific studies, however, have produced no proof of definite genetic racial divisions. Race is a cultural, political, and economic concept, not a biological one. Genetic differences do exist between populations but they do not define historical lineages, and are minimal compared to the genetic variation between individuals. Most anthropologists today, therefore, completely reject the concept of race, and social scientists tend to prefer the term 'ethnic group' (see *ethnicity).

Isolation in *Homo sapiens* has never lasted long enough for the establishment of the isolating mechanisms that prevent interbreeding and lead to speciation. Humans do, however, follow many of the rules that apply to animals; for example, pigmentation is more intense in the humid tropics than in arid, cooler regions. Body extremities and body surface as a whole are reduced in animals in very cold climates; this principle is demonstrated by the Inuit. It has proved impossible to measure mental differences between groups in an objective way, and there is no acceptable scientific evidence to suggest that one race is superior to others. The attempt to categorize human types, as in South Africa for

the purposes of segregation, is inevitably doomed by the absence of any straightforward distinction. Since humans can all interbreed to produce fertile offspring, they must all belong to the same genetic species.

radial drainage see *drainage pattern.

radiation 1. the transfer of heat and other energy by means of electromagnetic waves, as in the transfer of solar energy from the Sun to the Earth. This radiation enters the Earth's atmosphere in short wavelength forms (as ultraviolet and X-rays) whilst the loss of heat from the atmosphere occurs via the longer wavelength forms (as infra-red and microwaves.) See *insolation, *albedo. 2. (Nuclear physics) the emission of alpha-, beta- and gamma particles from radioactive particles. Prolonged exposure to alpha particles and beta particles can cause serious damage to living tissue, while gamma particles (commonly associated with nuclear reactors and bombs) can lead to *radiation sickness* in humans involving skin cancers, loss of hair, destruction of bone marrow and ultimately, death. 3. (Evolution) *adaptive radiation* occurs when primitive ancestors evolve into many divergent forms, each adapted for survival under specific conditions. Adaptive radiation is often found to have occurred on isolated islands such as the Galapagos Islands where for example, an ancestral finch has evolved to fill 14 separate modes of life.

radiation fog a type of fog formed by the rapid nocturnal cooling of the ground which chills the overlying layers of air below their *dew point and leads to the condensation of atmospheric water vapour. Radiation fogs frequently form in valley bottoms and are associated with a moist atmosphere, calm conditions and cloudless skies. This type of fog occurs in most latitudes and can develop at any time of year. In summer, early morning radiation fogs are quickly evaporated by the heat of the Sun but in winter, such fogs may persist for several days.

radiation monitoring system   network of monitors to detect any rise in background gamma radiation and to warn of a major nuclear accident within minutes of its occurrence. The accident at Chernobyl in Ukraine in 1986 prompted several Western European countries to begin installation of such systems locally, and in 1994 work began on a pilot system to provide a **gamma curtain**, a dense net of radiation monitors, throughout Eastern and Western Europe.

radioactive decay   process of disintegration undergone by the nuclei of radioactive elements, such as radium and various isotopes of uranium and the transuranic elements, in order to produce a more stable nucleus. The three most common forms of radioactive decay are alpha, beta, and gamma decay.

    In **alpha decay** (the loss of a helium nucleus–two protons and two neutrons) the atomic number decreases by two and a new nucleus is formed, for example, an atom of uranium isotope of mass 238, on emitting an alpha particle, becomes an atom of thorium, mass 234. In **beta decay** the loss of an electron from an atom is accomplished by the transformation of a neutron into a proton, thus resulting in an increase in the atomic number of one. For example, the decay of the carbon-14 isotope results in the formation of an atom of nitrogen (mass 14, atomic number 7) and the emission of a high-energy electron. **Gamma emission** usually occurs as part of alpha or beta emission. In gamma emission high-speed electromagnetic radiation is emitted from the nucleus, making it more stable during the loss of an alpha or beta particle. Certain lighter, artificially created isotopes also undergo radioactive decay. The associated radiation consists of alpha rays, beta rays, or gamma rays (or a combination of these), and it takes place at a constant rate expressed as a specific half-life, which is the time taken for half of any mass of that particular isotope to decay completely. Less commonly occurring decay forms include heavy-ion emission, electron capture, and spontaneous fission (in each of these the atomic number decreases). The original nuclide is known as the parent substance, and the product is a daughter nuclide (which may or may not be radioactive). The final product in all modes of decay is a stable element.

Radioactive Incident Monitoring Network   RIMNET, monitoring network at 46 (to be increased to a total of about 90) Meteorological Office sites throughout the UK. It feeds into a central computer, and was installed 1989 to record contamination levels from nuclear incidents such as the *Chernobyl disaster 1986.

radioactive tracer   any of various radioactive isotopes added to fluids in order to monitor their flow and therefore identify leaks or blockages.

radioactive waste   any waste that emits radiation in excess of the background level. See *nuclear waste.

radioactivity   spontaneous change of the nuclei of atoms, accompanied by the emission of radiation. Such atoms are called radioactive. It is the property exhibited by the radioactive isotopes of stable elements and all isotopes of radioactive elements, and can be either natural or induced. A radioactive material decays by releasing radiation, and transforms into a new substance. The energy is released in the form of alpha particles and beta particles or in the form of high-energy electromagnetic waves known as gamma radiation. Natural radioactive elements are those with an atomic number of 83 and higher. Artificial radioactive elements can also be formed. See *radioactive decay.

radiocarbon cycle   production and recycling of the radioisotope carbon-14 (^{14}C). The radioisotope occurs when a neutron flux, caused by cosmic radiation bombarding the upper atmosphere, reacts efficiently with nitrogen present. Carbon-14 intake by living organisms eventually returns to the atmosphere when dead vegetation or animal flesh decomposes, except when it is locked in preserved organic artefacts and remains. Radioactive decay occurs, forming the basis of the radiocarbon dating method.

Initially the concentration of carbon-14 is nonuniform (levels being higher over polar regions where the Earth's magnetic field is least effective in deflecting cosmic radiation) but air currents at about 10 km soon redistribute the newly formed carbon-14 as part of carbon dioxide gas. Around 7.5 kg of carbon-14 is added to the Earth's carbon reservoir each year and distributed throughout the oceans, the biosphere, and the atmosphere, although variations in the magnetic field and sunspot activity can alter the intensity of cosmic radiation, affecting carbon-14 production.

radiometric dating   method of dating rock by assessing the amount of *radioactive decay of naturally occurring isotopes. The dating of rocks may be based on the gradual decay of uranium into lead. The ratio of the amounts of 'parent' to 'daughter' isotopes in a sample gives a measure of the time it has been decaying, that is, of its age. Different elements and isotopes are used depending on the isotopes present and the age of the rocks to be dated. Once-living matter can often be dated by radiocarbon dating, employing the half-life of the isotope carbon-14, which is naturally present in organic tissue.

Radiometric methods have been applied to the decay of long-lived isotopes, such as potassium-40, rubidium-87, thorium-232, and uranium-238, which are found in rocks. These isotopes decay very slowly and this has enabled rocks as old as 3800 million years to be dated accurately. Carbon dating can be used for material between 1000 and 100,000 years old. **Potassium** dating is used for material more than 100,000 years old, **rubidium** for rocks more than 10 million years old, and **uranium** and **thorium** dating is suitable for rocks older than 20 million years.

rain   form of *precipitation in which separate drops of water fall to the Earth's surface from clouds. The drops are formed by the accumulation of fine droplets that condense from water vapour in the air. The *condensation is usually brought about by rising and subsequent cooling of air.

Rain can form in three main ways: frontal (or cyclonic) rainfall, orographic (or relief) rainfall, and convectional rainfall. **Frontal rainfall** takes place at the boundary, or *front, between a mass of warm air from the tropics and a mass of cold air from the poles. The water vapour in the warm air is chilled and condenses to form clouds and rain. **Orographic rainfall** occurs when an airstream is forced to rise over a mountain range. The air becomes cooled and precipitation takes place. In the UK, the Pennine hills, which

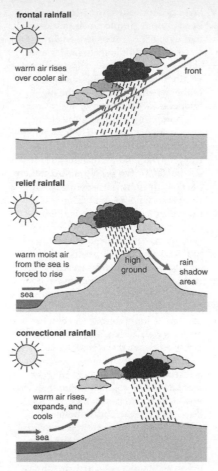

frontal rainfall

warm air rises
over cooler air   front

relief rainfall

warm moist air   high   rain
from the sea is   ground   shadow
forced to rise   area
sea

convectional rainfall

warm air rises,
expands, and
cools
sea

rain The three patterns of rainfall. Frontal
(or cyclonic) rain is caused by warm air rising
over cold air in a low pressure area. Relief
(or orthographic) rainfall occurs when warm,
moist air cools as it is forced to rise over hills or
mountains. Convectional rainfall is caused when
the surface of the Earth has been warmed by the
Sun and the air in contact with it rises and meets
colder layers. It is usually accompanied by a
thunderstorm and is common in tropical regions.

extend southwards from Northumbria
to Derbyshire in northern England,
interrupt the path of the prevailing
southwesterly winds, causing

orographic rainfall. Their presence
is partly responsible for the west of
the UK being wetter than the east.
Convectional rainfall, associated with
hot climates, is brought about by rising
and abrupt cooling of air that has been
warmed by the extreme heat of the
ground surface. The water vapour
carried by the air condenses,
producing heavy rain. Convectional
rainfall is usually accompanied by a
thunderstorm, and it can be intensified
over urban areas due to higher
temperatures there (see *heat island).

rainbow arch in the sky displaying
the colours of the spectrum formed
by the refraction and reflection of
the Sun's rays through rain or mist.
Its cause was discovered by Theodoric
of Freiburg in the 14th century.

rainfall gauge instrument used to
measure *precipitation, usually rain.
It consists of an open-topped cylinder,
inside which there is a close-fitting
funnel that directs the rain to a
collecting bottle inside a second, inner
cylinder. The gauge may be partially
embedded in soil to prevent spillage.
The amount of water that collects
in the bottle is measured every day,
usually in millimetres.

When the amount of water collected
is too little to be measured, **trace
rainfall** is said to have taken place.
Snow falling into the gauge must be
melted before a measurement is taken.

rainfall interception the process by
which plants and trees break the force
of precipitation. Rain lands on the
surfaces of the vegetation and fills the
hollows. It then runs down the trunks
and stems and drips from the leaves
on to the ground, where it can sink
in gently, instead of running off the
surface.

The density and type of vegetation
in an area are important in
determining the speed at which
rainwater moves through the

landscape. Different types of vegetation intercept water to a greater or lesser degree. Coniferous forest, for example, intercepts 58% of the rain falling upon it; tall grasses intercept about 27%.

Where large tracts of natural vegetation have been removed from an area, splash erosion and sheet erosion occur (see *soil erosion). Where soil on a slope is not consolidated by a network of roots, it can be loosened by rain and slip off, often with tremendous force, as a landslide. This has been a big problem in the Himalayas and areas of tropical rainforest, due to *deforestation.

In urban areas, the lack of vegetation is compensated for by an artificial drainage system.

rainforest dense forest usually found on or near the *Equator where the climate is hot and wet. Moist air brought by the converging trade winds rises because of the heat and produces heavy rainfall. More than half the tropical rainforests are in Central and South America, primarily the lower Amazon and the coasts of Ecuador and Columbia. The rest are in Southeast Asia (Malaysia, Indonesia, and New Guinea) and in West Africa and the Congo.

Tropical rainforests once covered 14% of the Earth's land surface, but are now being destroyed at an increasing rate as their valuable timber is harvested and the land cleared for agriculture, causing problems of *deforestation. Although by 1991 over 50% of the world's rainforests had been removed, they still comprise about 50% of all growing wood on the planet, and harbour at least 40% of the Earth's species (plants and animals).

The vegetation in tropical rainforests typically includes an area of dense forest called **selva**; a **canopy** formed by high branches of tall trees providing shade for lower layers; an intermediate layer of shorter trees and tree roots; lianas; and a ground cover of mosses and ferns. The lack of **seasonal rhythm** causes adjacent plants to flower and shed leaves simultaneously. Chemical weathering and leaching take place in the iron-rich soil due to the high temperatures and humidity.

Rainforests comprise some of the most complex and diverse *ecosystems on the planet, deriving their energy from the Sun and photosynthesis. In a hectare (10,000 sq m) of rainforest there are an estimated 200–300 tree species compared with 20–30 species in a hectare of temperate forest. The trees are the main **producers**. Herbivores such as insects, caterpillars, and monkeys feed on the plants and trees and in turn are eaten by the carnivores, such as ocelots and puma. Fungi and bacteria, the primary **decomposers**, break down the dead material from the plants, herbivores, and carnivores with the help of heat and humidity. This decomposed material provides the **nutrients** for the plants and trees.

The rainforest ecosystem helps to regulate global weather patterns – especially by taking up CO_2 (carbon dioxide) from the atmosphere – and stabilizes the soil. Rainforests provide most of the oxygen needed for plant and animal respiration. When deforestation occurs, the microclimate of the mature forest disappears; soil erosion and flooding become major problems since rainforests protect the deep tropical soils. Once an area is cleared it is very difficult for shrubs and bushes to re-establish because soils are poor in nutrients. This causes problems for plans to convert rainforests into agricultural land – after two or three years the crops fail and the land is left bare. Clearing of the rainforests may lead to *global warming of the atmosphere, and contribute to the *greenhouse effect.

Tropical rainforests are characterized by a great diversity of species (*biodiversity), usually of tall broad-leafed evergreen trees, with many climbing vines and ferns, some of which are a main source of raw materials for medicines. A tropical forest, if properly preserved, can yield medicinal plants, oils (from cedar, juniper, cinnamon, sandalwood), spices, gums, resins (used in inks, lacquers, linoleum), tanning and dyeing materials, forage for animals, beverages, poisons, green manure,

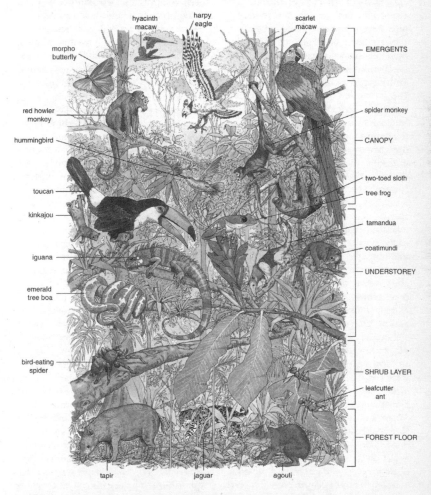

rainforest Cross section of a South American rainforest, showing the different horizontal layers of vegetation that make up the rainforest, and the animals and plants found in each layer. To a certain extent the separate layers provide unique microhabitats, with animals and plants found only in one stratum and being adapted to life there, so for example the tapir occupies only the forest floor and some butterfly species always fly at specific heights. Other animals and plants move across strata, for example many of the monkeys and birds.

rubber, and animal products (feathers, hides, honey). Other types of rainforest include montane, upper montane or cloud, mangrove, and subtropical.

Traditional ways of life in tropical rainforests are disappearing. The practice of **shifting cultivation** (*slash and burn), in which small plots of forest are cultivated and abandoned after two or three harvests, is being replaced by cultivation on such a large scale that the rainforests cannot regenerate. As a result hunting and gathering as a way of life is also becoming less viable. In the last 30 years, Central America has lost almost two-thirds of its rainforests to cattle ranching.

rainshadow see *orographic rainfall.

raised beach beach that has been raised above the present-day shoreline and is therefore no longer washed by the sea. It is an indication of a fall in sea level (eustatic) or of a rise in land level (isostatic).

Raised beaches are quite common in the north of Scotland.

ranching the extensive commercial grazing of large numbers of cattle or sheep either on open range or on large areas fenced by barbed wire. Ranching is less widespread now than a century ago as *arable farming has spread onto the humid margins of the semi-arid areas of the mid-latitude steppes and tropical savannas that were once its preserve. These areas include the Great Plains from Texas to the Canadian prairies, the llanos of Venezuela, the sertao of Brazil, the Karoo of South Africa, and the Australian interior. Ranching also occurs in the more humid environments of the Argentine pampas and the high country of South Island, New Zealand.

Ranches, or stations as they are known in Australia, occur in areas of sparse population with low land values, have few farm buildings, but may feature numerous windmills to pump ground-

water to the surface. Stock carrying capacities are low, at best perhaps 3 hectares per head of cattle, but in the more arid areas sometimes no more than 2 cattle per 100 hectares. Ranches in western Texas may exceed 8000 hectares, while some in Arizona are twice that size. Australian sheep stations average 8000 hectares, while some northern cattle stations cover 20,000 hectares or more. In recent times, however, ranching has tended towards smaller and more diversified units with some cultivation.

rancho Venezuelan makeshift housing or *shanty town.

random sample *sample taken from a group at random, with no attempt to emulate the make-up of the larger group (as in a quota sample). Random sampling is difficult to achieve in practice and requires special sampling techniques to avoid bias. In the case of market research the sample is drawn from the electoral register.

range in physical geography, a line of mountains (such as the Alps or Himalayas). In human geography, the distance that people are prepared to travel (often to a *central place) to obtain various goods or services. In mathematics, the range of a set of numbers is the difference between the largest and the smallest number; for example, 5, 8, 2, 9, 4 = 9 − 2 = 7; this sense is used in terms like 'tidal range' and 'temperature range'. Range is also a name for an open piece of land where cattle are ranched.

rapid an area of broken, turbulent water in a river channel, caused by a stratum of resistant *rock that dips downstream. The softer rock immediately upstream and downstream erodes more quickly, leaving the resistant rock sticking up and creating an obstacle to the flow of the water. Compare *waterfall.

Rare Breeds Survival Trust British organization dedicated to ensuring the survival of endangered breeds of livestock and poultry. The Trust was formed 1973, after a major survey of traditional breeds of farm animals had shown that many were in danger of extinction.

An official list of rare breeds was drawn up, and owners were advised on methods of increasing stocks, especially by seeking out animals of different blood lines in order to avoid in-breeding.

raw materials the *resources supplied to industries for subsequent manufacturing processes, for example agricultural products, minerals and timber are *raw materials. Many *primary industry products are used as raw materials.

recycling processing of industrial and household waste (such as paper, glass, and some metals and plastics) so that the materials can be reused. This saves expenditure on scarce raw materials, slows down the depletion of *non-renewable resources, and helps to reduce pollution. Aluminium is frequently recycled because of its value and special properties that allow it to be melted down and re-pressed without loss of quality, unlike paper and glass, which deteriorate when recycled.

The USA recycles only around 25% of its waste (1998), compared with around 33% in Japan. However, all US states encourage or require local recycling programmes to be set up. It has been estimated that around 33% of newspapers, 22% of office paper, 64% of aluminium cans, 3% of plastic containers, and 20% of all glass bottles and jars were recycled.

Red Data List report published by the *World Conservation Union and regularly updated that lists animal species by their conservation status. Categories of risk include **extinct in the wild**, **critically endangered**, **endangered**, **vulnerable**, and **lower risk** (divided into three subcategories).

reef underwater ridge or mound composed of the skeletal remains of colonies of *corals and calcareous algae. Types of reefs include atolls, barrier reefs, and fringing reefs.

refugee according to international law, a person fleeing from oppressive or dangerous conditions (such as political, religious, or military persecution) and seeking refuge in a foreign country. In 1995 there were an estimated 27 million refugees worldwide; their resettlement and welfare is the responsibility of the United Nations High Commission for Refugees (UNHCR). An estimated average of 10,000 people a day become refugees. Women and children make up 75% of all refugees and displaced persons. Many more millions are 'economic' or 'environmental' refugees, forced to emigrate because of economic circumstances, lack of access to land, or environmental disasters.

regeneration 1. in geography, the process of improving run-down inner city areas by investing money, improving transport links, providing grants and loans for buildings, improving the local environment, and providing good housing and social facilities. The aim is to attract business and to encourage people to live and work in the regenerated area.
2. renewed growth of, for example, forest after felling. Forest regeneration is crucial to the long-term stability of many *resource systems, from *bush fallowing to commercial forestry.

region an area of land which has marked boundaries or unifying internal characteristics. Geographers may identify regions according to physical, climatic, political, economic or other factors. The Sahara desert is an example of a climatic region, while southeast England (London and the Home Counties) is an economic region.

regional development a planning policy made by a country to overcome regional differences in income, wealth, education, medical facilities, transport, etc. Such divisions between regions must be minimized if countries are to achieve balanced development. Regional development plans can take many forms; for example, improvements in infrastructure (i.e. the construction of facilities that lay the foundations for further agricultural or industrial development – new roads, power supplies, etc.), resettlement schemes (such as the Indonesian transmigration scheme), or resource development (e.g. mineral exploitation).

Regional inequalities tend to be much more pronounced in countries in the *developing world, therefore such plans tend to be on a larger and more complex scale than those of developed countries.

regolith surface layer of loose material that covers most bedrock. It consists of eroded rocky material, volcanic ash, river alluvium, vegetable matter, or a mixture of these known as *soil.

rejuvenation in earth science, the renewal of a river's powers of downward erosion. It may be caused by a fall in sea level or a rise in land level, by climate change, by changes in vegetation cover (deforestation), or by the increase in water flow that results when one river captures another (*river capture).

Several river features are formed by rejuvenation. For example, as a river cuts down into its channel it will leave its old flood plain perched up on the valley side to form a *river terrace. Meanders (bends in the river) become deeper and their sides steeper, forming *incised meanders, and *waterfalls and rapids become more common.

relative humidity the relationship between the actual amount of water vapour in the air and the amount of vapour the air could hold at a particular temperature. This is usually expressed as a percentage. Relative humidity gives a measure of dampness in the *atmosphere, and this can be determined by a *hygrometer.

relief the differences in height between any parts of the Earth's surface. Hence a relief map will aim to show differences in the height of land by, for example, *contour lines or by a colour key.

relief rainfall see *orographic rainfall.

remote sensing gathering and recording information from a distance. Aircraft and satellites can observe a planetary surface or atmosphere, and space probes have sent back photographs and data about planets as distant as Neptune. Remote sensing usually refers to gathering data of the electromagnetic spectrum (such as visible light, ultraviolet light, and infrared light). In archaeology, surface survey techniques provide information without disturbing subsurface deposits.

Remote sensing is most commonly taken to refer to the process of photographing the Earth's surface with orbiting satellites. Satellites such as *Landsat* have surveyed all of the Earth's surface from orbit. Computer processing of data obtained by their scanning instruments, and the application of so-called false colours (generated by computer), have made it possible to reveal surface features invisible in ordinary light. This has proved valuable in agriculture, forestry, and urban planning, and has led to the discovery of new deposits of minerals.

rendzina calcareous rego black soil, rendoll or **derncarbonate soil**, a grey, brown or black soil of medium texture developed above parent material with a high calcium carbonate content such as limestone. Rendzinas are usually shallow (no more than 30 cm), are excessively drained and contain fragments of poorly weathered calcium carbonate. The typically dark colour of the rendzina is due to the abundance of humus. The combination of shallowness and excessive drainage makes this soil type unsuited for agricultural use.

renewable energy power from any source that can be replenished. Most renewable systems rely on solar energy directly or through the weather cycle as wave power, hydroelectric power, wind power via wind turbines, or solar energy collected by plants (alcohol fuels, for example). In addition, the gravitational force of the Moon can be harnessed through tidal power stations, and the heat trapped in the centre of the Earth is used via geothermal energy systems. Other examples are energy from biofuel and fuel cells. Renewable energy resources have the advantage of being non-polluting. However, some (such as wind energy) can be unreliable and therefore lose their effectiveness in providing a constant supply of energy.

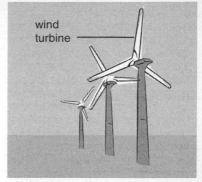

wind power

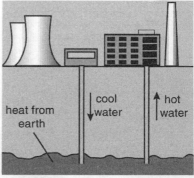

geothermal energy

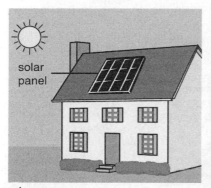

solar power

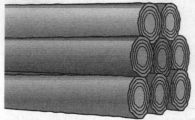

biomass

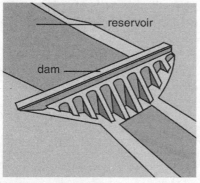

hydroelectric power

renewable energy Some examples of renewable energy. Panels of solar cells can be used to convert the light of the Sun into electrical energy. Wind turbines are connected to generators and turn to generate electricity. In a typical hydroelectric plant, water is stored in a reservoir. The water's potential energy is converted to kinetic energy as it is piped into turbines that are connected to generators. Geothermal energy harnesses the heat of the Earth to provide heating and electricity. Biomass is organic matter that can be converted to provide heat or electricity.

renewable resource natural resource that is replaced by natural processes in a reasonable amount of time. Soil, water, forests, plants, and animals are all renewable resources as long as they are properly conserved. Solar, wind, wave, and geothermal energies are based on renewable resources.

representative fraction the fraction of real size to which objects are reduced on a map; for example, on a 1:50,000 map, any object is shown at 1/50,000 of its real size.

republic (Latin *res publica* 'the state'; from *res* 'affair', and *publica* 'public') country where the head of state is not a monarch, either hereditary or elected, but usually a president, whose role may or may not include political functions.

research and development R&D, process undertaken by a business organization before the launch of a product. Research is usually scientific research into materials and production processes. Product development is application of that research to the development of existing or new products.

reserve currency in economics, a country's holding of internationally acceptable means of payment (major foreign currencies or gold); central banks also hold the ultimate reserve of money for their domestic banking sector. On the asset side of company balance sheets, undistributed profits are listed as reserves.

reserves 1. monies retained in case of emergency or to be used at a later date. Gold bullion and foreign currency reserves are held by a central bank such as the Bank of England. They are used to intervene in the foreign exchange market to change the exchange rate. If the Bank of England buys pounds sterling by selling some of its foreign currency reserves, the value of the pound should rise, and vice versa. In business, undistributed

profits are listed as reserves on the asset side of company balance sheets. 2. resources which are available for future use. The world has reserves of *fossil fuels which, if used with careful management, could last for many years. However, many of our fossil fuel and mineral reserves have been used up at a rapid rate by the developed countries, and more controlled use is required in the future to conserve reserves.

reservoir natural or artificial lake that collects and stores water for community water supplies and for *irrigation.

resettlement scheme policy for moving people (usually *squatters) out of their homes to other areas where they are provided with basic dwellings. This is common in urban areas of low-income countries; for example, in Mumbai (formerly Bombay), India. Resettlement on a large scale took place in Indonesia in the 1980s, and was known as transmigration.

resources materials that can be used to satisfy human needs. Because human needs are varied and extend from basic physical requirements, such as food and shelter, to spiritual and emotional needs that are hard to define, resources cover a vast range of items. The intellectual resources of a society – its ideas and technologies – determine which aspects of the environment meet that society's needs, and therefore become resources. For example, in the 19th century uranium was used only in the manufacture of coloured glass. Today, with the development of nuclear technology, it is a military and energy resource. Resources are often divided into **human resources**, such as labour, supplies, and skills, and **natural resources**, such as climate, *fossil fuels, and water. Natural resources are divided into *non-renewable resources and *renewable resources.

respiration   the release of energy from food in the cells of all living organisms (plants as well as animals). The process normally requires oxygen and releases carbon dioxide. It is balanced by *photosynthesis.

retail   sale of goods and services to a consumer. The retailer is the last link in the distribution chain. A retailer's purchases are usually made from a wholesaler, who in turn buys from a manufacturer.

reuse   multiple use of a product (often a form of packaging), by returning it to the manufacturer or processor each time. Many such returnable items are sold with a deposit which is reimbursed if the item is returned. Reuse is usually more energy- and resource-efficient than *recycling unless there are large transport or cleaning costs.

revenue   money received from taxes or the sale of a product. **Total revenue** can be calculated by multiplying the average price received by the total quantity sold. **Average revenue** is the average price received and is calculated by dividing total revenue by total quantity sold. **Marginal revenue** is the revenue gained from the sale of an additional unit of output.

revisionism   political theory derived from Marxism that moderates one or more of the basic tenets of Karl Marx, and is hence condemned by orthodox Marxists.

revolution   the passage of the Earth around the sun; one revolution is completed in 365.25 days. Due to the tilt of the Earth's axis ($23^1/2°$ from the vertical), revolution results in the sequence of seasons experienced on the Earth's surface.

rhyolite   *igneous rock, the fine-grained volcanic (extrusive) equivalent of granite.

ria   long narrow sea inlet, usually branching and surrounded by hills. A ria is deeper and wider towards its mouth, unlike a *fjord. It is formed by the flooding of a river valley due to either a rise in sea level or a lowering of a landmass.
There are a number of rias in the UK – Salcombe and Dartmouth in Devon are both situated on rias, for example.

ribbon development   another term for *linear development, housing that has grown up along a route.

ribbon lake   long, narrow lake found on the floor of a *glacial trough.
A ribbon lake will often form in an elongated hollow carved out by a glacier, perhaps where it came across a weaker band of rock. Ribbon lakes can also form when water ponds up behind a terminal moraine or a landslide. The English Lake District is named after its many ribbon lakes, such as Lake Windermere and Coniston Water.

Richter scale   quantitative scale of earthquake magnitude based on the measurement of seismic waves, used to indicate the magnitude of an *earthquake at its epicentre. The Richter scale is logarithmic, so an earthquake of 6.0 is ten times greater than one of 5.0. The magnitude of an earthquake differs from its intensity, measured by the *Mercalli scale, which is qualitative and varies from place to place for the same earthquake. The scale is named after US seismologist Charles Richter.
An earthquake's magnitude is a function of the total amount of energy released, and each point on the Richter scale represents a thirtyfold increase in energy over the previous point. One of the greatest earthquakes ever recorded, in 1920 in Gansu, China, measured 8.6 on the Richter scale.

ridge of high pressure   elongated area of high atmospheric pressure extending from an anticyclone.
On a synoptic weather chart it is shown as a pattern of lengthened isobars. The weather under a ridge of high pressure is the same as that under an anticyclone.

The UK is often under the influence of a ridge of high pressure – in summer it originates from the Azores, and in winter from Scandinavia.

riffle see *deposition, *pools and riffles.

rift valley valley formed by the subsidence of a block of the Earth's *crust between two or more parallel *faults. Rift valleys are steep-sided and form where the crust is being pulled apart, as at *mid-ocean ridges, or in the Great Rift Valley of East Africa. In cross-section they can appear like a widened gorge with steep sides and a wide floor.

rill a shallow channel often occurring in the middle of a slope which is formed by the concentration of thin sheets of surface water into flows capable of eroding the soil. Rills can reach up to 30 cm in width and depth and may gradually coalesce and enlarge downslope as the increased concentration of surface runoff results in gully erosion.

rime an accumulation of ice formed on the windward side of objects such as cars, fences and trees. Rime results when supercooled water droplets freeze when blown by a slight wind into contact with any exposed surface.

river large body of water that flows down a slope along a channel restricted by adjacent banks and *levees. A river starts at a point called its **source**, and enters a sea or lake at its **mouth**. Along its length it may be joined by smaller rivers called **tributaries**; a river and its tributaries are contained within a *drainage basin. The point at which two rivers join is called the *confluence.

Rivers are formed and moulded over time chiefly by the processes of *erosion, and by the **transport** and **deposition** of *sediment. Rivers are able to work on the landscape through erosion, transport, and deposition. The amount of potential energy available to a river is proportional to its initial height above sea level. A river follows the path of least resistance downhill, and deepens, widens and lengthens its channel by erosion. Up to 95% of a river's potential energy is used to overcome friction.

One way of classifying rivers is by their stage of development. An upper course is typified by a narrow V-shaped valley with numerous *waterfalls, *lakes, and rapids. Because of the steep gradient of the topography and the river's height above sea level, the rate of erosion is greater than the rate of deposition, and downcutting occurs by **vertical corrasion** (erosion or abrasion of the bed or bank caused by the load carried by the river).

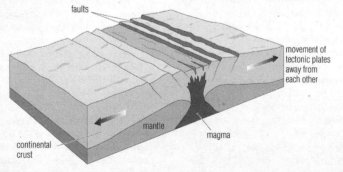

rift valley The subsidence of rock resulting from two or more parallel rocks moving apart is known as a graben. When this happens on a large scale, with tectonic plates moving apart, a rift valley is created.

middle course
The river flows through a broad valley floored with sediments and changes its course quite frequently. It cuts into the bank on the outsides of the curves where the current flows fast and deep. Along the inside of the curves, sand and gravel deposits build up. When the river washes against a valley spur it cuts it back into a steep bank, or bluff.

upper course
The river begins its descent through a narrow V-shaped valley. Falling steeply over a short distance, it follows a zig-zag course and produces interlocking spurs.

Loops and oxbow lakes form where the changing course of a river cuts off a meander.

lower course
The river meanders from side to side across a flat plain on which deep sediments lie; often the water level is higher than that of the plain. This is caused by the deposition of sediment forming high banks and levees particularly at times of flood.

Sand and mud deposited at the river mouth form sand banks and may produce a delta.

river The course of a river from its source of a spring or melting glacier, through to maturity where it flows into the sea.

In the middle course of a river, the topography has been eroded over time and the river's course has a shallow gradient. Such a river is said to be graded. Erosion and deposition are delicately balanced as the river meanders (gently curves back and forth) across the extensive *flood plain. **Horizontal corrasion** is the main process of erosion. The flood plain is an area of periodic *flooding along the course of a river valley where fine silty material called *alluvium is deposited by the flood water. Features of a

mature river (or the **lower course** of a river) include extensive *meanders, *oxbow lakes, and braiding.

Many important flood plains, such as the Nile flood plain in Egypt, occur in arid areas where their exceptional fertility is very important to the local economy. However, using flood plains as the site of towns and villages involves a certain risk, and it is safer to use flood plains for other uses, such as agriculture and parks. Water engineers can predict when flooding is likely and take action to prevent it by studying

*hydrographs, which show how the
*discharge of a river varies with time.

Major rivers of the world include the
Ganges, the Mississippi, and the Nile,
the world's longest river.

river basin   the area drained by a river
and its tributaries, sometimes referred
to as a *catchment area.

river capture   the diversion (capture)
of the headwaters of one river into a
neighbouring river. River capture occurs
when a stream is carrying out rapid
*headward erosion (backwards erosion
at its source). Eventually the stream
will cut into the course of a neighbouring
river, causing the headwaters of that
river to be diverted, or 'captured'.

The headwaters will then flow down
to a lower level (often making a sharp
bend, called an elbow of capture) over
a steep slope, called a knickpoint. A
waterfall will form here. *Rejuvenation
then occurs, causing rapid downwards
erosion. An excellent example is the
capture of the River Burn by the
River Lyd, Devon, England.

river cliff   or **bluff**, steep slope forming
the outer bank of a *meander (bend
in a river). It is formed by the
undercutting of the river current,
which at its fastest when it sweeps
around the outside of the meander.

river terrace   part of an old *flood
plain that has been left perched on
the side of a river valley. It results from
*rejuvenation, a renewal in the erosive
powers of a river. River terraces may
be fertile and are often used for
farming. They are also commonly
chosen as sites for human settlement
because they are safe from flooding.
Many towns and cities throughout
the world have been built on terraces,
including London, which is built on
the terraces of the River Thames.

Other terraces are formed as a result
of glacial outwash. This deposits large
amounts of sand and gravel, which are
later eroded by rivers. The uneroded

parts remain as terraces. In southern
England, the Harnborough, Wolvercote,
Summertown, and Radley terraces
near Oxford are areas of high ground
above the flood plains of the River
Thames and River Cherwell.

Roaring Forties   nautical expression
for regions south of latitude 40° S, in
the southern oceans, where strong
westerly winds prevail.

roche moutonnée   outcrop of tough
bedrock having one smooth side and
one jagged side, found on the floor
of a *glacial trough (U-shaped valley).
It may be up to 40 m high. A roche
moutonnée is a feature of glacial erosion
– as a glacier moved over its surface,
ice and debris eroded its upstream side
by corrasion, smoothing it and creating
long scratches or striations. On the
sheltered downstream side fragments
of rock were plucked away by the ice,
causing it to become steep and jagged.

rock   constituent of the Earth's crust
composed of *minerals or materials of
organic origin that have consolidated
into hard masses. There are three basic
types of rock: *igneous, *sedimentary,
and *metamorphic. Rocks are composed
of a combination (or aggregate) of
minerals, and the property of a rock
will depend on its components. Where
deposits of economically valuable
minerals occur they are termed *ores.
As a result of *weathering, rock breaks
down into very small particles that
combine with organic materials from
plants and animals to form *soil.
In *geology the term 'rock' can also
include unconsolidated materials such
as *sand, mud, *clay, and *peat.

The change of one type of rock
to another is called the *rock cycle.

rock studies   The study of the Earth's
crust and its composition fall under
a number of interrelated sciences
(generally known as geology), each
with its own specialists. Among these
are geologists, who identify and survey

rock formations and determine when and how they were formed, petrologists, who identify and classify the rocks themselves, and mineralogists, who study the mineral contents of the rocks. Palaeontologists study the fossil remains of plants and animals found in rocks.

applications of rock studies Data from these studies and surveys enable scientists to trace the history of the Earth and learn about the kind of life that existed here millions of years ago. The data are also used in locating and mapping deposits of *fossil fuels such as coal, oil, and natural gas, and valuable mineral-containing ores providing metals such as aluminium, iron, lead, and tin, and radioactive elements such as radium and uranium. These deposits may lie close to the Earth's surface or deep underground, often under oceans. In some regions, entire mountains are composed of deposits of iron or copper ores, while in other regions rocks may contain valuable non-metallic minerals such as borax and graphite, or precious gems such as diamonds and emeralds.

rock as construction material and building stone In addition to the mining and extraction of fuels, metals, minerals, and gems, rocks provide useful building and construction materials. Rock is mined through quarrying, and cut into blocks or slabs as building stone, or crushed or broken for other uses in construction work. For instance, cement is made from limestone and, in addition to its use as a bonding material, it can be added to crushed stone, sand, and water to produce strong, durable concrete, which has many applications, such as the construction of roads, runways, and dams.

Among the most widely used building stones are granite, limestone, sandstone, marble, and slate. Granite provides one of the strongest building stones and is resistant to weather, but its hardness makes it difficult to cut and handle. Limestone is a hard and lasting stone that is easily cut and shaped and is widely used for public buildings. The colour and texture of the stone can vary with location; for instance, Portland stone from the Jurassic rocks of Dorset is white, even-textured and durable, while Bath stone is an oolitic limestone that is honey-coloured and more porous. Sandstone varies in colour and texture; like limestone, it is relatively easy to quarry and work and is used for similar purposes. Marble is a classic stone, worked by both builders and sculptors. Pure marble is white, streaked with veins of black, grey, green, pink, red, and yellow. Slate is fine-grained rock that can be split easily into thin slabs and used as tiles for roofing and flooring. Its colour varies from black to green and red.

rock identification Rocks can often be identified by their location and appearance. For example, sedimentary rocks lie in stratified, or layered, formations and may contain fossils; many have markings such as old mud cracks or ripple marks caused by waves. Except for volcanic glass, all igneous rocks are solid and crystalline. Some appear dense, with microscopic crystals, and others have larger, easily seen crystals. They occur in volcanic areas, and in intrusive formations that geologists call batholiths, laccoliths, sills, dikes, and stocks. Many metamorphic rocks have characteristic bands, and are easily split into sheets or slabs. Rock formations and strata are often apparent in the cliffs that line a seashore, or where rivers have gouged out deep channels to form gorges and canyons. They are also revealed when roads are cut through hillsides, or by excavations for quarrying and mining. Rock and fossil collecting has been a popular activity since the 19th century and such sites can provide a treasure trove of finds for the collector.

rock cycle the recycling of the Earth's outer layers. *Rocks are continually being formed, destroyed, and re-formed in an endless cycle of change that takes millions of years. The processes involved include the formation of *igneous rock from *magma (molten rock); surface *weathering and *erosion; the compaction and cementation of sediments into *sedimentary rock; and metamorphism, chemical and physical changes brought about by heat and pressure, producing *metamorphic rock.

formation of igneous rock The newest rocks found on the Earth's surface are those caused by active *volcanoes. When *magma cools and solidifies, the minerals within it crystallize, and the magma becomes igneous rock. When a volcano erupts it may throw out molten magma that has built up below the surface in an **igneous intrusion**. Magma that flows out of the volcano is known as *lava.

*Extrusive rocks** are those that form from the molten magma above ground. As the magma cools rapidly on the surface, the minerals in the rock crystallize quickly to form small interlocking crystals. A common type of extrusive igneous rock is *basalt. Sometimes magma cools and solidifies before it reaches the surface, forming *intrusive rocks**. Intrusive igneous rock has a much larger crystal structure because the molten magma cools more slowly underground, allowing the crystals more time to grow. An example of intrusive igneous rock is *granite.

surface weathering and erosion Rocks exposed on the surface of the Earth are very slowly broken down into fragments through the action of sun, wind, rain, and ice, in a combination of *physical weathering and *chemical weathering. The broken pieces of rock are transported by gravity, water (rivers and streams), wind, or ice (glaciers), and are further broken up and smoothed during the process. The movement of the fragments causes further *erosion of the rock surfaces. In a river, the river bed is broken down and the smallest particles (clay and sand) are deposited at the river mouth forming an *estuary, or in lakes. The sea is also an important agent of erosion, causing *coastal erosion.

rock particles to sedimentary rock Rock particles are deposited as *sediments, often at the bottom of a lake or sea. Layers of sediment build up, sometimes trapping animals and plants that will later become *fossils. At this point the sediments are described as unconsolidated because the mixture is loosely packed together, for example as *gravel beds. Gradually the layers get thicker as more material is deposited. The transformation of sediments into sedimentary rocks is called *lithification. The layers of sediments are compressed (compacted) by the weight of the material above. Water is squeezed out and the grains are fused together. In certain cases the material is bound together by new *minerals, such as calcium carbonate, which crystallize around the rock particles. This process is known as *cementation** (not to be confused with the cement used in the building industry). *Sedimentary rocks are composed of grains, often in layers, and may contain fossils.

formation of metamorphic rock *Plate tectonics (movements of the plates that form the Earth's outermost layer) cause further change. Large areas of sedimentary rock may be folded and uplifted to form mountain ranges, such as the Himalayas, while in the active subduction zones of the tectonic plates, where one plate dips under another, the temperature and pressure will rise causing the rock structure to change or metamorphose,

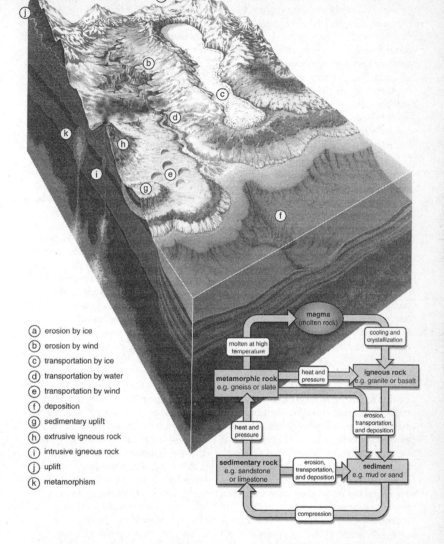

(a) erosion by ice

(b) erosion by wind

(c) transportation by ice

(d) transportation by water

(e) transportation by wind

(f) deposition

(g) sedimentary uplift

(h) extrusive igneous rock

(i) intrusive igneous rock

(j) uplift

(k) metamorphism

rock The rock cycle. Rocks are not as permanent as they seem but are being constantly destroyed and renewed. When a rock becomes exposed on the Earth's surface, it starts to break down through weathering and erosion. The resulting debris is washed or blown away and deposited, for example in sea or river beds, or in deserts, where it eventually becomes buried by yet more debris. Over time, this debris is compressed and compacted to form sedimentary rock, which may in time become exposed and eroded once more. Alternatively the sedimentary rock may be pushed further towards the Earth's centre where it melts and solidifies to form igneous rock or is heated and crushed to such a degree that its mineral content alters and it becomes metamorphic rock. Igneous and metamorphic rock may also become exposed and eroded by the same processes as sedimentary rock, and the cycle continues.

forming metamorphic rock. The huge pressures and high temperatures cause chemical changes in the minerals of the rock, and the new minerals to be compressed and aligned in a parallel pattern. Rocks buried deep in the crust will be heated by about 30°C per kilometre depth. *Metamorphic rock will also form next to an intrusion of *magma (molten rock), often found near a *volcano where the magma heats the surrounding rock. This type of change is called thermal or **contact metamorphism**. Examples of metamorphic rock are *slate and *marble.

completion of the rock cycle If the rock is heated until it melts then magma is formed. The material that once formed a mountain peak may now become the molten *lava forced out of an active volcano, and the rock cycle is complete. The timespan for these changes is enormous. Some of the oldest known rocks come from the *Precambrian era, and are at least 570 million years old.

The energy needed to power the rock cycle comes from the heat generated by the radioactive decay of elements in the material that makes up the *mantle (the zone between the Earth's crust and core). The *convection currents generated cause the semi-solid material of the mantle to rise, moving the tectonic plates, which causes fractures allowing volcanoes to form at the plate edges (see *plate tectonics).

Rostow model of development model of economic growth suggesting that all countries pass through a series of stages of development as their economies grow. US economist Walt Rostow presented this model in 1960 following a mainly European-based study.

Rostow described the first stage of development as **traditional society**. This is defined as subsistence economy based mainly on farming with very limited technology or capital to process raw materials or develop services and industries. **Preconditions for take-off**, the second stage, are said to take place when the levels of technology within a country develop and the development of a transport system encourages trade. During the next stage, **take-off**, manufacturing industries grow rapidly, airports, roads, and railways are built, and growth poles emerge as investment increases. Stage four is termed the **drive to maturity** during which growth should be self-sustaining, having spread to all parts of the country, and leading to an increase in the number and types of industry. During this stage more complex transport systems and manufacturing expand as transport develops, rapid *urbanization occurs, and traditional industries may decline. In Rostow's final stage, the **age of mass consumption**, rapid expansion of tertiary industries occurs alongside a decline in manufacturing.

rotation in astronomy, movement of a planet rotating about its own axis. For the *Earth, one complete rotation takes 23 hours and 56 minutes. The Earth rotates in an anticlockwise direction (as viewed looking along its axis from north to south), leading to the appearance from Earth of the Sun moving from east to west in the daily cycle. The rotation of the Earth produces a surface speed at the Equator of about 1600 kph. This speed decreases further north and south of the Equator. Artificial satellites use the Earth's natural rotation to orbit the Earth.

route the course taken between starting point and destination. For example, the route of the M1 motorway from London to Leeds is via Leicester and Sheffield. The study of routeways can be important to geographers in establishing the *accessibility of an area or examining the effects of a new road on the *environment.

day   night

Sun's rays

N

S

rotation The Earth rotates in an anticlockwise direction (looking along its axis from north to south) and takes approximately 24 hours to rotate on its own axis. Day and night depend on the Earth's rotation; the half of the Earth facing the Sun is in daylight and the half facing away from the Sun is in darkness.

ruby   red transparent gem variety of the mineral *corundum Al_2O_3, aluminium oxide. Small amounts of chromium oxide, Cr_2O_3, substituting for aluminium oxide, give ruby its colour. Natural rubies are found mainly in Myanmar (Burma), but rubies can also be produced artificially and such synthetic stones are used in lasers.

run-off   water that falls as rain and flows off the land, into streams or rivers.

rural depopulation   loss of people from remote country areas to cities; it is an effect of migration, due to the *pull factors of urban areas, such as employment opportunities, health care, and schools. In poor countries, large-scale migration to urban core regions may deplete the countryside of resources and workers. The population left behind will be increasingly aged, and agriculture declines. Rural depopulation may result in reduced village services.

Rural Development Commission UK government agency concerned with maintaining vigorous rural life and work. Its principal aim is the encouragement of employment and essential services in the countryside.

rural–urban continuum   a concept associated with the sociological typology of communities relating their social characteristics to the size, density and environment of a settlement. The term implies that in each kind of settlement in a continuum from the rural village to the urban city there is a characteristic way of life with different degrees of social stratification, stability, homogeneity and integration, and that the continuum indicates a process of social change.

The view that settlement patterns are associated with social characteristics and lifestyles has been attacked for failing to recognize that societies with essentially rural characteristics may be found in cities while those with urban values or attributes may occur in villages. Similarly, the process of social change may be evident within rural communities without any significant increase in settlement size. With only partial empirical support for the concept, it has taken on a wider if looser geographical meaning to represent the physical difficulty in recognizing a clear distinction between town and country. Thus, the idea of a rural-urban continuum supports the notion of a rural-urban fringe in which essentially urban ways of life mingle with, and merge into, rural ones.

rural–urban migration   the movement of people from rural to urban areas. See *migration and *rural depopulation. Compare *counter-urbanization.

rutile   titanium oxide mineral, TiO_2, a naturally occurring ore of titanium. It is usually reddish brown to black, with a very bright (adamantine) surface lustre. It crystallizes in the tetragonal system. Rutile is common in a wide range of igneous and metamorphic rocks and also occurs concentrated in sands; the coastal sands of eastern and western Australia are a major source. It is also used as a pigment that gives a brilliant white to paint, paper, and plastics.

S *abbreviation for* **south**.

SAARC *abbreviation for* **South Asian Association for Regional Cooperation**.

sabkha flat shoreline zone in arid regions above the high-water mark in which the sediments include large amounts of *evaporites. These occur in the form of nodules, crusts, and crystalline deposits of halite, anhydrite, and gypsum, as well as mineral grains of various sorts. Some of the evaporites form from rapid evaporation of marine waters soaking through from the bordering tidal flats, but some can be derived also from sediment-laden continental waters coming down from adjoining highlands.

Sahel (Arabic *sahil* 'coast') marginal area to the south of the Sahara, from Senegal to Somalia, which mostly has a thin natural vegetation of savannah grasses and shrubs and whose people live by herding, often nomadic, and by the cultivation of millet and ground nuts. The Sahel experiences desert-like conditions during periods of low rainfall and there are signs of progressive *desertification, partly due to climatic fluctuations but also caused by the pressures of a rapidly expanding population, which has led to overgrazing and the destruction of trees and scrub for fuelwood. In recent years many famines have taken place in the area.

The average rainfall in the Sahel ranges from 100–500 mm per year, but the rainfall over the past 30 years has been significantly below average. The resulting famine and disease are further aggravated by civil wars. The areas most affected are Ethiopia and the Sudan.

salinization accumulation of salt in water or soil; it is a factor in *desertification.

saltation in earth science, bouncing of rock particles along a river bed. It is the means by which *bedload (material that is too heavy to be carried in suspension) is transported downstream. The term is also used to describe the movement of sand particles bounced along by the wind.

salt, common or **sodium chloride**; NaCl, white crystalline solid, found dissolved in seawater and as rock salt (the mineral halite) in large deposits and salt domes. Common salt is used extensively in the food industry as a preservative and for flavouring, and in the chemical industry in the making of chlorine and sodium.

salt marsh wetland with halophytic vegetation (tolerant to seawater). Salt marshes develop around *estuaries and on the sheltered side of sand and shingle *spits. They are formed by the deposition of mud around salt-tolerant vegetation. This vegetation must tolerate being covered by seawater as well as being exposed to the air. It also traps mud as the tide comes in and out. This helps build up the salt marsh. Salt marshes usually have a network of creeks and drainage channels by which tidal waters enter and leave the marsh.

Typical plants of European salt marshes include salicornia, or saltwort, which has fleshy leaves like a succulent; sea lavender, sea pink, and sea aster. Geese such as brent, greylag, and bean are frequent visitors to salt marshes in winter, feeding on plant material.

salt pan a layer of salts formed by salinization caused by excessive and wrongly timed irrigation water, followed by a lengthy hot, dry phase. These conditions result in a substantial upward movement of salts which accumulate to form a salt pan. In the worst examples (as occurred soon after the expansion of irrigation in Egypt following the construction of the Aswan High Dam) the salt pan forms a toxic layer which kills agricultural crops.

sample small number of items drawn from a larger group and intended to be representative of that group. In market research, small groups of people are chosen to be representatives of a particular demographic group. Samples may be selected by quota, in order to reflect accurately the composition of the groups or markets they were drawn from, or they may be selected randomly. The term can also describe the small amount of a product given to potential retailers and consumers.

sample size size of a *sample. The sample size affects the accuracy of the results given by the sample. The larger the sample, the more accurate the results; this is because the sample is approaching more closely the group it was drawn from. It is important to remember when drawing conclusions from a sample that there is always a degree of error involved in sample selection.

sampling in statistics, taking a small number of cases and assuming that they are representative of the total number. In survey work in *market research, for example, it is usually not practical to survey every customer. It is also important in quality control to avoid the expense of testing every item produced. Instead, a sample is taken. A **random sample** is one where the sample is chosen in a random way.

San Andreas Fault geological fault stretching for 1125 km northwest–southeast through the state of California, USA. It marks a conservative plate margin, where two plates slide past each other (see *plate tectonics).

Friction is created as the coastal Pacific plate moves northwest, rubbing against the American continental plate, which is moving slowly southeast. The relative movement is only about 5 cm a year, which means that Los Angeles will reach San Francisco's latitude in 10 million years. The friction caused by the tectonic movement gives rise to frequent, destructive *earthquakes. For example, in 1906 an earthquake originating from the fault almost destroyed San Francisco. Contemporary official figures put the death toll at 700–800, although later estimates put the figure closer to 3000 deaths caused by the quake and its aftermath.

sanction economic or military measure taken by a state or number of states to enforce international law. The first use of sanctions, as a trade *embargo, was the attempted economic boycott of Italy 1935–36 during the Abyssinian War by the League of Nations.

sand loose grains of rock, 0.0625–2.00 mm in diameter, consisting most commonly of *quartz, but owing their varying colour to mixtures of other minerals. Sand is used in cement-making, as an abrasive, in glass-making, and for other purposes.

Sands are classified into marine, freshwater, glacial, and terrestrial. Some 'light' soils contain up to 50% sand. Sands may eventually consolidate into *sandstone.

sandbar ridge of sand built up by he currents across the mouth of a river or bay. A sandbar may be entirely underwater or it may form an elongated island that breaks the surface. A sandbar stretching out from a headland is a **sand spit**.

Coastal bars can extend across estuaries to form **bay bars**.

sand dune see *dune.

sandstone *sedimentary rocks formed from the consolidation of sand, with sand-sized grains (0.0625–2 mm) in a matrix or cement. Their principal component is quartz. Sandstones are commonly permeable and porous, and may form freshwater *aquifers. They are mainly used as building materials.

Sandstones are classified according to the matrix or cement material (whether derived from clay or silt).

sandur see *outwash plain.

Santa Ana periodic warm Californian *wind.

sapphire deep-blue, transparent gem variety of the mineral *corundum Al_2O_3, aluminium oxide. Small amounts of iron and titanium give it its colour. A corundum gem of any colour except red (which is a ruby) can be called a sapphire; for example, yellow sapphire.

sard yellow or red-brown variety of *chalcedony.

satellite any small body that orbits a larger one. **Natural satellites** that orbit planets are called **moons**. The first **artificial satellite**, Sputnik 1, was launched into orbit around the Earth by the USSR in 1957. Artificial satellites can transmit data from one place on Earth to another, or from space to Earth. *Satellite applications include science, communications, weather forecasting, and military use.

satellite applications uses to which artificial satellites are put. These include:

reconnaissance, land resource, and mapping applications Apart from military use and routine mapmaking, the US Landsat, the French Satellite Pour l'Observation de la Terre, and equivalent Russian satellites have provided much useful information about water sources and drainage,

vegetation, land use, geological structures, oil and mineral locations, and snow and ice.

weather monitoring The US National Oceanic and Atmospheric Administration series of satellites, and others launched by the European Space Agency, Japan, and India, provide continuous worldwide observation of the atmosphere.

navigation The US Global Positioning System (GPS) uses 24 Navstar satellites that enable users (including walkers and motorists) to find their position to within 100 m/328 ft. The US military can make full use of the system, obtaining accuracy to within 1.5 m/4 ft 6 in. The Transit system, launched in the 1960s, with 12 satellites in orbit, locates users to within 100 m/328 ft.

communications A complete worldwide communications network is now provided by satellites such as the US-run Intelsat system.

scientific experiments and observation Many astronomical observations are best taken above the disturbing effect of the atmosphere. Satellite observations have been carried out by the Infrared Astronomy Satellite (1983) which made a complete infrared survey of the skies, and Solar Max (1980), which observed solar flares. The Hipparcos satellite, launched in 1989, measured the positions of many stars. The Röntgen Satellite, launched in 1990, examined ultraviolet and X-ray radiation. In 1992, the Cosmic Background Explorer satellite detected details of the Big Bang that mark the first stage in the formation of galaxies. Medical experiments have been carried out aboard crewed satellites, such as the Russian space station *Mir* and the US *Skylab*.

satellite image an image giving information about an area of the Earth or another planet, obtained from a satellite. Satellite images can provide a variety of information, including vegetation patterns, sea surface temperature, weather, and geology.

Landsat 4, launched in 1982, orbits at 705 km above the Earth's surface. It completes nearly 15 orbits per day, and can survey the entire globe in 16 days. Instruments on Landsat continually scan the Earth and sense the brightness of reflected light. When the information is sent back to Earth, computers turn it into *false colour images* in which built-up areas appear in one colour (perhaps blue), vegetation in another (often red), bare ground in a third, and water in a fourth colour, making it easy to see their distribution and to monitor any changes, such as *deforestation. Satellite photography can be used to measure the depth of shallow water, to monitor the health and development of forestry and agricultural crops, and as an aid in oil and mineral exploration.

saturated adiabatic lapse rate (SALR) the rate of heat loss which occurs from a rising, saturated air mass. The SALR can vary depending upon the temperature of the air mass and the associated water vapour content. A minimum value of 0.4°C per 100 m to a maximum of 0.9°C per 100 m can be recorded. See *dry adiabatic lapse rate.

savannah or **savanna**, extensive open tropical grasslands, with scattered trees and shrubs. Savannahs cover large areas of Africa, North and South America, and northern Australia. The soil is acidic and sandy and generally considered suitable only as pasture for low-density grazing.

A new strain of rice suitable for savannah conditions was developed in 1992. It not only grew successfully under test conditions in Colombia but also improved pasture quality so that grazing numbers could be increased twentyfold.

scale the size ratio represented by a map; for example, on a map of scale 1:25000 the real landscape is portrayed at 1/25000 of its actual size.

scapolite group of white or greyish minerals, silicates of sodium, aluminium, and calcium, common in metamorphosed limestones and forming at high temperatures and pressures.

scarcity in economics, insufficient availability of *resources to satisfy wants. The use of scarce resources has an *opportunity cost.

scarp and dip in geology, the two slopes that comprise an escarpment. The scarp is the steep slope and the dip is the gentle slope. Such a feature is common when sedimentary rocks are uplifted, folded, or eroded, the scarp slope cuts across the bedding planes of the sedimentary rock while the dip slope follows the direction of the strata. An example is Salisbury Crags in Edinburgh, Scotland.

scatter diagram a graph showing the scores of a set of individuals or items on two variables. The *line of best fit* runs through the points so that the sum of the squares of the vertical distances (or offsets) from the points on to the line is minimal. The purpose of the graph is to indicate visually the general trend of the data, i.e. the degree of relationship, or correlation, between the two variables applying to that set of individuals or items.

When a unique position for the line cannot be found, because of the purely random distribution of points, then no relationship exists and the correlation value is zero.

The correlation may be positive: the scores on both variables increase or decrease together; or negative: the scores on one of the variables increase while those of the other decrease.

scavenger in ecology, animal that feeds on plant or animal remains. Crows, woodlice, and vultures are examples.

Rhinoceros
Vulture
Gazelle
Baobob tree
Wildebeest
Acacia tree

Giraffe
Dik dik
Zebra
Termite mound
Termite
Elephant
Lion
Buffalo

savannah Composite to show the diversity of wildlife found on the African savannah. The savannah habitat supports large herds of grazers and their large carnivorous predators.

Schengen Group association of states in Europe that in theory adhere to the ideals of the Schengen Convention, notably the abolition of passport controls at common internal borders and the strengthening of external borders. The Convention, which first came into effect in March 1995, was signed by Belgium, France, Germany, Luxembourg, and the Netherlands in 1985; Italy in November 1990; Portugal and Spain in 1991; Greece in 1992; Austria in 1995; and Denmark, Sweden, Finland, and the first two non-European Union (EU) countries, Iceland and Norway in 2001.

schist *metamorphic rock containing *mica; its crystals are arranged in parallel layers or bands. Schist may contain additional minerals such as *garnet.

The *sedimentary rocks mudstone and shale become *slate under stress. When slate is subjected to higher temperatures and pressure schist is formed. The temperature causes recrystallization of new minerals, such as mica, and the pressure causes the new minerals to be aligned in bands.

science park an area developed for the use of firms engaged in research and development in advanced technology. Science parks tend to be located on *greenfield and/or landscaped sites. Commercial development is closely linked to fundamental research taking place in universities and the provision of science parks in university towns is common.

The provision of these parks has been an attempt to facilitate a trend observed in the USA since the early 1970s (especially in 'Silicon Valley' in California) for high-tech industries to cluster near advanced research establishments. In the UK, science parks developed spontaneously in areas of high growth such as Cambridge well before financial incentives in the form of government grants were made available to encourage the setting up of science parks in less favoured locations. Compare *business park.

scientific management or **Taylorism**, approach to management asserting that measurement improves productivity. Scientific management is associated with its originator, industrial engineer and inventor Frederick Winslow Taylor. Taylor, chief engineer at the Midvale Steel Works at the beginning of the 20th century, noticed that his fellow workers were 'soldiering', that is deliberately working at a slow pace. The workers were only able to do this because no one had thought to study the process of work and how long the tasks the men were carrying out should actually take. Armed with a stopwatch, Taylor set about measuring the time individual tasks should take, making him a favourite of managers, and the enemy of workers across the globe. Employees were, according to Taylor, nothing more than components in a machine.

Scottish Natural Heritage Scottish nature conservation body formed in 1991 after the break-up of the national *Nature Conservancy Council. It is government-funded.

scree pile of rubble and sediment that collects at the foot of a mountain range or cliff. The rock fragments that form scree are usually broken off by the action of frost (*freeze–thaw weathering).

With time, the rock-waste builds up into a heap or sheet of rubble that may eventually bury even the upper cliffs, and the growth of the scree then stops. Usually, however, erosional forces remove the rock waste so that the scree stays restricted to lower slopes.

seabird wreck the washing ashore of significantly larger numbers of seabirds than would be expected for the time of year.

In February 1994 around 75,000 birds were washed up along the east coast of Britain. They were mostly fish-eating species, such as guillemots, shags, and razorbills, and appeared to have died of starvation. The cause of seabird wrecks is unknown though overfishing could be a contributory factor.

sea breeze gentle coastal wind blowing off the sea towards the land. It is most noticeable in summer when the warm land surface heats the air above it and causes it to rise. Cooler air from the sea is drawn in to replace the rising air, so causing an onshore breeze. At night and in winter, air may move in the opposite direction, forming a *land breeze.

seafloor spreading growth of the ocean *crust outwards (sideways) from ocean ridges. The concept of seafloor spreading has been combined with that of continental drift and incorporated into *plate tectonics.

Seafloor spreading was proposed in 1960 by US geologist Harry Hess (1906–69), based on his observations of ocean ridges and the relative youth of all ocean beds. In 1963 British geophysicists Fred Vine and Drummond Matthews observed that the floor of the Atlantic Ocean was made up of rocks that could be

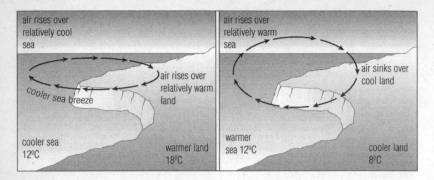

air rises over relatively cool sea

air rises over relatively warm land

cooler sea breeze

cooler sea 12°C

warmer land 18°C

air rises over relatively warm sea

air sinks over cool land

warmer sea 12°C

cooler land 8°C

sea breeze The direction of a sea breeze can vary depending on which is warmer, the land or the sea. Warm air rises and is replaced by cool air. When the land is warmer than the sea, the warm air rises drawing in the cooler air from the sea, creating an onshore breeze. The opposite happens when the sea is warmer than the land, drawing cooler air from the land creating an offshore or land breeze.

arranged in strips, each strip being magnetized either normally or reversely (due to changes in the Earth's polarity when the North Pole becomes the South Pole and vice versa, termed *polar reversal). These strips were parallel and formed identical patterns on both sides of the ocean ridge. The implication was that each strip was formed at some stage in geological time when the magnetic field was polarized in a certain way. The seafloor magnetic-reversal patterns could be matched to dated magnetic reversals found in terrestrial rock. It could then be shown that new rock forms continuously and spreads away from the ocean ridges, with the oldest rock located farthest away from the midline. The observation was made independently in 1963 by Canadian geologist Lawrence Morley, studying an ocean ridge in the Pacific near Vancouver Island.

Confirmation came when sediments were discovered to be deeper further away from the oceanic ridge, because the rock there had been in existence longer and had had more time to accumulate sediment.

sea level average height of the surface of the oceans and seas measured throughout the tidal cycle at hourly intervals and computed over a 19-year period. It is used as a datum plane from which elevations and depths are measured.

Factors affecting sea level include temperature of seawater (warm water is less dense and therefore takes up a greater volume than cool water) and the topography of the ocean floor.

Rising sea levels and erosion have claimed large portions of the British coastline; in 1996 it was reported that 29 villages had disappeared along the Yorkshire coast since 1926 as a result of tidal battering. Rising sea levels present a particularly severe problem on Britain's east coast, where the land has been sinking by half a centimetre a year.

seamount in earth science, isolated volcanic mount on the sea bed at least 700 m tall. Seamounts vary from smallish conical peaks to large flat-topped masses (*guyots) and may occur singly or in chains or groups.

season period of the year having a characteristic climate. The change in seasons is mainly due to the Earth's axis being tilted in relation to the Sun, and hence the position of the Sun in the sky at a particular place changes as the Earth orbits the Sun.

When the northern hemisphere is tilted away from the Sun (winter) the Sun's rays have further to travel through the atmosphere (they strike the Earth at a shallower angle) and so have less heating effect, resulting in colder weather. The days are shorter and the nights are longer. At the same time, the southern hemisphere is tilted towards the Sun (summer) and experiences warmer weather, with longer days and shorter nights. The opposite occurs when the northern hemisphere is tilted towards the Sun and the southern hemisphere away from the Sun.

In temperate latitudes four seasons are recognized: spring, summer, autumn (fall), and winter. Tropical regions have two seasons – the wet and the dry. Monsoon areas around the Indian Ocean have three seasons: the cold, the hot, and the rainy.

The differences between the seasons are more marked inland than near the coast, where the sea has a moderating effect on temperatures. In polar regions the change between summer and winter is abrupt; spring and autumn are hardly perceivable. In tropical regions, the belt of rain associated with the trade winds moves north and south with the Sun, as do the dry conditions associated with the belts of high pressure near the tropics. The monsoon's three seasons result from the influence of the Indian Ocean on the surrounding land mass of Asia in that area.

seasonal demand   demand for products which varies according to the time of year. For many products in the UK, demand is highest at the Christmas period. For cars, demand is highest in August when the registration letter changes. Seasonal demand poses problems for businesses because they have to build up stocks for sale during the period of peak demand. Stocks are expensive to keep. Firms are also caught with too many

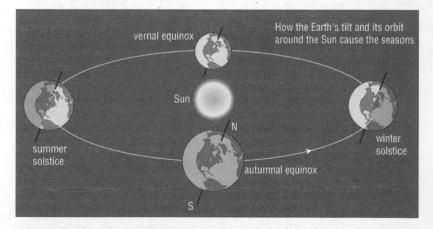

season The cause of the seasons. As the Earth orbits the Sun, its axis of rotation always points in the same direction. This means that, during the northern hemisphere summer solstice (usually 21 June), the Sun is overhead in the northern hemisphere. At the northern hemisphere winter solstice (usually 22 December), the Sun is overhead in the southern hemisphere.

stocks if peak demand sales fail to live up to expectation, but sales are lost if peak demand exceeds expectations.

seasonal unemployment unemployment arising from the seasonal nature of some economic activities. An example is agriculture, which uses a smaller labour force in winter.

seawater water of the seas and oceans, covering about 70% of the Earth's surface and comprising about 97% of the world's water (only about 3% is freshwater). Seawater contains large numbers of electrically charged particles, or ions. These may be positively or negatively charged. The most common positive ions are sodium, potassium, magnesium, and calcium. These positive ions are balanced by negative ions, for example chloride, iodide, fluoride, bromide, sulphates, carbonates, bicarbonates, phosphates, nitrates, and others. As a result the physical chemistry of seawater is extremely complex.

As seawater evaporates, the positive and negative ions are attracted to one another to form crystals of various types, the most common of these being sodium chloride (common salt). Seawater also contains a large amount of dissolved carbon dioxide, and thus the oceans act as carbon 'sinks' that may help to reduce the greenhouse effect.

secession Latin *secessio*, in politics, the withdrawal from a federation of states by one or more of its members, as in the secession of the Confederate states from the Union in the USA 1860, Singapore from the Federation of Malaysia 1965, and Croatia and Slovenia from the Yugoslav Federation 1991.

secondary consumer in ecology, animals (primary carnivores) that eat herbivores.

secondary industry manufacturing from raw materials; see *industrial sector.

secondary market market for resale of purchase or shares, bonds, and commodities outside of organized stock exchanges and primary markets.

secondary source a supply of information or data that has been researched or collected by an individual or group of people and made available for others to use; *census data is an example of this. A *primary source* of data or information is one collected at first hand by the researcher who needs it; for example, a traffic count in an area, undertaken by a student for his or her own project.

Second World former term for the industrialized *communist countries of the Soviet Union and Eastern bloc, used by the West during the Cold War, alongside the terms First World (industrialized free-market countries of the West) and Third World (non-aligned, developing nations). Originally denoting political alignment, the classifications later took on economic connotations. The terms have now lost their political meaning, and are considered derogatory.

sector model or *Hoyt model A model of urban structure developed by H. Hoyt in 1939 and based on an analysis of the land-use patterns of 142 American cities.

The model differs from the *Burgess model in allowing sectorial development along major lines of communication. Thus, zone 2 (industry) and its related workers' housing zone (3) will develop along, for example, a river valley which favoured canal and railway expansion. Zone 5, high-quality housing, will extend along ridges of high ground or other pleasant environmental corridors. Whilst Hoyt's model is a step closer to reality, the *multiple-nuclei theory is more so. No model will ever be an entirely successful simulation of real urban land-use patterns, as the local factors responsible for the structure of a given city will be unique to that location. Compare *shanty town.

sector theory   model of urban land use in which the various land-use zones are shaped like wedges radiating from the central business district. According to sector theory, the highest prices for land are found along transport routes (especially roads), and once an area has gained a reputation for a particular type of land use (such as industry), it will attract the same land users as the city expands outward.

secular variation   changes in the position of Earth's magnetic poles measured with respect to geographical positions, such as the North Pole, throughout geological time.

Security Council   the most important body of the United Nations; see *United Nations.

sediment   any loose material that has 'settled' after deposition from suspension in water, ice, or air, generally as the water current or wind speed decreases. Typical sediments are, in order of increasing coarseness: clay, mud, silt, sand, gravel, pebbles, cobbles, and boulders.

Sediments differ from *sedimentary rocks, in which deposits are fused together in a solid mass of rock by a process called *lithification (solidification). Pebbles are cemented into *conglomerates; sands become sandstones; muds become mudstones or shales; peat is transformed into coal.

sedimentary rock   rock formed by the accumulation and *cementation of deposits that have been laid down by water, wind, ice, or gravity. Sedimentary rocks cover more than two-thirds of the Earth's surface and comprise three major categories: clastic, chemically precipitated, and organic (or biogenic). Clastic sediments are the largest group and are

composed of fragments of pre-existing rocks; they include clays, sands, and gravels.

Chemical precipitates include some limestones and evaporated deposits such as gypsum and halite (rock salt). Coal, oil shale, and limestone made of fossil material are examples of organic sedimentary rocks.

Most sedimentary rocks show distinct layering (stratification), because they are originally deposited as more or less horizontal layers.

Sedimentary rocks are categorized by the grain sizes of the particles: *conglomerate rocks may contain rock pieces over 2 mm in diameter; *sandstone particles are up to 0.2 mm; siltstone up to 0.02 mm; and claystone less than 0.02 mm in diameter.

sedimentology   branch of geology dealing with the structure, composition, and evolution of *sedimentary rocks, and what they can tell us about the history of the Earth.

seed bank   repository for seeds. The seeds of crop plants and wild species of plants threatened with extinction are kept in cold storage, forming a reference collection that is an invaluable source of genetic material for breeding and conservation programmes. The world's largest seed bank is connected to Kew Gardens, the popular name for the Royal Botanic Gardens at Kew, Surrey, in England.

segmentation   see *market segmentation.

seiche   pendulous movement seen in large areas of water resembling a *tide. It was originally observed on Lake Geneva and is created either by the wind, earth tremors or other atmospheric phenomena.

seif dune   or **longitudinal dune**, in earth science, long narrow ridge of sand with sinuous crests, aligned for up to 10 km along the direction of the prevailing wind.

seismic gap theory theory that along faults that are known to be seismically active, or in regions of high seismic activity, the locations that are more likely to experience an *earthquake in the relatively near future are those that have not shown seismic activity for some time. When records of past earthquakes are studied and plotted onto a map, it becomes possible to identify **seismic gaps** along a fault or plate margin. According to the theory, an area that has not had an earthquake for some time will have a great deal of stress building up, which must eventually be released, causing an earthquake.

Although the seismic gap theory can suggest areas that are likely to experience an earthquake, it does not enable scientists to predict when that earthquake will occur.

seismic wave energy wave generated by an *earthquake or an artificial explosion. There are two types of seismic waves: **body waves** that travel through the Earth's interior; and **surface waves** that travel through the surface layers of the crust and can be felt as the shaking of the ground, as in an earthquake.

Seismic waves show similar properties of reflection and refraction as light and sound waves. Seismic waves change direction and speed as they travel through different densities of the Earth's rocks.

body waves There are two types of body waves: P-waves and S-waves, so-named because they are the primary and secondary waves detected by a seismograph. **P-waves**, or compressional waves, are longitudinal waves (wave motion in the direction the wave is travelling), whose compressions and rarefactions resemble those of a sound wave. **S-waves** are transverse waves or shear waves, involving a back-and-forth shearing motion at right angles to the direction the wave is travelling.

Because liquids have no resistance to shear and cannot sustain a shear wave, S-waves cannot travel through liquid material. The Earth's outer core is believed to be liquid because S-waves disappear at the mantle-core boundary, while P-waves do not.

surface waves Surface waves travel in the surface and subsurface layers of the crust. **Rayleigh waves** travel along the free surface (the uppermost layer) of a solid material. The motion of particles is elliptical, like a water wave, creating the rolling motion often felt during an earthquake. **Love waves** are transverse waves trapped in a subsurface layer due to different densities in the rock layers above and below. They have a horizontal side-to-side shaking motion transverse (at right angles) to the direction the wave is travelling.

seismogram or **seismic record**, trace, or graph, of ground motion over time, recorded by a seismograph. It is used to determine the magnitude and duration of an earthquake.

seismograph instrument used to record ground motion. A heavy inert weight is suspended by a spring and attached to this is a pen that is in contact with paper on a rotating drum. During an earthquake the instrument frame and drum move, causing the pen to record a zigzag line on the paper; held steady by inertia, the pen does not move.

seismology study of *earthquakes, the seismic waves they produce, the processes that cause them, and the effects they have. By examining the global pattern of waves produced by an earthquake, seismologists can deduce the nature of the materials through which they have passed. This leads to an understanding of the Earth's internal structure.

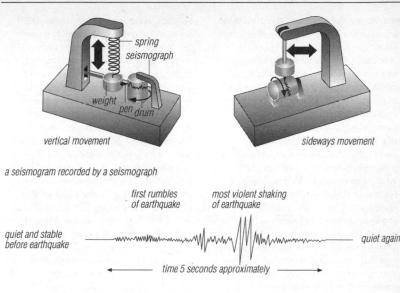

vertical movement   sideways movement

a seismogram recorded by a seismograph

first rumbles
of earthquake

most violent shaking
of earthquake

quiet and stable
before earthquake

quiet again

◄──────────── *time 5 seconds approximately* ────────────►

seismograph A seismogram, or recording made by a seismograph. Such recordings are used to study earthquakes and in prospecting.

On a smaller scale, artificial earthquake waves, generated by explosions or mechanical vibrators, can be used to search for subsurface features in, for example, oil or mineral exploration. Earthquake waves from underground nuclear explosions can be distinguished from natural waves by their shorter wavelength and higher frequency.

self-help project any scheme for a community to help itself under official guidance. The most popular self-help projects in the developing world are aimed at improving conditions in *shanty towns. Organized building lots are commonly provided, together with properly laid-out drains, water supplies, roads, and lighting. *Squatters are expected to build their own homes on the prepared sites, perhaps with loans provided by the government or other agencies. An example is the Arumbakkam scheme in Madras (now Chennai), India, begun 1977. Alternatively, 'basic shell' housing may be provided, as in parts of São Paulo, Brazil, and Colombia.

self-sufficiency situation where an individual or group does not rely on outsiders. Economic self-sufficiency means that no trade takes place between the individual or group and others. If an economy were self-sufficient, it would not export or import. For a family to be self-sufficient, for example, it would have to grow all its own food, make its own clothes, and provide all its own services. In a modern economy, there is very little self-sufficiency because *specialization enables individuals to enjoy a much higher standard of living than if they were self-sufficient.

selva equatorial rainforest, such as that in the Amazon basin in South America.

serac a pinnacle of ice formed by the tumbling and shearing of a *glacier at an *ice fall, i.e. the broken ice associated with a change in *gradient of the valley floor.

sere plant *succession developing in a particular habitat. A **lithosere** is a succession starting on the surface of bare rock. A **hydrosere** is a succession in shallow freshwater, beginning with planktonic vegetation and the growth of pondweeds and other aquatic plants, and ending with the development of swamp. A **plagiosere** is the sequence of communities that follows the clearing of the existing vegetation.

serial entrepreneur entrepreneur who, after successfully creating one company, moves on to create another, and so on. The ranks of serial entrepreneurs include Internet pioneers such as US computer scientist Jim Clark, US systems developer Marc Andreessen, and Indian businessman Sabeer Bhatia, who co-founded Hotmail.

serpentine member of a group of minerals, hydrous magnesium silicate, $Mg_3Si_2O_5(OH)_4$, occurring in soft *metamorphic rocks and usually dark green. The fibrous form **chrysotile** is a source of *asbestos; other forms are **antigorite** and **lizardite**. Serpentine minerals are formed by hydration of ultramafic rocks during metamorphism. Rare snake-patterned forms are used in ornamental carving.

service industry or **tertiary industry**, sector of the economy that supplies services such as retailing, banking, and education.

set-aside scheme policy introduced by the European Community (now the European Union) in the late 1980s, as part of the Common Agricultural Policy, to reduce overproduction of certain produce. Farmers are paid not to use land but to keep it fallow. The policy may bring environmental benefits by limiting the amount of fertilizers and pesticides used.

settlement collection of dwellings forming a community. There are many different types of settlement and most owe their origin to historical and geographical factors. The growth and development of a settlement is greatly influenced by its location, site, situation, and function. Human settlements can be identified as centres that function as marketplaces, administrative centres, and social and cultural meeting places serving surrounding hinterlands.

The first settlements probably developed between 13,000 and 10,000 years ago, when groups of nomadic hunter-gatherers in the Middle East adopted more settled ways of life and developed social organizations based on larger, more formal groups. These societies all developed in areas where food supplies such as fish and easily gathered food plants were readily available, and the social changes may have come about in part because some

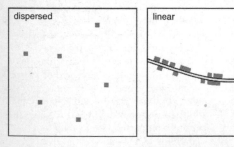
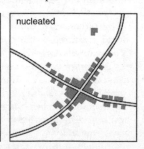

settlement Three basic shapes of rural settlements. Nucleated settlements have a defined centre, such as a market square or crossroads. Linear settlements, also known as ribbon settlements, are long and narrow and often grow up along a road. The buildings in a dispersed settlement tend to be scattered over a wide area with undeveloped land in between.

of the grain plants became more plentiful towards the end of the last ice age, about 10,000 years ago. The technology and social organizations of some of these more advanced societies served as a foundation for later farming settlements. It is thought that the first settlements in Britain probably originated as 'base-camp' sites for small clans or family groups of Mesolithic hunter-gatherers who, in the course of a year, progressed round a circuit that formed the framework of their territory. Within this area they would have known where particular food resources could be located according to the season.

settlement functions   the main activities of a *settlement. As a settlement grows in size it often develops specific functions. These relate to economic and social development. Functions may change over time and the growth of a settlement may have been based on an activity that no longer exists, such as coalmining. Many towns and cities in more developed countries are multifunctional.

settlement hierarchy   the arrangement of settlements within a given area in order of importance. Hamlets form the base of the hierarchy followed by villages, small towns, large towns, and cities. A country's capital or largest city forms the top.

The range and number of services in a settlement is proportionate to the size of the population and may determine a settlement's *sphere of influence.

sewage disposal   the disposal of human excreta and other waterborne waste products from houses, streets, and factories. Conveyed through sewers to sewage works, sewage has to undergo a series of treatments to be acceptable for discharge into rivers or the sea, according to various local laws. Raw sewage, or sewage that has

not been treated adequately, is one serious source of water pollution and a cause of *eutrophication.

In the industrialized countries of the West, most industries are responsible for disposing of their own wastes. Government agencies establish industrial waste-disposal standards. In most countries, sewage works for residential areas are the responsibility of local authorities. The solid waste (sludge) may be spread over fields as a fertilizer or, in a few countries, dumped at sea. A significant proportion of bathing beaches in densely populated regions have unacceptably high bacterial content, largely as a result of untreated sewage being discharged into rivers and the sea. This can, for example, cause stomach upsets in swimmers.

In Europe and North America 30–60% of sludge is spread on agricultural land. The use of raw sewage as a fertilizer (long practised in China) has the drawback that disease-causing micro-organisms can survive in the soil and be transferred to people or animals by consumption

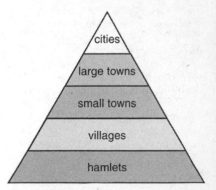

settlement hierarchy The size and range of functions (services, transport, and so on) of a settlement determine its position in the hierarchy. Hamlets are small in size and have a very small range of functions while cities are large in size and have a wide range of facilities, goods, and services.

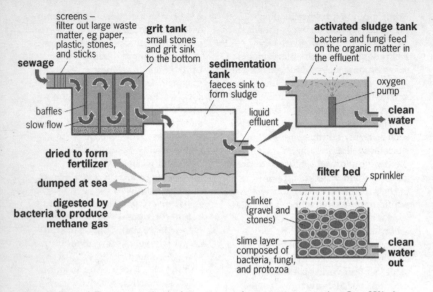

sewage disposal The processes involved in treatment of sewage at a sewage plant. Over 95% of sewage is water and once the solids have been removed and the water cleaned, it can be discharged into rivers.

of subsequent crops. Sewage sludge is safer, but may contain dangerous levels of heavy metals and other industrial contaminants.

shading map see *choropleth map.

shaft mine a vertical mine. Compare *drift mine.

shale fine-grained and finely layered *sedimentary rock composed of silt and clay. It is a weak rock, splitting easily along bedding planes to form thin, even slabs (by contrast, mudstone splits into irregular flakes). Oil shale contains kerogen, a solid bituminous material that yields *petroleum when heated.

shanty town group of unplanned shelters constructed from cheap or waste materials (such as cardboard, wood, and cloth). Shanty towns are commonly located on the outskirts of cities in poor countries, or within large cities on derelict land or near rubbish tips.

Land available for shanties is often of poor quality (for example, too steep or poorly drained). This makes them vulnerable to natural disasters such as the mudslides in Rio de Janeiro, Brazil, in 1988. Shanty areas often lack such services as running water, electricity, and sanitation. They are crowded areas, with high-density *populations of squatters (people who have no right to the land). In the developing world, shanty towns result from mass migration from rural areas in response to *pull factors, especially the perceived prospects of employment. One solution to uncontrolled shanty development is *self-help projects.

Shanty towns are sometimes referred to by words in or derived from the local language: in Brazil, *favela*; in India, especially around Kolkata (formerly Calcutta), *basti*; in Peru, *colonia proletaris*; in Tunisia, *gourbi* or *bidonville*; and in Venezuela, *rancho*.

In Cairo, Egypt, squatters have taken over the City of the Dead, which was originally built as a burial ground.

sharecropping a form of land tenure in which the tenant pays the landlord a share of the crop produced rather than a monetary payment for the right to use the land. The landlord generally provides much of the movable capital such as farm equipment, seed and stock, as well as the fixed capital of the land and buildings, while the tenant farmer provides labour.

Efforts to reform or regulate sharecropping have focused on the issues of security of tenure and limiting the share paid to the landowner. The system is characterized by short-term leases, or as in many parts of the developing world, by no more than oral agreements; tenants thus have little incentive to invest in the land or engage in long-term land use planning. Maximum shares have become fixed in law in most developed countries and in areas where effective land reform has been implemented.

shelf sea relatively shallow sea, usually no deeper than 200 m, overlying the continental shelf around the coastlines. Most fishing and marine mineral exploitations are carried out in shelf seas.

shell in zoology, hard outer covering of a wide variety of invertebrates. The covering is usually mineralized, normally with large amounts of calcium. The shell of birds' eggs is also largely made of calcium.

shield in geology, alternative name for *craton, the ancient core of a continent.

shield volcano broad, flat *volcano formed at a *constructive margin between tectonic plates or over a *hot spot. The *magma (molten rock) associated with shield volcanoes is usually basalt – thin and free-flowing. An example is Mauna Loa on the Pacific island of Hawaii. A *composite

volcano, on the other hand, is formed at a destructive margin.

Shield volcanoes are found along the rift valleys and ocean ridges of constructive plate margins. The lava flows for some distance over the surface before it sets, so forming low, broad volcanoes. The lava of a shield volcano is not ejected violently, but simply flows over the crater rim.

shifting cultivation see *bush fallowing.

shift work method of organizing work so that the same machinery or equipment (capital) is used at least twice during a 24-hour period by different groups or shifts of workers.

shoreface terrace a bank of *sediment accumulating at the change of slope which marks the limit of a marine *wave-cut platform.

Material removed from the retreating cliff base is transported by the *undertow off the wave-cut platform to be deposited in deeper water offshore.

sial in geochemistry and geophysics, the substance of the Earth's continental *crust, as distinct from the *sima of the ocean crust. The name, now used rarely, is derived from silica and alumina, its two main chemical constituents. Sial is often rich in granite.

Sierra Club US environmental organization, founded in 1892. Its mission is to protect the environment through congressional lobbying, direct action, and education. Campaigns have been run on US issues such as urban development, the commercial logging of forests, and water contamination, and global concerns such as global warming, population growth, and responsible trade.

The Sierra Club also organizes outings to wild areas in order to promote environmental awareness, produces books on environmental issues, and runs local action groups throughout the USA.

silage any *fodder crop harvested whilst still green. The crop is kept succulent by partial fermentation in a *silo*. It is used as animal feed during the winter.

silica silicon dioxide, SiO_2, the composition of the most common mineral group, of which the most familiar form is quartz. Other silica forms are *chalcedony, chert, opal, tridymite, and cristobalite.

Common sand consists largely of silica in the form of quartz.

silicate one of a group of minerals containing silicon and oxygen in tetrahedral units of SiO_4, bound together in various ways to form specific structural types. Silicates are the chief rock-forming minerals. Most rocks are composed, wholly or in part, of silicates (the main exception being limestones). Glass is a manufactured complex polysilicate material in which other elements (boron in borosilicate glass) have been incorporated.

sill 1. sheet of igneous rock created by the intrusion of magma (molten rock) between layers of pre-existing rock. (A *dyke, by contrast, is formed when magma cuts *across* layers of rock.) An example of a sill in the UK is the Great Whin Sill, which forms the ridge along which Hadrian's Wall was built.

A sill is usually formed of **dolerite**, a rock that is extremely resistant to erosion and weathering, and often forms ridges in the landscape or cuts across rivers to create *waterfalls.
2. (also called threshold) the lip of a *corrie.

sillimanite aluminium silicate, Al_2SiO_5, a mineral that occurs either as white to brownish prismatic crystals or as minute white fibres. It is an indicator of high-temperature conditions in metamorphic rocks formed from clay sediments.

Andalusite, kyanite, and sillimanite are all polymorphs of Al_2SiO_5.

silt sediment intermediate in coarseness between clay and sand; its grains have a diameter of 0.002–0.02 mm. Silt is usually deposited in rivers, and so the term is often used generically to mean a river deposit, as in the silting-up of a channel.

Silurian Period period of geological time 439–409 million years ago, the third period of the Palaeozoic era. Silurian sediments are mostly marine and consist of shales and limestone. Luxuriant reefs were built by coral-like organisms. The first land plants began to evolve during this period, and there were many ostracoderms (armoured jawless fishes). The first jawed fishes (called acanthodians) also appeared.

sima in geochemistry and geophysics, the substance of the Earth's oceanic *crust, as distinct from the *sial of the continental crust. The name, now used rarely, is derived from **si**lica and **ma**gnesia, its two main chemical constituents.

Single European Act act signed in 1986 (and in force from July 1987) to establish a *single European market, defined as an area without frontiers in which free movement of goods, services, people, and capital is ensured.

single European currency former name for the *euro.

single European market single market within the *European Union. Established under the *Single European Act, it was the core of the process of European economic integration, involving the removal of obstacles to the free movement of goods, services, people, and capital between member states of the EU. It covers, among other benefits, the elimination of customs barriers, the liberalization of capital movements, the opening of public procurement markets, and the mutual recognition of professional qualifications. It came into effect on 1 January 1993.

single-party state   another name for *one-party state.

sink hole   funnel-shaped hollow in an area of limestone. A sink hole is usually formed by the enlargement of a joint, or crack, by *carbonation (the dissolving effect of water). It should not be confused with a *swallow hole, or swallet, which is the opening through which a stream disappears underground when it passes onto limestone.

sirocco   hot, normally dry and dust-laden wind that blows from the deserts of North Africa across the Mediterranean into southern Europe. It occurs mainly in the spring. The name 'sirocco' is also applied to any hot oppressive wind.

site of special scientific interest SSSI, in the UK, land that has been identified as having animals, plants, or geological features that need to be protected and conserved. From 1991 these sites were designated and administered by English Nature, Scottish Natural Heritage, and the Countryside Council for Wales.

slash and burn   or **shifting cultivation**, simple agricultural method whereby natural vegetation is cut and burned, and the clearing then farmed for a few years until the soil loses its fertility, whereupon farmers move on and leave the area to regrow. Although this is possible with a small, widely dispersed population, as in the Amazon rainforest for example, it becomes unsustainable with more people and is now a cause of *deforestation.

slate   fine-grained, usually grey metamorphic rock that splits readily into thin slabs along its *cleavage planes. It is the metamorphic equivalent of *shale.

Slate is highly resistant to atmospheric conditions and can be used for writing on with chalk (actually gypsum). Quarrying slate takes such skill and time that it is now seldom used for roof and sill material except in restoring historic buildings.

sleet   precipitation consisting of a mixture of water and ice.

slip   the amount of vertical displacement of *strata at a *fault.

slip-off slope   gentle slope forming the inner bank of a *meander (bend in a river). It is formed by the deposition of fine silt, or alluvium, by slow-flowing water.

As water passes round a meander the fastest current sweeps past the outer bank, eroding it to form a steep river cliff. Water flows more slowly past the inner bank, and as it reduces speed the material it carries is deposited around the bank to form a slip-off slope.

slum   area of poor-quality *housing. Slums are typically found in parts of the *inner city in rich countries and in older parts of cities in poor countries.

Slum housing is usually densely populated, in a bad state of repair, and has inadequate services (poor sanitation, for example). Its occupants are often poor with low rates of literacy.

slump   the downward sliding of soil and rock debris under the influence of gravity. Slumps are usually considered a form of landslide and are characterized by their rotational movement on a curved slip plane.

slurry   form of manure composed mainly of liquids. Slurry is collected and stored on many farms, especially when large numbers of animals are kept in factory units. When slurry tanks are accidentally or deliberately breached, large amounts can spill into rivers, killing fish and causing *eutrophication. Some slurry is spread on fields as a fertilizer.

Small Is Beautiful   book by E F Schumacher, published in 1973, which

SMALL- OR MEDIUM-SIZED ENTERPRISE

argues that the increasing scale of corporations and institutions, concentration of power in fewer hands, and the overwhelming priority being given to economic growth are both unsustainable and disastrous to environment and society.

small- or medium-sized enterprise see *SME.

SME *abbreviation for* small- or medium-sized enterprise, enterprise defined under European Union legislation as one with fewer than 250 employees. Medium-sized companies have between 50 and 250 employees, and small companies have fewer than 50.

smog natural fog containing impurities, mainly nitrogen oxides (NO_x) and volatile organic compounds (VOCs) from domestic fires, industrial furnaces, certain power stations, and internal-combustion engines (petrol or diesel). It can cause substantial illness and loss of life, particularly among chronic bronchitics, and damage to wildlife.

photochemical smog is mainly prevalent in the summer as it is caused by chemical reaction between strong sunlight and vehicle exhaust fumes. Such smogs create a build-up of ozone and nitrogen oxides which cause adverse symptoms, including coughing and eye irritation, and in extreme cases can kill.

The London smog of 1952 lasted for five days and killed more than 4000 people from heart and lung diseases. The use of smokeless fuels, the treatment of effluent, and penalties for excessive smoke from poorly maintained and operated vehicles can be effective in reducing smog but it still occurs in many cities throughout the world.

smoker hot fissure in the ocean floor, known as a *hydrothermal vent.

smokestack industry a term used to describe any traditional heavy industry

– iron and steel, shipbuilding – as opposed to newer, cleaner industries such as electronics.

snout in earth science, the front end of a *glacier, representing the furthest advance of the ice at any one time. Deep cracks, or *crevasses, and ice falls are common.

Because the snout is the lowest point of a glacier it tends to be affected by the warmest weather. Considerable melting takes place, and so it is here that much of the rocky material carried by the glacier becomes deposited. Material dumped just ahead of the snout may form a *terminal moraine.

The advance or retreat of the snout depends upon the glacier budget – the balance between *accumulation (the addition of snow and ice to the glacier) and *ablation (their loss by melting and evaporation).

snow precipitation in the form of soft, white, crystalline flakes caused by the condensation in air of excess water vapour below freezing point. Light reflecting in the crystals, which have a basic hexagonal (six-sided) geometry, gives snow its white appearance.

snow line the altitude above which permanent snow exists, and below which any snow that falls will not persist during the summer months. The altitude of the snow line varies with *latitude: at the equator it is approximately 5000 metres, in the Alps it is approximately 3000 metres, and at the Poles the snow line is at sea level.

soapstone compact, massive form of impure *talc.

social anthropology another term for *cultural anthropology.

social costs and benefits in economics, the costs and benefits to society as a whole that result from economic decisions. These include private costs (the financial cost of production incurred by firms) and benefits (the profits made by firms and the value to people of consuming goods and services) and external costs and benefits (affecting those not directly involved in production or consumption); pollution is one of the external costs.

social fabric the make-up of an area in terms of its social geography, such as class, ethnic composition, employment, education, and values.

socialism movement aiming to establish a classless society by substituting public for private ownership of the means of production, distribution, and exchange. The term has been used to describe positions as widely apart as anarchism and social democracy.

social mobility movement of groups and individuals up and down the social scale in a classed society. The extent or range of social mobility varies in different societies. Individual social mobility may occur through education, marriage, talent, and so on; group mobility usually occurs through change in the occupational structure caused by new technological or economic developments.

society the organization of people into communities or groups. Social science, in particular sociology, is the study of human behaviour in a social context, or setting. Various aspects of society are discussed under *class, *community, *culture, *kinship, norms, role, socialization, and status.

socio-economic group classification of buyers according to social and economic characteristics. A classification commonly used in marketing research is the A to E system, which classifies households according to the profession of the head of household, or principal wage earner. The grades are as follows: A (upper middle class) – professional or at director level; B (middle class) – senior management; C1 (lower middle class) – junior management and

clerical; C2 (working class) – skilled; D (working class) – unskilled, manual labour; and E – those reliant on the state, such as pensioners and the long-term unemployed.

socio-technical system theory suggesting that the organization of work has both an economic and a social element. The theory is based largely on work done by the Tavistock Institute in London. Researchers at the Tavistock Institute examined the effects on employees in the mining industry of a shift from team working to mass production. They found that disruption to the social systems had an adverse effect on employee motivation and concluded that companies should take into consideration not only economic but also social factors when implementing change in an organization.

soil loose covering of broken rocky material and decaying organic matter overlying the bedrock of the Earth's surface. It is composed of minerals (formed from *physical weathering and *chemical weathering of rocks), organic matter (called *humus) derived from decomposed plants and organisms, living organisms, air, and water. Soils differ according to climate, parent material, rainfall, relief of the bedrock, and the proportion of organic material. The study of soils is **pedology**.

Soils influence the type of agriculture employed in a particular region – light well-drained soils favour arable farming, whereas heavy clay soils give rise to lush pasture land. Plant roots take in nutrients (in the form of ions) dissolved in the water in soil. The main elements that plants need to absorb through their roots are nitrogen, phosphorus, and potassium.

Soil is formed by the weathering of rocks. *Physical weathering breaks the rock into small pieces. *Chemical weathering releases the minerals that plants need for growth.

A soil can be described in terms of its **soil profile**, a vertical cross-section from ground-level to the bedrock on which the soils sits. The profile is divided into layers called horizons. The A horizon, or topsoil, is the uppermost layer, consisting primarily of humus and living organisms and some mineral material. Most soluble material has been leached from this layer or washed down to the B horizon. The B horizon, or subsoil, is the layer where most of the nutrients accumulate and is enriched in clay minerals. The C horizon is the layer of weathered parent material at the base of the soil.

Two common soils are the *podzol and the chernozem soil. The **podzol** is common in coniferous forest regions where precipitation exceeds evaporation. The A horizon consists of a very thin litter of organic material producing a poor humus. Coniferous tree leaves (needles) take a long time to decompose. The relatively heavy precipitation causes *leaching of minerals, as nutrients are washed downwards.

Chernozem soils are found in grassland regions, where evaporation exceeds precipitation. The A horizon is rich in humus due to decomposition of a thick litter of dead grass at the surface. Minerals and moisture migrate upwards due to evaporation, leaving the B and A horizons enriched.

The organic content of soil is widely variable, ranging from zero in some desert soils to almost 100% in *peats.

Soil Association pioneer British ecological organization founded in 1946, which campaigns against pesticides and promotes organic farming.

soil creep gradual movement of soil down a slope in response to gravity. This eventually results in a mass downward movement of soil on the slope.

Evidence of soil creep includes the formation of terracettes (steplike ridges along the hillside), leaning walls and telegraph poles, and trees that grow in a curve to counteract progressive leaning.

soil depletion decrease in soil quality over time. Causes include loss of nutrients caused by overfarming, erosion by wind, and chemical imbalances caused by acid rain.

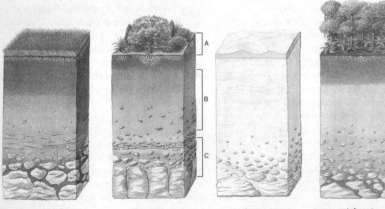

chernozem   podzol   desert   rainforest

soil Common types of soil profile.

soil erosion wearing away and redistribution of the Earth's soil layer. It is caused by the action of water, wind, and ice, and also by improper methods of agriculture. If unchecked, soil erosion results in the formation of deserts (*desertification). It has been estimated that 20% of the world's cultivated topsoil was lost between 1950 and 1990.

If the rate of erosion exceeds the rate of soil formation (from rock and decomposing organic matter), then the land will become infertile. The removal of forests (*deforestation) or other vegetation often leads to serious soil erosion, because plant roots bind soil, and without them the soil is free to wash or blow away, as in the American *dust bowl. The effect is worse on hillsides, and there has been devastating loss of soil where forests have been cleared from mountainsides, as in Madagascar.

Improved agricultural practices such as contour ploughing are needed to combat soil erosion. Windbreaks, such as hedges or strips planted with coarse grass, are valuable, and organic farming can reduce soil erosion by as much as 75%.

Soil degradation and erosion are becoming as serious as the loss of the rainforest. It is estimated that more than 10% of the world's soil lost a large amount of its natural fertility during the latter half of the 20th century. Some of the worst losses are in Europe, where 17% of the soil is damaged by human activity such as mechanized farming and fallout from acid rain. In Mexico and Central America, 24% of soil is highly degraded, mostly as a result of deforestation.

soil mechanics branch of engineering that studies the responses of soils with different water and air contents, to loads. Soil is investigated during construction work to ensure that it has the mechanical properties necessary to support the foundations of dams, bridges, and roads.

soil profile the sequence of layers or *horizons usually seen in an exposed *soil section. Topsoil is the uppermost layer and subsoil is the layer between topsoil and bedrock.

solar constant the rate at which the Sun's energy is received per unit area on a horizontal surface at the outer margins of the Earth's atmosphere. The term 'constant' is misleading since its value fluctuates slightly; however, the average solar constant is 1388 Wm^{-2}. See also *radiation, *albedo, *insolation.

solar cycle variation of activity on the *Sun over an 11-year period indicated primarily by the number of *sunspots visible on its surface. The next period of maximum activity is expected around 2011.

solar energy 1. the radiant energy which originates from the Sun. The Sun can be considered a large thermonuclear device with a total energy output estimated to be 200,000 million, million times that of the largest nuclear reactor on Earth, and yet our planet intercepts only one fifty-millionth of this energy. This amounts, on average, to 15.3 × 10^8 cal/m^2/year (equivalent to 40,000 kW of electrical energy for every human on Earth). Due to latitudinal effects and cloud cover, the pattern of incoming solar radiation shows great variation. The UK, for example, receives on average only 2.5 × 10^8 cal/m^2/year.

Of the solar energy which reaches the outer layer of the Earth's atmosphere between 49% and 51% is lost through scattering or absorption within the atmosphere. Of the energy which reaches the surface, some 95–99% is absorbed by the oceans and is used to drive the *hydrological cycle. The remaining 1–5% is used for *photosynthesis in green plants upon which all animal life is dependent.

2. the capture of the Sun's rays to provide a renewable source of energy for human use. Large-scale use of solar energy as a power source is still confined to experimental and research projects. Initial experiments in the USA and Israel which have shown promising results include those where concave mirrors positioned in hot desert areas are used to focus the Sun's rays to raise the temperature of water sufficient to raise steam and generate electricity from steam turbines. Elsewhere and on a smaller scale, solar power has been successfully used to heat domestic water via solar panels, to charge batteries in locations remote from fixed power lines, to operate photo-electric cells, and to provide on-board power for satellite equipment. Solar energy may also be harnessed indirectly using **solar cells** (photovoltaic cells) made of panels of *semiconductor material (usually silicon), which generate electricity when illuminated by sunlight.

solar flare brilliant eruption on the Sun above a *sunspot, thought to be caused by release of magnetic energy. Flares reach maximum brightness within a few minutes, then fade away over about an hour. They eject a burst of atomic particles into space at up to 1,000 kps. When these particles reach Earth they can cause radio blackouts, disruptions of the Earth's magnetic field, and *auroras.

solar pond natural or artificial 'pond', such as the Dead Sea, in which salt becomes more soluble in the Sun's heat. Water at the bottom becomes saltier and hotter, and is insulated by the less salty water layer at the top. Temperatures at the bottom reach about 100°C and can be used to generate electricity.

solar power heat radiation from the sun converted into electricity or used directly to provide heating.

Solar power is an example of a renewable source of energy (see *renewable resources).

solar radiation radiation given off by the Sun, consisting mainly of visible light, *ultraviolet radiation, and *infrared radiation, although the whole spectrum of *electromagnetic waves is present, from radio waves to X-rays. High-energy charged particles, such as electrons, are also emitted, especially from *solar flares. When these reach the Earth, they cause magnetic storms (disruptions of the Earth's magnetic field), which interfere with radio communications.

solar time time of day as determined by the position of the *Sun in the sky.

sol brun or **sol brun acide**, the term extensively used by European soil classification systems for a brown earth soil.

solifluction movement (flow) of topsoil that has become saturated with water. It is an important process of transportation in *periglacial environments where water released by the spring thaw cannot percolate downwards because the ground below is permanently frozen (permafrost). The saturated topsoil may then flow slowly downhill to form a **solifluction lobe**.

solstice either of the days on which the Sun is farthest north or south of the celestial equator each year. In the northern hemisphere, the **summer solstice**, when the Sun is farthest north, occurs around 21 June and the **winter solstice** around 22 December; see *season.

solution or **dissolution**, in earth science, process by which the minerals in a rock are dissolved in water. Solution is one of the processes of *erosion as well as *weathering (in which the breakdown of rock occurs without transport of the dissolved material). An example of this is when weakly acidic rainfall dissolves calcite.

Solution commonly affects
*limestone and *chalk, which are both
formed of calcium carbonate. It can
occur in coastal environments along
with corrasion and hydraulic action,
producing features like the white cliffs
of Dover, as well as fluvial (river)
environments, like the one that formed
Cheddar Gorge.

Solution is also responsible for the
weathering of many buildings,
monuments and other structures.

South American geology   South
America can be divided into three
major geological provinces. The oldest
rocks occur in the east of the continent,
in the plateaus of the **Guyana and
Brazilian Shields**, separated by the
Amazon basin. To the west and north
is a **Palaeozoic area** of broad plains,
underlain by Palaeozoic rocks, and to
the west again is the **Andean mobile
belt**, lying along the length of the
Pacific coast and extending along the
northern Caribbean coast. This belt is
of Mesozoic to Tertiary age.

the Guyana Shield This is a high
plateau area of old rocks, occupying
Guyana, Suriname, French Guiana,
and parts of Venezuela, Colombia,
and Brazil. The oldest rocks are highly
metamorphosed basement plutonic
complexes, mainly of middle to late
Precambrian age, but including some
remnants of early Precambrian
complexes, together with a variety
of less highly metamorphosed
sedimentary and volcanic successions.
These represent the remains of mobile
belts which reached the final stages of
their activity 2000 million years ago.
These old assemblages are
unconformably overlain by a flat-lying
sedimentary formation known as the
Roraima Formation. This comprises
several thousand metres of sandstones,
conglomerates, and shales, some
containing detrital diamonds. The

sediments are intruded by large
masses of basic igneous rock, forming
thick sheets of gabbro, sills, and dykes.
These were intruded in late
Precambrian times, 2000–1800 million
years ago.

the Brazilian Shield This shield area
underlies the whole of Brazil and
stretches through Uruguay and into
the adjoining states to the west.
Although covered in part by younger
sediments, the old Precambrian
crystalline basement is well exposed
in northeastern Brazil. Most of the
rocks are of late Precambrian age,
and represent deposits laid down in
geosynclinal zones parallel to the
present Atlantic coast. A basement of
older rocks has been largely reworked
during late Precambrian mobile belt
activity. Two major cycles can be seen;
during the first cycle the rocks of the
Pre-Minas Series were laid down,
consisting of sediments, volcanics,
and some thick ironstone formations.
These were folded, metamorphosed,
and invaded by granites during an
orogenic phase (see *orogeny) about
1350–1100 million years ago. The
second cycle includes the rocks of the
Minas Series, and makes up the wide
coastal zone known as the Brazilides.
The Minas Series consists of psammites,
pelites, laminated ironstones,
manganiferous beds, and quartzites.
These were folded and metamorphosed
650–500 million years ago, and late
granites were then intruded. This belt
in turn was covered by an
unconformable series of sediments.

In many parts of the Brazilian Shield
the Precambrian metamorphic rocks
are overlain by glacial deposits of
very late Precambrian age, termed
Infracambrian. These are similar to
glacial deposits of the same age found
in other continents.

Palaeozoic area Rocks of Palaeozoic
age cover a large area between the old

shield areas of the east and the Andean belt to the west. They consist of shallow-water sediments laid down on a platform fringing the ancient shield areas. Deposition appears to have started in the west, where Cambro-Ordovician sediments are seen in Bolivia; elsewhere the sedimentary sequence begins with rocks of Silurian or Devonian age. The Devonian sediments include glacial tillites in a number of areas, indicating a further period of glaciation in Devonian times.

Towards the end of the Upper Palaeozoic, important changes took place in the super-continent of *Gondwanaland, of which South America was still a part. Beds of a continental facies accumulated in new basins of deposition at the margins of the old shield and within the shield. At the base of this succession are glacial deposits of Permo-Carboniferous age. In the Santa Catarina system of the Paraná basin these terrestrial glacial deposits reach a maximum thickness of 1600 m, and they are followed by coal measures, shales, sandstones, and volcanics, spanning a period from late Carboniferous to late Triassic times. The whole sequence forms a pile of continental sediments 4 km thick.

The first phase in the break-up of Gondwanaland is seen in late Jurassic to early Cretaceous times, when plateau basalts outpoured over much of the Paraná and Amazon basins. Associated with the continental disruption are the marine evaporites and carbonates deposited in mid-Cretaceous times along the Atlantic coast. At the same time the diamond-bearing kimberlite pipes were intruded in the Minas Gerais area of the Brazilian Shield. From late Mesozoic times onwards the southern continents moved apart by a process of *seafloor spreading in the South Atlantic.

the Andean mobile belt This belt at the western margin of the continent was active through most of Phanerozoic time. Thick accumulations of Palaeozoic and Mesozoic rocks built up in a geosynclinal environment, and there were episodes of orogenic activity in mid- and late Palaeozoic times, and again in late Mesozoic times, when huge granite masses were intruded. This was followed by widespread volcanic activity and uplift of the mountain chain of the Andes. The continued westward movement of South America is evident from the deep-seated earthquakes occurring below the Andes, and the deep ocean trench off the coast. Late orogenic volcanic activity continued from late Cretaceous times up to the present day, with the eruption of andesite lavas. Copper deposits are found in Chile associated with granites intruded in Palaeogene to Neogene times. In the late Tertiary, uplift was accompanied by explosive volcanic activity, resulting in vast sheets of ignimbrite and acid pyroclastic rocks, especially in eastern Chile.

Outside the Andean belt, thick continental sandstones and conglomerates were deposited from the rising mountain chain during Tertiary and Quaternary times. In the extreme north, adjoining the Caribbean mobile belt, thick deposits of marine Cretaceous, Tertiary, and recent sediments were laid down in the eastern Venezuela basin. This contains important oil deposits.

South Asian Association for Regional Cooperation SARCC, organization established in 1985 by India, Pakistan, Bangladesh, Nepal, Sri Lanka, Bhutan, and the Maldives to cover agriculture, telecommunications, health, population, sports, art, and culture. In 1993 a preferential trade agreement was adopted to reduce tariffs on intraregional trade. At the May 1997 summit it was agreed to establish a regional free trade zone by 2001.

Southern African Development Community   SADC, organization of countries in the region working together initially to reduce their economic dependence on South Africa and harmonize their economic policies, but from 1995 to promote the creation of a free-trade zone by 2000. It was established in 1980 as the **Southern African Development Coordination Conference (SADCC)**, adopting its present name in 1992, and focuses on transport and communications, energy, mining, and industrial production. The member states are Angola, Botswana, Lesotho, Malawi, Mauritius, Mozambique, Namibia, South Africa, Swaziland, Tanzania, Zambia, and Zimbabwe; headquarters in Gaborone, Botswana.

Southern Cone Common Market alternative name for *Mercosur.

Southern Ocean   corridor linking the Pacific, Atlantic, and Indian oceans, all of which receive cold polar water from the world's largest ocean surface current, the Antarctic Circumpolar Current, which passes through the Southern Ocean.

South Pacific Bureau for Economic Cooperation   SPEC, organization founded 1973 for the purpose of stimulating economic cooperation and the development of trade in the region. The headquarters of SPEC are in Suva, Fiji Islands.

South Pacific Commission   SPC, former name, until February 1998, of the *Pacific Community.

South Pacific Forum   former name, until October 2000, of *Pacific Islands Forum.

sovereignty   absolute authority within a given territory. The possession of sovereignty is taken to be the distinguishing feature of the state, as against other forms of community. The term has an internal aspect, in that it refers to the ultimate source of authority within a state, such as a parliament or monarch, and an external aspect, where it denotes the independence of the state from any outside authority.

space   the geographer's term for area; the context within which distributions and patterns occur.

spaghetti organization   extremely flexible organizational structure. The concept of the spaghetti organization was introduced by Danish businessman Lars Kolind at the company Oticon. Kolind took over the company when it had stagnated and restructured it along radical lines. He rethought the organization, placing interaction, collaboration, and connectivity of people, customers, suppliers, and ideas, at its heart. Kolind called it, 'a spaghetti organization of rich strands in a chaotic network'.

spatial analysis   the description and explanation of distributions of people and their activities in *space.

spatial distribution   the pattern of locations of, for example, population or *settlement in a region.

specialization   in economics, a method of organizing production where economic units such as households or nations are not self-sufficient but concentrate on producing certain goods and services and trading the surplus with others. Specialization of workers is known as the *division of labour.

speleology   scientific study of caves, their origin, development, physical structure, flora, fauna, folklore, exploration, mapping, photography, cave-diving, and rescue work. **Potholing**, which involves following the course of underground rivers or streams, has become a popular sport.

Speleology first developed in France in the late 19th century, where the Société de Spéléologie was founded in 1895.

sphalerite   mineral composed of zinc sulphide with a small proportion of iron, formula (Zn,Fe)S. It is the chief ore of zinc. Sphalerite is brown with a non-metallic lustre unless an appreciable amount of iron is present (up to 26% by weight). Sphalerite usually occurs in ore veins in limestones, where it is often associated with galena. It crystallizes in the cubic system but does not normally form perfect cubes.

sphere of influence   the area surrounding a settlement or service (for example a leisure complex) that is affected by the activities of that settlement or service. The sphere of influence may be mapped by looking at the catchment areas of various services or by considering local-newspaper circulation, delivery areas, and public-transport destinations.

spinel   any of a group of 'mixed oxide' minerals consisting mainly of the oxides of magnesium and aluminium, $MgAl_2O_4$ and $FeAl_2O_4$. Spinels crystallize in the cubic system, forming octahedral crystals. They are found in high-temperature igneous and metamorphic rocks. The aluminium oxide spinel contains gem varieties, such as the ruby spinels of Sri Lanka and Myanmar (Burma).

spit   ridge of sand or shingle projecting from the land into a body of water. It is formed by a combination of *longshore drift, tides, river currents, and/or a bend in the coastline. The decrease in wave energy causes more material to be deposited than is transported down the coast, building up a finger of sand that points in the direction of the longshore drift. Deposition in the brackish water behind a spit may result in the formation of a *salt marsh.

spodosol   the term used in the American soil classification system for a *podzol-type soil.

spring   in geology, a natural flow of water from the ground, formed at the point of intersection of the water table and the ground's surface. The source of water is rain that has percolated through the overlying rocks. During its underground passage, the water may have dissolved mineral substances that may then be precipitated at the spring (hence, a mineral spring).

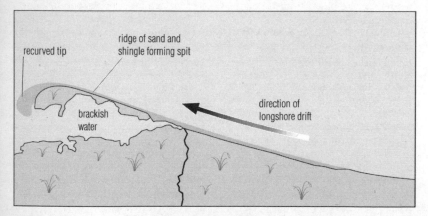

spit Longshore drift carries sand and shingle up coastlines. Deposited material gradually builds up over time at headlands forming a new stretch of land called a spit. A spit that extends across a bay is known as a bar.

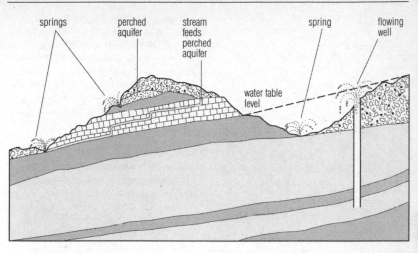

spring Springs occur where water-laden rock layers (aquifers) reach the surface. Water will flow from a well whose head is below the water table.

A spring may be continuous or intermittent, and depends on the position of the water table and the topography (surface features).

spring line geological feature where water springs up in several places along the edge of a permeable rock escarpment. Springline settlements may become established around this.

spring tides see *tide.

sprite in earth science, rare thunderstorm-related luminous flash. Sprites occur in the mesosphere, at altitudes of 50–90 km. They are electrical, like lightning, and arise when the electrical field that occurs between the thunder cloud top and the ionosphere (ionized layer of the Earth's atmosphere) draws electrons upwards from the cloud. If the air is thin and this field is strong the electrons accelarate rapidly, transferring kinetic energy to molecules they collide with. The excited molecules then discharge this energy as a light flash.

spur ridge of rock jutting out into a valley or plain. In mountainous areas rivers often flow around *interlocking spurs because they are not powerful enough to erode through the spurs. Spurs may be eroded away by large and powerful glaciers to form *truncated spurs.

squatter person illegally occupying someone else's property; for example, some of the urban homeless in contemporary Britain making use of vacant houses. Squatters commit a criminal offence if they take over property where there is a 'residential occupier'; for example, by moving in while the owner is on holiday.

squatter settlement an area of peripheral urban settlement in which the residents occupy land to which they have no legal title. See *shanty town.

SSSI *abbreviation for* *Site of Special Scientific Interest.

stack isolated pillar of rock that has become separated from a headland by *coastal erosion. It is usually formed by the collapse of an *arch. Further erosion will reduce it to a stump, which is exposed only at low tide.

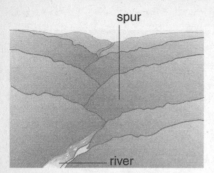

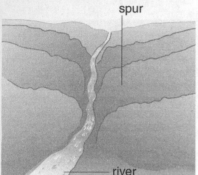

spur Interlocking spurs (above) and truncated spurs (below). Interlocking spurs are formed by meandering rivers; truncated spurs are interlocking spurs eroded by glaciation.

stalactite and stalagmite cave structures formed by the deposition of calcite dissolved in ground water. **Stalactites** grow downwards from the roofs or walls and can be icicle-shaped, straw-shaped, curtain-shaped, or formed as terraces. **Stalagmites** grow upwards from the cave floor and can be conical, fir-cone-shaped, or resemble a stack of saucers. Growing stalactites and stalagmites may meet to form a continuous column from floor to ceiling.

Stalactites are formed when ground water, hanging as a drip, loses a proportion of its carbon dioxide into the air of the cave. This reduces the

amount of calcite that can be held in solution, and a small trace of calcite is deposited. Successive drips build up the stalactite over many years. In stalagmite formation the calcite comes out of the solution because of agitation – the shock of a drop of water hitting the floor is sufficient to remove some calcite from the drop. The different shapes result from the splashing of the falling water.

standard atmosphere alternative term for *atmosphere, a unit of pressure.

standard deviation in statistics, a measure (symbol σ or s) of the spread of data. The deviation (difference) of each of the data items from the mean is found, and their values squared. The mean value of these squares is then calculated. The standard deviation is the square root of this mean.

standard of living in economics, the measure of consumption and welfare of a country, community, class, or person. Individual standard-of-living expectations are heavily influenced by the income and consumption of other people in similar jobs.

standard time time in any of the 24 time zones, each an hour apart, into which the Earth is divided. The respective times depend on their distances, east or west of Greenwich, England. In North America the eight zones (Atlantic, Eastern, Central, Mountain, Pacific, Alaska, Hawaii-Aleutian, and Samoa) use the mean solar times of meridians 15° apart, starting with 60° longitude. (See also *time.)

standing crop in ecology, the total number of individuals of a given species alive in a particular area at any moment. It is sometimes measured as the weight (or *biomass) of a given species in a sample section.

staple diet the basic foodstuff which comprises the daily meals of a given people. In South East Asia rice is the staple food, while, in many parts of Africa the staple food is maize.

star dune   see **dune.

state   territory that forms its own domestic and foreign policy, acting through laws that are typically decided by a government and carried out, by force if necessary, by agents of that government. It can be argued that the growth of regional international bodies such as the European Union (formerly the European Community) means that states no longer enjoy absolute *sovereignty.

staurolite   silicate mineral, $(Fe,Mg)_2(Al,Fe)_9Si_4O_{20}(OH)_2$. It forms brown crystals that may be twinned in the form of a cross. It is a useful indicator of medium grade (moderate temperature and pressure) metamorphism in metamorphic rocks formed from clay sediments.

steel industry   see **iron and steel industry.

steppe   temperate grasslands of Europe and Asia. The term is sometimes used to refer to other temperate grasslands and semi-arid desert edges.

Stevenson screen   box designed to house weather-measuring instruments such as thermometers. It is kept off the ground by legs, has louvred sides to encourage the free passage of air, and is painted white to reflect heat radiation, since what is measured is the temperature of the air, not of the sunshine.

stishovite   highest density, highest pressure polymorph (substance with the same chemical composition but different crystal structure) of silica, SiO_2, in which the crystal structure is the same as that of the mineral *rutile, titanium dioxide. Rare in nature it is found primarily near meteor impact craters. It is thought to be a rare constituent of the lower *mantle.

stock   materials, unfinished goods, or work-in-progress, and finished goods that businesses hold. They need to hold materials and unfinished goods because they are needed for production. Finished goods are held until sold. Companies usually try to keep the value of their stock as low as possible, while still being able to make enough finished goods to supply their markets on time.

stock exchange   institution for the buying and selling of stocks and shares (securities). The world's largest stock exchanges are London, New York (Wall Street), and Tokyo. The oldest stock exchanges are Antwerp (1460), Hamburg (1558), Amsterdam (1602), New York (1790), and London (1801). The former division on the London Stock Exchange between brokers (who bought shares from jobbers to sell to the public) and jobbers (who sold them only to brokers on commission, the 'jobbers' turn') was abolished in 1986.

stone stripes   see **patterned ground.

storm   extreme weather condition characterized by strong winds, rain, hail, thunder, and lightning.

storm hydrograph   a graph showing the *discharge of a stream and its response to a period of rainfall. There is a time lag between peak rainfall and peak discharge. This is because it takes time for the rainwater to pass through the various 'storage units', i.e. vegetation, soil and rock. A series of storm hydrographs can be plotted over several days to give a picture of the river's *regime* or pattern of flow.

storm surge   abnormally high tide brought about by a combination of a deep atmospheric *depression (very low pressure) over a shallow sea area, high spring tides, and winds blowing from the appropriate direction. It can be intensified by snowmelt and/or river flooding. A storm surge can cause severe flooding of lowland coastal regions and river estuaries.

Bangladesh is particularly prone to surges, because it is sited on a low-lying *delta where the Indian Ocean funnels into the Bay of Bengal. In May 1991, 125,000 people were killed there in such a disaster. In February 1953 more than 2000 people died when a North Sea surge struck the Dutch and English coasts.

strata singular **stratum**, layers or *beds of sedimentary rock.

strategic islands islands (Azores, Canary Islands, Cyprus, Iceland, Madeira, and Malta) of great political and military significance likely to affect their stability; they held their first international conference in 1979.

stratigraphy branch of geology that deals with *sedimentary rock layers (strata) and their sequence of formation. Its basis was developed by English geologist William Smith. The basic principle of superposition establishes that upper layers or deposits accumulated later in time than the lower ones.

Stratigraphy involves both the investigation of sedimentary structures to determine past environments represented by rocks, and the study of fossils for identifying and dating particular beds of rock. A body of rock strata with a set of unifying characteristics indicative of an environment is called a *facies.

Stratigraphic units can be grouped in terms of time or lithology (rock type). Strata that were deposited at the same time belong to a single **chronostratigraphic unit** but need not be the same lithology. Strata of a specific lithology can be grouped into a **lithostratigraphic unit** but are not necessarily the same age.

Stratigraphy in the interpretation of archaeological excavations provides a relative chronology for the levels and the artefacts within rock beds. It is the principal means by which the context of archaeological deposits is evaluated.

stratocumulus a grey or white low-altitude cloud mass of globular visual appearance usually of extensive form and associated with the passage of a warm front. This cloud type is often associated with continuous drizzle.

stratopause layer in the Earth's atmosphere at an altitude of 50–55 km, at the top of the stratosphere.

stratosphere that part of the atmosphere 10–40 km from the Earth's surface, where the temperature slowly rises from a low of –55°C to around 0°C. The air is rarefied and at around 25 km much *ozone is concentrated.

stratovolcano another term for *composite volcano, a type of volcano made up of alternate ash and lava layers.

stratus layer-cloud of uniform grey appearance, often associated with the warm sector of a *depression. Stratus is a type of low *cloud which may hang as mist over mountain tops; broken stratus is referred to as *fractostratus*.

striation scratch formed by the movement of a glacier over a rock surface. Striations are caused by the scraping of rocky debris embedded in the base of the glacier (*corrasion), and provide an useful indicator of the direction of ice flow in past *ice ages.

They are common features of *roche moutonnées.

strike compass direction of a horizontal line on a planar structural surface, such as a fault plane, bedding plane, or the trend of a structural feature, such as the axis of a fold. Strike is 90° from *dip.

strip cropping a method of *soil conservation whereby different crops are planted in a series of strips, often following *contours around a hillside. The purpose of such a sequence of cultivation is to arrest the downslope movement of soil, especially if one of the crops (e.g. maize) tends to expose the soil surface. Alternate strips may be planted with grass which effectively binds the soil and acts as a brake on run-off. See *soil erosion.

strike-slip fault   common name for a lateral fault in which the motion is sideways in the direction of the *strike of the fault.

stromatolite   mound produced in shallow water by mats of algae that trap mud particles. Another mat grows on the trapped mud layer and this traps another layer of mud and so on. The stromatolite grows to heights of a metre or so. They are uncommon today but their fossils are among the earliest evidence for living things – over 2000 million years old.

structural functionalism anthropological theory formulated by Alfred Radcliffe-Brown, which argued that social structures arise and are maintained in order to facilitate the smooth and harmonious functioning of society as a whole.

structural geology   branch of geology dealing with the physical shape of rock formations, from large-scale features such as mountain belts and rift valleys, to microscopic features such as deformed mineral grains. Structural geologists deal with the brittle and ductile deformation of planetary surfaces and the processes that cause deformation.

structural unemployment unemployment resulting from the decline of an industry, often as a result of technological innovation, or a change in demand. In the UK, the demise of the coal mining, ship building, and steel industries have all resulted in structural unemployment.

structure plan   a statutory planning document that describes the spatial allocation of land for various uses within its stated area in forthcoming years. The written strategy is accompanied by a map or *key diagram* portraying the spatial implications of future land use. The structure plan provides the framework within which more detailed decisions about particular sites are made. The pattern of land use in the structure plan is the design for the infrastructure required to house the projected future population, to provide social services, and to provide opportunities for economic development.

stump   low outcrop of rock formed by the erosion of a coastal *stack. Unlike a stack, which is exposed at all times, a stump is exposed only at low tide. Eventually it is worn away completely.

subduction zone   in *plate tectonics, a region where two plates of the Earth's rigid *lithosphere collide, and one plate descends below the other into the weaker *asthenosphere. Subduction results in the formation of ocean trenches, most of which encircle the Pacific Ocean.

    Ocean trenches are usually associated with volcanic *island arcs and deep-focus earthquakes (more than 300 km below the surface), both the result of disturbances caused by the plate subduction.

subglacial   beneath a glacier. Subglacial rivers are those that flow under a glacier; subglacial material is debris that has been deposited beneath glacier ice. Features formed subglacially include *drumlins and *eskers.

subsidence   downward movement of a block of rock. Subsidence is usually due to the removal of material from below the surface, and can be caused by faults, *erosion, or by human activities such as mining.

subsidiarity   devolution of decision-making within the European Union from the centre to the lowest level possible. Since the signing of the *Maastricht Treaty on European union 1991, which affirms that, wherever possible, decisions should be 'taken as closely as possible to the citizens', subsidiarity has been widely debated as a means of countering trends towards excessive centralization.

subsidiary cone   a volcanic cone which develops within the *caldera of a previous, larger cone which was blown away during an eruption.

subsistence agriculture   a system of *agriculture in which farmers produce exclusively for their own consumption, in contrast to *commercial agriculture where farmers produce purely for sale at the market.

subsoil   see *soil profile.

suburb   outer part of an urban area. Suburbs generally consist of residential housing and shops of a low order (newsagent, small supermarket), which act as *central places for the local community. Often, suburbs are the most recent growth of an urban area, and their end marks the *urban fringe. Their growth may result in *urban sprawl. Increasingly, *out-of-town shopping centres and retail parks are locating on the urban fringe.

suburbanized village   or **commuter village** or **dormitory village**, village settlement affected by *counter-urbanization. The reversal of movement from the centre of large urban areas in many parts of the developed world has led to many people moving into the villages surrounding these areas. This has changed the character of these settlements in relation to form, function, size, and population structure.

As villages become increasingly suburbanized a number of physical, social, and economic changes take place. Housing, population structure, employment, transport, services, and the environment are all affected.

succession   in ecology, a series of changes that occur in the structure and composition of the vegetation in a given area from the time it is first colonized by plants (**primary succession**), or after it has been disturbed by fire, flood, or clearing (**secondary succession**).

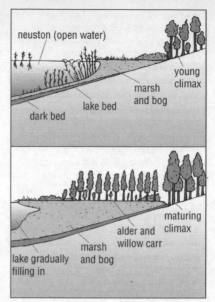

succession The succession of plant types along a lake. As the lake gradually fills in, a mature climax community of trees forms inland from the shore. Extending out from the shore, a series of plant communities can be discerned with small, rapidly growing species closest to the shore.

If allowed to proceed undisturbed, succession leads naturally to a stable *climax community (for example, oak and hickory forest or savannah grassland) that is determined by the climate and soil characteristics of the area.

Sun   star at the centre of our Solar System. It is about 5 billion years old, with a predicted lifetime of 10 billion years; its diameter is 1.4 million km; its temperature at the surface (the *photosphere) is about 5530°C, and at the centre about 15 million°C. It is composed of about 70% hydrogen and 30% helium, with other elements making up less than 1%. The Sun's energy is generated by nuclear fusion reactions that turn hydrogen into helium, producing large amounts of light and heat that sustain life on Earth.

Space probes to the Sun have included NASA's series of Orbiting Solar Observatory satellites, launched between 1963 and 1975, the Ulysses space probe, launched in 1990, and Genesis, launched in 2001.

sunspot   dark patch on the surface of the Sun, actually an area of cooler gas, thought to be caused by strong magnetic fields that block the outward flow of heat to the Sun's surface. Sunspots consist of a dark central **umbra**, about 3700°C, and a lighter surrounding **penumbra**, about 5200°C. They last from several days to over a month, ranging in size from 2000 km to groups stretching for over 100,000 km.

sunshine recorder   device for recording the hours of sunlight during a day. The **Campbell-Stokes sunshine recorder** consists of a glass sphere that focuses the Sun's rays on a graduated paper strip. A track is burned along the strip corresponding to the time that the Sun is shining.

superimposed drainage   a river drainage system originally developed on a land surface since removed by erosion and subsequently imposed on older underlying rocks. Superimposed drainage often bears little relation to the nature and structure of the present land surface. Examples of superimposed drainage are found in the English Lake District and in the Appalachians in the USA. See also *drainage pattern.

supermarket   large self-service shop selling food and household goods. The first, Piggly-Wiggly, was introduced by US retailer Clarence Saunders in Memphis, Tennessee, in 1919.

Supermarkets have a high turnover and are therefore able to buy goods in bulk. This cuts down the unit cost and, in turn, the price, which further encourages business. Cut-price supermarkets have in some places led to the closure of small local shops.

superpower   state that through disproportionate military or economic strength can dominate smaller nations. The term was used to describe the USA and the USSR from the end of World War II, when they emerged as significantly stronger than all other countries. With the collapse of the Soviet Union in 1991, the USA is, arguably, now the world's sole superpower.

supply   in economics, the production of goods or services for a market in anticipation of an expected *demand. The level of supply is determined by the price of the product, the cost of production, the level of technology available for production, and the price of other goods. There is no guarantee that supply will match actual demand.

supply and demand   one of the fundamental approaches to economics, which examines and compares the supply of a good with its demand (usually in the form of a graph of supply and demand curves plotted against price). For a typical good, the supply curve is upward-sloping (the higher the price, the more the manufacturer is willing to sell), while the demand curve is downward-sloping (the cheaper the good, the more demand there is for it). The point where the curves intersect is the equilibrium price at which supply equals demand.

supply chain management management of two-way flows of materials, finance, people, and information along the chain from raw material to the customer. At the centre of modern ideas about supply chain management are speed and flexibility. The concept is founded on the image of a chain with continual links between the different stages. Effective supply chain management can lead to processes, people, and material working more efficiently, and can have a significant impact on costs. Removing a link in the supply chain, usually through the use of technology, often leads to improved services at highly competitive prices, something that computer manufacturer Dell achieved by electing to sell computers directly to the public.

supply curve diagrammatic illustration of the relationship between the price of the good and the quantity that producers will supply at that price. It is said to be upward-sloping because the higher the price, the more profitable existing production becomes, attracting new companies into the industry and thus increasing the quantity supplied.

supply-side economics school of economic thought advocating government policies that allow market forces to operate freely, such as privatization, cuts in public spending and income tax, reductions in trade-union power, and cuts in the ratio of unemployment benefits to wages. Supply-side economics developed as part of the monetarist (see *monetarism) critique of *Keynesian economics.

supraglacial on top of a glacier. A supraglacial stream flows over the surface of the glacier; supraglacial material collected on top of a glacier may be deposited to form *lateral and *medial moraines.

surface run-off overland transfer of water after a rainfall. It is the most rapid way in which water reaches a river. The amount of surface run-off increases given:
(1) heavy and prolonged rainfall;
(2) steep gradients;
(3) lack of vegetation cover;
(4) saturated or frozen soil.

A *hydrograph can indicate the time the run-off takes to reach the river. Throughflow is another way water reaches a river.

surge abnormally high tide; see *storm surge.

surging glacier glacier that has periods, generally lasting one to four

years, of very rapid flow (up to several metres per hour compared to normal glaciers which move at a rate of several metres per year) followed by periods of stagnation lasting up to ten years. Surging glaciers are heavily crevassed.

survey means of finding out information by posing questions of individuals or organizations. Surveys may be carried out by post, telephone, or personal interview. Most surveys only involve a sample of respondents.

surveying accurate measuring of the Earth's crust, or of land features or buildings. It is used to establish boundaries, and to evaluate the topography for engineering work. The measurements used are both linear and angular, and geometry and trigonometry are applied in the calculations.

Survival International organization formed in 1969 to support tribal peoples and their right to decide their own future, and to help them protect their lands, environment, and way of life. It operates in more than 60 countries worldwide. Its headquarters are in London, England.

suspended load see *load.

suspension in earth science, the *sediment that is carried by a river, or a wave, and is kept off the floor (bed) of the river, or shore. The material is carried within the body of water and is held up by turbulent flow. Most of the particles are less than 0.2 mm in diameter. In a glacier, most of the load is held in suspension by layers of ice.

sustainable capable of being continued indefinitely. For example, the sustainable yield of a forest is equivalent to the amount that grows back. Environmentalists made the term a catchword, in advocating the sustainable use of resources.

sustainable development the ability of a country to maintain a level of economic development, thus enabling the majority of the population to have a reasonable standard of living.

sustained-yield cropping in ecology, the removal of surplus individuals from a *population of organisms so that the population maintains a constant size. This usually requires selective removal of animals of all ages and both sexes to ensure a balanced population structure. Taking too many individuals can result in a population decline, as in overfishing.

Excessive cropping of young females may lead to fewer births in following years, and a fall in population size. Appropriate cropping frequencies can be determined from an analysis of a life table.

swallet alternative name for a *swallow hole.

swallow hole or **swallet**, hole, often found in limestone areas, through which a surface stream disappears underground. It will usually lead to an underground network of caves. Gaping Gill in North Yorkshire, England, is an example.

swamp region of low-lying land that is permanently saturated with water and usually overgrown with vegetation; for example, the everglades of Florida, USA. A swamp often occurs where a lake has filled up with sediment and plant material. The flat surface so formed means that run-off is slow, and the water table is always close to the surface. The high humus content of swamp soil means that good agricultural soil can be obtained by draining.

swash advance of water and sediment up a beach as a *wave breaks. Swash plays a significant role in the movement of beach material by *longshore drift, and is responsible for throwing shingle and pebbles up a beach to create ridges called *berms.

sweatshop workshop or factory where employees work long hours under substandard conditions for low wages. Exploitation of labour in this way is associated with unscrupulous employers, who often employ illegal immigrants or children in their labour force.

swell see *wave.

syenite grey, crystalline, plutonic (intrusive) *igneous rock, consisting of feldspar and hornblende; other minerals may also be present, including small amounts of quartz.

syncline geological term for a fold in the rocks of the Earth's crust in which the layers or *beds dip inwards, thus forming a trough-like structure with a sag in the middle. The opposite structure, with the beds arching upwards, is an *anticline.

synoptic chart weather chart in which symbols are used to represent the weather conditions experienced over an area at a particular time. Synoptic charts appear on television and newspaper forecasts, although the symbols used may differ.

taiga or **boreal forest**, Russian name for the forest zone south of the *tundra, found across the northern hemisphere. Here, dense forests of conifers (spruces and hemlocks), birches, and poplars occupy glaciated regions punctuated with cold lakes, streams, bogs, and marshes. Winters are prolonged and very cold, but the summer is warm enough to promote dense growth.

The varied fauna and flora are in delicate balance because the conditions of life are so precarious. This ecology is threatened by mining, forestry, and pipeline construction.

talc $Mg_3Si_4O_{10}(OH)_2$, mineral, hydrous magnesium silicate. It occurs in tabular crystals, but the massive impure form, known as **steatite** or **soapstone**, is more common. It is formed by the alteration of magnesium compounds and is usually found in metamorphic rocks. Talc is very soft, ranked 1 on the Mohs scale of hardness. It is used in powdered form in cosmetics, lubricants, and as an additive in paper manufacture.

tarn the postglacial lake which often occupies a *corrie.

Taylorism see *scientific management.

TBT *abbreviation for* tributyl tin, chemical used in antifouling paints that has become an environmental pollutant.

technocracy society controlled by technical experts such as scientists and engineers. The term was invented by US engineer W H Smyth (1855–1940) 1919 to describe his proposed 'rule by technicians', and was popularized by James Burham (1905–87) in *Managerial Revolution* 1941.

tectonics in geology, the study of the movements of rocks on the Earth's surface. On a small scale tectonics involves the formation of *folds and *faults, but on a large scale *plate tectonics deals with the movement of the Earth's surface as a whole.

telecottage centre, particularly in rural areas, where people come together to work remotely using technology such as fax machines, telephones, and the Internet. The telecottage supplies the office machinery and facilities for a fee.

temperate climate a climate typical of mid-latitudes. The British Isles, for example, have a temperate climate. Such a climate is intermediate between the extremes of hot (tropical) and cold (polar) climates. Compare *extreme climate. See also *maritime climate.

temperate deciduous forest type of vegetation mass which covers over 4.5% of the Earth's land surface and is found mainly in Europe, the USA, and China. Common tree species are oak, maple, and beech and the annual leaf fall means that soils are rich in nutrients.

The overall area of temperate deciduous forest has declined rapidly since the agrarian revolution. In Western Europe reafforestation schemes are gradually leading to the expansion of forest area.

temperate grasslands a vegetation type, dominated by deep-rooted native grasses, found in areas which have a dry continental climate with a summer drought. Temperate grasslands cover 6% of the Earth's land surface and are found in the USA (*prairies), Central Asia (*steppe), and South America (pampas).

temperature   measure of how hot an object is. It is temperature difference that determines whether heat transfer will take place between two objects and in which direction it will flow, that is from warmer object to cooler object. The temperature of an object is a measure of the average kinetic energy possessed by the atoms or molecules of which it is composed. The SI unit of temperature is the kelvin (symbol K) used with the Kelvin scale. Other measures of temperature in common use are the *Celsius scale and the *Fahrenheit scale. The Kelvin scale starts at absolute zero (0 K = −273°C). The Celsius scale starts at the freezing point of water (0°C = 273 K). 1 K is the same temperature interval as 1°C.

temperature inversion   see *lapse rate.

temporal niche diversification   in ecology, sharing of a niche by different species separated in time, so that only one species occupies the niche for a particular period. In this way the resources of a habitat are fully exploited but inter-specific competition is reduced.

tenure   employment terms and conditions. Security of tenure is often granted to the judiciary, civil servants, educators, and others in public office, where impartiality and freedom from political control are considered necessary.

terminal moraine   linear, slightly curved ridge of rocky debris deposited at the front end, or snout, of a glacier. It represents the furthest point of advance of a glacier, being formed when deposited material (till), which was pushed ahead of the snout as it advanced, became left behind as the glacier retreated.

tephigram   see *meteorology.

terms of trade   in international trade, the ratio of export prices to import prices. An improvement in the terms of trade (there is an increase in the

value of the ratio) should mean that the country is better off, having to give foreigners fewer exports for the same number of imports as before. *Devaluation of the currency leads to a deterioration of the terms of trade.

terrace   1. see *river terrace.
2. a type of housing in which all the dwelling units are joined together, as opposed to detached or semi-detached housing.

terracette   a small terrace on a slope formed by *solifluction or *slumps. Terracettes may be up to 30 cm in height and width and usually occur in parallel groups following the contours of a slope. Cattle and sheep often use terracettes as paths across slopes and this may lead to the growth of the feature.

terracing   a means of *soil conservation and land utilization whereby steep hillsides are engineered into a series of flat ledges which can be used for *agriculture, held in places by stone banks to prevent *soil erosion. Terracing reaches its most sophisticated and intricate form in the fertile volcanic hills of Java, Indonesia, where a subsistence rice economy prevails.

terrane   tract of land with a distinct geological character. The term **exotic terrane** is commonly used to describe a rock mass that has a very different history from others near by. The exotic terranes of the Western *Cordillera of North America represent old island chains that have been brought to the North American continent by the movements of plate tectonics, and welded to its edge.

terra rosa   a red clay-rich soil which develops on parent material with a high calcium carbonate content. Terra rosa soils comprise the insoluble residue products of the chemical weathering of limestone and as such they are usually very old. They are most common in semi-arid regions which border the Mediterranean Basin and in parts of Australia. The red colouration arises from the oxidized (ferric) state of the iron compounds in the parent material. They form fertile agricultural soils.

terrigenous derived from or pertaining to the land. River sediment composed of weathered rock material and deposited near the mouth of the river on the ocean's continental shelf (the shallow ledge extending out from the continent) is called **terrigenous sediment**.

Tertiary period period of geological time 65–1.64 million years ago, divided into five epochs: Palaeocene, Eocene, Oligocene, Miocene, and Pliocene. During the Tertiary period, mammals took over all the ecological niches left vacant by the extinction of the dinosaurs, and became the prevalent land animals. The continents took on their present positions, and climatic and vegetation zones as we know them became established. Within the geological time column the Tertiary follows the Cretaceous period and is succeeded by the Quaternary period.

tertiary sector that sector of the economy which provides services such as transport, finance and retailing, as opposed to the *primary industry sector which provides *raw materials, the secondary sector which processes and manufactures products, and the quaternary sector which provides information and expertise. In highly developed economies the tertiary sector is the dominant employer, though in recent years numbers employed in the quaternary sector have gone up due to increased use of computers and electronic information. See *industrial sector.

Test Ban Treaty agreement signed by the USA, the USSR, and the UK on 5 August 1963 contracting to test nuclear weapons only underground. All nuclear weapons testing in the atmosphere, in outer space, and under water was banned. In the following two years 90 other nations signed the treaty, the only major nonsignatories being France and China, which continued underwater and ground-level tests. In January 1996 France announced the ending of its test programme, and supported the implementation of a universal test ban. The treaty did not restrict or regulate underground testing, or the possession or use of nuclear weapons during wartime.

Tethys Sea sea that in the Mesozoic era separated *Laurasia from *Gondwanaland. The formation of the Alpine fold mountains caused the sea to separate into the Mediterranean, the Black, the Caspian, and the Aral seas.

textile industry the manufacture of cloth and related materials. Traditionally the textile industry used wool, cotton, flax and other natural products as its *raw materials, but this sector has declined with the rise in use of *artificial fibres derived from oil and other hydrocarbons. Nylon, rayon and polyester are examples of such fibres

theocracy political system run by priests, as was once found in Tibet. In practical terms it means a system where religious values determine political decisions. The closest modern examples have been Iran during the period when Ayatollah Khomeini was its religious leader, 1979–89, and Afghanistan under the Islamic fundamentalist Taliban regime, 1996–2001. The term was coined by the historian Josephus in the 1st century AD.

theory of three worlds view expounded by Chinese communist leader Deng Xiaoping at the United Nations General Assembly in 1974. He maintained that the two superpowers the USA and the USSR – were seeking world hegemony and that China, as a developing socialist country, should oppose this by making firmer links with other countries in the developing world.

therblig system that divides a task into a number of individual steps. The term was coined by US pioneers of the scientific management movement Frank Gilbreth and Lilian Gilbreth, who invented and refined the system between 1908 and 1924. The origins of the system were the motion studies the Gilbreths carried out on bricklayers.

thermal column of warm air, which is of a lower density than its surroundings, and contains rising currents of air.

thermal power station an electricity-generating plant which burns *coal, oil or natural gas to produce steam to drive turbines.

thermometer instrument for measuring temperature. There are many types, designed to measure different temperature ranges to varying degrees of accuracy. Each makes use of a different physical effect of temperature. Expansion of a liquid is employed in common **liquid-in-glass thermometers**, such as those containing mercury or alcohol. The more accurate **gas thermometer** uses the effect of temperature on the pressure of a gas held at constant volume. A **resistance thermometer** takes advantage of the change in resistance of a conductor (such as a platinum wire) with variation in temperature. Another electrical thermometer is the thermocouple. Mechanically, temperature change can be indicated by the change in curvature of a **bimetallic strip** (as commonly used in a thermostat).

thermosphere layer in the Earth's *atmosphere above the mesosphere and below the exosphere. Its lower level is about 80 km above the ground, but its upper level is undefined. The ionosphere is located in the thermosphere. In the thermosphere the temperature rises with increasing height to several thousand degrees Celsius. However, because of the thinness of the air, very little heat is actually present.

Third World former term used to describe countries of the *developing world, now considered derogatory. The classifications First (western industrialized free-market), Second (eastern Communist bloc), and Third (developing or non-aligned) worlds arose during the Cold War, but began to lose their political meaning as the Cold War came to an end in the late 1980s.

Three Gorges Dam dam being built to harness the power of the Chang Jiang. It will create a reservoir 600 km long and its turbines will have a 18,000 megawatt capacity (eight times the capacity of Egypt's Aswan Dam). Construction began in December 1994.

In November 1997 the last gap in the main channel of the Chang Jiang River was blocked, and the water forced into a man-made channel.

The Three Gorges Dam will be the world's largest dam and hydroelectric power station, and will save the central and lower river valley from annual flooding. The dam will be 2309 m long and 175 m high, raising the water level in the Three Gorges area by 40–50 m. The major technical problem to be overcome is siltation, as 530 million tonnes of silt is carried down in the river every year, and much of this will be trapped behind the dam walls. The project is scheduled for completion by about 2009 and will cost $22–34 billion. More than a million homes will be lost during the process, which will involve the resettlement of up to 1.8 million people in the Hubei and Sichuan provinces.

THRESHOLD

threshold see *sill (sense 2).

threshold population in geography, the minimum number of people necessary before a particular good or service will be provided in an area. Typically a low-order shop (such as a grocer or newsagent) may require only 800 or so customers, whereas a higher-order store such as Marks and Spencer may need a threshold of 70,000 to be profitable, and a university may need 350,000 to be viable.

Thresholds may also be linked to the spending power of customers; this is most obvious in periodic markets in poor countries, where wages are so low that people can buy the goods or services only once in a while.

throughflow the lateral and downslope subsurface movement of water through the soil. The erosive effect of throughflow is usually limited due to its low velocity. Throughflow often becomes concentrated in natural 'pipes' in the soil to form *percolines* which may flow into rivers or cause springs.

thunderstorm severe storm of very heavy rain, thunder, and lightning. Thunderstorms are usually caused by the intense heating of the ground surface during summer. The warm air rises rapidly to form tall cumulonimbus clouds with a characteristic anvil-shaped top. Electrical charges accumulate in the clouds and are discharged to the ground as flashes of lightning. Air in the path of lightning becomes heated and expands rapidly, creating shock waves that are heard as a crash or rumble of thunder.

The rough distance between an observer and a lightning flash can be calculated by timing the number of seconds between the flash and the thunder. A gap of three seconds represents about a kilometre; five seconds represents about a mile.

tidal limit the point upstream of which there is no tidal rise and fall in river level.

tidal range the mean difference in water level between high and low tides at a given location. See *tide.

tidal wave common name for a *tsunami.

tide rhythmic rise and fall of the *sea level in the Earth's oceans and their inlets and estuaries due to the gravitational attraction of the Moon and, to a lesser extent, the Sun, affecting regions of the Earth unequally as it rotates. Water on the side of the Earth nearest to the Moon feels the Moon's pull and accumulates directly below the Moon, producing a high tide.

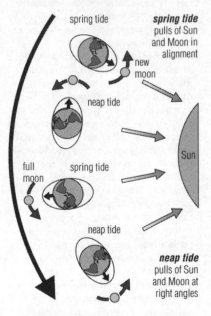

spring tide

spring tide pulls of Sun and Moon in alignment

new moon

neap tide

full moon

spring tide

Sun

neap tide

neap tide pulls of Sun and Moon at right angles

tide The gravitational pull of the Moon is the main cause of the tides. Water on the side of the Earth nearest the Moon feels the Moon's pull and accumulates directly under the Moon. When the Sun and the Moon are in line, at new and full Moon, the gravitational pull of Sun and Moon are in line and produce a high spring tide. When the Sun and Moon are at right angles, lower neap tides occur.

High tide occurs at intervals of 12 hr 24 min 30 sec. The maximum high tides, or spring tides, occur at or near new and full Moon when the Moon and Sun are in line and exert the greatest combined gravitational pull. Lower high tides, or neap tides, occur when the Moon is in its first or third quarter and the Moon and Sun are at right angles to each other.

tied cottage   a house on a farm or estate provided for a worker as part of his or her terms of employment. It is either free or let at a very low rent.

till   or **boulder clay**, deposit of clay, mud, gravel, and boulders left by a *glacier. It is unsorted, with all sizes of fragments mixed up together, and shows no stratification; that is, it does not form clear layers or *beds.

time zone   longitudinal strip of the Earth's surface, stretching from pole to pole and sharing the same time of day or night. In a 24-hour period the Earth makes one complete rotation on its axis; thus the direct rays of the Sun pass through one degree of longitude every 4 minutes. To allow for time changes on an hourly basis, each time zone covers 15 degrees of longitude in width. In practice, however, zone boundary lines are adjusted to accommodate political units. (See also *international date line).

tin ore   mineral from which tin is extracted, principally cassiterite, SnO_2. The world's chief producers are Malaysia, Thailand, and Bolivia.

The UK was a major producer in the 19th century but today only a few working mines remain, and production is small.

titanium ore   any mineral from which titanium is extracted, principally ilmenite ($FeTiO_3$) and rutile (TiO_2). Brazil, India, and Canada are major producers. Both these ore minerals are found either in rock formations or concentrated in heavy mineral sands.

topaz   mineral, aluminium fluorosilicate, $Al_2(F_2SiO_4)$. It is usually yellow, but pink if it has been heated, and is used as a gemstone when transparent. It ranks 8 on the Mohs scale of hardness.

tombolo   a *spit which extends to join an island to the mainland, as in the case of Chesil Beach, Portland Island, southern England.

topography   surface shape and composition of the landscape, comprising both natural and artificial features, and its study. Topographical features include the relief and contours of the land; the distribution of mountains, valleys, and human settlements; and the patterns of rivers, roads, and railways.

topsoil   the uppermost layer of *soil, more rich in organic matter than the underlying subsoil. See *soil profile.

tor   isolated mass of rock, often granite, left upstanding on a hilltop after the surrounding rock has been broken down. Weathering takes place along the joints in the rock, reducing the outcrop into a mass of rounded blocks.

tornado   extremely violent revolving storm with swirling, funnel-shaped clouds, caused by a rising column of warm air propelled by strong wind. A tornado can rise to a great height, but with a diameter of only a few hundred metres or less. Tornadoes move with wind speeds of 160–480 kph, destroying everything in their path. They are common in the central USA and Australia.

A series of tornadoes killed 47 people, destroyed 2000 homes, and caused $500 million worth of damage in Oklahoma, Nebraska, Kansas, and Texas in May 1999.

totalitarianism   government control of all activities within a country, openly political or otherwise, as in fascist or communist dictatorships. Examples of totalitarian regimes are

Italy under Benito Mussolini 1922–45; Germany under Adolf Hitler 1933–45; the USSR under Joseph Stalin from the 1930s until his death in 1953; and more recently Romania under Nicolae Ceausescu 1974–89.

tourism travel and visiting places for pleasure, often involving sightseeing and staying in overnight accommodation. Regarded as an industry, tourism can increase the wealth and job opportunities in an area, although the work is often seasonal and low paid. Among the negative effects of tourism are traffic and people congestion, and damage to the environment.

Tourism is the world's largest industry. It sustained 120 million jobs in 1995, accounting for 7% of the global workforce. It is estimated that the number of international travellers in 1994 will double to 1 billion by 2010, and 80% of tourists come from the 20 richest countries.

tourmaline hard, brittle mineral, a complex silicate of various metals, but mainly sodium aluminium borosilicate.

Small tourmalines are found in granites and gneisses. The common varieties range from black (schorl) to pink, and the transparent gemstones may be colourless (achromatic), rose pink (rubellite), green (Brazilian emerald), blue (indicolite, verdelite, Brazilian sapphire), or brown (dravite).

toxic waste *hazardous waste, especially when it has been dumped.

trace element chemical element necessary in minute quantities for the health of a plant or animal. For example, magnesium, which occurs in chlorophyll, is essential to photosynthesis, and iodine is needed by the thyroid gland of mammals for making hormones that control growth and body chemistry.

trade exchange of commodities between groups, individuals, or countries. Direct trade is usually known as barter, whereas indirect trade is carried out through a medium such as money.

trade cycle another term for *business cycle.

trade wind prevailing wind that blows towards the Equator from the northeast and southeast. Trade winds are caused by hot air rising at the Equator and the consequent movement of air from north and south to take its place. The winds are deflected towards the west because of the Earth's west-to-east rotation.

The unpredictable calms known as the *doldrums lie at their convergence.

The trade-wind belts move north and south about 5° with the seasons. The name is derived from the obsolete expression 'blow trade' meaning to blow regularly, which indicates the trade winds' importance to navigation in the days of cargo-carrying sailing ships.

traditional economy economy based on subsistence agriculture where small family groups or tribes produce nearly all of what they need themselves. There is therefore very little trade, and *barter rather than money is used for any trade that does take place. In a traditional economy, people are generally averse to risk, preferring to keep to traditional modes of production and avoiding change.

TRAFFIC the arm of the *World Wide Fund for Nature (WWF) that monitors trade in endangered species.

transfer earnings earnings which a factor of production, such as a worker, could earn in its next highest paid use or occupation. For example, if the maximum a footballer who earns £50,000 a year could earn in any other job was £20,000 as a sports centre manager, then his transfer earnings would be £30,000.

transform margin or **transform boundary**, in *plate tectonics, a boundary between two lithospheric plates in which the plates move past each other rather than towards or away from one another. Transform boundaries are characterized by large *strike-slip faults called **transform faults**, and powerful and destructive earthquakes.

transhumance the practice whereby herds of farm animals are moved between regions of different climates. Pastoral farmers (see *pastoral farming) take their herds from valley pastures in the winter to mountain pastures in the summer. Frequently farmers will live in mountain huts during the summer in mountainous regions such as the Himalayas and the Alps.

translocation the movement of materials, often in solution, through a *soil profile. Clay particles, plant nutrients and dissolved iron compounds are common examples of substances that can be translocated between different soil horizons. In climates in which *precipitation exceeds the rate of *evapotranspiration, translocation will result in the washing down of materials and may result in the formation of an ironpan or claypan (see *hardpan). When evapotranspiration exceeds precipitation the dominant movement through the horizons will be upwards and salts may accumulate at the surface in a *salt pan. Translocation incorporates a variety of processes including *leaching, *eluviation, *illuviation, *calcification, *salinization.

transnational corporation concept developed by business school academics Christopher Bartlett and Sumantra Ghoshal in the late 1980s and early 1990s to describe an emergent organizational model that transcended national boundaries. In their work they identified four forms of multinational organization: multinational (decentralized with strong local presence), global (more centralized, achieves cost advantages through centralized production), international (exploits parent company's research and development and systems capability), and transnational. The fourth form, the transnational, combines local responsiveness with global efficiency.

transpiration the process whereby plants give off water vapour via the stomata of their leaves. Water taken up by roots is thus returned to the *atmosphere.

transportation the movement of soil and rock fragments by the action of rivers, glaciers, sea and wind. See *slump, *creep, *solifluction, *mudflow, *earthflow, *fluvial or glacial transportation.

transport network or **communication network**, an interconnected system of fixed lines of movement between *activity nodes. The main means of carrying traffic are roads, footpaths, railways, and inland *waterways. Sea-travel routes and air corridors also form parts of transport networks but are neither fixed in position nor visible. Each section of fixed line forms a link in a national network, connecting junction points at which travellers change mode of travel or terminate a journey.

trawling a method of fishing in which a large bag-shaped net is dragged or trawled at deep levels behind a specially adapted boat. The older form of trawl net had a 10 m wooden beam across its mouth to hold it open. The modern net has a mouth of approximately 25 m and has large flat plates (*otter boards*) attached to either side of it; as the net is dragged along, the water pressure forces the otter boards apart and so keeps the net open. Trawling is the single most important method of fishing in terms of catch.

tree line or **timber line**, the altitudinal limit of tree growth. Below the tree line, trees are erect but often as the tree line is approached the trees become stunted and distorted by the wind and may eventually assume a horizontal, prostrate, mat-like form known as *krummholz*. Above the treeline, exposure, frost damage and snow cover all prevent tree colonization.

trellis drainage see *drainage pattern.

tremor minor *earthquake.

trench see *ocean trench.

Triassic Period period of geological time 245–208 million years ago, the first period of the Mesozoic era. The present continents were fused together in the form of the world continent *Pangaea. Triassic sediments contain remains of early dinosaurs and other animals now extinct. By late Triassic times, the first mammals had evolved.

There was a mass extinction of 95% of plants at the end of the Triassic possibly caused by rising temperatures.

The climate was generally dry; desert sandstones are typical Triassic rocks.

tribal society way of life in which people govern their own affairs as independent local communities of families and clans without central government organizations or states. They are found in parts of Southeast Asia, New Guinea, South America, and Africa.

As the world economy expands, natural resources belonging to tribal peoples are coveted and exploited for farming or industrial use and the people are frequently dispossessed. Pressure groups such as *Survival International and Cultural Survival have been established in some Western countries to support the struggle of tribal peoples for property rights as well as civil rights within the borders of the countries of which they are technically a part.

tributyl tin TBT, chemical used in antifouling paints on ships' hulls and other submarine structures to deter the growth of barnacles. The tin dissolves in sea water and enters the food chain. It can cause reproductive abnormalities such as exposed female whelks developing penises; the use of TBT has therefore been banned in many countries, including the UK.

troglodyte ancient Greek term for a cave dweller, designating certain pastoral peoples of the Caucasus, Ethiopia, and the southern Red Sea coast of Egypt.

troilite FeS, iron sulphide, mineral abundant in meteorites and probably present in the Earth's core.

trophic level in ecology, the position occupied by a species (or group of species) in a *food chain. The main levels are **primary producers** (photosynthetic plants), **primary consumers** (herbivores), **secondary consumers** (carnivores), and **decomposers** (bacteria and fungi).

tropical cyclone another term for *hurricane.

tropical storm intense low pressure system which forms over the warm oceans of the world's tropical areas, occurring in late summer and early autumn. Warm, moist air rises in a spiral to form a storm that can measure several kilometres across. The weather conditions vary considerably throughout the duration of the storm. Coastal areas are usually the worst affected by tropical storms, which can bring about both human and financial damage. Areas affected by a tropical storm often lose their electricity, water, and sewerage facilities. Tropical storms in different parts of the world are known by various names: *hurricane, *tropical cyclone, or *typhoon.

As a storm approaches, temperature and pressure both fall rapidly, whilst cloud cover and wind speed increase. Rainfall gradually increases until it is torrential, often with thunder and lightning. Wind speeds can reach up to 160 km per hour. The passage of the calm 'eye' of the storm may take two or three hours. During this time, the skies clear and the temperature and pressure rise. Once the eye has passed the storm begins again with renewed force. This time the winds blow from the opposite direction. The whole storm moves at about 25 km per hour and takes several hours finally to move away.

tropics   area between the tropics of Cancer and Capricorn, defined by the parallels of latitude approximately 23°30' north and south of the Equator. They are the limits of the area of Earth's surface in which the Sun can be directly overhead. The mean monthly temperature is over 20°C.

Climates within the tropics lie in parallel bands. Along the Equator is the *intertropical convergence zone, characterized by high temperatures and year-round heavy rainfall. Tropical rainforests are found here. Along the tropics themselves lie the tropical high-pressure zones, characterized by descending dry air and desert conditions. Between these, the conditions vary seasonally between wet and dry, producing the tropical grasslands.

tropopause   thin layer in the Earth's atmosphere whose altitude varies from about 7 km at the poles to 28 km at the equator. It is located at the top of the troposphere.

troposphere   lower part of the Earth's *atmosphere extending about 10.5 km from the Earth's surface, in which temperature decreases with height to about –60°C except in local layers of temperature inversion. The **tropopause** is the upper boundary of the troposphere, above which the temperature increases slowly with height within the atmosphere. All of the Earth's weather takes place within the troposphere.

trough   an area of low pressure, not sufficiently well-defined to be regarded as a depression.

trough end wall   the steep rear wall of a *U-shaped valley, formed where coalescing *corrie *glaciers cause an increase in *erosion and a consequent deepening of the glacial valley.

truncated spur   blunt-ended ridge of rock jutting from the side of a *glacial trough, or valley. As a glacier moves down a former river valley it is unable to flow around the *interlocking spurs that project from either side, and so it erodes straight through them, shearing away their tips and forming truncated spurs.

trust territory   country or area placed within the United Nations trusteeship system and, as such, administered by a UN member state on the UN's behalf. A trust territory could be one of three types: one administered under a mandate given by the UN, or its predecessor, the League of Nations; a territory which was removed from an enemy state, namely Germany, Italy, or Japan, at the end of World War II; or a territory which had been placed voluntarily within the trusteeship system by a member state responsible for its administration. The last territory remaining under the UN trusteeship system, the Republic of Palau, became independent in 1994.

tsunami   (Japanese 'harbour wave') ocean wave generated by vertical movements of the sea floor resulting from *earthquakes or volcanic activity or large submarine landslides. Unlike waves generated by surface winds, the entire depth of water is involved in the wave motion of a tsunami. In the open ocean the tsunami takes the form of several successive waves, rarely in excess of 1 m in height but travelling at speeds of 650–800 kph. In the coastal shallows tsunamis slow down and build up producing huge swells over 15 m high in some cases and over 30 m in rare instances. The waves sweep inland causing great loss of life and damage to property.

Before each wave there may be a sudden withdrawal of water from the beach. Used synonymously with tsunami, the popular term 'tidal wave' is misleading: tsunamis are not caused by the gravitational forces that affect *tides.

tufa or **travertine**, soft, porous, *limestone rock, white in colour, deposited from solution from carbonate-saturated ground water around hot springs and in caves.

tuff volcanic ash or dust which has been consolidated into *rock.

Tullgren funnel in biology, device used to extract mites, springtails, fly larvae, and other small invertebrates from a sample of soil.

Soil resting on a net inside a funnel is illuminated from above, so that the top layers of the soil warm and begin to dry out. The invertebrates, trying to escape the desiccating effect of the heat burrow downwards and in the process drop out of the soil and through the net, to be collected in a beaker resting below the funnel.

tundra region in high latitudes with almost no trees – they cannot grow because the ground is permanently frozen (*permafrost). The vegetation consists mostly of grasses, sedges, heather, mosses, and lichens. Tundra stretches in a continuous belt across northern North America and Eurasia. Tundra is also used to describe similar conditions at high altitudes.

The term was originally applied to the topography of part of northern Russia, but is now used for all such regions.

tungsten ore either of the two main minerals, wolframite (FeMn)WO$_4$ and scheelite, CaWO$_4$, from which tungsten is extracted. Most of the world's tungsten reserves are in China, but the main suppliers are Bolivia, Australia, Canada, and the USA.

turbidity current gravity-driven current in air, water, or other fluid resulting from accumulation of suspended material, such as silt, mud, or volcanic ash, and imparting a density greater than the surrounding fluid. Marine turbidity currents originate from tectonic movement, storm waves, tsunamis (tidal waves) or earthquakes and move rapidly downward, like underwater avalanches, leaving distinctive deposits called **turbidites**. They are thought to be one of the mechanisms by which submarine canyons are formed.

turnkey project large-scale project, such as the building of a manufacturing plant, where the supplier builds, installs, and tests production processes before handing the plant over. The plant is then ready to start 'at the turn of a key'. In computing, the term is used to describe an IT system that is delivered and installed ready to run.

turquoise mineral, hydrous basic copper aluminium phosphate, CuAl$_6$(PO$_4$)$_4$(OH)$_8$5H$_2$O. Blue-green, blue, or green, it is a gemstone. Turquoise is found in Australia, Egypt, Ethiopia, France, Germany, Iran, Turkestan, Mexico, and southwestern USA. It was originally introduced into Europe through Turkey, from which its name is derived.

24/7 business conducting business 24 hours a day, 7 days a week. This concept has become more prevalent since the focus of companies on customer service, for example permanently open grocery stores.

twilight period of faint light that precedes sunrise and follows sunset, caused by the reflection of light from the upper layers of the atmosphere. The limit of twilight is usually regarded as being when the Sun is 18° below the horizon. The length of twilight depends on the latitude – at the tropics, it only lasts a few minutes; near the poles, it may last all night.

typhoon violent revolving storm, a *hurricane in the western Pacific Ocean.

ultrabasic   in geology, an igneous rock with a lower silica content than basic rocks (less than 45% silica). Part of a system of classification based on the erroneous concept of silica acidity and basicity. Once used widely it has now been largely replaced by the term **ultramafic**.

ultraviolet radiation   electromagnetic radiation of wavelengths from about 400–10 nanometres (where the X-ray range begins). Physiologically, ultraviolet radiation is extremely powerful, producing sunburn and causing the formation of vitamin D in the skin. Ultraviolet radiation is invisible to the human eye, but its effects can be demonstrated.

UN   *abbreviation for* *United Nations.

unconformity   surface of erosion or nondeposition eventually overlain by younger *sedimentary rock strata and preserved in the geologic record. A surface where the *beds above and below lie at different angles is called an **angular unconformity**. The boundary between older igneous or metamorphic rocks that are truncated by erosion and later covered by younger sedimentary rocks is called a **nonconformity**.

UNCTAD   acronym for **United Nations Conference on Trade and Development**.

underdevelopment   term generally referring to countries which have not gone through the phase of industrialization (the development of mass production in factories) and have not reached post-industrialization (where the proportion of output and employment devoted to services rather than goods is expanding).

Underdevelopment has been subject to changing definitions, but countries which are referred to as less developed are those with a relatively high proportion of output and employment devoted to agriculture, and a low national income per head. Under these criteria, the majority of countries in Africa, Asia, and Central and South America would qualify as underdeveloped. There is disagreement over the extent to which economic development constitutes 'progress' or improvement in the quality of life.

underemployment   inefficient use of the available workforce. Underemployment exists when many people are performing tasks that could easily be done by fewer workers, or when a large percentage of the workforce is unemployed.

undernourishment   condition that results from consuming too little food over a period of time. Like **malnutrition** – the result of a diet that is lacking in certain nutrients (such as protein or vitamins) – undernourishment is common in poor countries. Both lead to a reduction in mental and physical efficiency, a lowering of resistance to disease in general, and often to deficiency diseases such as beriberi or anaemia. In the developing world, lack of adequate food is a common cause of death.

In 1996, an estimated 195 million children under the age of five were undernourished in the world. Undernourishment is not just a problem of the developing world: there were an estimated 12 million children eating inadequately in the USA in 1992. According to UN figures there were 200 million Africans suffering from undernourishment in 1996.

underpopulation too few people for the resources available in an area (such as food, land, and water). Underpopulated countries, like Canada and Australia, have a vast wealth of resources that could be exploited, such as food, energy, and minerals, and are able to export these commodities whilst maintaining a high standard of living. They have well-developed economic infrastructures and relatively highly skilled populations with generally high incomes.

undertow the counter-current to water breaking onshore as waves. The undertow is responsible for the removal of eroded material from the *wave-cut platform, to be deposited as the *shoreface terrace. The undertow operates at a much larger scale than *backwash.

unemployment lack of paid employment. The unemployed are usually defined as those out of work who are available for and actively seeking work. Unemployment is measured either as a total or as a percentage of those who are available for work, known as the working population, or labour force. Periods of widespread unemployment in Europe and the USA in the 20th century include 1929–39, and the years since the mid-1970s. According to a report released by the UN's International Labour Organization November 1995, nearly 1 billion people, about 30% of the global workforce, were out of work or underemployed. The reduction in job opportunities was said to be due to lower growth rates in industrialized countries since 1973, and the failure of most developing nations to recover fully from the economic crisis of the early 1980s. The ILO argued that despite increasing worldwide competition, the 1996 jobless figures were neither politically nor socially sustainable.

Unemployment in industrialized countries (the members of the *Organization for Economic Cooperation and Development (OECD)) in 2003 averaged 7.1%, and in the European Union (EU) 8.8% in 2003. Within the OECD group the country with the lowest percentage of unemployed in 2003 was Korea (3.6%) and the highest was Poland (19.2%).

UNEP acronym for **United Nations Environmental Programme**.

UNESCO acronym for United Nations Educational, Scientific, and Cultural Organization, specialized agency of the United Nations, established in 1946, to promote international cooperation in education, science, and culture, with its headquarters in Paris.

UNHCR *abbreviation for* **United Nations High Commission for Refugees**.

uniformitarianism in geology, the principle that processes that can be seen to occur on the Earth's surface today are the same as those that have occurred throughout geological time. For example, desert sandstones containing sand-dune structures must have been formed under conditions similar to those present in deserts today. The principle was formulated by Scottish geologists James Hutton and expounded by Charles Lyell.

unilateralism in politics, support for **unilateral nuclear disarmament**: scrapping a country's nuclear weapons without waiting for other countries to agree to do so at the same time.

unitary authority administrative unit of Great Britain. Since 1996 the two-tier structure of local government has ceased to exist in Scotland and Wales, and in some parts of England, and has been replaced by unitary authorities, responsible for all local government services.

United Nations UN, association of states for international peace, security, and cooperation, with its headquarters in New York City. The UN was established on 24 October 1945 by 51 states as a successor to the League of Nations. Its Charter, whose obligations member states agree to accept, sets out four purposes for the UN: to maintain international peace and security; to develop friendly relations among nations; to cooperate in solving international problems and in promoting respect for human rights; and to be a centre for the harmonizing the actions of nations. The UN has played a role in development assistance, disaster relief, cultural cooperation, aiding refugees, and peacekeeping. Its membership in 2004 stood at 191 states, and the total budget for 2002–03 (raised by the member states) was US$2.9 billion, supporting more than 50,000 staff.

The UN has six principal institutions. Five are based in New York: the General Assembly, the Security Council, the Economic and Social Council, the Trusteeship Council, and the Secretariat. The sixth, the International Court of Justice, is located at the Peace Palace in the Hague, Netherlands. Kofi Annan, from Ghana, became secretary general in 1997, and was re-elected for a second term in 2001. In January 1998, Louise Fréchette, a Canadian, was elected its first deputy secretary general. In October 2001, Annan and the UN itself were awarded the 2001 Nobel Prize for Peace. There are six official working languages: English, French, Russian, Spanish, Chinese, and Arabic. The name 'United Nations' was coined by US President Franklin D Roosevelt.

unleaded petrol petrol manufactured without the addition of antiknock. It has a slightly lower octane rating than leaded petrol, but has the advantage of not polluting the atmosphere with lead compounds. Many cars can be converted to run on unleaded petrol by altering the timing of the engine, and most new cars are designed to do so. Cars fitted with a catalytic converter must use unleaded fuel.

Aromatic hydrocarbons and alkenes are added to unleaded petrol instead of lead compounds to increase the octane rating. After combustion the hydrocarbons produce volatile organic compounds. These have been linked to cancer, and are involved in the formation of photochemical smog. A low-lead fuel is less toxic than unleaded petrol for use in cars that are not fitted with a catalytic converter.

Unrepresented Nations' and Peoples' Organization UNPO, international association founded in 1991 to represent ethnic and minority groups unrecognized by the United Nations and to defend the right to self-determination of oppressed peoples around the world. The founding charter was signed by representatives of Tibet, the Kurds, Turkestan, Armenia, Estonia, Georgia, the Volga region, the Crimea, the Greek minority in Albania, North American Indians, Australian Aborigines, West Irians, West Papuans, the minorities of the Cordillera in the Philippines, and the non-Chinese in Taiwan. UNPO is based in the Netherlands.

upper course see *river.

uraninite uranium oxide, UO_2, an ore mineral of uranium, also known as **pitchblende** when occurring in massive form. It is black or brownish-black, very dense, and radioactive. It occurs in veins and as massive crusts, usually associated with granite rocks.

uranium ore material from which uranium is extracted, often a complex mixture of minerals. The main ore is uraninite (or pitchblende), UO_2, which is commonly found with sulphide minerals. The USA, Canada, and South Africa are the main producers in the West.

urban decay urban decay the process of deterioration in the *infrastructure of parts of the city – especially in the old industrial cities of, for example, Northern England and the Midlands. It is the result of long-term shifts in patterns of economic activity and residential location.

Parts of the inner city are especially decayed: old Victorian *terraced housing, mills and other traditional industrial installations such as disused canals and warehouses.

Much of the decay results from neglect, as the focus of the urban system moves away from these areas – for example, towards new peripheral locations for industry, newer housing in the outer suburbs, and new *communications bypassing the city. In order to reverse the problem, many cities have undertaken 'face-lift' schemes to improve the decayed *environments, either by demolition and landscaping, or by renovating old property for new uses. See *comprehensive redevelopment, *inner city.

urban development corporation
UDC, in the UK, an organization set up by the central government to coordinate rapid improvements within depressed city areas. UDCs were first introduced 1981 in the London Docklands and Merseyside. Their aims were typically:
(1) to improve the local environment, making it more attractive to business;
(2) to give cash grants to firms setting up or expanding within the area;
(3) to renovate and reuse buildings;
(4) to offer advice and practical help to firms considering moving to the location.

urban ecology study of the *ecosystems, animal and plant communities, soils and microclimates found within an urban landscape.

Parks are important for many organisms, such as song birds, while birds of prey (the kestrel being a notable example) find ample food in the wasteland around estates and offices. Mammals, including foxes and badgers, are regular visitors, especially if there is an undisturbed corridor penetrating into the town, such as a disused railway.

urban fringe boundary area of a town or city, where new building is changing land use from rural to urban (see *urban sprawl). It is often a zone of planning conflict.

urbanization process by which the proportion of a population living in or around towns and cities increases through migration and natural increase. The growth of urban concentrations in the USA and Europe is a relatively recent phenomenon, dating back only about 150 years to the beginning of the Industrial Revolution (although the world's first cities were built more than 5000 years ago). The UN Population Fund reported in 1996 that within ten years the majority of the world's population would be living in urban conglomerations. Almost all urban growth will occur in the developing world, creating ten large cities a year.

Urbanization has had a major effect on the social structure of industrial societies, affecting not only where people live but how they live.

In England, about 705,000 sq km of former agricultural land was lost to housing, industrial development, and road building between 1945 and 1992.

The UN Centre for Human Settlements (UNCHS) reported in 1996 that the growth of most cities was slower in the 1980s than in any of the previous three decades, owing to many people moving out of, rather than into, cities during that time.

urban land-use model in the social sciences, a simplified pattern of the land use (such as industry, housing,

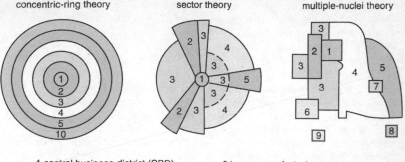

concentric-ring theory   sector theory   multiple-nuclei theory

1 central business district (CBD)   6 heavy manufacturing
2 wholesale light manufacturing   7 outlying business district
3 low-cost housing   8 residential suburb
4 medium-cost housing   9 industrial suburb
5 high-cost housing   10 commuter zone

urban land-use model Three models for looking at the pattern of land use in older towns and cities. The concentric model, suggested by E W Burgess in 1925, displays a series of concentric circles with the central business district (CBD) at the centre and the commuter zone on the outer ring. The sector theory, put forward by Homer Hoyt in 1939, has industrial and residential zones in sectors radiating from the CBD at the centre. The multiple-nuclei theory was developed by D Harris and E L Ullman in 1945. In this model, the different zones grow in several independent areas rather than solely around the CBD.

and commercial activity) that may be found in towns and cities. These models are based on an understanding of the way in which these areas have grown. The most common ways of looking at urban land use are: *concentric-ring theory, *sector theory, and *multiple-nuclei theory. Each results in different shapes of land-use areas. In practice, factors such as *topography, land fertility, and culture vary from one city to another and affect their final form.

urban sprawl   outward spread of built-up areas caused by their expansion. This is the result of *urbanization. Unchecked urban sprawl may join cities into *conurbations; green belt policies are designed to prevent this.

URUPABOL   organization formed 1981 by Bolivia, Paraguay, and Uruguay to foster economic and commercial cooperation.

U-shaped valley   another term for a *glacial trough, a valley formed by a glacier.

valley a long, linear depression sloping downwards towards the sea or an inland drainage basin. Types of valleys include the *V-shaped valley, *U-shaped valley, *hanging valley, *dry valley, *misfit valley, *asymmetric valley, and *rift valley.

Valley of Ten Thousand Smokes valley in southwestern Alaska, on the Alaska Peninsula, where in 1912 Mount Katmai erupted in one of the largest volcanic explosions ever known, although without loss of human life since the area was uninhabited. The valley was filled with ash to a depth of 200 m. It was dedicated as the Katmai National Monument in 1918. Thousands of fissures on the valley floor continue to emit steam and gases.

valley wind see *anabatic wind, katabatic wind.

variation in biology, a difference between individuals of the same species, found in any sexually reproducing population. Variations may be almost unnoticeable in some cases, obvious in others, and can concern many aspects of the organism. Typically, variations in size, behaviour, biochemistry, or colouring may be found. The cause of the variation is genetic (that is, inherited), environmental, or more usually a combination of the two. Some variation is the result of the environment modifying inherited characteristics. The origins of variation can be traced to the recombination of the genetic material during the formation of the gametes, and, more rarely, to mutation.

variance in business, the difference between expected or predicted costs, and actual costs. It is used in budgeting and costing control.

variance analysis looking at variances in budgeting and costings and determining their causes. Variances are not referred to as positive or negative but as favourable or adverse, that is contributing to profits or reducing them.

variety in biology, a stable group of organisms within a single species, clearly different from the rest of the species. Such a group would generally be called a variety for plants and a breed for animals. The differences lie in their genetic make-up and could have arisen naturally–through *natural selection or as a result of selective breeding by humans. Most varieties have been produced by selective breeding, for example 'Cox's', 'Golden Delicious', and 'Bramley' apple varieties.

Variscan see *European geology.

varve in geology, a pair of thin sedimentary beds, one coarse and one fine, representing a cycle of thaw followed by an interval of freezing, in lakes of glacial regions.

varve analysis method of archaeological dating using annual varve (glacial deposit) thickness patterns formed in lakes near the edge of a glacier's retreat. A hot summer results in a thick varve, owing to a greater meltwater rate and discharge of gravel and clay, whereas a cold summer results in a thin one. By analysis of thickness patterns and cross-linking to adjacent regions, a chronology has been developed back to the end of the last Ice Age.

vegetation plant life of a particular area, such as shrubs, trees, and grasses.

The type of vegetation depends on the soil, which may become impoverished by deforestation; climate, especially the amount of moisture available and the possibility of frost, since these factors influence the types of vegetation that can survive; and human activity, such as overgrazing or the destruction of grasslands.

veldt   subtropical grassland in South Africa, equivalent to the Pampas of South America.

vermilion   HgS, red form of mercuric sulphide; a scarlet that occurs naturally as the crystalline mineral *cinnabar.

vesiculation   in earth science, process by which the expansion of gases in molten extrusive igneous rocks forms small spherical cavities (vesicles), which may later become filled with various minerals.

Vesuvius   Italian **Vesuvio**, active volcano in Campania, Italy, 15 km southeast of Naples, Italy; height 1277 m (1969; the height of the main cone changes with each eruption). In AD 79 it destroyed the cities of Pompeii, Herculaneum, and Stabiae. It is the only active volcano on the European mainland.

Vesuvius is a composite *volcano at the **convergent plate margin** where the African plate is subducting beneath the Eurasian plate. Its lava is *andesite in composition and consequently very viscous, giving rise to explosive eruptions. Vesuvius is comprised of two cones. Monte Somma (1277 m), the remnant of a massive wall which once enclosed a huge cone in prehistoric times, is now a semicircular girdle of cliff to the north and east, separated from the main eruptive cone by the valley of Atrio di Cavallo. Layers of lava, scoriae, ashes, and pumice make up the mountain.

The surprising fertility of the volcano's slopes, especially for the cultivation of grapes and production of 'Lacrimae Christi' wine, explains why the environs of Vesuvius remain densely populated in spite of the constant threat of eruption.

eruptions The eruption on 24 August AD 79 ended a dormant period so long that the volcano had been presumed extinct. This, the earliest recorded eruption, buried Pompeii, Herculaneum, and Stabiae under ashes and mud, and was described by Pliny the Younger. During the eruptions of 472 and 1631 particles of dust are said to have landed in Constantinople (modern Istanbul). Other years of great activity were 1794, 1822, 1855, 1871, 1906, 1929, and 1944. There has been no eruption since 1944.

veto   (Latin 'I forbid') exercise by a sovereign, branch of legislature, or other political power, of the right to prevent the enactment or operation of a law, or the taking of some course of action.

vicious cycle of poverty   the poverty trap in which much of the population of the *Third World finds itself. Poor farmers cannot afford to invest in their land through improved seed or fertilizer: *yields thus remain low, there is no surplus for sale at market, and so the poverty continues. In the absence of credit or grant aid, there is no way in which the farmer can break out of the cycle.

Given that poor yields also lead to a food shortage, the farmer will suffer poor nutrition and may even have to borrow money to secure enough food, let alone invest in the land. It is not only necessary to provide the financial means for the farmer to improve his or her livelihood, it is also important to provide the correct institutional context for progress – for example, *land reform may be necessary to ensure security of tenure.

viscous lava   *lava that resists the tendency to flow. It is sticky, flows slowly and congeals rapidly. *Nonviscous* lava is very fluid, flows quickly and congeals slowly.

Visegrad Group association of the four neighbouring central European states of the Czech Republic, Hungary, Poland, and the Slovak Republic. Originally the 'Visegrad Three', the group was formed in 1991 when Czechoslovakia, Hungary, and Poland signed a cooperation treaty at the Hungarian town of Visegrad in the wake of their recent democratization. The treaty was extended in 1992 by a Central European Free Trade Agreement (CEFTA) which was later joined by Slovenia and, in 1997, Romania. With the division of Czechoslovakia into the Czech Republic and the Slovak Republic in 1994, the 'Visegrad Four' was created. The 'Visegrad Four' press for early membership of the European Union and have voiced concern over rising Russian nationalism.

viticulture a form of *horticulture concerned with the cultivation of the grapevine. Viticulture had its origins in western Asia, was known to the ancient Egyptians 6000 years ago, and spread out across the Mediterranean Basin as Greek settlements spread westwards. The vineyards of Europe outside the Mediterranean region are the oldest areas of specialized horticulture, the Romans having been influential in the northward spread of viticulture. The spread of European settlement has seen the growth of viticulture in California, Argentina, Chile, South Africa, Australia and New Zealand.

volcanic rock or *extrusive rock a category of *igneous rock which comprises those rocks formed from *magma which has reached the Earth's surface (This is to be contrasted with *plutonic rock which forms below the surface.) *Basalt is an example of a volcanic rock, as are all solidified lavas.

volcano crack in the Earth's crust through which hot *magma (molten rock) and gases well up. The magma is termed **lava** when it reaches the surface. A volcanic mountain, usually cone-shaped with a crater on top, is formed around the opening, or vent, by the build-up of solidified lava and ash (rock fragments). Most volcanoes occur on plate margins (see *plate tectonics), where the movements of plates generate magma or allow it to rise from the mantle beneath. However, a number are found far from plate-margin activity, on *hot spots where the Earth's crust is thin, for example in Hawaii. There are two main types of volcano: *composite volcanoes and *shield volcanoes.

The type of volcanic activity also depends on the age of the volcano. The first stages of an eruption are usually vigorous as the magma forces its way to the surface. As the pressure drops and the vents become established, the main phase of activity begins. Composite volcanoes emit *pyroclastic debris, while shield volcanoes produce lava flows. When the pressure from below ceases, due to exhaustion of the magma chamber, activity wanes and is confined to the emission of gases, and in time this also ceases. The volcano then enters a period of quiescence, after which activity may resume after a period of days, years, or even thousands of years. Only when the root zones of a volcano have been exposed by erosion can a volcano be said to be truly extinct.

Many volcanoes are submarine and occur along *mid-ocean ridges. The main volcanic regions are around the Pacific rim (Cape Horn to Alaska); the central Andes of Chile (with the world's highest active volcano, Guallatiri, 6063 m); North Island, New Zealand; Hawaii; Japan; and Antarctica. There are more than 1300 potentially active volcanoes on Earth.

composite
volcano

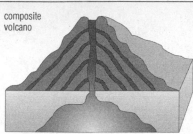

cinder
cone

shield volcano

volcano There are two main types of volcano, but three distinctive cone shapes. Composite volcanoes emit a stiff, rapidly solidifying lava which forms high, steep-sided cones. Volcanoes that regularly throw out ash build up flatter domes known as cinder cones. The lava from a shield volcano is not ejected violently, flowing over the crater rim forming a broad low profile.

Volcanism has also helped shape other members of the Solar System, including the Moon, Mars, Venus, and Jupiter's moon Io.

There are several methods of monitoring volcanic activity. They include seismographic instruments on the ground, aircraft monitoring, and monitoring from space using remote-sensing satellites.

volcanology   study of *volcanoes, the lava, rocks, and gases that erupt from them, and the geological phenomena that cause them.

Von Thünen theory   J. H. Von Thünen's early-19th century model of the distribution of agricultural land use around a town is still the basis for many investigations of such land-use patterns. Von Thünen envisaged a single market town, surrounded by a region supplying agricultural produce. Physical and economic characteristics of this agricultural area were regarded as uniform.

Such conditions of course do not exist in reality, but the value of a model like Von Thünen's is that it sweeps away the clutter of reality and allows us to observe basic processes at work. Von Thünen, given his simplifying assumptions, postulated that there were five zines of land use around a town: zone 1, market gardening; zone 2, dairying; zone 3, arable farming; zone 4, rough grazing; and zone 5, wilderness. The zones 1-5 were derived from bid-rent curves (*see bid-rent theory), with rents decreasing the greater the distance from the town.

The implication of the theory is that the distribution of land uses around the town will depend upon a variety of factors: perishability, transport costs and land area requirements.

Highly perishable produce such as salad crops must be grown close to town to ensure freshness at market and to minimize the transport costs necessitated by frequent marketing. The location of market gardens close to the town is also explained by the fact that land area requirements are small, and therefore the bid rent per unit area can be high. Market gardeners can achieve high productivity from small areas by the intensive nature of their farming, and through investment in greenhouses, fertilizers and other inputs. Land uses with progressively larger land area requirements must locate further away from the town;

a farmer needing many units of land will only be able to bid a lower rent per unit area. Land uses at progressively greater distance from town are also characterized by a decreasing frequency of marketing of products, and by decreasing in the investment land. In reality such patterns are constrained by variations in the physical *environment and in human behaviour; government policy also affects farming practices and the distribution of land uses.

V-shaped valley   river valley with a V-shaped cross-section. Such valleys are usually found near the source of a river, where the steeper gradient means that there is a great deal of *corrasion (grinding away by rock particles) along the stream bed, and erosion cuts downwards more than it does sideways. However, a V-shaped valley may also be formed in the lower course of a river when its powers of downward erosion become renewed by a fall in sea level, a rise in land

V-shaped valley

V-shaped valley Cross section of a V-shaped valley. V-shaped valleys are formed through erosion by water.

level, or the capture of another river (see *rejuvenation).

The angle of the V-shaped cross-section depends on the rate of uplift of rock, the type of rock, the erosive ability of the river, the type of climate, and the stage of the river.

vulcanicity   a collective term for those processes which involve the intrusion of *magma into the *crust, or the extrusion of such molten material onto the Earth's surface.

wadi in arid regions of the Middle East, a steep-sided valley containing an intermittent stream that flows in the wet season.

Wafd (Arabic 'deputation') the main Egyptian nationalist party between World Wars I and II. Under Nahas Pasha it formed a number of governments in the 1920s and 1930s. Dismissed by King Farouk in 1938, it was reinstated by the British in 1941. The party's pro-British stance weakened its claim to lead the nationalist movement, and the party was again dismissed by Farouk in 1952, shortly before his own deposition. Wafd was banned in January 1953.

Waldsterben (German 'forest death') tree decline related to air pollution, common throughout the industrialized world. It appears to be caused by a mixture of pollutants; the precise chemical mix varies between locations, but it includes acid rain, ozone, sulphur dioxide, and nitrogen oxides.

want in economics, the desire of consumers for material goods and services. Wants are argued to be infinite, meaning that consumers can never be satisfied with their existing standard of living but would always like to consume more goods and services. Infinite wants mean that *resources have to be allocated.

warm front see *depression.

warm sector see *depression.

warm temperate (subtropical) climate climate classified into two areas: *Mediterranean climate and the Eastern margin climate. **Eastern margin** climates, which are found in southeast and eastern Asia, are dominated by the *monsoon. Temperature figures and rainfall distributions are similar to those of tropical continental areas although annual amounts of rain are much higher.

Washington Convention alternative name for *CITES, the international agreement that regulates trade in endangered species.

waste materials that are no longer needed and are discarded. Examples are household waste, industrial waste (which often contains toxic chemicals), medical waste (which may contain organisms that cause disease), and nuclear waste (which is radioactive). By *recycling, some materials in waste can be reclaimed for further use. In 1990 the industrialized nations generated 2 billion tonnes of waste. In the USA, 40 tonnes of solid waste are generated annually per person, roughly twice as much as in Europe or Japan.

There has been a tendency to increase the amount of waste generated per person in industrialized countries, particularly through the growth in packaging and disposable products, creating a 'throwaway society'.

In Britain, the average person throws away about ten times their own body weight in household refuse each year. Collectively the country generates about 50 million tonnes of waste per year. In principle, over 50% of UK household waste could be recycled, although less then 5% is currently recovered. The European Commission has set targets for a much greater percentage of recycling in the 21st century.

waste disposal depositing of waste. Methods of waste disposal vary according to the materials in the waste and include incineration, burial at designated sites, and dumping at sea. Organic waste can be treated and reused as fertilizer (see *sewage disposal). Nuclear waste and *toxic waste are usually buried or dumped at sea, although this does not negate the danger.

environmental pollution Waste disposal is an increasing problem. Environmental groups, such as Greenpeace and Friends of the Earth, are campaigning for more recycling, a change in lifestyle so that less waste (from packaging and containers, to nuclear materials) is produced, and safer methods of disposal.

Although incineration cuts down on landfill and can produce heat as a useful by-product it is still a wasteful method of disposal in comparison with recycling. For example, recycling a plastic bottle saves twice as much energy as is obtained by burning it.

waste disposal, USA The USA burns very little of its rubbish as compared with other industrialized countries. Most of its waste, 80%, goes into landfills. Many of the country's landfill sites had to close in the 1990s because they did not meet standards to protect groundwater.

waste disposal, UK The average Briton throws away about ten times his or her own body weight in household refuse each year. Collectively the country generates about 50 million tonnes of waste per annum. Methods of waste disposal vary, although the use of landfill sites has been the preferred option for many years; up to 90% of domestic rubbish is disposed of in this fashion. In principle, over 50% of household waste could be recycled, although less then 5% is currently recovered The industrial waste dumped every year by the UK in the North Sea includes 550,000 tonnes of fly ash from coal-fired power stations. The British government agreed in 1989 to stop North Sea dumping from 1993, but dumping in the heavily polluted Irish Sea would

continue. Industrial pollution is suspected of causing ecological problems, including an epidemic that killed hundreds of seals in 1989.

water chemical compound of hydrogen and oxygen elements – H_2O. It can exist as a solid (ice), liquid (water), or gas (water vapour). Water is the most common compound on Earth and vital to all living organisms. It covers 70% of the Earth's surface, and provides a habitat for large numbers of aquatic organisms. It is the largest constituent of all living organisms–the human body consists of about 65% water. It is found in all cells and many chemicals involved in processes such as respiration and photosynthesis need to be in solution in water in order to react. Pure water is a colourless, odourless, tasteless liquid which freezes at 0°C, and boils at 100°C. Natural water in the environment is never pure and always contains a variety of dissolved substances. Some 97% of the Earth's water is in the oceans; a further 2% is in the form of snow or ice, leaving only 1% available as freshwater for plants and animals. The recycling and circulation of water through the biosphere is termed the *water cycle, or 'hydrological cycle'; regulation of the water balance in organisms is termed osmoregulation. Water becomes more dense when it cools but reaches maximum density at 4°C. When cooled below this temperature the cooler water floats on the surface, as does ice formed from it. Animals and plants can survive under the ice.

water-borne disease disease associated with poor water supply. In developing world countries four-fifths of all illness is caused by water-borne diseases, with diarrhoea being the leading cause of childhood death. Malaria, carried by mosquitoes that are dependent on stagnant water for breeding, affects 400 million people every year and kills 5 million. Polluted water is also a problem in industrialized nations, where industrial dumping of chemical, hazardous, and radioactive wastes causes a range of medical conditions from headache to cancer.

water-cooled reactor   a type of *nuclear reactor in which water is used as a core coolant and also, in most cases, as a moderator. Water-cooled reactors were pioneered in North America, first in military and later in commercial programmes. Water-cooled reactors are of two types:
(a) *pressurized water reactor (PWR).* Heat generated in the nuclear core is removed by water which circulates at high pressure through a primary circuit. Heat is transferred from the primary to a secondary circuit via a heat exchanger, thereby generating steam in the secondary circuit which can be used to power turbines. PWRs are relatively inefficient due to the technical difficulties in working at pressures greater than 1.5×10^7 Pa (2000 psi). About 60% of the heat

energy generated escapes as the coolant is 'dumped' usually into the sea, lake or river alongside which PWRs are inevitably built.
(b) *boiling water reactor (BWR).* These differ from PWRs in that the pressurized water acts not only as a coolant and moderator but also is allowed to boil within the reactor itself. The BWR works at 7.5×10^6 Pa (1000 psi) and the steam, having passed through a turbine, is condensed and the water re-used, thus avoiding some of the problems of 'dumping' large volumes of very hot water into the environment.

water cycle   or **hydrological cycle**, natural circulation of water through the upper part of the Earth. It is a complex system involving a number of physical and chemical processes (such as evaporation, *precipitation, and *infiltration) and stores (such as rivers, oceans, and soil). The cycle is powered by solar radiation which provides the energy to maintain the flow through the processes of evaporation, transpiration, precipitation, and run-off.

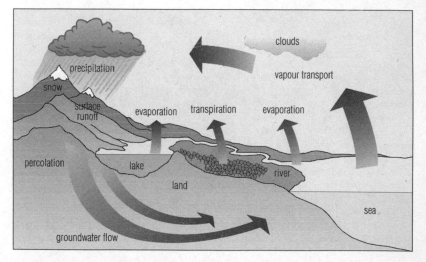

water cycle About one-third of the solar energy reaching the Earth is used in evaporating water. About 380,000 cubic km is evaporated each year. The entire contents of the oceans would take about one million years to pass through the water cycle.

Water is lost from the Earth's surface to the atmosphere by evaporation caused by the Sun's heat on the surface of lakes, rivers, and oceans, and through the transpiration of plants. This atmospheric water is carried by the air moving across the Earth, and **condenses** as the air cools to form clouds, which in turn deposit moisture on the land and sea as *precipitation. The water that collects on land flows to the ocean overland – as streams, rivers, and glaciers – or through the soil (*infiltration) and rock (*groundwater). The boundary that marks the upper limit of groundwater is called the *water table.

The oceans, which cover around 70% of the Earth's surface, are the source of most of the moisture in the atmosphere.

waterfall cascade of water in a river or stream. It occurs when a river flows over a bed of rock that resists erosion; weaker rocks downstream are worn away, creating a steep, vertical drop and a plunge pool into which the water falls. Over time, continuing erosion causes the waterfall to retreat upstream forming a deep valley, or *gorge. Good examples of waterfalls include Victoria Falls (Zimbabwe/Zambia), Niagara Falls (USA/Canada), and Angel Falls (Venezuela).

water pollution any addition to fresh or seawater that interferes with biological processes or causes a health hazard. Common pollutants include nitrates, pesticides, and sewage (resulting from poor *sewage disposal methods), although a huge range of industrial contaminants, such as chemical by-products and residues created in the manufacture of various goods, also enter water – legally, accidentally, and through illegal dumping.

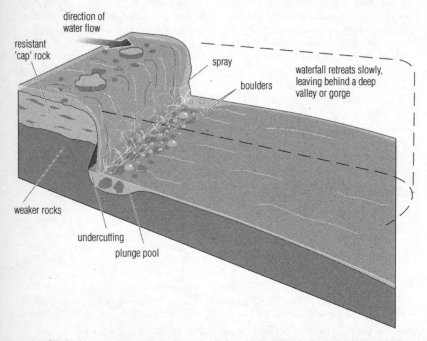

direction of water flow

resistant 'cap' rock

spray

boulders

waterfall retreats slowly, leaving behind a deep valley or gorge

weaker rocks

undercutting

plunge pool

waterfall When water flows over hard rock and soft rock, the soft rocks erode creating waterfalls. As the erosion processes continue, the falls move backwards, in the opposite direction of the water.

In 1980 the UN launched the 'Drinking Water Decade', aiming for cleaner water for all by 1990. However, in 1994 it was estimated that roughly half of all people in the developing world did not have safe drinking water. A 1995 World Bank report estimated that some 10 million deaths in developing countries were caused annually by contaminated water.

A 1999 study by Swiss chemists revealed that European rain contains high levels of dissolved *pesticides which lead to unacceptable pesticide levels in drinking water.

In the UK, water pollution is controlled by the Environment Agency (formerly the National Rivers Authority, NRA) and, for large industrial plants, Her Majesty's Inspectorate of Pollution. In 1989 the regional water authorities of England and Wales were privatized to form ten water and sewerage companies. Following concern that some of the companies were failing to meet EU drinking-water standards on nitrate and pesticide levels, the companies were served with enforcement notices by the government Drinking Water Inspectorate.

Concern is growing over the amount of medication in drinking water, as up to 90% of a dose of antibiotics is excreted in the urine. Other drugs, from aspirin to oestrogen from the contraceptive pill, are also excreted and end up in the environment. Inevitably, some of these drugs find their way back into our drinking water. A 1995 German study found anticholesterol drugs and analgesics such as ibuprofen, in Berlin drinking water. An earlier and overlooked British study estimated that a London river contained more than 170 drugs with a concentration of up to one part per billion (over a tonne per year).

watershed   boundary formed by height differences in the land, situated between two rivers or *drainage basins.

waterspout   funnel-shaped column of water and cloud that is drawn from the surface of the sea or a lake by a *tornado.

water supply   distribution of water for domestic, municipal, or industrial consumption. Water supply in sparsely populated regions usually comes from underground water rising to the surface in natural springs, supplemented by pumps and wells. Urban sources are deep artesian wells, rivers, and reservoirs, usually formed from enlarged lakes or dammed and flooded valleys, from which water is conveyed by pipes, conduits, and aqueducts to filter beds. As water seeps through layers of shingle, gravel, and sand, harmful organisms are removed and the water is then distributed by pumping or gravitation through mains and pipes.

   water treatment Often other substances are added to the water, such as chlorine and fluoride; aluminium sulphate, a clarifying agent, is the most widely used chemical in water treatment. In towns, domestic and municipal (road washing, sewage) needs account for about 135 litres per head each day. In coastal desert areas, such as the Arabian peninsula, desalination plants remove salt from sea water. The Earth's waters, both fresh and saline, have been polluted by industrial and domestic chemicals, some of which are toxic and others radioactive (see *water pollution).

   drought A period of prolonged dry weather can disrupt water supply and lead to drought. The area of the world subject to serious droughts, such as the Sahara, is increasing because of destruction of forests, overgrazing, and poor agricultural practices. A World Bank report in 1995 warned that a global crisis was imminent:

chronic water shortages were experienced by 40% of the world's population, notably in the Middle East, northern and sub-Saharan Africa, and central Asia. 1.4 billion people (25 % of the population) had no access to safe drinking water in 1997.

In the UK, drought is defined as the passing of 15 days with less than 0.2 mm of rain. The amount of water available to consumers in England and Wales has fallen by 5% over the period 1994–98, owing to drought.

In 1992 the town of Cgungungo in the Atacama Desert began using a system to convert water from fog as a public water supply. The system supplies 11,000 litres of water per day.

water supply since 1989 ten privatized water companies have statutory responsibility for water supply, sewerage, and sewerage treatment in England and Wales, under the supervision of the industry's regulatory body, Ofwat. In Scotland the responsibility for all water and sewerage services rests with three public water authorities, while in Northern Ireland the Department of the Environment holds this brief. About 75% of Britain's water supplies are obtained from mountain lakes, reservoirs, and rivers; the rest comes from underground sources. 1995–97 was the driest two-year period in over 200 years and, as a result, there was a severe depletion of Britain's water supplies. The Environment Agency, which is responsible for monitoring and protecting water resources, has since introduced statutory efficiency measures which the water companies are now obliged to follow.

water consumption The water supply companies in England and Wales provide almost 4.6 million household with water; over 75% of commercial and industrial customers pay for water and sewerage services on the basis of their metered consumption. 1% of total water provided by these companies is used for drinking; 2% for cooking; 3% for gardening; and 49% in bathrooms. Following concern that some of the water companies were were failing to meet EU drinking-water standards on nitrate and pesticide levels, the companies were served with enforcement notices by the government Drinking Water Inspectorate.

water table upper level of *groundwater (water collected underground in porous rocks). Water that is above the water table will drain downwards; a spring forms where the water table meets the surface of the ground. The water table rises and falls in response to rainfall and the rate at which water is extracted, for example for irrigation and industry.

In many irrigated areas the water table is falling due to the extraction of water. Below northern China, for example, the water table is sinking at a rate of 1 m a year. Regions with a high water table and dense industrialization have problems with *pollution of the water table. In the USA, New Jersey, Florida, and Louisiana have water tables that are contaminated by both industrial *wastes and saline seepage from the ocean.

water vapour a colourless, ordourless gaseous form of water which mixes perfectly with the other gases of the air. Water vapour enters the atmosphere following the evaporation of fresh and salt water, condensing into clouds and fog before being released from the atmosphere as *precipitation, see *water cycle. Water vapour effectively absorbs the Sun's radiant energy and as such the presence of water vapour in the *troposphere stabilizes the temperature of the Earth. *Humidity is the concentration of water vapour in the air and is of prime importance as an environmental factor. The water vapour content of the atmosphere varies greatly from as low as 0.02% in hot deserts to 1.8% in humid equatorial regions.

waterway a natural or artificial network of water channels used for travel or transport. These are mostly river or lake systems but include the canals which were built to interconnect towns for freight transport purposes in the 19th century. The Great Lakes and the St Lawrence Seaway of North America represents one of the most important waterways in the world. Since the 1960s inland water-borne recreation has developed significantly, and stretches of abandoned canal have been rehabilitated to create long distance routes for recreational cruising. Disused canals have become important havens for many plant and animal species which have been ousted from their traditional habitats by intensification of agriculture and the expansion of urban land use.

wave in the oceans, a ridge or swell formed by wind or other causes. The power of a wave is determined by the strength of the wind and the distance of open water over which the wind blows (the fetch). Waves are the main agents of *coastal erosion and deposition: sweeping away or building up beaches, creating *spits and *berms, and wearing down cliffs by their *hydraulic action and by the *corrosion of the sand and shingle that they carry. A *tsunami (misleadingly called a 'tidal wave') is formed after a submarine earthquake.

As a wave approaches the shore it is forced to break as a result of friction with the seabed. When it breaks on a beach, water and sediment are carried up the beach as **swash**; the water then drains back as **backwash**.

A **constructive wave** causes a net deposition of material on the shore because its swash is stronger than its backwash. Such waves tend be low and have crests that spill over gradually as they break. The backwash of a **destructive wave** is stronger than its swash, and therefore causes a net removal of material from the shore. Destructive waves are usually tall and have peaked crests that plunge downwards as they break, trapping air as they do so.

If waves strike a beach at an angle the beach material is gradually moved

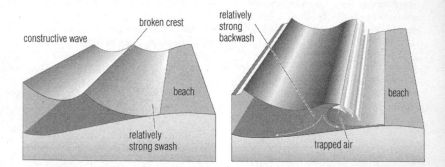

wave The low, gentle crests of a constructive wave, with the energy of the wave flowing up the beach in a strong swash and depositing material, contrasts with the high, steep-crested, more forceful motions of destructive waves which crash in at an angle to the beach directing all their energy into plunging waves which tear up the sand and shingle and carry it out with the strong backwash.

along the shore (*longshore drift), causing deposition of material in some areas and erosion in others.

Atmospheric instability caused by the *greenhouse effect and *global warming appears to be increasing the severity of Atlantic storms and the heights of the ocean waves. Waves in the South Atlantic are shrinking – they are on average half a metre smaller than in the mid-1980s – and those in the northeast Atlantic have doubled in size over the last 40 years. As the height of waves affects the supply of marine food, this could affect fish stocks, and there are also implications for shipping and oil and gas rigs in the North Atlantic, which will need to be strengthened if they are to avoid damage.

Freak or 'episodic' waves form under particular weather conditions at certain times of the year, travelling long distances across the Atlantic, Indian, and Pacific oceans. They are considered responsible for the sudden unexplained disappearance of many ships.

Freak waves become extremely dangerous when they reach the shallow waters of the continental shelves at 100 fathoms (180 m), especially when they meet currents: for example, the Agulhas Current to the east of South Africa, and the Gulf Stream in the North Atlantic. A wave height of 34 m has been recorded.

wave-cut platform gently sloping rock surface found at the foot of a coastal cliff. Covered by water at high tide but exposed at low tide, it represents the last remnant of an eroded headland (see *coastal erosion). Wave-cut platforms have a gentle gradient (usually less than 1°) and can be up to about 1 km in length. Over time, wave-cut platforms are lengthened and protect the cliffs behind them.

wave power power obtained by harnessing the energy of water waves.

Various schemes have been advanced since 1973 when oil prices rose dramatically and an energy shortage threatened. In 1974 the British engineer Stephen Salter developed the 'duck'– a floating boom, the segments of which nod up and down with the waves. The nodding motion can be used to drive pumps and spin generators. Another device, developed in Japan, uses an oscillating water column to harness wave power. A major technological breakthrough will be required if wave power is ever to contribute significantly to the world's energy needs, although several ideas have reached prototype stage.

wave refraction distortion of waves as they reach the coast, due to variations in the depth of the water and shape of the coastline. It is particularly evident where there are headlands and bays.

The bending of a wave crest as it approaches a headland concentrates the energy of the wave in the direction of that headland, and increases its power of erosion. By contrast, the bending that a wave crest experiences when it moves into a bay causes its energy to be dissipated away from the direction of the shore. As a result the wave loses its erosive power and becomes more likely to deposit sediment on the shore.

wealth in economics, the wealth of a nation is its stock of physical capital, human capital, and net financial capital owned overseas. Physical capital is the stock of buildings, factories, offices, machines, roads, and so on. Human capital is the workforce; not just the number of workers, but also their stock of education and training which makes them productive. Net financial capital is the difference between the money value of assets owned by foreigners in the domestic economy and the assets owned by the country abroad.

For individuals, the most significant wealth they have is themselves and their ability to generate an income by working.

After that, the largest item of wealth is likely to be their house. Possessions, money, and insurance policies are other examples of individual wealth.

weather   variation of atmospheric conditions at any one place over a short period of time. Such conditions include humidity, precipitation, temperature, cloud cover, visibility, and wind. Weather differs from *climate in that the latter is a composite of the average weather conditions of a locality or region over a long period of time (at least 30 years). *Meteorology is the study of short-term weather patterns and data within a particular area; climatology is the study of weather over longer timescales on a zonal or global basis. *Weather forecasts are based on current meterological data, and predict likely weather for a particular area.

weather area   any of the divisions of the sea around the British Isles for the purpose of weather forecasting for shipping. The areas are used to indicate where strong or gale-force winds are expected.

weather chart   a map or chart of an area giving details of *weather experienced at a particular time of day, such as 0600 hrs. The information is gathered at weather stations throughout the country, and whenall observations have been collected *isobars can be drawn and the *depressions, *anticyclones, *fronts, etc. can be identified. Weather charts are sometimes called *synoptic charts*, as they give a synopsis of the weather at a particular time.

weather forecast   prediction for changes in the *weather. The forecast is based on several different types of information from several sources. Weather stations are situated around the world, on land, in the air (on aircraft and air balloons), and in the sea (on ships, buoys, and oilrigs). Forecasters also use satellite photographs which show cloud patterns, and information from the *Meteorological Office, such as a *synoptic chart which shows *atmospheric pressure and changes in *air masses (known as a weather *front). The information is used to predict the weather over a 24-hour period, and to suggest likely changes over a period of three or four days.

Forecasts may be short-range (covering a period of one or two days), medium-range (five to seven days), or long-range (a month or so). Weather observations are made on an hourly basis at meteorological recording stations – there are more than 3500 of these around the world. More than 140 nations participate in the exchange of weather data through the World Weather Watch programme, which is sponsored by the World Meteorological Organization (WMO), and information is distributed among the member nations by means of a worldwide communications network. Incoming data is collated at weather centres in individual countries and plotted on weather maps, or charts. The weather map uses internationally standardized symbols to indicate barometric pressure, cloud cover, wind speed and direction, precipitation, and other details reported by each recording station at a specific time. Points of equal atmospheric pressure are joined by lines called *isobars and from these the position and movement of weather fronts and centres of high and low pressure can be extrapolated. The charts are normally compiled on a three-hourly or six-hourly basis – the main synoptic hours are midnight, 0600, 1200 and 1800 – and predictions for future weather are drawn up on the basis of comparisons between current charts and previous charts. Additional data received from weather balloons and satellites help to complete and corroborate the picture obtained from the weather map.

weather in Britain   the climate of Britain is notoriously variable and changeable from day to day. Weather is generally cool to mild with frequent cloud and rain, but occasional settled spells of weather occur at all seasons. Visitors are often surprised by the long summer days, which are a consequence of the northerly latitude; in the north of Scotland in midsummer the day is 18 hours long and twilight lasts all night. Conversely, winter days are short.

While the south is usually a little warmer than the north and the west wetter than the east, the continual changes of weather mean that, on occasions, these differences may be reversed. Extremes of weather are rare but they do occur. For example, in December 1981 and January 1982, parts of southern and central England experienced for a few days lower temperatures than central Europe and Moscow. During the long spells of hot, sunny weather in the summers of 1975 and 1976 parts of Britain were drier and warmer than many places in the western Mediterranean.

The greatest extremes of weather and climate occur in the mountains of Scotland, Wales, and northern England. Here at altitudes exceeding 600 m conditions are wet and cloudy for much of the year with annual rainfall exceeding 1500 mm and in places reaching as much as 5000 mm. These are among the wettest places in Europe. Winter conditions may be severe with very strong winds, driving rain, or snow blizzards.

In spite of occasional heavy snowfalls on the Scottish mountains, conditions are not reliable for skiing and there has been only a limited development of winter sports resorts. Because of the severe conditions which can arise very suddenly on mountains, walkers and climbers who go. unprepared face the risk of exposure or even frostbite. Conditions may be vastly different from those suggested by the weather at lower levels.

Virtually all permanent settlement in Britain lies below 300 m, and at these levels weather conditions are usually much more congenial. As a general rule the western side of Britain is cloudier, wetter, and milder in winter, than the east, with cooler summers. The eastern side of Britain is drier, with a tendency for summer rain to be heavier than that of winter. Much of central England has very similar weather to that of the east and south of the country. Southwestern England shares the greater summer warmth of southern England but experiences rather milder and wetter winters than the east of the country.

The average number of hours of sunshine is greatest in the south and southeast of England. Western Scotland, Wales, and Northern Ireland have rather less sunshine than most of England. Generally, daily sunshine hours range from between one and two in midwinter to between five and seven in midsummer. Winter sunshine is much reduced because of frequent fogs and low cloud. This is a consequence of winds from the Atlantic and seas surrounding Britain, which bring high humidity. For the same reason the mountain areas are particularly cloudy and wet.

Snow may occur anywhere in Britain in winter or even spring but, except on the hills, it rarely lies for more than a few days. In some winters there may be very little snow, but every 15 or 20 years it may lie for some weeks during a prolonged cold spell.

Northern Ireland shares with the rest of the British Isles a mild, changeable climate with very rare extremes of heat or cold. Ireland is even more influenced by the warm waters of the North Atlantic than England and, consequently, its climate is a little wetter the year round, milder in winter and cooler and cloudier in summer. This mild, rainy climate is particularly favourable to the growth of grass and moss and for this reason Ireland has been called the Emerald Isle.

In the wetter west of the country rain is frequent but on many days it is very light and in the form of drizzle. The sunniest parts of the country are the east and south coasts, with sunshine hours averaging from two a day in winter to six in midsummer. Over most of Ireland spring is the driest time of the year and May is the sunniest month. Except in the extreme east around Dublin, autumn and winter are the wettest seasons. Occasional severe weather in winter takes two forms: storms and gales which particularly affect the west; and rare spells with frost and snow when cold easterly or northerly winds bring severe weather to the whole British Isles.

weathering process by which exposed rocks are broken down on the spot (in situ) by the action of rain, frost, wind, and other elements of the weather. It differs from *erosion in that no movement or transportation of the broken-down material takes place. Two types of weathering are recognized: *physical (or mechanical) weathering and *chemical weathering. They usually occur together, and are an important part of the development of landforms.

weather station a place where all elements of the weather are measured and recorded. Each station will have a *Stevenson screen and a variety of instruments such as a maximum and minimum thermometer, a *hygrometer, a *rainfall gauge, a *wind vane, an *anemometer and a *sunshine recorder.

weather vane instrument that shows the direction of wind. Wind direction is always given as the direction from which the wind has come – for example, a northerly wind comes from the north.

Weber's theory see *industrial location.

weedkiller or **herbicide**, chemical that kills some or all plants. Selective herbicides are effective with cereal crops because they kill all broad-leaved plants without affecting grasslike leaves. Those that kill all plants include sodium chlorate and paraquat. The widespread use of weedkillers in agriculture has led to an increase in crop yield but also to pollution of soil and water supplies and killing of birds and small animals, as well as creating a health hazard for humans.

Wegener, Alfred see *continental drift.

weighted average average that reflects the relative importance of its components. Each element is allotted a proportion of 100% according to its relative importance. Its contribution to the total is then reduced by that percentage. The average is then calculated using the adjusted total.

West African Economic Community international organization established in 1975 to end barriers in trade and to achieve cooperation in development. Members include Burkina Faso, Côte d'Ivoire, Mali, Mauritania, Niger, and Senegal; Benin and Togo have observer status.

westerlies prevailing winds from the west that occur in both hemispheres between latitudes of about 35° and 60°. Unlike the *trade winds, they are very variable and produce stormy weather.

The westerlies blow mainly from the southwest in the northern hemisphere and the northwest in the southern hemisphere, bringing moist weather to the west coast of the landmasses in these latitudes.

wet and dry bulb thermometer see *hygrometer.

wetland permanently wet land area or habitat. Wetlands include areas of *marsh, fen, *peat bog, flood plain, and shallow coastal areas. Wetlands are extremely fertile. They provide warm, sheltered waters for fisheries, lush vegetation for grazing livestock, and an abundance of wildlife. Estuaries and seaweed beds are more than 16 times as productive as the open ocean.

The term is often more specifically applied to a naturally flooding area that is managed for agriculture or wildlife. A water meadow, where a river is expected to flood grazing land at least once a year thereby replenishing the soil, is a traditional example.

In the UK, the Royal Society for the Protection of Birds (RSPB) manages 2800 hectares of wetland, using sluice gates and flood-control devices to produce sanctuaries for wading birds and wild flowers.

Approximately 60% of the world's wetlands are peat, and in May 1999 the Ramsar Convention on the Conservation of Wetlands approved a peatlands action plan that should have a major impact on the conservation of peat bogs.

wet-rice cultivation a type of agriculture involving the cultivation of rice in paddies. In contrast to upland or dry rice, which is grown like any other grain crop, wet-rice must be covered by water for much of the growing season. The need for level land, as in deltas and river flood plains or on artificially constructed *terracing restricts the geographical distribution of wet-rice cultivation, although its economic importance in Asian

agriculture is unmatched by other crops.

Most areas of wet-rice cultivation support a high density of peasant farmers intensively operating small fragmented paddy-fields within a largely subsistence farming economy. Rice is often the only crop grown, and soil fertility is maintained by green manuring, by the application of night soil, animal droppings and crop residues, and by the paddy's unique microenvironment. In rain-fed areas without supplementary irrigation, a crop other than rice may be taken from the drained field during the dry season. Labour inputs are invariably high, with high marginal returns to labour, that is, the system continues to reward ever-greater inputs of labour long after other types of farming would have collapsed from over-intensification.

wetted perimeter length of that part of a river's cross-section that is in contact with the water. The wetted perimeter is used to calculate a river's hydraulic radius, a measure of its *channel efficiency.

whaling the hunting of whales. Whales have been killed by humans since at least the middle ages. There were hundreds of thousands of whales at the beginning of the 20th century, but the invention of the harpoon in 1870 and improvements in ships and mechanization have led to the near-extinction of several species of whale. Commercial whaling was largely discontinued in 1986, although Norway and Japan have continued the practice.

whirlwind rapidly rotating column of air, often synonymous with a *tornado. On a smaller scale it produces the dust-devils seen in deserts.

white-collar worker a worker who is not a manual worker and who does not work in 'dirty conditions'.

The term 'white-collar' derives from the white shirts worn by e.g. clerical workers and professional people whose jobs do not make their clothes dirty. Compare *blue-collar worker.

white ice   ice from which air has not been totally expelled (see *corrie). Contrast this with *blue ice which is found at greater depth in a corrie icefield and from which air has been expelled by compression.

whiteout   'fog' of grains of dry snow caused by strong winds in temperatures of between –18°C and –1°C. The uniform whiteness of the ground and air causes disorientation in humans.

WHO   acronym for *World Health Organization, an agency of the United Nations established to prevent the spread of diseases.

wholesale   the business of selling merchandise to anyone other than the final customer. Most manufacturers or producers sell in bulk to a wholesale organization which distributes the smaller quantities required by retail outlets.

wilderness   area of uninhabited land that has never been disturbed by humans, usually located some distance from towns and cities. According to estimates by US group Conservation International, 52% (90 million sq km) of the Earth's total land area was still undisturbed in 1994.

wildlife corridor   passage between habitats. See *corridor, wildlife.

wildlife trade   international trade in live plants and animals, and in wildlife products such as skins, horns, shells, and feathers. The trade has made some species virtually extinct, and whole ecosystems (for example, coral reefs) are threatened. Wildlife trade is to some extent regulated by *CITES (Convention on International Trade in Endangered Species).

will-o'-the-wisp   light sometimes seen over marshy ground, believed to be burning gas containing methane from decaying organic matter.

willy-willy   Australian Aboriginal term for a cyclonic whirlwind.

wilting point   the point at which a plant becomes unable to extract sufficient water from the soil via its roots to compensate for the loss of moisture through its leaves by *transpiration. Wilting point is reached at variable times following rainfall depending on the soil texture which determines a soil's water retention capacity. Water still remains in the soil even though the wilting point has been reached and is called *hygroscopic water; it is unavailable for plant growth because it is retained in the pore spaces by a considerable surface tension force.

wind   lateral movement of the Earth's atmosphere from high-pressure areas (anticyclones) to low-pressure areas (depressions). Its speed is measured using an anemometer or by studying its effects on, for example, trees by using the *Beaufort scale. Although modified by features such as land and water, there is a basic worldwide system of *trade winds, *westerlies, and polar easterlies.

A belt of low pressure (the *doldrums) lies along the Equator. The trade winds blow towards this from the horse latitudes (areas of high pressure at about 30° north and south of the Equator), blowing from the northeast in the northern hemisphere, and from the southeast in the southern hemisphere. The westerlies (also from the horse latitudes) blow north of the Equator from the southwest, and south of the Equator from the northwest.

Cold winds blow outwards from high-pressure areas at the poles. More local effects result from landmasses heating and cooling faster than the adjacent sea, producing onshore winds in the daytime and offshore winds at night.

The *monsoon is a seasonal wind of southern Asia, blowing from the southwest in summer and bringing the rain on which crops depend. It blows from the northeast in winter.

Famous warm winds include the **chinook** of the eastern Rocky Mountains, North America; the **föhn** of Europe's Alpine valleys; the **sirocco** (Italy)/**khamsin** (Egypt)/**sharav** (Israel) – spring winds that bring warm air from the Sahara and Arabian deserts across the Mediterranean; and the **Santa Ana**, a periodic warm wind from the inland deserts that strikes the California coast.

The dry northerly **bise** (Switzerland) and the **mistral** (which strikes the Mediterranean area of France) are unpleasantly cold winds.

The fastest wind speed ever measured on Earth, 512 kph, occurred on 3 May 1999 in a *tornado that struck the suburbs of Oklahoma City, Oklahoma, USA.

wind-chill factor or **wind-chill index**, estimate of how much colder it feels when a wind is blowing. It is arrived at by combining the actual temperature and wind speed and is given as a different temperature.

Wind chill can be calculated in Celsius using the formula: wind chill = $0.323(18.97 \times \sqrt{V} - V + 37.62) \times (33 - T)$ where V is the wind speed in kilometres per hour and T is the temperature in degrees Celsius.

wind gap a *dry valley, for example in a chalk *escarpment, now standing above the *water table, but formed at a time when the water table was higher or when the ground was frozen.

wind power power produced from the harnessing of wind energy. The wind has long been used as a source of energy: sailing ships and windmills are ancient inventions. After the energy crisis of the 1970s *wind turbines began to be used to produce electricity on a large scale.

wind turbine windmill of advanced aerodynamic design connected to a generator producing electrical energy and used in wind-power installations. Wind is a form of *renewable energy that is used to turn the turbine blades of the windmill. Wind turbines can be either large propeller-type rotors mounted on a tall tower, or flexible metal strips fixed to a vertical axle at top and bottom.

The world's largest wind turbine is on Hawaii, in the Pacific Ocean. It has two blades 50 m long on top of a tower 20 storeys high. An example of a propeller turbine is found at Tvind in Denmark; it has an output of some 2 MW. Other machines use novel rotors, such as the 'egg-beater' design developed at Sandia Laboratories in New Mexico, USA.

wind vane an instrument used to indicate wind direction. It consists of a rotating arm which always points in the direction from which the wind blows. The wind is named after this direction.

wold (Old English *wald* 'forest') open, hilly country. The term refers specifically to certain areas in England, notably the Yorkshire and Lincolnshire Wolds and the Cotswold Hills.

wolframite iron manganese tungstate, (Fe,Mn)WO$_4$, an ore mineral of tungsten. It is dark grey with a submetallic surface lustre, and often occurs in hydrothermal veins in association with ores of tin.

woodland area in which trees grow more or less thickly; generally smaller than a *forest. Temperate climates, with four distinct seasons a year, tend to support a mixed woodland habitat, with some conifers but mostly broad-leaved and deciduous trees, shedding their leaves in autumn and regrowing them in spring. In the Mediterranean region and parts of the southern hemisphere, the trees are mostly evergreen.

Temperate woodlands grow in the zone between the cold coniferous forests and the tropical forests of the hotter climates near the Equator. They develop in areas where the closeness of the sea keeps the climate mild and moist.

Old woodland can rival tropical rainforest in the number of species it supports, but most of the species are hidden in the soil. A study in Oregon, USA, in 1991 found that the soil in a single woodland location contained 8000 arthropod species (such as insects, mites, centipedes, and millipedes), compared with only 143 species of reptile, bird, and mammal in the woodland above.

Woodland covers about 2.4 million hectares in the UK: about 8% of England, 15% of Scotland, 12% of Wales, and 6% of Northern Ireland. This is over 10% of the total land area – well below the 25% average in the rest of Europe. An estimated 33% of ancient woodland has been destroyed in Britain since 1945. The Forestry Commission aims to double the area of woodlands in England by the middle of this century.

worker cooperative business owned and controlled by its workers rather than outside shareholders. In some worker cooperatives each member worker has one vote at meetings, however many shares he or she owns. There are relatively few worker cooperatives in the UK; they are far more popular in Europe and Japan.

working capital current assets minus current liabilities of a business organization. It is the assets which are left free, after liabilities have been covered, for the business to use or put to work if it feels that it should take that risk.

working conditions the physical environment in which a worker has to work, as well as the way workers are expected to complete their tasks.

For example, working conditions include heat and light in the workplace, availability of toilets, noise levels, the degree of danger in the job, hours of work, the size of the team of workers, how repetitive a task is, and whether or not the worker is able to vary the pattern of work. Working conditions are important in determining the motivation of workers.

working population the number of people in work, both the employed and the self-employed, and people out of work, the unemployed.

World Bank officially the **International Bank for Reconstruction and Development**, specialized agency of the United Nations that borrows in the commercial market and lends on commercial terms. It was established in 1945 under the 1944 Bretton Woods agreement, which also created the International Monetary Fund (IMF). The **International Development Association** is an arm of the World Bank.

world city a city widely recognised as a centre of economic and political power within the capitalist world economy. These cities are important financial and business centres and usually boast a high concentration of headquarters of *multinational corporations. London, England, New York, USA, and Tokyo, Japan, are examples of world cities.

world conservation strategy in ecology, plan prepared in 1980 by the United Nations Environmental Programme (UNEP), the then World Wildlife Fund (now the World Wide Fund for Nature, WWF), and the International Union for the Conservation of Nature and Natural Resources (IUCN), which established an agenda that protected the environment while allowing the exploitation of its resources for human use. The plan has been updated and aims to influence politicians whose decisions affect the environment and its natural resources.

World Conservation Union or **International Union for the Conservation of Nature (IUCN)**, organization established by the *United Nations to promote the conservation of wildlife and habitats as part of the national policies of member states.

World Health Organization WHO, specialized agency of the United Nations established in 1946 to prevent the spread of diseases and to eradicate them. For 2004–05 it had a budget of US$881 million. Its headquarters are in Geneva, Switzerland. The WHO's greatest achievement to date has been the eradication of smallpox.

World Meteorological Organization agency, part of the United Nations since 1950, that promotes the international exchange of weather information through the establishment of a worldwide network of meteorological stations. It was founded as the International Meteorological Organization in 1873 and its headquarters are now in Geneva, Switzerland.

World Trade Organization WTO, specialized, rules-based, member-driven agency of the United Nations, world trade monitoring body established in January 1995, on approval of the Final Act of the Uruguay round of the *General Agreement on Tariffs and Trade (GATT). Under the Final Act, the WTO, a permanent trading body with a status comparable with that of the International Monetary Fund or the World Bank, effectively replaced GATT. The WTO oversees and administers agreements to reduce barriers to trade, such as tariffs, subsidies, quotas, and regulations which discriminate against imported products. Other functions of the WTO include: handling trade disputes, offering a forum for trade negotiations, technical assistance and training for developing countries, and monitoring national trade policies.

World Wide Fund for Nature WWF; formerly the **World Wildlife Fund**, international organization established in 1961 to raise funds for conservation by public appeal. Projects include conservation of particular species, for example, the tiger and giant panda, and special areas, such as the Simen Mountains, Ethiopia.

World Wildlife Fund former and US name of the *World Wide Fund for Nature.

WWF *abbreviation for* *World Wide Fund for Nature (formerly World Wildlife Fund).

xerophyte   any plant with special
characteristics such that it can survive
in climates with a pronounced dry
phase. Vegetation which grows in very
cold climates may also display
xeromorphic features.

   Xerophytes can display some but
rarely all of the following features:
reduction of leaf size; thickening of the
cuticular covering of leaves and stems;
thick corky bark; leaves which are
protected by hairs; stomata that are
sunken and protected by enlarged
guard cells; stomatal openings which
may be filled with a wax-like
substance; leaves that can be rotated to
remain 'edge-on' to the Sun (*heliotrophic*).

xerosere   any vegetation succession
developed on dry soil as opposed to
rock, as for example the succession
of plants on sand dunes.

X-ray diffractometer   instrument
used to determine the crystalline
structure (or the atomic structure)
of a material such as a mineral.
X-ray diffractometers are used to
identify finely crystalline minerals
such as clays, and to determine the
crystal structure of newly discovered
minerals.

yardang   ridge formed by wind erosion
from a dried-up riverbed or similar
feature, as in Chad, China, Peru, and
North America. On the planet Mars
yardangs occur on a massive scale.

yield   the productivity of land as
measured by the weight or volume
of produce per unit area. Agricultural
yields are usually expressed per
hectare.

Yixian formation   in palaeontology,
a Chinese geological formation in the
the rural province of Liaoning that is
yielding a wealth of extraordinarily
well-preserved fossils. The fossils
date from 150–120 million years ago
(late Jurassic or early Cretaceous)
and include those of hundreds of early
birds, such as *Confuciusornis*, with
feathers, lizards with full skin, and
mammals with hair.

Z

zeolite   any of the hydrous aluminium silicates, also containing sodium, calcium, barium, strontium, or potassium, chiefly found in igneous rocks and characterized by a ready loss or gain of water. Zeolites are used as 'molecular sieves' to separate mixtures because they are capable of selective absorption. They have a high ion-exchange capacity and can be used to separate petrol, benzene, and toluene from low-grade raw materials, such as coal and methanol.

zero population growth (ZPG)   a situation in which the birth rate is constant and equal to the death rate, the *age structure* (the proportion of the population at each age group) is constant, and the growth rate is zero.

Zeugen   *pedestal rocks* in arid regions; wind *erosion is concentrated near the ground where *corrasion by wind-borne sand is most active. This leads to undercutting and the pedestal profile emerges.

zinc ore   mineral from which zinc is extracted, principally sphalerite $(Zn,Fe)S$, but also zincite, ZnO_2, and smithsonite, $ZnCO_3$, all of which occur in mineralized veins. Ores of lead and zinc often occur together, and are common worldwide; Canada, the USA, and Australia are major producers.

zircon   zirconium silicate, $ZrSiO_4$, a mineral that occurs in small quantities in a wide range of igneous, sedimentary, and metamorphic rocks. It is very durable and is resistant to erosion and weathering. It is usually coloured brown, but can be other colours, and when transparent may be used as a gemstone.

APPENDIX A – Useful Data

IMPERIAL AND METRIC CONVERSION FACTORS

To convert from Imperial	Multiply by	To convert from Metric	Multiply by
Length			
inches to millimetres	25.4	millimetres to inches	0.0393701
feet to metres	0.3048	metres to feet	3.28084
yards to metres	0.9144	metres to yards	1.09361
furlongs to kilometres	0.201168	kilometres to furlongs	4.97097
miles to kilometres	1.609344	kilometres to miles	0.621371
Area			
square inches to square centimetres	6.4516	square centimetres to square inches	0.1550
square feet to square metres	0.092903	square metres to square feet	10.7639
square yards to square metres	0.836127	square metres to square yards	1.19599
square miles to square kilometres	2.589988	square kilometres to square miles	0.386102
acres to square metres	4,046.856422	square metres to acres	0.000247
acres to hectares	0.404685	hectares to acres	2.471054
Volume/capacity			
cubic inches to cubic centimetres	16.387064	cubic centimetres to cubic inches	0.061024
cubic feet to cubic metres	0.028317	cubic metres to cubic feet	35.3147
cubic yards to cubic metres	0.764555	cubic metres to cubic yards	1.30795
cubic miles to cubic kilometres	4.1682	cubic kilometres to cubic miles	0.239912
fluid ounces (imperial) to millilitres	28.413063	millilitres to fluid ounces (imperial)	0.035195
fluid ounces (US) to millilitres	29.5735	millilitres to fluid ounces (US)	0.033814
pints (imperial) to litres	0.568261	litres to pints (imperial)	1.759754
pints (US) to litres	0.473176	litres to pints (US)	2.113377
quarts (imperial) to litres	1.136523	litres to quarts (imperial)	0.879877
quarts (US) to litres	0.946353	litres to quarts (US)	1.056688
gallons (imperial) to litres	4.54609	litres to gallons (imperial)	0.219969
gallons (US) to litres	3.785412	litres to gallons (US)	0.364172
Mass/weight			
ounces to grams	28.349523	grams to ounces	0.035274
pounds to kilograms	0.453592	kilograms to pounds	2.20462
stones (14 lb) to kilograms	6.350293	kilograms to stones (14 lb)	0.157473
tons (imperial) to kilograms	1,016.046909	kilograms to tons (imperial)	0.000984
tons (US) to kilograms	907.18474	kilograms to tons (US)	0.001102
tons (imperial) to metric tonnes	1.016047	metric tonnes to tons (imperial)	0.984207
tons (US) to metric tonnes	0.907185	metric tonnes to tons (US)	1.10231
Speed			
miles per hour to kilometres per hour	1.609344	kilometres per hour to miles per hour	0.621371
feet per second to metres per second	0.3048	metres per second to feet per second	3.28084
Force			
pounds-force to newtons	4.44822	newtons to pounds-force	0.224809
		newtons to kilograms-force	0.101972
		kilograms-force to newtons	9.80665
Pressure			
pounds-force per square inch to kilopascals	6.89476	kilopascals to pounds-force per square inch	0.145038
tons-force per square inch (imperial) to megapascals	15.4443	megapascals to tons-force per square inch (imperial)	0.064779
atmospheres to newtons per square centimetre	10.1325	newtons per square centimetre to atmospheres	0.098692
pounds-force per square inch to atmospheres	0.068948		

TABLE OF EQUIVALENT TEMPERATURES

Celsius and Fahrenheit temperatures can be interconverted as follows: $C = (F - 32) \times 100/180$;
$F = (C \times 180/100) + 32$.

°C	°F	°C	°F	°C	°F	°C	°F
100	212.0	70	158.0	40	104.0	10	50.0
99	210.2	69	156.2	39	102.2	9	48.2
98	208.4	68	154.4	38	100.4	8	46.4
97	206.6	67	152.6	37	98.6	7	44.6
96	204.8	66	150.8	36	96.8	6	42.8
95	203.0	65	149.0	35	95.0	5	41.0
94	201.2	64	147.2	34	93.2	4	39.2
93	199.4	63	145.4	33	91.4	3	37.4
92	197.6	62	143.6	32	89.6	2	35.6
91	195.8	61	141.8	31	87.8	1	33.8
90	194.0	60	140.0	30	86.0	0	32.0
89	192.2	59	138.2	29	84.2	−1	30.2
88	190.4	58	136.4	28	82.4	−2	28.4
87	188.6	57	134.6	27	80.6	−3	26.6
86	186.8	56	132.8	26	78.8	−4	24.8
85	185.0	55	131.0	25	77.0	−5	23.0
84	183.2	54	129.2	24	75.2	−6	21.2
83	181.4	53	127.4	23	73.4	−7	19.4
82	179.6	52	125.6	22	71.6	−8	17.6
81	177.8	51	123.8	21	69.8	−9	15.8
80	176.0	50	122.0	20	68.0	−10	14.0
79	174.2	49	120.2	19	66.2	−11	12.2
78	172.4	48	118.4	18	64.4	−12	10.4
77	170.6	47	116.6	17	62.6	−13	8.6
76	168.8	46	114.8	16	60.8	−14	6.8
75	167.0	45	113.0	15	59.0	−15	5.0
74	165.2	44	111.2	14	57.2	−16	3.2
73	163.4	43	109.4	13	55.4	−17	1.4
72	161.6	42	107.6	12	53.6	−18	−0.4
71	159.8	41	105.8	11	51.8	−19	−2.2

LARGEST COUNTRIES BY POPULATION SIZE

Source: *State of the World Population 1999*; Population Division and Statistics
Division of the United Nations Secretariat

Countries with a population of over 100 million, 2000 and 2050.

Rank	Country	Population (millions)	% of world population
2000			
1	China	1,278	21.11
2	India	1,014	16.75
3	United States	278	4.59
4	Indonesia	212	3.50
5	Brazil	170	2.81
6	Pakistan	156	2.58
7	Russian Federation	147	2.43
8	Bangladesh	129	2.13
9	Japan	127	2.10
10	Nigeria	112	1.85
	World total	6,055	
2050 (projected)			
1	India	1,529	17.16
2	China	1,478	16.58
3	United States	349	3.91
4	Pakistan	346	3.88
5	Indonesia	312	3.50
6	Nigeria	244	2.73
7	Brazil	244	2.73
8	Bangladesh	213	2.39
9	Ethiopia	170	1.90
10	Congo, Democratic Republic of	160	1.79
11	Mexico	147	1.65
12	Philippines	131	1.47
13	Vietnam	127	1.42
14	Russian Federation	122	1.42
15	Iran	115	1.29
16	Egypt	115	1.29
17	Japan	105	1.17
18	Turkey	101	1.13
	World total	8,909	

Useful Data

Useful Data

LARGEST DESERTS IN THE WORLD

Desert	Location	Area[1] sq km	sq mi
Sahara	northern Africa	9,065,000	3,500,000
Gobi	Mongolia/northeastern China	1,295,000	500,000
Patagonian	Argentina	673,000	260,000
Rub al-Khali	southern Arabian peninsula	647,500	250,000
Kalahari	southwestern Africa	582,800	225,000
Chihuahuan	Mexico/southwestern USA	362,600	140,000
Taklimakan	western China	362,600	140,000
Great Sandy	northwestern Australia	338,500	130,000
Great Victoria	southwestern Australia	338,500	130,000
Kyzyl Kum	Uzbekistan/Kazakhstan	259,000	100,000
Thar	India/Pakistan	219,000	84,556
Sonoran	Mexico/southwestern USA	181,300	70,000
Simpson	Australia	103,600	40,000
Mojave	southwestern USA	65,000	25,000

[1] Desert areas are very approximate because clear physical boundaries may not occur.

LARGEST LAKES IN THE WORLD

Lake	Location	Area sq km	sq mi
Caspian Sea	Azerbaijan/Russian Federation/Kazakhstan/Turkmenistan/Iran	370,990	143,239
Superior	USA/Canada	82,071	31,688
Victoria	Tanzania/Kenya/Uganda	69,463	26,820
Aral Sea	Kazakhstan/Uzbekistan	64,500	24,903
Huron	USA/Canada	59,547	22,991
Michigan	USA	57,735	22,291
Tanganyika	Tanzania/Democratic Republic of Congo/Zambia/Burundi	32,880	12,695
Baikal	Russian Federation	31,499	12,162
Great Bear	Canada	31,316	12,091
Malawi (or Nyasa)	Malawi/Tanzania/Mozambique	28,867	11,146
Great Slave	Canada	28,560	11,027
Erie	USA/Canada	25,657	9,906
Winnipeg	Canada	25,380	9,799
Ontario	USA/Canada	19,010	7,340
Balkhash	Kazakhstan	18,421	7,112
Ladoga	Russian Federation	17,695	6,832
Chad	Chad/Cameroon/Nigeria	16,310	6,297
Maracaibo	Venezuela	13,507	5,215

MAJOR OCEANS AND SEAS OF THE WORLD

Ocean/sea	Area[1]		Average depth	
	sq km	sq mi	m	ft
Pacific Ocean	166,242,000	64,186,000	3,939	12,925
Atlantic Ocean	86,557,000	33,420,000	3,575	11,730
Indian Ocean	73,429,000	28,351,000	3,840	12,598
Arctic Ocean	13,224,000	5,106,000	1,038	3,407
South China Sea	2,975,000	1,149,000	1,464	4,802
Caribbean Sea	2,754,000	1,063,000	2,575	8,448
Mediterranean Sea	2,510,000	969,000	1,501	4,926
Bering Sea	2,261,000	873,000	1,491	4,893
Sea of Okhotsk	1,580,000	610,000	973	3,192
Gulf of Mexico	1,544,000	596,000	1,614	5,297
Sea of Japan	1,013,000	391,000	1,667	5,468
Hudson Bay	730,000	282,000	93	305
East China Sea	665,000	257,000	189	620
Andaman Sea	565,000	218,000	1,118	3,667
Black Sea	461,000	178,000	1,190	3,906
Red Sea	453,000	175,000	538	1,764
North Sea	427,000	165,000	94	308
Baltic Sea	422,000	163,000	55	180
Yellow Sea	294,000	114,000	37	121
Gulf	230,000	89,000	100	328
Gulf of California	153,000	59,000	724	2,375
English Channel	90,000	35,000	54	177
Irish Sea	89,000	34,000	60	197

[1] All figures are approximate as boundaries of oceans and seas cannot be exactly determined.

LONGEST RIVERS IN THE WORLD

River	Location	Approximate Length	
		km	mi
Nile	Africa	6,695	4,160
Amazon	South America	6,570	4,083
Chang Jiang (Yangtze)	China	6,300	3,915
Mississippi–Missouri–Red Rock	USA	6,020	3,741
Huang He (Yellow River)	China	5,464	3,395
Ob–Irtysh	China/Kazakhstan/Russian Federation	5,410	3,362
Amur–Shilka	Asia	4,416	2,744
Lena	Russian Federation	4,400	2,734
Congo	Africa	4,374	2,718
Mackenzie–Peace–Finlay	Canada	4,241	2,635
Mekong	Asia	4,180	2,597
Niger	Africa	4,100	2,548
Yenisei	Russian Federation	4,100	2,548
Paraná	Brazil	3,943	2,450
Mississippi	USA	3,779	2,348
Murray–Darling	Australia	3,751	2,331
Missouri	USA	3,726	2,315
Volga	Russian Federation	3,685	2,290
Madeira	Brazil	3,241	2,014
Purus	Brazil	3,211	1,995
São Francisco	Brazil	3,199	1,988
Yukon	USA/Canada	3,185	1,979
Rio Grande	USA/Mexico	3,058	1,900
Indus	Tibet/Pakistan	2,897	1,800
Danube	central and eastern Europe	2,858	1,776
Japura	Brazil	2,816	1,750
Salween	Myanmar/China	2,800	1,740
Brahmaputra	Asia	2,736	1,700
Euphrates	Iraq	2,736	1,700
Tocantins	Brazil	2,699	1,677
Zambezi	Africa	2,650	1,647
Orinoco	Venezuela	2,559	1,590
Paraguay	Paraguay	2,549	1,584
Amu Darya	Tajikistan/Turkmenistan/Uzbekistan	2,540	1,578
Ural	Russian Federation/Kazakhstan	2,535	1,575
Kolyma	Russian Federation	2,513	1,562
Ganges	India/Bangladesh	2,510	1,560
Arkansas	USA	2,344	1,459
Colorado	USA	2,333	1,450
Dnieper	Russian Federation/Belarus/Ukraine	2,285	1,420
Syr Darya	Asia	2,205	1,370
Irrawaddy	Myanmar	2,152	1,337
Orange	South Africa	2,092	1,300

APPENDIX B – Web Resources

There is a wealth of information available on the internet to help you find out more about many aspects of geography and on related topics from current affairs to statistics. A selected listing is given below. Of particular value are the gateway sites, which give easy access to a wide range of resources. We would welcome further suggestions for inclusion, which can be sent via the Collins website, Collins.co.uk

1. Gateway websites

Earth Sciences Gateway PSIgate (Physical Sciences Information Gateway) is the physical sciences hub of the Resource Discovery Network (RDN).

www.psigate.ac.uk/newsite/earth-gateway.html

Geography Network A global network of geographic information users and providers through which you can access many types of geographical content, including dynamic maps, downloadable data, and more advanced Web services.

www.geographynetwork.com

Geography websites A site that offers links to a very wide range of websites in the following areas: ecosystems, the environment, climate change, pollution, agriculture, development, population and settlement, tourism, transport, coasts, earthquakes, geology, glaciers, rivers, volcanoes, weather charts, climate, satellite images and more. Compiled and maintained by Radley College Geography Department.

http://atschool.eduweb.co.uk/radgeog/websites.html

Geography World Links to sites about maps, weather, population, conservation, and other geographical subjects.

http://members.aol.com/bowermanb/101.html

Geo-Information Gateway An index of on-line geographical resources.

www.geog.le.ac.uk/cti/info.html

Georesources A wide range of geographical resources and information for school and college students.

www.georesources.co.uk

GEOSOURCE Gateway Web resources for human geography, physical geography, planning, geoscience and environmental science.

www.library.uu.nl/geosource

GEsource An information resource for geography and the environment.

www.gesource.ac.uk/home

Internet Resources for Physical Geography Links to sites covering all aspects of physical geography.

www.uwsp.edu/geo/internet/geog_geol_resources

SOSIG: Geography The Social Science Information Gateway, a comprehensive site of subject specific information and links.

www.sosig.ac.uk/geography

2. General geographical websites

Geography in the news A resource site provided by the Royal Geographical Society.

www.geographyinthenews.rgs.org

Global Earthquake Response Centre A site for information on earthquakes.

www.earthquake.org/core.html

National Geographic online The website of the National Geographic Society.

www.nationalgeographic.com

Oceanography Virtual Library Gateway to many oceanographic websites worldwide.

www.mth.uea.ac.uk/ocean/vl

This Dynamic Earth Plate tectonics explained.

http://pubs.usgs.gov/publications/text/dynamic.html

Volcano World Comprehensive site on volcano information.

http://volcano.und.nodak.edu

3. Geographical organisation websites

Association for Geographic Information (AGI)

www.agi.org.uk

Association of American Geographers
www.aag.org
British Cartographic Society
www.cartography.org.uk
British Geological Survey
www.bgs.ac.uk
Centre for Ecology and Hydrology
www.nwl.ac.uk/ih
Geography Departments Worldwide   Links to 1044 university geographical
departments worldwide.
http://univ.cc/geolinks
Hydrographic Office, The
www.hydro.gov.uk
Royal Geographical Society with the Institute of British Geographers
www.rgs.org
Royal Scottish Geographical Society
www.geo.ed.ac.uk/~rsgs/
Society of Cartographers
www.soc.org.uk
The Geographical Association
www.geography.org.uk
The Geological Society of London
www.geolsoc.org.uk
US Geological Survey
www.usgs.gov.uk

4. Environmental websites

Antarctica Online   The Australian Antarctic programme.
www.antdiv.gov.au
EarthTrends: Environmental Information Portal   A source of global environmental
and sustainable development information. You can search the World Resources
Institute's database, maps, feature articles and country profiles for information and
statistics on key environmental trends and issues.
www.earthtrends.wri.org
English Nature   The main government funded English conservation organisation.
www.english-nature.org.uk
Environmental data from NERC   The Natural Environment Research Council in the UK.
www.nerc.ac.uk/data
European Environment Agency   A source for Europe-wide information.
www.eea.eu.int
Forestry Commission   The UK government organisation responsible for the
protection and expansion of Britain's forests and woodlands.
www.forestry.gov.uk
Friends of the Earth   A major environmental campaigning organisation.
www.foe.org.uk
Greenpeace   A major environmental campaigning organisation.
www.greenpeace.org/international_en
Macaulay Land Use Research Institute   For research on the use of rural land
resources for the benefit of people and the environment.
www.mluri.sari.ac.uk
United Nations Environment Programme   Information on this UN programme and
on its Global Environmental Outlook project.
www.grida.no/geo
US Environmental Protection Agency   The main US environmental agency.
www.epa.gov
World Wide Fund for Nature   Details of its conservation work and campaigns.
www.panda.org

5. Development issues websites

Eldis   A gateway to information on development issues, providing easy access to wide range of online resources.
    www.eldis.org
Food and Agriculture Organisation, UN   Information on the programmes and publications of this UN agency.
    www.fao.org
InfoRurale   Consolidates information on rural development from around the world.
    www.inforurale.org.uk/inforurale
Institute of Development Studies   A leading centre for research and teaching on international development.
    www.ids.ac.uk/ids
New Internationalist   Website of popular monthly magazine devoted to development issues.
    www.newint.org
OECD Online   The site of the OECD, whose work covers economic and social issues from macroeconomics, to trade, education, development, and science and innovation.
    www.oecd.org
State of the World's Cities Report 2001   A report from the United Nations Human Settlement Programme.
    www.unchs.org/Istanbul+5/statereport.html
The World Bank   Information on World Bank programmes and on economic development.
    www.worldbank.org
United Nations   Site contains information on worldwide economic and social development, population, environment, trade and more.
    www.un.org
United Nations Children's Fund   Information on the work of UNICEF.
    www.unicef.org
World Resources Institute   Information on development issues and other data.
    www.wri.org/wri/index.html

6. Statistical websites

Australian Bureau of Statistics   The official Australian statistics site.
    www.abs.gov.au
Fedstats   The gateway to statistics from over 100 US Federal agencies.
    www.fedstats.gov
International Data Base   Source of global demographic and socio-economic statistics, maintained by the US Census.
    www.census.gov/ipc/www/idbnew.html
National Statistics   The official UK statistics site.
    www.statistics.gov.uk/
UK agricultural statistics
    http://statistics.defra.gov.uk/esg
United States Census Bureau
    www.census.gov
World statistics   Statistical resources on the web from the University of Michigan.
    www.lib.umich.edu/govdocs/stats.html

7. Cartographic websites

CTI – Geographical Information Systems   A gateway to a range of GIS websites
    www.geog.le.ac.uk/cti/gis.html
GIS Dictionary   Produced by the Association for Geographic Information.
    www.geo.ed.ac.uk/agidict/welcome.html
Historical Maps
    www.lib.utexas.edu/maps/map_sites/hist_sites.html

Web Resources

Millennium Mapping Project
    www.millennium-map.com
Multimap
    http://uk8.multimap.com/map/places.cgi
NASA's Visible Earth   A searchable collection of Earth images.
    http://visibleearth.nasa.gov
National Geographic Map Machine
    http://plasma.nationalgeographic.com/mapmachine
Ordnance Survey
    www.ordnancesurvey.co.uk/oswebsite
Remote Sensing and Photogrammetry Society
    www.rspsoc.org
Satellite Views of the Earth
    www.fourmilab.ch/cgi-bin/uncgi/Earth/action?opt=-p
World Atlas.com
    www.graphicmaps.com/aatlas/world.htm

8. Countries of the world websites

Atlapedia: Countries A to Z
    www.atlapedia.com/online/country_index.htm
Background Notes   U.S. State Department provides information on countries and
international organizations.
    www.state.gov/r/pa/ei/bgn
BBC News Country Profiles   Profiles of every country in the world, regularly
updated.
    http://news.bbc.co.uk/1/hi/country_profiles
CIA – The World Factbook   Up to date data on any country in the world.
    www.odci.gov/cia/publications/factbook
City Population
    www.citypopulation.de
Country Studies   On-line versions of books previously published in hard copy by the
Federal Research Division of the Library of Congress and studies political, economic,
social and cultural aspects of 102 countries.
    http://lcweb2.loc.gov/frd/cs
European Union Online
    www.europa.eu.int/index_en.htm
GeoHive
    www.geohive.com
World Gazetteer
    www.gazetteer.de/home.htm

9. Climatology websites

BBC Weather Website
    www.bbc.co.uk/weather
Met Office   The site of the UK's official meteorological organisation.
    www.met-office.gov.uk
Meteorology Site, World Data Center   World Data Center for Meteorology.
    http://lwf.ncdc.noaa.gov/oa/wmo/wdcamet.html
NASA's Earth Observation System Project
    http://eospso.gsfc.nasa.gov
National Climatic Data Center   The world's largest active archive of weather data.
    http://lwf.ncdc.noaa.gov/oa/about/about.html
Online Weather   World weather reports.
    www.onlineweather.com
The Tornado Project   For information on tornados.
    www.tornadoproject.com/